Results of Researches of the International Geophysical Projects

Collection of Articles
No. 16

METEOROLOGICAL RESEARCHES
(Meteorologicheskie issledovaniya)

B.L. Dzerdzeevskii and Kh. P. Pogosyan, Editors

GENERAL CIRCULATION OF THE ATMOSPHERE

(Obshchaya tsirkulyatsiya atmosfery)

Izdatel'stvo "Nauka"
Moskva 1968

Translated from Russian

Israel Program for Scientific Translations
Jerusalem 1971

TT 70-50052
Published Pursuant to an Agreement with
THE NATIONAL OCEANIC AND ATMOSPHERIC ADMINISTRATION,
U.S. DEPARTMENT OF COMMERCE
and
THE NATIONAL SCIENCE FOUNDATION, WASHINGTON, D.C.

IPST Cat. No. 5754

SBN 7065 1088 7

Translated by D. Lederman
Edited by P. Greenberg

Printed in Jerusalem by Keter Press
Binding: Wiener Bindery Ltd., Jerusalem

Available from the
U.S. DEPARTMENT OF COMMERCE
National Technical Information Service
Springfield, Va. 22151

III/6/3

Table of Contents

FOREWORD ix

THEORY OF THE GENERAL CIRCULATION OF THE ATMOSPHERE

Mashkovich, S. A. Hydrodynamic theory of the general circulation of the atmosphere and numerical experiments 1

Yudin, M. I. Factors responsible for unsteadiness of the general circulation of the atmosphere 11

Marchuk, G. I. and T. A. Temnoeva. The solution of certain quasi-steady problems of dynamic meteorology 20

DESCRIPTION AND STUDY OF THE GENERAL CIRCULATION USING METHODS OF MATHEMATICAL STATISTICS

Obukhov, A. M. and M. I. Fortus. Statistical description of the general circulation of the atmosphere 26

Kolesnikova, V. N. and A. S. Monin. The spectra of micrometeorological, synoptic and climatic fluctuations of meteorological fields 30

Gruza, G. V. Integral characteristics of the general circulation of the atmosphere 62

ENERGETICS OF THE GENERAL CIRCULATION, CONSIDERATION OF RADIATIVE HEAT INFLUX, UTILIZATION OF SATELLITE OBSERVATIONS

Borisenkov, E. P. Energetics of the general circulation of the atmosphere 76

Feigel'son, E. M. Certain radiation problems related to the problem of the general circulation 95

Rakipova, L. R. and B. E. Shneerov. Consideration of radiative heat influx in problems of dynamic meteorology 106

Malkevich, M. S. Determination of meteorological elements from satellites 129

SYNOPTICOSTATISTICAL METHODS IN THE STUDY OF THE GENERAL CIRCULATION OF THE ATMOSPHERE. FLUCTUATIONS OF THE GENERAL CIRCULATION OF THE ATMOSPHERE AND OF THE CLIMATE

Pogosyan, Kh. P., A. A. Pavlovskaya, and M. V. Shabel'nikova. Nature of the interaction between air circulation in the troposphere and stratosphere of different regions of the northern hemisphere 144

Khrgian, A.Kh., V.I.Bekoryukov, V.M.Berezin, and Ya.A.Shafrin. Problems of ozone circulation in the terrestrial atmosphere 174
Dzhordzhio, V.A. Jet streams as indicators of the general circulation of the atmosphere 185
Astapenko, P.D. and S.S.Gaigerov. Some features of the atmospheric circulation over the southern hemisphere 191
Dobryshman, E.M. Some features of pressure and wind fields in the near-equitorial belt 205
Dubentsov, V.R. Some features of air circulation in the stratosphere and mesosphere 222
Petrosyants, M.A. Effect of orography on the general circulation of the atmosphere 235
Drozdov, O.A. The general circulation of the atmosphere and the hydrologic cycle 269
Dzerdzeevskii, B.L. Long-term fluctuations of the general circulation of the atmosphere and of the climate and the major regularities of planetary circulation 283
Girs, A.A. Epochal variations in atmospheric circulation in the northern hemisphere and corresponding changes in characteristics of the atmosphere and hydrosphere 305

ATMOSPHERE—OCEAN INTERACTION

Laikhtman, D.L., B.A.Kagan, and V.V.Timonov. Methods of studying ocean—atmosphere interaction 323
Kitaigorodskii, S.A. and Yu.A.Volkov. Some problems of the physics of the water-adjoining layer of the atmosphere 338

LABORATORY MODELING OF THE GENERAL CIRCULATION OF THE ATMOSPHERE

Bonchkovskaya, T.V. Experience in modeling certain schemes of atmospheric circulation and heat exchange 372

ATMOSPHERIC CIRCULATION OF OTHER PLANETS OF THE SOLAR SYSTEM

Golitsyn, G.S. and V.I.Moroz. Problems of general circulation in planetary atmospheres 382

BRIEF COMMUNICATIONS

Dmitriev, A.A. Objective determination of circulation types 390
Bulinskaya, N.A. Analysis of atmospheric circulation for forecasting purposes 393
Serlapov, S.T. Comparison of maps of maximum wind strength in the upper troposphere with ground-surface synoptic maps 396

Baidal, M.Kh. Dynamics of the "circulation pole" and their synoptico-climatological significance 397
Karol', I.L. The use of tracers in the study of the general circulation of the atmosphere 398
Khrabrov, Yu.B. Modeling of typhoons and thunderclouds 400
Usmanov, R.F. Underwater trenches in oceans as factors amplifying atmospheric processes 401

UDC 551.5

The collection contains reports about the present state of knowledge of the general (planetary) circulation of the earth atmosphere.
The following subjects are treated in detail: the hydrodynamic theory of the general circulation and numerical experiments; the application of modern mathematical statistics to the study and description of the general circulation; energetics of the general circulation, accounting for the radiative heat influx in solving problems of the general circulation of the atmosphere and the use of information on radiative processes in the atmosphere obtained from artificial satellites; the synopticostatistical aspects of the study of the general circulation using empirical data; the relationship between general-circulation fluctuations and climatic variations; studies of interaction between the atmosphere and the ocean; use of laboratory methods in modeling general circulation models in atmospheres of other planets.

EXPLANATORY LIST OF USSR ABBREVIATIONS APPEARING IN THIS BOOK

Abbreviation	Full name (transliterated)	Translation
AANII	Arkticheskii i Antarkticheskii Nauchno-Issledovatel'skii Institut	Arctic and Antarctic Scientific Research Institute
AN SSSR	Akademiya Nauk SSSR	Academy of Sciences of the USSR
GGI	Gosudarstvennyi Gidrologicheskii Institut	State Hydrological Institute
GGO	Glavnaya Geofizicheskaya Observatoriya	Main Geophysical Observatory
GOIN	Gosudarstvennyi Okeanograficheskii Institut	State Oceanographic Institute
GUGMS	Glavnoe Upravlenie Gidrometeorologicheskoi Sluzhby SSSR	Main Administration of the Hydrometeorological Service of the USSR
IFA	Institut Fiziki Atmosfery	Institute of Atmospheric Physics
IOAN	Institut Okeanologii Akademii Nauk SSSR	Institute of Oceanology of the Academy of Sciences of the USSR
LGMI	Leningradskii Gidrometeorologicheskii Institut	Leningrad Hydrometeorological Institute
LGU	Leningradskii Gosudarstvennyi Universitet	Leningrad State University
LVIMU	Leningradskoe Vysshee Inzhenernoe Morekhodnoe Uchilishche	Leningrad Higher School of Naval Engineering
MGU	Moskovskii Gosudarstvennyi Universitet	Moscow State University
MMTs	Mezhdunarodnyi Meteorologicheskii Tsentr	World Meteorological Center

Abbreviation	Full name (transliterated)	Translation
NIGMI	Nauchno-Issledovatel'skii Gidrometeorologicheskii Institut	Hydrometeorological Scientific Research Institute
NIIAK	Nauchno-Issledovatel'skii Institut Aeroklimatologii	Scientific Research Institute of Aeroclimatology
NIU	Nauchnyi Institut po Udobreniyam	Scientific Institute for Fertilizers
SAE	Sovetskaya Antarkticheskaya Ekspeditsiya	Soviet Antarctic Expedition
SANIGMI	Sredneaziatskii Nauchno-Issledovatel'skii Gidrometeorologicheskii Institut	Central Asian Hydrometeorological Scientific Research Institute
TsAO	Tsentral'naya Aerologicheskaya Observatoriya	Central Aerological Observatory
TsIP	Tsentral'nyi Institut Prognozov	Central Weather Forecasting Institute
UGMS	Upravlenie Gidrometeorologicheskoi Sluzhby	Hydrometeorological Service Administration
VNMS	Vsesoyuznoe Nauchno-Meteorologicheskoe Soveshchanie	All-Union Scientific Meteorological Conference

FOREWORD

The present book contains comprehensive and somewhat revised reports read at the conference on "The General Circulation of the Earth Atmosphere" at the end of 1964. This conference was organized by the Institute of Atmospheric Physics and the Institute of Geography of the USSR Academy of Sciences. Two hundred and eighty representatives of 42 scientific organizations from 14 cities of the USSR participated in the conference. Twenty-four reports and a number of brief communications were read, and more than 70 people took part in editing them.

The conference was planned and organized along topical lines. The subjects of all the reports were preassigned, and each of the scientists invited by the organizing committee undertook to respect the agreed plan. The topics were selected to elucidate all the principal aspects of this complex problem, viz., the study of the general circulation of the atmosphere. Each report is thus an integral part of one of the seven parts of this book, while the entire book may be regarded as a single monograph written by a number of authors.

Was this endeavor completely successful? Apparently it was not. Several topics are not mentioned, and there are departures in uniformity of format by a number of articles. This is due to the extent to which the particular questions were studied and to the unavailability, in a number of cases, of the information required. Nevertheless, the book in general does clarify the various aspects of the state and prospects for development of the study of the general circulation, even if it is not an exhaustive treatment of the subject.

The complete proceedings of the conference were too voluminous to be included in a single book. The book was reduced to its present size by excluding conference resolutions, the summing up remarks of session chairmen, and discussion. We shall therefore mention here briefly the topics which attracted the greatest interest.

The conference participants noted the ever increasing interest in the general circulation of the earth atmosphere, which is closely related to problems of long-range weather forecasting, to the theory of the climate, and to questions of nature modification. It was noted particularly that, besides "traditional" topics of synopticostatistical studies of circulation, a great deal of attention is being paid to new problems, such as the study of circulation energy (using satellite observations), laboratory modeling of circulation, and the study of circulation in atmospheres of other planets.

The conference participants emphasized the wide range and high level of theoretical studies of the general circulation of the atmosphere. The significance of statistical processing of empirical data and the necessity of bringing these into closer agreement with numerical methods were noted.

The great benefit derived from publications of initial data and the poor practical implementation of this principle were pointed out. The ever increasing role of natural orthogonal functions in the study of atmospheric processes was noted. In particular, it is now possible to estimate the degree of accuracy and the reliability of the statistical features in the classification of atmospheric process, found from empirical data.

A great deal of attention was paid to the absolute necessity of further analyzing the role played by the surface (with its inhomogeneities) in the formation of meteorological fields. The study of radiative processes carried out for these purposes has advanced significantly. However, here also there are a number of important, as yet unsolved problems (for example, the use of radiation measurements for estimating the state of the atmosphere, the solution of unsteady radiative-transfer problems, etc.). The consideration of air humidity is more complicated, and research is needed in this direction.

Analysis of the spectra of meteorological fields is both interesting and productive, but it has only been carried out for individual locations, while energy relationships in the atmosphere have been analyzed for short periods only. The extent of these and other studies must be greatly increased, and this can only be achieved by the participation of more workers and by consolidating the efforts of scientists from different organizations.

The conference emphasized the value of studying synchronous relationships between stratospheric and tropospheric processes in different parts of a hemisphere and between the northern and southern hemispheres.

Reports read at the conference stressed the great importance of implementing the IGY and IQSY programs to develop studies of the general circulation of the atmosphere.

Unfortunately, it took a long time to prepare the book for print. This was due to the delay, by some authors, in submitting the final drafts of articles and to subsequent technical difficulties. Invaluable assistance in overcoming the technical difficulties was rendered by the staff of the Climatology Department of the Institute of Geography of the USSR Academy of Sciences, as well as by some staff members of the Institute of Atmospheric Physics of the USSR Academy of Sciences and of the Hydrometeorological Center of the Main Administration of the Hydrometeorological Service. The editors wish to express their particular thanks to E. M. Dobryshman and E. M. Feigel'son, Doctors of Physical and Mathematical Sciences, and to K. V. Kuvshinov, Candidate of Physical and Mathematical Sciences, for their participation in editing the book.

S. A. Mashkovich

HYDRODYNAMIC THEORY OF THE GENERAL CIRCULATION OF THE ATMOSPHERE AND NUMERICAL EXPERIMENTS

A survey is presented of researches on the hydrodynamic theory of the climate and of the general circulation of the atmosphere. The most significant simplifications and limitations introduced in solving the problem are noted. The importance of numerical experiments in the study of large-scale atmospheric processes is emphasized. Some problems requiring further investigation are formulated.

Lately, increased attention is being paid to the hydrodynamic theory of the general circulation of the atmosphere. The importance of theoretical studies of this problem is beyond doubt. In fact, while a physicist wishing to study some phenomenon is usually able to set up an experiment, the meteorologist, who is concerned with large-scale atmospheric processes, is as a rule unable to do so, and must be satisfied with data of actual observations in the atmosphere. Analysis of these data yields a number of features of the atmospheric structure and enables important regularities to be established. However, in many cases the study of observational data alone is insufficient. This is partly due to the fact that not all the quantities of interest are measured directly.

Systematic studies into the hydrodynamic theory of the general circulation of the atmosphere and of the climate started in the 1930's with the basic works of Kochin /15, 14/. Kochin applied boundary-layer theory to the study of the general circulation. His main result was that the troposphere should be regarded as the planetary boundary layer.

Kochin simplified the equation of motion as a result of his estimates, and was thus able to find the velocity field for the zonal circulation and the zonal pressure on the assumption that the temperature field in the entire space and the earth-surface pressure are known. Kochin's results were further developed in /11, 30, 1/.

A solution to the problem of determining the nonzonal pressure field was obtained by Blinova /2, 4/. She was the first to apply the linearized vorticity equation to the solution of problems of this kind, and she used it to calculate the steady-state nonzonal pressure field at the average atmospheric level /2, 33/. The temperature field at this level was regarded as known. Later Blinova /4/ obtained a solution for the three-dimensional problem (the component containing the vertical velocity was retained in the vorticity equation), which made it possible to calculate the steady-state nonzonal pressure field at any level, in particular at sea level, as long as the temperature distribution in the atmosphere was known.

Subsequently Chekirda /29/ calculated the ground-level pressure using Blinova's solution.

The above studies demonstrated the feasibility of calculating the steady-state pressure distribution and the field of fluxes from a known temperature field. The next problem was that of simultaneously finding the temperature, pressure, and wind, i.e., of simultaneously solving the equations of motion and of heat influx.

The first attempts to obtain theoretically the temperature distribution in the atmosphere involved the assumption of radiation equilibrium (Gold, Emden, Hulbert, Milankovich, Kuznetsov, and others). The temperatures found are in this case relatively close to those actually observed. However, the calculated lapse rates exceeded appreciably the actual ones; this shortcoming is inherent in all studies based on the assumption of radiation equilibrium and not allowing for air motion and atmospheric turbulence. The calculated zonal temperature distribution along the meridian yielded overestimated values for southern latitudes and underestimated values for northern latitudes.

A solution of the problem of temperature distribution in the troposphere with consideration of eddy as well as radiative transport of heat (within the framework of the gray-body radiation model) was proposed by Kibel' /13/. The calculated data were in satisfactory agreement with the observed. Kibel' was the first to obtain the lapse rates and inversion in the surface layer which are typical of the troposphere. The Kibel' model was subsequently refined by taking into account the diffusivity of the propagation of long-wave radiation /32/.

The problem of the zonal-temperature distribution in the northern hemisphere was solved by Blinova /3/, who supplemented the Kibel' computational model /13/ by introducing into the heat-influx equation a term describing the effect of horizontal mixing. The solution thus obtained allowed the yearly average vertical and meridional distribution of the zonal temperature to be found; the computed results were close to climatic data. Reference /3/ served as the basis for a number of studies pertaining to the distribution of zonal temperature over the globe, the yearly march of zonal temperature, and wind fields /10, 17, 20, 21/. Here the "gray" atmosphere model was used to describe radiative processes. This model only describes approximately radiative heat transfer (for example, values of the parameters used yield an underestimated outgoing radiation). Hence one problem in the theory of thermal-field formation is to define the radiative processes more exactly.

The theory of steady-state zonal temperature distribution was further developed by Rakipova /28/. She refined the theoretical model by taking into account selectivity in the absorption and diffusivity in the propagation of radiation as well as by introducing a relationship between the density of the radiation-absorbing substance [atmospheric air] and the latitude, and by considering the heat expended in evaporation and condensation. This made it possible to obtain a more exact temperature distribution in the atmosphere. The merit of her work is the estimate of contributions made by various factors to the thermal regime of the atmosphere and calculation of the heat budget components at different latitudes and levels. For example, she showed that the temperature contrast at the pole and the equator increase by 13° when the latitudinal variation in the albedo is considered. She also showed that consideration of large-scale eddy transports decreases this difference by 32° at the earth surface and by 16° at an elevation of 10 km, and that consideration of the meridional change in the optical mass of the atmosphere increases it by 20—25%.

In the above, purely zonal models of the atmosphere, the problem is appreciably simplified from a mathematical point of view, since the heat influx equation does not contain terms describing the advection of temperature and the effect of vertical motions. As a result, it becomes possible to obtain a solution independently of the equations of motion. In other words, the problem is broken up into two stages: first, the heat influx equation is solved and the zonal temperature field is determined; and then equations of statics are used to calculate the gradient wind from this temperature field.

Studies on the theory of zonal-temperature formation provided the basis for the simultaneous solution of the equations of motion and of heat influx, and for formulating the problem of the formation of nonzonal fields of meteorological elements.

The problem of steady-state nonzonal temperature distribution was solved by Blinova /3/, who considered simultaneously the vorticity equation, and the equations of heat influx and of "gray" radiative transfer. These were linearized relative to the zonal state, the motion was regarded as quasisolenoidal, and vertical fluxes were regarded as nonexistent. Later Blinova /5/ generalized the above results to a periodic process and solved the problem of the yearly march of the nonzonal temperature field and the stream function. Kurbatkin /16/ used this model to calculate the temperature field at sea level in January, and obtained satisfactory agreement with the observed pattern. Musaelyan /26/ studied the effect of the orography on the formation of the steady nonzonal baric field in the middle troposphere and examined the formation of long stationary waves of orographic origin. He showed /27/ that the pressure perturbance produced by large mountainous obstacles decreases with altitude in accordance with a power law; here three waves are produced along the parallel.

This series of studies on the theory of formation of climatic fields of meteorological elements made it possible to accumulate experience in the use of hydrodynamic and thermodynamic equations for solving problems of dynamic meteorology. Of greatest importance is the development of methods for taking into account radiative processes in the atmosphere. These processes were either not considered at all in non-Soviet studies until very recently, or were taken into account only very approximately.

The above studies were restricted to considering steady or periodic processes in the atmosphere. However, the study of unsteady large-scale processes is an important aspect of the theory of the general circulation of the atmosphere. It should be noted that the consideration of these processes in the atmosphere brings studies on the theory of the general circulation significantly closer to prognostic studies.

The dynamics of long waves in the atmosphere is an important subject in the study of unsteady large-scale processes, since these waves are an important factor in the formation of large-scale processes. The development of long waves is accompanied by significant energy transformations, by the transfer of heat, by angular momentum, etc. At the same time the model of large waves is a very convenient apparatus for research purposes.

The dynamics of long waves has been studied by many investigators. Without discussing these studies in detail, we note only that very interesting and important results were obtained in the study of the stability of these waves. It was established, for example, that for certain values of vertical wind shear, as well as in the presence of horizontal shift, the waves may be found to be unstable (their amplitude may, with time, increase without limit). In the majority of cases long and short waves are stable, while waves of intermediate size are unstable. It should, however, be remembered that these conclusions were obtained within the framework of the linear problem. The behavior of long waves in the atmosphere was also considered in a number of studies pertaining to the hydrodynamic theory of long-range forecasting. Blinova /33, 8, 9/ has surveyed these studies. The above studies on the theory of the general circulation and of the climate allowed significant advances to be made in explaining the formation of planetary meteorological fields. The role of certain factors in the formation of the atmospheric temperature field could be estimated. A quantitative theory of the climatic temperature distribution for both the zonal and the horizontal atmospheric models was constructed. In the latter case methods were given for the simultaneous solution of the equations of dynamics and thermodynamics for large-scale atmospheric processes, and the effect of temperature advection in the study of the formation of the nonzonal temperature field was taken into account. The theoretical results were found to be in quite satisfactory quantitative agreement with observational data. At the same time it should be noted that the complexity of the problem imposed a number of limitations and simplifications, among which are the following.

The system of equations describing atmospheric processes is nonlinear. Solution of equations of this kind involves great mathematical difficulties, for which reason these equations were linearized. The majority of studies used the model of the "gray" atmosphere, which provided a relatively approximate description of radiative processes.

Insufficient attention was paid to a consideration of differences in the properties of various types of surfaces. In many cases it was necessary to use averaged values of parameters which vary perceptibly in space. Heat expended in evaporation and condensation was ignored and the atmospheric hydrological cycle was not considered. This list of simplifications, as well as of problems to be considered, can obviously be enlarged.

New prospects in the development of the theory of the general circulation of the atmosphere and of the climate appeared with the advent of high-speed computers. It became possible to use more effective mathematical methods to solve more complex problems, and consequently to take into account many more weather-producing factors. The use of electronic computers allowed one to solve steady-state and periodic problems in the theory of the general circulation of the atmosphere as well as unsteady processes, and to study developing large-scale processes. An important point here is the feasibility of the extensive use of numerical experiments as a means of studying atmospheric macroprocesses.

Studies in this direction were initiated by Phillips /36/, who solved the hydrodynamic equations for a model with a number of typical features of the general circulation. He considered the quasigeostrophic motion of the "β-plane."* The equations were written for a two-level atmospheric model. The effect of dissipative forces was considered and the heat influx was

* $\beta = \frac{dl}{dy}$ describes the variation in l, the Coriolis factor, with latitude; the y-axis is aligned with the meridian.

taken into account (as a linear function of y). First the formation of zonal flux from the state of rest is considered. When the zonal circulation had formed, random disturbances were imposed on it. Subsequent calculations (for 40 days) demonstrated the formation and evolution of waves, which are similar in many respects to baric formations on aerological charts, presence of a jet stream, formation of three meridional circulation cells, etc. Phillips' experiment allowed quite a detailed analysis of energy transformation to be carried out.

Among subsequent studies is the work of Mintz /35/, who compared the development of long waves within the frameworks of the nonlinear and the linear problems, the study by Lorenz /34/ of the nonlinear interaction between the zonal current and nonzonal formations, and calculations of the evolution of large-scale processes over a long period of time, carried out by Smagorinsky, Mintz, and others. The work of Smagorinsky /37/ is a direct continuation and generalization of the work by Phillips /36/. The generalization consists primarily in the use of "complete" equations, from which the external gravitational waves were "filtered out," and in the consideration of the kinematics of motion over the globe. The motion in a channel with smooth walls located at the equator and 64.4°N is considered within the framework of the two-level model of the atmosphere. Dissipative forces are taken into account more precisely (the ground-surface friction is regarded as proportional to the square of the wind speed at the ground; the internal convective stresses are proportional to the vertical wind shear; consideration is given to mixing with a variable exchange coefficient) and the heat influx is allowed for. The latter is represented as the difference between the averaged value of short-wave solar radiation absorbed by the earth and the atmosphere, and the radiation which is proportional to the temperature at the mid-level of the atmosphere.

Water vapor condensation is taken into account by reducing the static stability. As in /36/, the formation of zonal circulation is considered first. After this circulation forms, temperature disturbances of amplitude 2.5° are imposed on it; the area integral due to the temperature disturbances is equal to zero. After ten days the small details of the initial state are damped out, waves with numbers 5 and 6 are formed, and cyclical evolution of the synoptic state starts. Of greatest interest is the formation of an index cycle with a variation period of 11 days. A detailed analysis is carried out of various factors which play a role in the development of the process (energy transformation, angular momentum transfer, heat transfer, etc.).

The work of Smagorinsky /37/, as well as numerous experiments carried out by other investigators, showed the effectiveness of this approach to the study of the general circulation of the atmosphere. It was shown that one can integrate the equations of motion over a large time interval (tens of days and months), obtain a number of physical conclusions as to the manner in which the processes develop, estimate the role of individual factors, and so on.

At the same time it is obvious that the models for describing large-scale atmospheric processes used in these studies were quite approximate. The use of a number of substantial simplifications (for example, the rough approximation of the vertical atmospheric structure, simplified specification of heat influx, assumption of homogeneity of surface properties, artificial boundary conditions, the use of unreal initial fields of meteorological elements, etc.) resulted in excessive schematization of the processes, so that the models used were far from conditions actually prevailing.

The first steps are only now being made in the use of numerical experiments in atmospheric studies. At present there exist quite refined physical models for studying atmospheric macroprocesses, and methods for solving the problems have been developed. One should mention primarily the work of Blinova /6, 7/, who has formulated and solved the three-dimensional nonlinear problem of the simultaneous determination of climatic as well as of unsteady values of meteorological elements. She takes into account the turbulent and the radiative heat influxes. The quasi-solenoidal property is not assumed, and a solution is sought for the stream function, velocity potential, vertical velocity, temperature, and geopotential. The motion is regarded as occurring over a spherical globe. The variation of the solar energy influx in time and over the earth surface is considered. The boundary conditions at the surface are such that it is possible to describe the continuous heat exchange between the atmosphere and ocean or land.

The present author /22/ has suggested a theoretical model for the study of the general circulation of the atmosphere. He considers a quasi-solenoidal motion above a spherical globe. The nonlinear system of equations is solved for an arbitrary initial distribution of meteorological elements and for a time-varying solar energy influx. In addition to radiative and turbulent heat exchange, consideration is given to heat influx due to condensation, with the amount of condensed moisture determined as a part of the forecast. Allowance is made for differences in the surface properties (oceans and continents). The first numerical experiments checked on the basis of this model show that the steady zonal flux taken as the initial condition produces disturbances due to nonadiabatic factors; circulation factors come into play only after several hours.

The theoretical model suggested by Marchuk /18/ accounts very well for the various forms of energy. An important item is the inclusion of radiative processes which occur under cloudy skies. Numerical experiments on the basis of models of this kind may yield more information on large-scale atmospheric processes. It should, however, be noted that the implementation of solutions (e.g., those described in /6, 7, 22/) with the aid of electronic computers is very difficult. In addition, within the framework of their rather general statement it is frequently difficult to analyze the role of individual factors and processes. Therefore there is still much validity in considering simpler models and in studying individual processes and their interrelationships. Numerical experiments make possible the effective study of various important mechanisms, and the combination and study of their interaction. This approach appears to be very fruitful, since one of the central problems of the theory of the general circulation of the atmosphere is probably the study of the interrelationship of processes of various durations and scales. Reference is made here primarily to nonlinear interactions. As an example we cite the study of the interaction of baric formations of different size, in particular the interaction of the zonal flux and nonzonal disturbances. Reference /24/ considers the mutual effect of zonal circulation and of long waves. It considers quasisolenoidal motion of a barotropic atmosphere above a spherical earth. The evolution of long waves is described by the vorticity equation linearized relative to the zonal flux, whose velocity is an arbitrary function of the latitude and time. Changes in the zonal flux are described by a nonlinear equation of motion, averaged over the latitude circle. The solution of these equations is sought simultaneously in time steps. The changes in the zonal velocity are determined from nonzonal values of the stream function, found at the preceding step, while the vorticity equation for the nonzonal disturbance is solved with allowance for the zonal velocity profile, also obtained at the preceding step. In this manner the nonlinear interaction is partially taken into account. It was calculated that even in this approximate calculation one may obtain substantial differences in the development of the process (the amplitude of the disturbance is smaller, the meridional distribution of the amplitude of the disturbance and of the zonal velocity is different from the case when interaction is not considered, natural oscillations set in and an index cycle forms).

The study of energy conversion in the atmosphere is important for the study of the interaction of different-scale processes. A deeper study of macroprocesses can then be undertaken and the relative significance of various factors in their development evaluated. However, this involves a number of difficulties. For example, one of the problems not yet solved is the analysis of dissipative forces. Although the main factors leading to dissipation are known and the total energy dissipation in the atmosphere can be estimated, a quantitative description of the individual mechanism and clarification of the role of energy dissipation devolving upon each of these mechanisms involves certain difficulties. These difficulties also arise when carrying out numerical experiments in which the derivatives are approximated by finite differences. The introduction of finite differences actually removes from consideration objects of study whose dimensions do not exceed the resolving power of the grid. Consequently, the energy transfer between these small-scale disturbances and between formations whose size exceeds the grid spacing are not considered. An estimate of resulting errors, as well as an estimate of the dissipation, can be obtained only in the course of appropriate numerical experiments.

One of the more important problems of the general circulation theory is the determination of the "sensitivity" of large-scale processes to changes in various factors, and the clarification of the reaction of the given processes on the operation of individual mechanisms. Resolution of this problem will enable one to determine which of the factors should be taken into account and the amount of detail needed in carrying out the calculations. Some studies are being carried out in this direction. In particular, some results can be quoted as to the role of the radiative heat influx in the formation of large-scale processes. For example, the problem is solved in /23/ for the case of zonal circulation, where effects of eddy and radiative heat exchange were taken into account in the heat influx equation. An estimate of the heat-influx effect was made by Shvets and Shneerov /31/ and by Meleshko /25/ for the case of nonzonal circulation. These studies are discussed in greater detail in the article by Rakipova and Shneerov.

Many experiments in the study of the general circulation of the atmosphere can be carried out only after developing algorithms of the solution which will ensure computational stability over a long period of time. It is also important that these algorithms be amenable to the introduction of additional factors. The "uncoupling" method, suggested by Marchuk /19/, shows great promise.

References

1. Berlyand, O.S. Raspredelenie atmosfernogo davleniya na poverkhnosti zemli v sluchae statsionarnoi zonal'noi tsirkulyatsii atmosfery (Distribution of the Terrestrial-Surface Atmospheric Pressure in the Case of Steady Zonal Atmospheric Circulation). — Izvestiya AN SSSR, seriya geograficheskaya i geofizicheskaya, 14(3). 1950.
2. Blinova, E.N. Gidrodinamicheskaya teoriya voln davleniya, temperaturnykh voln i tsentrov deistviya atmosfery (The Hydrodynamic Theory of Pressure and Temperature Waves and of Atmospheric Action Centers). — Doklady AN SSSR, 39(7). 1943.
3. Blinova, E.N. K voprosu o srednem raspredelenii temperatury v zemnoi atmosfere (The Average Temperature Distribution in the Terrestrial Atmosphere).— Izvestiya AN SSSR, seriya geograficheskaya i geofizicheskaya, No. 1. 1947.
4. Blinova, E.N. K voprosu ob opredelenii davleniya na urovne morya (The Determination of Pressure at Sea Level). — Doklady AN SSSR, 92(3). 1953.
5. Blinova, E.N. K teorii godovogo khoda nezonal'noi tsirkulyatsii atmosfery zemli (The Theory of the Annual Variation in the Nonzonal Circulation of the Terrestrial Atmosphere). — Trudy Instituta Fiziki Atmosfery AN SSSR, No. 2. 1958.
6. Blinova, E.N. Gidrodinamicheskaya teoriya klimata i dolgosrochnogo prognoza pogody (The Hydrodynamic Theory of the Climate and of Long-Range Weather Forecasting). — Doklady AN SSSR, 140(3). 1961.
7. Blinova, E.N. Gidrodinamicheskoe issledovanie nestatsionarnykh atmosfernykh protsessov planetarnogo masshtaba (A Hydrodynamic Study of Planetary-Scale Unsteady Atmospheric Processes). — Doklady AN SSSR, 140(2). 1961.
8. Blinova, E.N. Sostoyanie i perspektivy razvitiya gidrodinamicheskoi teorii klimata i dolgosrochnogo prognoza pogody (The State and Prospects for Development of the Hydrodynamic Theory of Climate and of Long-Range Weather Forecasting). — Trudy MMTs, No. 2. 1964.
9. Blinova, E.N. Sostoyanie i blizhaishie zadachi gidrodinamicheskoi teorii klimata i dolgosrochnogo prognoza pogody (The State and the Immediate Problems of the Hydrodynamic Theory of Climate and of Long-Range Weather Forecasting). — Izvestiya AN SSSR, seriya geofizicheskaya, No. 1. 1964.
10. Blinova, E.N. and G.I. Marchuk. K teorii godovogo khoda chisto zonal'noi tsirkulyatsii (The Theory of the Annual Variation of Purely Zonal Circulation). — Trudy Instituta Fiziki Atmosfery AN SSSR, No. 2. 1958.
11. Dorodnitsyn, A.A. Raspredelenie atmosfernogo davleniya po poverkhnosti zemli v statsionarnoi zonal'noi tsirkulyatsii atmosfery (Distribution of Atmospheric Pressure at the Earth Surface in a Steady Zonal Atmospheric Circulation). — Trudy GGO, No. 18. 1937.
12. Dorodnitsyn, A.A., B.I. Izvekov, and M.E. Shvets. Matematicheskaya teoriya obshchei tsirkulyatsii (The Mathematical Theory of the General Circulation). — Meteorologiya i gidrologiya, No. 4. 1939.

13. Kibel', I.A. Opredelenie temperatury v zemnoi atmosfere (Determination of Temperature in the Terrestrial Atmosphere). — Doklady AN SSSR, 39(1): 1943.
14. Kochin, N.E. Ob uproshcheniyakh uravnenii gidromekhaniki dlya sluchaya obshchei tsirkulyatsii atmosfery (Simplification of Hydromechanical Equations for the General Circulation of the Atmosphere). — Trudy GGO, No. 4. 1935.
15. Kochin, N.E. Postroenie modeli zonal'noi tsirkulyatsii atmosfery (Construction of a Model of Zonal Atmospheric Circulation). — Trudy GGO, No. 10. 1936.
16. Kurbatkin, G.P. Vliyanie krupnykh orograficheskikh prepyatstvii na zapadno-vostochnyi potok (Effect of Large Terrain Obstacles on the West—East Current). — Izvestiya AN SSSR, seriya geofizicheskaya, No. 2. 1958.
17. Marchuk, G.I. Godovoi khod indeksa tsirkulyatsii (Annual Variation of the Circulation Index). — Trudy Instituta Fiziki Atmosfery AN SSSR, No. 2. 1958.
18. Marchuk, G.I. Teoreticheskaya model' prognoza pogody maloi zablagovremennosti (A Theoretical Model of Short-Range Weather Forecasting). — Izvestiya AN SSSR, seriya geofizicheskaya, No. 5. 1964.
19. Marchuk, G.I. Novyi podkhod k chislennomu resheniyu zadachi prognoza pogody (A New Approach to the Numerical Solution of Weather Forecasting Problems). — Meteorologiya i gidrologiya, No. 5. 1964.
20. Mashkovich, S.A. Raspredelenie temperatury po vysote na raznykh shirotakh severnogo i yuzhnogo polushariya (The Vertical Temperature Distribution at Different Latitudes of the Northern and Southern Hemispheres). — Trudy TsIP, 21(48). 1950.
21. Mashkovich, S.A. O sezonnykh izmeneniyakh struinykh techenii (Seasonal Variations of Jet Streams). — Meteorologiya i gidrologiya, No. 6. 1956.
22. Mashkovich, S.A. Teoreticheskaya model' dlya izucheniya formirovaniya obshchei tsirkulyatsii atmosfery i klimaticheskikh polei meteoelementov (A Theoretical Model for the Study of the Formation of the General Circulation of the Atmosphere and of Climatic Fields of Meteorological Elements). — Trudy TsIP, No. 78. 1958.
23. Mashkovich, S.A. O formirovanii zonal'noi tsirkulyatsii (The Formation of Zonal Circulation). — Trudy TsIP, No. 78. 1958.
24. Mashkovich, S.A. O vzaimodeistvii zonal'noi tsirkulyatsii i nalozhennykh na nee nezonal'nykh vozmushchenii (Interaction between Zonal Circulation and Nonzonal Disturbances Imposed upon it). — Izvestiya AN SSSR, No. 6. 1964.
25. Meleshko, V.P. K voprosu o chislennom prognoze oblachnosti i osadkov s uchetom transformatsii vozdushnykh mass (Numerical Forecasting of Clouds and Precipitation with Consideration of Air-Mass Modification). — Trudy GGO, No. 151. 1964.
26. Musaelyan, Sh.A. Planetarnye orograficheskie volny v zapadnom potoke (Planetary Orographic Waves in the Western Current). — In: "Voprosy dinamicheskoi meteorologii," Izdatel'stvo AN SSSR. 1960.

27. Musaelyan, Sh. A. Prostranstvennaya zadacha obtekaniya nerovnostei zemnoi poverkhnosti s uchetom sferichnosti zemli (A Three-Dimensional Problem of Flow Past Terrestrial-Surface Obstacles with Consideration of Sphericity of the Earth). — Doklady AN SSSR, 103(5). 1955.
28. Rakipova, L.R. Teplovoi rezhim atmosfery (The Thermal Regime of the Atmosphere). — Gidrometeoizdat. 1957.
29. Chekirda, A.Z. K gidrodinamicheskoi teorii tsentrov deistviya atmosfery (The Hydrodynamic Theory of Atmospheric Action Centers). — Izvestiya AN SSSR, seriya geofizicheskaya, No. 7. 1957.
30. Shvets, M.E. Model' statsionarnoi tsirkulyatsii atmosfery (A Model of Steady-State Atmospheric Circulation). — Meteorologiya i gidrologiya, No. 1. 1939.
31. Shvets, M.E. and B.E. Shneerov. K uchetu rezul'tatov radiatsionnykh izmerenii so sputnikov v neadiabaticheskoi modeli atmosfernykh dvizhenii (Incorporation of Results of Satellite Radiation Measurements into a Nonadiabatic Model of Atmospheric Motions). — Doklady AN SSSR, 152(3). 1963.
32. Shekhter, F.N. K vychisleniyu luchistykh potokov tepla v atmosfere (Calculation of Radiative Heat Fluxes in the Atmosphere). — Trudy GGO, No. 22. 1950.
33. Blinova, E.N. Hydrodynamic Methods of Short- and Long-Range Weather Forecasting in the USSR, pt. 2. Long-Range Forecasting. — Tellus, 9(4). 1957.
34. Lorenz, E.N. Maximum Simplification of the Dynamic Equations. — Tellus, 12(3). 1960.
35. Mintz, Y. Numerical Studies of Planetary Circulations. — Dept. Meteorol., Univ. California, Los Angeles, Final Report, Contr. NAE, 19(604). 1963.
36. Phillips, N.A. The General Circulation of the Atmosphere: A Numerical Experiment. — Quart. J. Roy. Meteorol. Soc., 82(352). 1956.
37. Smagorinsky, J. General Circulation Experiments with the Primitive Equations, pt. 1. The Basic Experiment. — Monthly Weather Review, 91(3). 1963.

M. I. Yudin

FACTORS RESPONSIBLE FOR UNSTEADINESS OF THE GENERAL CIRCULATION OF THE ATMOSPHERE

The starting premise of the present article is representation of the atmosphere as a nonlinear oscillatory system with many degrees of freedom and in close interaction with the surface (continents and oceans).

The fundamental modes of oscillation are determined by expanding the meteorological fields in terms of natural orthogonal functions. Fields of these functions are treated as the relative amplitudes of oscillation of a standing wave. It is shown that the two modes of world weather suggested by Walker, viz., the North Atlantic and the Pacific fluctuations, are actually elements of a single principal oscillation, which is described by the first natural component of the baric field.

The time during which the sign of the oscillation remains constant (i.e., the relaxation time) is examined. It is shown that the first two terms of the expansion have a relaxation time of from 2 to 3 weeks, i.e., exceeding several-fold the relaxation time of atmospheric processes calculated with the aid of data from several randomly selected locations.

One of the primary tasks of meteorology is to determine the factors responsible for the basic features of the circulation of the atmosphere during seasons, months, and shorter periods of time for different years. This necessitates developing the hydrodynamic theory of planetary atmospheric processes, and the extensive use of numerical experiments and of data obtained from analysis of results of long-range hydrodynamic weather forecasts. New methods of statistical analysis of actual data must also be developed and utilized. Obviously, if one does not consider the relatively well investigated question of the annual variation of meteorological elements, then there are very few established facts which can shed light on the nature of the unsteady circulation of the atmosphere. To prevent this from slowing down the solution of the problem as a whole, one must know the methods for describing and empirically studying planetary atmospheric processes in accordance with the basic physical information on the general circulation. Statistical work should also be developed much more quickly. This is possible only on the basis of quantitative characteristics of atmospheric processes and by markedly increasing the volume of computer computations.

What additional requirements must be satisfied for an empirical analysis of planetary processes in the atmosphere? First of all it is obvious that any acceptable explanation of circulation changes during the same seasons of different years should take into account the interaction between the ocean and the atmosphere as well as that between the atmospheric and the active layer of the soil. It follows from this that in statistical studies of atmospheric processes of extended duration, as well as in hydrodynamic analysis, one must, in addition to the pressure field, also consider the characteristics of fields determining the magnitude of heat and moisture influx to the atmosphere.

A number of papers have recently been published /5, 7, 8, 13, 17/ which describe the relationship between the subsequent weather and the location of the boundary of the snow cover, heat content of water in the ocean, ice and cloud covered areas, values of radiative fluxes, soil moisture, etc. The validity of these relationships will no doubt be verified by future data, and the principal properties of the atmosphere, as a complex nonlinear oscillatory system with many degrees of freedom, will be better understood. In order to construct even a very approximate initial theory of processes developing within the system one must separate empirically the main modes of oscillation and write equations of the dynamics of the atmosphere and boundary conditions for this initially simplified model. In a certain sense this approach is similar to the procedure of "noise filtering," which was to develop a number of schemes of hydrodynamic numerical forecasting and is now a basic feature of the methods actually in use. From this point of view short-period local atmospheric processes represent noise, whose effect should be studied not individually, but statistically.

Let us state more specifically the idea that the principal modes of fluctuations should be detected on the basis of an empirical (i.e., statistical) analysis of real processes. In principle, it is more logical and correct to use another approach, viz., the modes are found by solving the equations of the dynamics of the atmosphere. In particular, the regularities of the vertical distribution of elements can be explained to a large extent by resorting to precisely these equations /6, 15/. However, the theoretical determination of the relationship between synchronous oscillations in different regions of the globe is at present extremely complex. If one were to wait for its solution, the time when one could formulate a starting model of planetary atmospheric processes which would be close to the actual picture would be delayed appreciably. It is probably simpler to give a theoretical explanation of the properties of oscillations using this starting model as an illustration. Hence the Dynamic Meteorology Department of the Main Geophysical Observatory studies unsteady circulation of the atmosphere by combining the hydrodynamic and statistical methods. In our opinion the development of the combined hydrodynamic and statistical approach is the most promising for developing reliable methods of long-range weather forecasting.

In keeping with this general approach, primary attention was first paid to preparing starting statistical data. The principal method used to describe meteorological fields was expansion in terms of natural orthogonal functions. It is known that this is the best of all the possible methods of expanding a random function $F(\vec{r}_k, t)$ in terms of a complete system of orthogonal functions $f_j(\vec{r}_k)$, in the sense that natural functions $f_j(\vec{r}_k)$ are obtained from the requirement that the rms error of the expansion be at a minimum. This is expressed by

$$D_n(F) = \sum_{k=1}^{N} \overline{\left[F(\vec{r}_k, t) - \sum_{j=1}^{n} T_j(t) f(\vec{r}_k)\right]^2} = \min, \qquad (1)$$

where $n = 1, 2, \ldots, N-1$. Here $F(\vec{r}_k, t)$ is a random function specified at N points $\vec{r}_k$ $(k = 1, 2 \ldots N)$; $f_j(\vec{r}_k)$ is a natural function $(j = 1, 2 \ldots n)$; $T_j(t)$ is the corresponding coefficient of expansion, representing a random function of time t; the bar denotes averaging with respect to time. The rms error

$D_n(F)$ is frequently called the residual variance. This quantity is easily found from the same standard programs used for calculating the modes of natural functions $f_j(\vec{r}_k)$. Hence, by specifying the allowable relative error of the expansion it is immediately possible to determine the number of terms to be retained. We shall not consider here the theory and techniques of expansion in terms of natural functions, since these are given in sufficient detail in /2, 4, 16/; a number of methodological problems are also examined in /3, 12, 20, etc./.

However, one very important property of expansion should be pointed out, viz., the mode of the function is uniquely related to its number. It is most convenient to explain this property for the case when one considers values of the function $F(\vec{r}_k, t)$, and by virtue of this fact, also of function $f_j(\vec{r}_h)$ at a given isobaric surface.

It is additionally assumed that the mathematical expectation of function F is zero, i.e., that the climatic norm was subtracted from the values of the corresponding meteorological element. Function F is represented in the form

$$F(\vec{r}_k) = \sum_{k=1}^{N} T_j(t) f_j(\vec{r}_k), \tag{2}$$

the value of each function $f_j(\vec{r}_h)$ at points $\vec{r}_k$ are plotted on charts, and isolines are drawn, while values of $T_j(t)$ are represented on time graphs. It is now easy to see that each term of the sum in equation (2) can be interpreted as the value of some element of a standing-wave type oscillatory motion.* Here the maxima and minima of the function $f_j(\vec{r}_k)$ correspond to antinodes of the oscillations, while the isoline at which $f_j(\vec{r}_k) = 0$ is a nodal line. In this graphical interpretation, the first natural functions (which, by definition, make the greatest contribution to the expansion) in almost all the cases considered are represented by relatively smooth large-scale fields.

If the expansion is carried out for an area commensurable with the dimensions of a planet (for example, the northern hemisphere), then it will be found that regions for which the natural functions are of the same sign are very large. This means that the first natural functions describe oscillatory processes on a planetary scale.

Increasing the number of the function decreases L_s, the spatial scale of the field, which is defined by the expressions

$$\frac{1}{L_{s,j}^2} = \frac{1}{L_{x,j}^2} + \frac{1}{L_{y,j}^2}; \quad \frac{1}{L_x} = N^{1/2} O\left[\frac{\partial f_j}{\partial x}\right], \quad \frac{1}{L_y} = N^{1/2} O\left[\frac{\partial f_j}{\partial y}\right]$$

(see /8/, Chapter II), so that functions $f_j(\vec{r}_k)$ with large numbers represent small-scale fields with many local maxima and minima. This means that when the field of a meteorological element is described by a small number of natural functions (terms of expansion), it is possible to filter out and not consider inconsequential small-scale disturbances, i.e., to solve the problem at hand.

* But, obviously, not of an elementary periodic oscillation, since $T_j(t)$ is not represented in the form of a simple sinusoid.

This approach is only possible if new statistical regularities can be detected. This can only be done by excluding "background noise," i.e., the effect of random small-scale disturbances.

We now present some results which verify the correctness of the above approach. Figure 1 shows the mode of the first natural function obtained by expanding the pressure field at sea level. The function $F(\vec{r}_k, t_m)$ was specified for 26 points located mainly at high and middle latitudes of the northern hemisphere. All the instants t_m $(m = 1, 2. . . 150)$ pertain to the winter season and were determined by random sampling. A more detailed description of the use of the network of locations and of the entire technique used in the investigation can be found in /12/. It should be emphasized that the average magnitude of the oscillation whose mode is described by the first natural function is quite appreciable and comprises more than 40% of the total rms pressure amplitude (averaged over all the points under study).

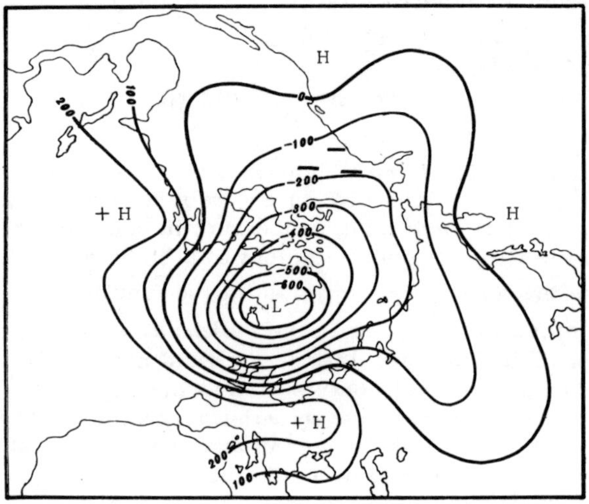

FIGURE 1. Mode of the first natural function obtained by expanding the pressure field at sea level. The plus sign is attached to points at which the pressure in the Walker equation is positive, while the minus sign denotes the opposite cases.

We now analyze the mode of oscillation depicted in Figure 1. It describes a planetary process under which the pressure in the entire polar region, as well as in a large part of the moderate belt, increases or decreases synchronously. The phase of the oscillation reverses itself in the lower latitudes. Thus, the oscillation being considered is related to the transport from south to north, or in the opposite direction, of large air masses. However, the iso-amplitudes of this oscillation are asymmetrical with respect to the pole. The antinode of the standing wave is located in the region of Greenland, while the nodal line has quite perceptible wavy inflections. In a number of regions the direction of the nodal line is almost parallel to the shoreline.

The large gradients of the function $f_1(\vec{r}_k)$ close to the shoreline also suggest that the mode of the first natural pressure oscillation is related to a certain degree to the effect of the underlying surface.

These preliminary conclusions obviously require verification. The oscillatory mode must be plotted more exactly by using a large number of points and varying synoptic situations.

We now consider differences in the oscillatory phases at high and low latitudes. This property recalls the conclusions of Walker /18, 19/, who detected this phase difference in two regions of the northern hemisphere.

Walker defined the North Atlantic and North Pacific oscillations as the principal forms of the world weather, and obtained formulas for describing the oscillatory phase on the basis of the average winter (December—February) temperature and pressure at a number of places.

We shall now attempt to compare the present results with those of Walker. The plus sign in Figure 1 is assigned to all points where the pressure carries a positive sign in the Walker formula, and the minus sign in the opposite cases. Naturally, the relationship between the temperature and pressure cannot be used. As can be seen from Figure 1, all the pluses and minuses belong to the corresponding regions of the first principal oscillation. This result, which could be expected a priori, yields a number of important conclusions.

1. Walker used data for 1875—1930, while the information in /11/ is for 1951—1960. The fact that the nature of the oscillations during these rather far removed periods is the same shows that this is a constant element of the general circulation of the atmosphere.

2. Walker obtained his conclusions by using very pronounced smoothing with respect to time (over the season). By expanding in terms of natural functions it is possible to go even further. It shows that the first oscillation is an appreciable component of pressure oscillations during individual days.

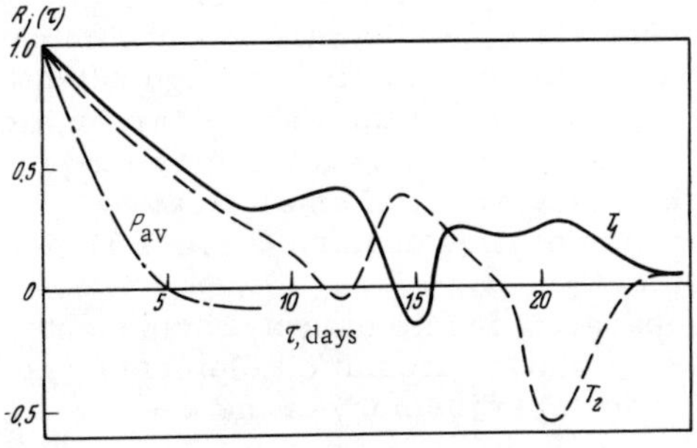

FIGURE 2. Graph of correlation functions of the form $R_j(\tau) = T_j(t)\,T_j(t+\tau)/T_j^2(t)$ for the first two terms of the expansion of the pressure at sea level.

3. This comparison also showed that the North Atlantic and North Pacific forms of the world weather discovered by Walker represent elements of a single fundamental oscillation, extending to at least the subtropic, moderate, and polar zones of the entire northern hemisphere.

Up to now the method of expanding the pressure field in terms of natural functions has not been applied to the tropical zone of both hemispheres, since a number of pertinent conclusions by Walker and others have not as yet been correlated for these areas.

We now consider the average duration during which the function $T_j(t)$ retains its sign for different j. This is a question of considerable importance for the theory of planetary processes and has great practical significance for long-range weather forecasting.

It is rather natural that the amplitudes and phases of oscillations for a complex, nonlinear self-oscillating system, such as the atmosphere, do not coincide from year to year. However, this general statement is insufficient for determining the main trend of work on long-range forecasting. One must know first of all to what extent the differences in the weather of the same seasons for different years are related to oscillations of different duration and which elements serve as the main storage units of "meteorological memory." In the opinion of this author, many-year fluctuations play a marked, but not principal role in the formation of the weather of a specific month and season. At the same time it is shown by direct analysis of correlations between values of meteorological elements for adjoining years that the relaxation time thus determined amounts to three to five, and in no case more than seven days. Hence the question as to whether a real correlation exists in the atmosphere for time periods of about a month or whether the observed circulations of a month and a season are mainly the result of the random superposition of relatively short-period processes still remains open. Baur's proof of nonrandomness of weather differences over extended periods of time is compatible with the hypothesis that the relaxation time does not exceed a week. A number of other proofs of this hypothesis do not stand up to statistical examination.

It could be expected, from the assumption that the characteristic period of an oscillation depends on its spatial scale, that expansion in terms of natural functions would make it possible to isolate long-period atmospheric fluctuations. This was in fact confirmed by calculations carried out by the Dynamic Meteorology Department of the Main Geophysical Observatory. Correlation functions were calculated /12/ in the form

$$R_j(\tau) = \frac{\overline{T_j(t)\, T_j(t+\tau)}}{\overline{T_j^2(t)}}.$$

Figure 2 shows graphs of these functions for the first two components of the expansion of the pressure at sea level. The correlation function of P_{av}, the average pressure for Europe, taken from /10/, is also plotted for comparison. It is seen from the figure that the characteristic time of sign retention for oscillations described by the first components of the field is two or three weeks. At the same time, even when pressure is averaged over a large area (Europe) it is still impossible to eliminate the effect of small-scale disturbances if the entire area under consideration does not lie in a region in which $f_1(\mathbf{r}_k)$ or $f_2(\mathbf{r}_k)$ possess the same sign.

Analysis of the pressure field by separating it into natural components thus makes it possible to isolate explicitly those circulation elements which serve as storage cells of a relatively long-term memory of the past circulation history.

When looking for memory elements of an even longer duration, one must consider elements describing the state of the surface and of the active layer of the water and soil, mentioned at the start of the article. Work on the natural expansion of these fields has just started. We shall consider here the description of the inertia of snow-cover boundaries. The numerical characteristic of the snow-cover boundary used in /1, 9/ is the latitude $\varphi(\lambda, t)$ at which the line connecting the points at which a snow cover was observed for five out of ten days intersected the meridian λ. The corresponding time unit was ten days. Subsequently, the function $\varphi(\lambda_j, t_m) - \overline{\varphi}(\lambda_j, t_m)$ was expanded in terms of natural components.

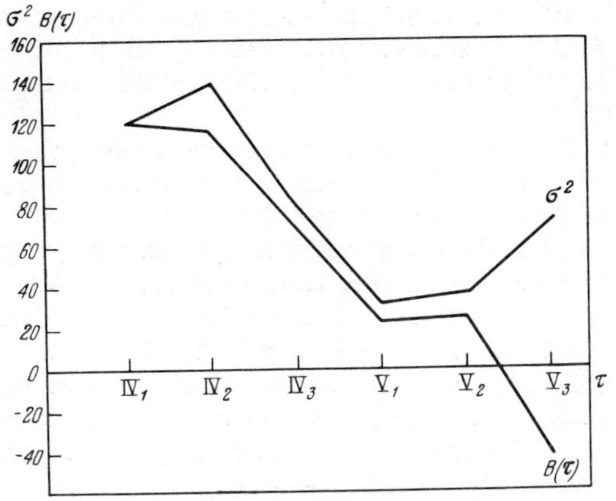

FIGURE 3. Graph of covariance and the annual variation in variance.

Figure 3 shows a graph of the covariance

$$\overline{T_{\varphi,1}(t_{0,k}+\tau)\,T_{\varphi,1}(t_{0,k})},$$

where $T_{\varphi,1}$ is the first coefficient of the expansion of function φ; $t_{0,k}$ is the first ten-day period in April of the k-th year. The figure also gives the variance of the first parameter of the expansion, which allows one to calculate the correlation factor. Unfortunately, the short period of observations (eight years) is not enough for definite conclusions to be made as to the relaxation time. It appears probable, however, that it is markedly greater than the relaxation time for the pressure field.

These statistical studies are continuing at present and the author intends to construct a starting dynamic-statistical model of circulation fluctuations which will produce results close to those actually observed.

References

1. Afanas'eva, V. B. and N. P. Isakova. O sposobakh kharakteristiki anomalii oblachnosti, snezhnogo pokrova i radiatsionnykh potokov (Methods for Describing Anomalies of Cloudiness, Snow Cover, and Radiative Fluxes).— Trudy GGO, No. 143. 1962.
2. Bagrov, N. A. Analiticheskoe predstavlenie posledovatel'nosti meteorologicheskikh polei posredstvom estestvennykh ortogonal'nykh sostavlyayushchikh (Analytical Representation of the Sequence of Meteorological Fields by Means of Natural Orthogonal Components).— Trudy TsIP, No. 74. 1959.
3. Meshcherskaya, A. V. and N. I. Yakovleva. Utochnenie estestvennykh funktsii polei geopotentsiala (davleniya) Atlantiko-Evropeiskogo sektora (Improving the Natural Functions of the Geopotential (Pressure) Fields of the Atlantic-European Sector).— Trudy GGO, No. 168. 1965.

4. Obukhov, A. M. O statisticheski ortogonal'nykh razlozheniyakh empiricheskikh funktsii (Statistically Orthogonal Expansions of Empirical Functions). — Izvestiya AN SSSR, seriya geofizicheskaya, No. 3. 1960.
5. Rakipova, L. R. O formirovanii srednemesyachnykh anomalii temperatury atmosfery (The Formation of Monthly Anomalies of the Temperature of the Atmosphere). — Trudy GGO, No. 99. 1959.
6. Rukhovets, L. V. Mnogourovennaya model' prognoza polya geopotentsiala, osnovannaya na malom chisle parametrov (Multilevel Model for Forecasting the Geopotential Field, Based on a Small Number of Parameters). — Trudy GGO, No. 151. 1964.
7. Semenov, V. G. Vliyanie Atlanticheskogo okeana na rezhim temperatury i osadkov na Evropeiskoi territorii SSSR (Effect of the Atlantic Ocean on the Temperature and Precipitation Regimes of the European USSR). — Gidrometeoizdat. 1960.
8. Yudin, M. I., N. P. Isakova, and V. B. Afanas'eva. Predvaritel'naya otsenka prognosticheskoi znachimosti informatsii, poluchaemoi s meteorologicheskikh sputnikov (Preliminary Estimate of the Forecasting Significance of Meteorological Satellite Information). — Trudy GGO, No. 166. 1964.
9. Yudin, M. I. Novye metody i problemy kratkosrochnogo prognoza pogody (New Methods and Problems of Short-Range Weather Forecasting). — Gidrometeoizdat. 1963.
10. Yaglom, A. M. Teoriya korrelyatsii nepreryvnykh protsessov i polei s prilozheniyami k zadache o statisticheskom ekstrapolirovanii vremennykh ryadov i k teorii turbulentnosti (The Theory of the Correlation of Continuous Processes and Fields with Applications to Statistical Extrapolation of Time Series and to the Theory of Turbulence) (Dissertation). — Geofizicheskii Institut AN SSSR. 1955.
11. Yakovleva, N. I. and A. V. Meshcherskaya. Analiz baricheskikh polei nad severnym polushariem metodom razlozheniya po estestvennym ortogonal'nym funktsiyam (Analysis of Baric Fields over the Northern Hemisphere by Expansion in Terms of Natural Orthogonal Functions). — Trudy GGO, No. 168. 1965.
12. Yakovleva, N. I., A. V. Meshcherskaya, and G. D. Kudashkin. Issledovanie polei davleniya (geopotentsiala) metodom razlozheniya po estestvennym sostavlyayushchim (Study of Pressure (Geopotential) Fields by Expansion in Terms of Natural Components). — Trudy GGO, No. 165. 1964.
13. Aden, J. The Physical Basis for the Numerical Prediction of Monthly and Seasonal Temperatures in the Troposphere Ocean-Continent System. — Monthly Weather Rev., 92(3). 1964.
14. Baur, F. Über die grundsätzliche Möglichkeit langfristiger Witterungsvorhersagen. — Ann. Hydrogr., Vol. 72. 1944.
15. Holmstrom, J. The Vertical Structure of the Atmosphere. — Tellus, 16(3). 1964.
16. Lorenz, E. N. Empirical Orthogonal Functions and Statistical Weather Prediction. — Sci. Rept. No. 1, Statis. Fest. Proj. Contr., NAF 19(604), 1566, MIT. 1956.

17. Namias, L. Influences of Abnormal Surface Heat Sources and Sinks on Atmospheric Behavior. — Proc. Internat. Sympos. Numerical Weather Prediction, Tokyo, 7—13 Nov. 1960. Meteorol. Soc. Japan. 1962.
18. Walker, G.T. and E.W. Bliss. World Weather. III. — Lond. Met. Roy. Meteorol. Soc. 2(17). 1928.
19. Walker, G.T. and E.W. Bliss. World Weather. V. — Lond. Met. Roy. Meteorol. Soc., 4(36). 1932.
20. White, R.M., R.C. Derby, D.S. Cooley, and F.A. Seaver. Hemispherical Prediction by Statistical Techniques. — J. Meteorol., 14(5). 1957.

G. I. Marchuk and T. A. Temnoeva

THE SOLUTION OF CERTAIN QUASISTEADY PROBLEMS OF DYNAMIC METEOROLOGY

A numerical method is applied, in conjunction with the method of successive approximations, for solving certain problems of dynamic meteorology. It is shown that the algorithms developed by the authors have advantages over others in solving problems of this kind.

When solving problems related to weather forecasting which involve the study of the general circulation of the atmosphere and the theory of the climate, one encounters mathematical formulations of quasisteady problems which incorporate only the basic factors continuously affecting the formation of fields of meteorological elements. Sometimes this approach markedly simplifies the picture of the hydrodynamic process.

This situation may be illustrated by solving the well-known quasisteady problem of weather forecasting. Although approaches to it in the quasi-geostrophic approximation are known and have been considered in detail, the problem has not yet been solved in the general form. As is known, the main difficulty consists in integrating the quasisteady system of Euler's equations.

For the simplest barotropic model for a given pressure field, this system has the form

$$\begin{aligned} u\frac{\partial u}{\partial x} + v\frac{\partial u}{\partial y} - lv &= -\frac{\partial H}{\partial x} \\ u\frac{\partial v}{\partial x} + v\frac{\partial v}{\partial y} + lu &= -\frac{\partial H}{\partial y}. \end{aligned} \tag{1}$$

The components (u, v) of the velocity vector are regarded as specified at the boundary (x, y) of a region G at points where the velocity vector is directed into the region (l is the Coriolis parameter).

The main problem consists in relating the components u and v to the geopotential field H or, more precisely, in finding u and v from a given field of H. As is well known, in the geostrophic approximation this relationship is specified in the form

$$u_g = -\frac{1}{l}\frac{\partial H}{\partial y}; \quad v_g = \frac{1}{l}\frac{\partial H}{\partial x}.$$

It would have been possible to construct a subsequent approximate expression, which takes into account the departure of the wind from geostrophic behavior by using nonlinear terms of the system of equations. However, the series thus constructed may be found to be divergent.

In the present paper an attempt is made to construct a solution for the problem. For this purpose we consider a new system of equations

$$\left.\begin{aligned} \frac{\partial u}{\partial t}+u\frac{\partial u}{\partial x}+v\frac{\partial u}{\partial y}-lV&=-\frac{\partial H}{\partial x}\\ \frac{\partial V}{\partial t}+u\frac{\partial V}{\partial x}+v\frac{\partial V}{\partial y}+lU&=-\frac{\partial H}{\partial y},\end{aligned}\right\} \tag{2}$$

where $u(x, y)$ and $v(x, y)$ are the exact solutions of the system of equations (1), while $H(x, y)$, as previously, is not a function of time. We use the same boundary conditions as for equation (1); the variable t represents the time.

The system of equations (2) is now solved with the following initial conditions:

$$U(x, y, 0)=U_0(x, y), \quad V(x, y, 0)=V_0(x, y). \tag{3}$$

The formal statement of the problem, unfortunately, assumes that the solutions of the exact problem, which are used as coefficients in this system of equations, are known. We shall subsequently resolve this problem using a special method of successive approximations.

It is seen by examining equations (2) and (3) that as $t\to\infty$ one generally would not need to solve steady-state problem (1) in order to solve these two systems of equations. In order to prove this it suffices to consider the simplest case of $u=0$, $v=0$, $H=0$. System of equations (2) then takes the form

$$\left.\begin{aligned} \frac{\partial U}{\partial t}-lV&=0\\ \frac{\partial V}{\partial t}+lU&=0\end{aligned}\right\} \tag{4}$$

with the condition that

$$U=U_0,\ V=V_0 \text{ for } t=0. \tag{5}$$

The solution of equations (4) and (5) for $l=\text{const}$ has the form

$$\left.\begin{aligned} U&=U_0\cos lt+V_0\sin lt\\ V&=V_0\cos lt-U_0\sin lt.\end{aligned}\right\} \tag{6}$$

However, the exact solution of the problem in this case will be $U_0=0$, $V_0=0$. Obviously, there will be no convergence when $U_0\neq 0$ and $V_0\neq 0$. Thus, the solution of the steady-state problem using equations (2) and (3) requires the use of new principles.

We assume further that the solution of the problem stated by equations (2) and (3) has been found. We now show that the solution of this unsteady problem can be used for obtaining a corresponding solution of the system of equations (1) using the smoothing operator

$$m=\frac{1}{T}\int_0^T dt. \tag{7}$$

In fact, let us apply operator (7) to the system of equations (2). Then, using initial conditions (3), we obtain for any given T

$$\left.\begin{aligned}\frac{U(T)-U_0}{T}+u\frac{\partial mU}{\partial x}+v\frac{dmU}{\partial x}-lmV=-\frac{\partial H}{\partial x}\\ \frac{V(T)-V_0}{T}+u\frac{\partial mV}{\partial x}+v\frac{\partial mV}{\partial y}+lmU=-\frac{\partial H}{\partial y},\end{aligned}\right\} \qquad (8)$$

where $U(T)\equiv U(x, y, m)$, $V(T)\equiv V(x, y, m)$.

Since solutions U and V are expected to be continuous functions, differentiable in the range $(0, \infty)$ with respect to all the variables, it follows that one can always find a T for which

$$\frac{U(T)-U_0}{T}\leqslant\varepsilon, \quad \frac{V(T)-V_0}{T}\leqslant\varepsilon.$$

In this case, to within infinitesimals of the order of ε, the system of equations (8) will approximate equations (1) and, consequently, the functions

$$mU=\frac{1}{T}\int_0^T U\,dt, \quad mV=\frac{1}{T}\int_0^T V\,dt$$

will be approximate solutions of (1). Naturally, when $T\to\infty$ the equality is exact, i.e.,

$$mU=U, \quad mV=V. \qquad (9)$$

The above equalities can be checked directly, for example, in the particular case of solution of equations (6). It is easy to see that for any specified T one has

$$\begin{aligned}mU&=\frac{U_0}{lT}\sin lT-\frac{V_0}{lT}(\cos lT-1)\\ mV&=\frac{V_0}{lT}\sin lT+\frac{U_0}{lT}(\cos lT-1),\end{aligned} \qquad (10)$$

from which it follows that when $T\to\infty$, $mU\to 0$ and $mV\to 0$.

We now remove the limitation imposed by the requirement that an exact solution is to be found. For this purpose we introduce a series of solutions $mU^{(n)}=u^{(n)}$ and $mV^{(n)} = V^{(n)}$, defined by the system of equations

$$\left.\begin{aligned}\frac{\partial U^{(n+1)}}{\partial t}+u^{(n)}\frac{\partial U^{(n+1)}}{\partial x}+v^{(n)}\frac{\partial U^{(n+1)}}{\partial y}-lV^{(n+1)}=-\frac{\partial H}{\partial x}\\ \frac{\partial V^{(n+1)}}{\partial t}+u^{(n)}\frac{\partial V^{(n+1)}}{\partial x}+v^{(n)}\frac{\partial V^{(n+1)}}{\partial y}+lU^{(n+1)}=-\frac{\partial H}{\partial y}\end{aligned}\right\} \qquad (11)$$

with the initial conditions

$$U^{(n+1)}(x, y, 0)=U_0, \quad V^{(n+1)}(x, y, 0)=V_0 \qquad (12)$$

and corresponding boundary conditions. It is convenient to use the geostrophic values of the coefficients $u^{(0)}$ and $v^{(0)}$ as the starting quantities. When using the method of successive approximations, equations (11) and (12) converge very rapidly, although, due to the complex nonlinear structure of the differentiated operators, no theoretical estimate of the convergence was obtained. One may, however, put forward a consideration emphasizing the advantage of this method. It consists in the fact that the method proper involves errors which are introduced into the coefficients of the differential equations, when all other methods, which involve the use of quasigeostrophic models and corrections to them, infringe upon the differential properties of the solutions, since they use approximate solutions of the problem.

Let us consider the above method from a somewhat different point of view. We again turn to (2) and (3) and examine them from the theoretical point of view. For given u and v, (2) and (3) are linear relative to functions U and V. Since the free terms in the equations are not functions of time, we shall seek the solution of system (2) in the form of the sum

$$\left.\begin{aligned} U &= u(x, y) + U_1(x, y, t) \\ V &= v(x, y) + V_1(x, y, t), \end{aligned}\right\} \tag{13}$$

where u and v are solutions of (1) with corresponding boundary conditions, while U_1 and V_1 are the solutions of

$$\left.\begin{aligned} \frac{\partial U_1}{\partial t} + u\frac{\partial U_1}{\partial x} + v\frac{dU_1}{\partial y} - lV_1 &= 0 \\ \frac{\partial V_1}{\partial t} + u\frac{\partial V_1}{\partial x} + v\frac{\partial V_1}{dy} + lU_1 &= 0. \end{aligned}\right\} \tag{14}$$

For zero boundary conditions, at points where the velocity vector $\vec{u} = \vec{u_i} + \vec{v_i}$ is directed into the region in which the solution is determined, and with the initial data

$$U_1(x, y, 0) = U_0, \quad V_1(x, y, 0) = V_0, \tag{15}$$

we set $U_0 = 0$ and $V_0 = 0$. We obtain a homogeneous equation for functions U_1 and V_1. It may be expected that the above homogeneous equation permits a single solution, which is trivial. In this case we arrive at an important result. The equations

$$\left.\begin{aligned} \frac{\partial U}{\partial t} + u\frac{\partial U}{\partial x} + v\frac{\partial U}{\partial y} - lV &= -\frac{\partial H}{\partial x} \\ \frac{\partial V}{\partial t} + u\frac{\partial V}{\partial t} + v\frac{\partial V}{\partial y} + lU &= -\frac{\partial H}{\partial y} \end{aligned}\right\} \tag{16}$$

with corresponding boundary conditions and zero initial conditions

$$U(x, y, 0) = 0, \quad V(x, y, 0) = 0 \tag{17}$$

approach uniformly, as $t \to \infty$, the solution of steady-state problem (1), which satisfies the corresponding boundary conditions.

This result is clear theoretically, although an approximate solution of equation (16) with conditions (17) involves errors in numerical computation which not only result in departure from uniform convergence and in deviation of the approximate from the exact solution, but do not yield a solution which converges even in the limit. It is true, however, that the scatter in the solution from one time step to another occurs only in the lower orders of the approximate solution. The steady-state solution is obtained by applying an averaging operator.

As a result we obtain the steady-state solution of the problem at hand by the shortest possible route. This, in a given sense, is the sought optimum algorithm. Obviously, all the above considerations on the method of successive approximation with respect to the coefficients are also valid in this case.

The above method is easily extended to the case of more complex problems of dynamic meteorology. We consider a model problem of the theory of climate. For simplicity, the earth is regarded as flat, since it is a rather trivial task to extend the algorithm to a spherical earth. In this case we have

$$\left.\begin{aligned} &u\frac{\partial u}{\partial x}+v\frac{\partial u}{\partial y}+\tau\frac{\partial u}{\partial P}-lv=-\frac{\partial H}{\partial x}\\ &u\frac{\partial v}{\partial x}+v\frac{\partial v}{\partial y}+\tau\frac{\partial v}{\partial P}+lu=-\frac{\partial H}{\partial y}\\ &u\frac{\partial T}{\partial x}+v\frac{\partial T}{\partial y}-\frac{\gamma_a-\gamma}{g}RT\frac{\tau}{P}=\frac{\varepsilon}{c_p}\\ &\frac{\partial u}{\partial x}+\frac{\partial v}{\partial y}+\frac{\partial \tau}{\partial P}=0,\quad T=-\frac{P}{R}\frac{\partial H}{\partial P}, \end{aligned}\right\}\tag{18}$$

where ε, the heat influx due to nonadiabatic factors, is treated as a known function of coordinates; R is the gas constant; c_p is the specific heat at constant pressure; g is the acceleration of gravity; γ is the lapse rate; γ_a is the adiabatic lapse rate; P is the pressure; T is the temperature; $\tau=\partial P/\partial t$ is the partial derivative of the pressure.

The conditions specified at the upper and lower boundaries of the atmosphere are similar to those which are usually used in weather forecasting problems. Functions u, v, and T are specified at the boundary (x, y) of a region Γ for the set of points $\gamma\leftarrow\Gamma$, where the velocity vector is directed into the region defined by the solution; at the same time function H is specified at all points of Γ.

In accordance with the above considerations, steady-state problem (18) is replaced by an unsteady problem of the form

$$\left.\begin{aligned} &\frac{\partial U}{\partial t}+u\frac{\partial U}{\partial x}+v\frac{\partial U}{\partial y}+\tau\frac{\partial U}{\partial P}-lV=-\frac{\partial H}{\partial x}\\ &\frac{\partial V}{\partial t}+u\frac{\partial V}{\partial x}+v\frac{\partial V}{\partial y}+\tau\frac{\partial V}{\partial P}+lU=-\frac{\partial H}{\partial y}\\ &\frac{\partial T}{\partial t}+u\frac{\partial T}{\partial x}+v\frac{\partial T}{\partial y}-\frac{\gamma_a-\gamma}{g}RT\frac{\tau}{P}=\frac{\varepsilon}{c_p}\\ &\frac{\partial U}{\partial x}+\frac{\partial V}{\partial y}+\frac{\partial \tau}{\partial P}=0\\ &T=-\frac{P}{R}\frac{\partial H}{\partial P} \end{aligned}\right\}\tag{19}$$

with boundary conditions of the corresponding steady-state problem and zero initial conditions

$$\left.\begin{aligned} U(x, y, P, 0)=0, V(x, y, P, 0)=0,\\ T(x, y, P, 0)=0. \end{aligned}\right\}\tag{20}$$

Since the coefficients of the equations, i.e., functions u, v, and τ, are unknown, the method of successive approximations must be used. As a first approximation we solve the set of equations

$$\left.\begin{aligned}
&\frac{\partial U}{\partial t}+U\frac{\partial U}{\partial x}+V\frac{\partial U}{\partial y}+\tau\frac{\partial U}{\partial P}-lV=-\frac{\partial H}{\partial x}\\
&\frac{\partial V}{\partial t}+U\frac{\partial V}{\partial x}+V\frac{\partial V}{\partial y}+\tau\frac{\partial V}{\partial P}+lU=-\frac{\partial H}{\partial y}\\
&\frac{\partial T}{\partial t}+U\frac{\partial T}{\partial x}+V\frac{\partial T}{\partial y}-\frac{\gamma_a-\gamma}{g}RT\frac{\tau}{P}=\frac{\varepsilon}{c_p}\\
&\frac{\partial U}{\partial x}+\frac{\partial V}{\partial y}+\frac{\partial \tau}{\partial P}=0\\
&T=-\frac{P}{R}\frac{\partial H}{\partial P}
\end{aligned}\right\} \qquad (21)$$

with the same boundary conditions and zero initial conditions (20). We use ordinary methods quite well developed in weather forecasting theory. The solution is continued until the computations result in a refinement and the result is found to lie near some average value. Integration of the solution with respect to t within some interval (T_1, T_2) yields

$$\Phi=\frac{1}{T_2-T_1}\int_{T_1}^{T_2}\varphi\,dt. \qquad (22)$$

Using the above expression one obtains approximate values for the coefficients of system (18) and the problem is refined further to the desired accuracy, which, as a rule, occurs very rapidly.

A.M. Obukhov and M.I. Fortus

STATISTICAL DESCRIPTION OF THE GENERAL CIRCULATION OF THE ATMOSPHERE

The expedience of using the statistical approach in describing the general circulation of the atmosphere is discussed. It is suggested that the statistical properties of the geopotential field, in particular its correlation structure, be studied first. Results are presented of calculations of the spatial correlation function of the geopotential field for the northern hemisphere on the assumption of azimuthal homogeneity. A theoretical model is described of the correlation structure of the geopotential field, based on the hypothesis of random distribution of the potential eddy distribution.

The general circulation of the atmosphere can be regarded as an ensemble of meteorological processes with different scales, generated by latitudinal inhomogeneity of the heat influx and the related global inhomogeneity of the temperature field. Meteorologists are usually interested only in the largest links in this chain, with the lower limit depending on the specific application of the results. Since the most important field of application in the given case is the theory of long-range weather forecasting, the spectrum of atmospheric disturbances in the study of the general circulation can be terminated in the region of scales of the order of thousands of kilometers. The ensemble of disturbances with smaller scales should be treated, in this approach, as a kind of "turbulence" (macroturbulence).

The application of statistical methods for the quantitative description of the general circulation as the common regime of motions of different scales arising in the atmosphere can be quite useful, and work has begun in this direction in a number of scientific institutions in the USSR and in the West. Apparently a program for this kind of research was first formulated in sufficiently general form by Keller /4/ on the basis of ideas due to Fridman.

The gist of this program is to find, alongside with average values of meteorological fields over a large part of the globe (which is the problem of "classical" climatology), also correlation characteristics of the field's disturbances at two different points, i.e., covariances. The consideration of covariances is of interest also in dynamic problems arising from the theory of long-range forecasting (Blinova /1/ as well as Smirnov /2/). The characteristics of large-scale meteorological processes (quasi-steadiness, quasigeostrophicity or quasisolenoidality and quasihorizontality of movements) are such that it is possible to describe the "dynamic state" of the atmosphere by a function of the coordinates only. For extratropical latitudes it is most convenient to describe the dynamics of the atmosphere in terms of the geopotential, regarded as a function of the horizontal coordinates and pressure. A statistical description of the general circulation is thus reduced to the application of statistical methods to the analysis of the baric field, utilizing general probability methods for describing the fields, introduced to meteorology by Fridman and Keller and subsequently developed primarily in conjunction with problems of the statistical theory of turbulence. During the past few years a relatively large volume of aerological data has been obtained for our planet, particularly for the northern hemisphere, and the possibilities for carrying out computations have been tremendously increased by the use of electronic computers. This makes it possible to find correlation functions primarily for the geopotential over a large part of the globe and in the near future for the entire earth.

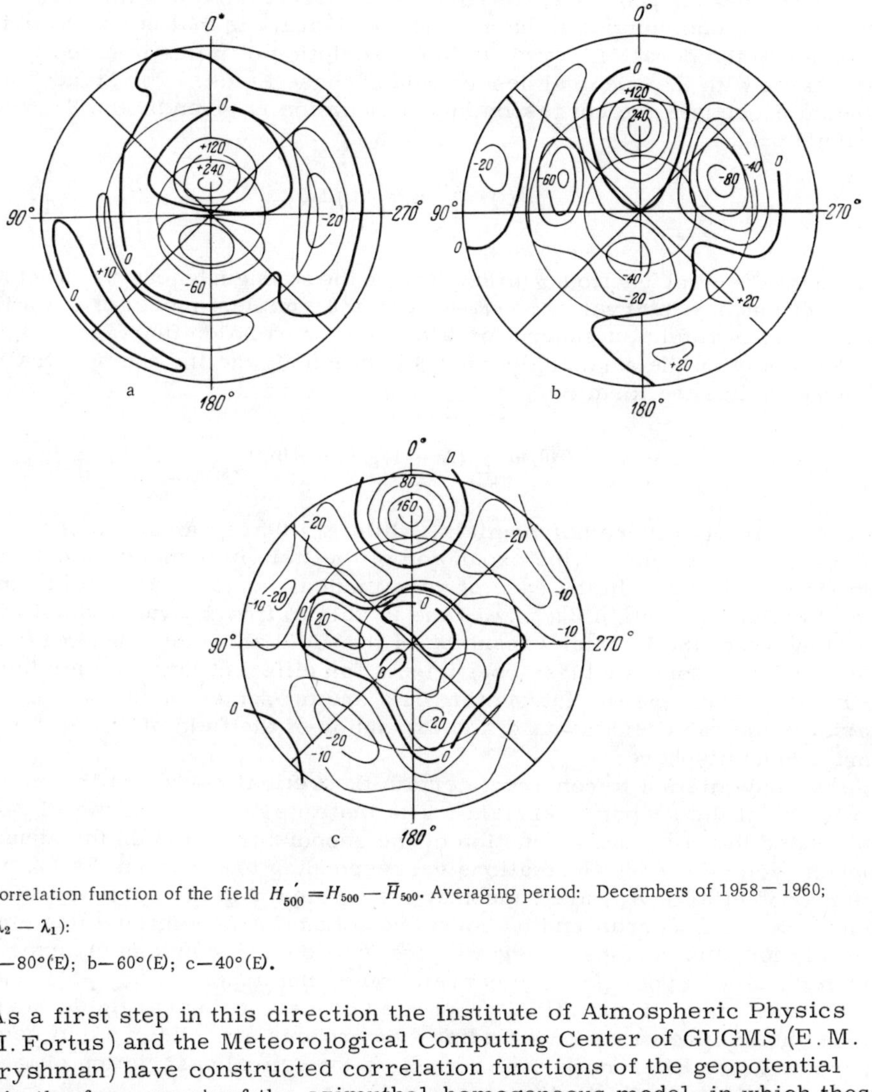

Correlation function of the field $H'_{500} = H_{500} - \overline{H}_{500}$. Averaging period: Decembers of 1958–1960; $(\lambda_2 - \lambda_1)$:

a—80°(E); b—60°(E); c—40°(E).

As a first step in this direction the Institute of Atmospheric Physics (M.I. Fortus) and the Meteorological Computing Center of GUGMS (E.M. Dobryshman) have constructed correlation functions of the geopotential within the framework of the azimuthal-homogeneous model, in which these functions depend only on the longitude difference $\lambda_2 - \lambda_1$ of the observational points M_1 and M_2 under consideration. Thus, the correlation functions of the geopotential

$$B(M_1M_2) = \langle H'(M_1)\,H'(M_2)\rangle$$

depend on three arguments: the two latitudes φ_1 and φ_2, and the longitude difference $\lambda_2 - \lambda_1$. Use was made here of data for the northern hemisphere bounded by the 30°N latitude. The most complete results were obtained for the 500 mb level for the three Decembers of 1958—1960 (see figure).

We wish to point out the clearly expressed regions of negative correlations, whose center is shifted by approximately 80° with respect to longitude, and the poor positive correlation for greater longitudinal shifts. The greatest contribution to baric-field variability is that of waves which fit twice along the parallel. Correlation description of fields is directly associated with the spectral description of these fields. For steady-state random processes the corresponding correlation is given by the Khinchin formula

$$B(\tau) = \int_0^\infty \cos \tau\omega g(\omega)\, d\omega, \qquad (1)$$

where the spectral function $g(\omega)\, d\omega$ defines the strength (energy) of fluctuations in the frequency interval $(\omega_1\omega + d\omega)$. For the simplest model of a statistically homogeneous field at a sphere for which the correlation function $B(M_1, M_2)$ depends only on the angular distance θ between M_1 and M_2, the spectral divergence has the form

$$B(\theta) = \sum_{n=1}^{\infty} (2n+1)\, g_n P_n(\cos\theta), \qquad (2)$$

where P_n are Legendre polynomials, while $g_n > 0$ are variances of the individual coefficients of the expansion of the field in terms of spherical functions $S_{n,m}, (\varphi, \lambda)$. In the case of an azimuthal-homogeneous field, knowing the correlation function $B(\varphi_1\varphi_2, \lambda_1\lambda_2)$ one is able to find the variance of the coefficients of the Fourier expansion of the field along any parallel and the correlation factors for these coefficients for different latitudes for the same wave numbers m. Interesting data are presented in /3/ on the variances of the coefficients of the expansion of the field of H_{500} in the northern hemisphere.

One may attempt to construct certain theoretical models of the correlation structure of the geopotential field. The Institute of Atmospheric Physics calculated the correlation function of the geopotential field on the assumption that the potential eddy fluctuations corresponding to the given distribution of the geopotential of H_{500} are randomly distributed with the specified variance density b_0. The corresponding correlation function is similar to an empirical correlation, the regions of negative correlation are very clearly expressed, but regions with poor positive correlation do not appear.

Further study of statistical properties of meteorological fields on a global scale should involve not only finding synchronous correlations between fluctuations at different points, but also a statistical description of the time evolutions of these fluctuations.

References

1. Blinova, E.N. K gidrodinamicheskoi teorii klimata i dolgosrochnogo prognoza pogody (The Theory of Hydrodynamic Climate and Long-Range Weather Forecasting). — In: "Gidrodinamicheskii dolgosrochnyi prognoz pogody," Moskva. 1964.
2. Smirnov, I.P. O raspredelenii momentov svyazi polya davleniya na srednem urovne (Distribution of the Pressure-Field Covariances at the Mean Level of the Atmosphere). — Ibid.
3. Jones, R.H. Stochastic Processes on a Sphere as Applied to Meteorological 500 mb Forecasts. — Proc. Sympos. Time Series. An., N.Y. — London. 1963.
4. Keller, L. Über die Ausstellung eines Systems der Charakteristiken der atmosphärischen Turbulenz. — Zhurnal geofiziki i meteorologii, 2(34). 1925.

V. N. Kolesnikova and A. S. Monin

THE SPECTRA OF MICROMETEOROLOGICAL, SYNOPTIC AND CLIMATIC FLUCTUATIONS OF METEOROLOGICAL FIELDS

The main concepts of the space and time analysis of meteorological fields are presented. A survey is made of data on the time spectra of the wind speed, temperature, pressure, and turbulent fluxes of the momentum and heat for intervals of seconds, months, and years. New data are given on the spectra of temperature fluctuations with periods of from 1 — 2 minutes to several months. An explanation is given of the inertial period in the synoptic domain and of the minimum in the mesometeorological range of spectra. Problems related to concepts of climate and climatic fluctuations, solar activity, and terrestrial weather are considered.

Many quantitative characteristics of the state of the atmosphere which are of interest to meteorological studies (atmospheric pressure, temperature, wind speed, specific humidity, momentum, heat, moisture fluxes, etc.) are functions of a point in space and of time, or, as these are called, **fields**. In other words, their values at a given time instant change from point to point, and the values of these quantities at a given point in space (or other functionals of appropriate spatial fields, for example, their average values over a given region in space, the so-called circulation indexes, etc.) vary with time. The study of these fluctuations of meteorological elements is a large part of the domain of atmospheric physics. In the present paper one of the methods used to describe meteorological fields statistically — **spectral analysis** — is considered, and an attempt is made to **explain** some regularities detected by this method.

DETERMINATION AND MEANING OF TIME SPECTRA

Spectral analysis consists, firstly, of representing the fluctuation under study as a sum of some elementary waves (which may contain a finite, enumerable number or even a continuum of components (in the latter case, obviously, the sum is replaced by an integral)). Secondly, it consists of constructing curves of the rms amplitudes of these waves as a function of their frequencies (or periods) if one refers to fluctuations in time, or of their wave numbers (or wavelengths) if one refers to oscillatory variations in space. Since here one refers to **average** rms amplitudes, the possibility of some kind of data averaging is inferred. Theoretically, one refers to averaging over some statistical ensemble of individual realizations of the oscillations under study under homogeneous "external conditions." Actually, one usually assumes the existence of ergodicity, i.e., the feasibility of replacing averaging over an ensemble of realizations by averaging over a sufficiently large time interval (or spatial domain).

The set of frequencies or periods (wave numbers or wavelengths) of elementary waves which form the oscillations under study is called the time (space) **spectrum** of these oscillations. The function describing the rms elementary-wave amplitudes as a function of points of the spectrum is called a **spectral function.**

In order to be able to carry out more specific quantitative formulations, the case of fluctuations in time of some meteorological characteristic $\xi=\xi(t)$, which will be most frequently referred to in the present article, will be considered first. The observed fluctuations in time of meteorological elements are frequently so irregular that it is completely hopeless trying to provide them with an exact analytical solution and it is more expedient to describe them statistically. This means, first of all, that the quantity $\xi=\xi(t)$ is regarded as a **random function** of time t, i.e., that for its values $\xi(t_1), \xi(t_2),\ldots,\xi(t_n)$ at any finite set of time instants $t_1, \ldots, t_n$ probability distributions exist (approximately correlated with one another). This makes it possible to determine the average value of any function of $\xi(t_1), \xi(t_2), \ldots,$ as the corresponding mathematical expectation (which shall be denoted by a bar over the symbols). As was noted, calculation of the mathematical expectation must, in the majority of cases, be replaced by averaging over a sufficiently long time interval.

The average values $\overline{\xi(t)}$ of meteorological elements are usually found to be relatively simple functions of t, which lend themselves easily to analytical description (for example, constant or periodic functions with a daily or annual period). Upon subtracting the regular component $\overline{\xi(t)}$ from the random function $\xi(t)$, we obtain the fluctuations $\xi'(t)=\xi(t)-\overline{\xi(t)}$, the average values of which are zero.

Fluctuations $\xi'(t)$ may be composed of fluctuations with any frequency $f(f=1/T$, where T is the period, so that the frequency is measured by the number of fluctuations per unit time; sometimes the angular frequency $\omega=2\pi f=2\pi/T$ expressed as radians per unit time is used). Upon subdividing the entire set of possible frequencies $0<f<\infty$ into sufficiently small intervals Δf, it becomes possible to represent function $\xi'(t)$ as a sum of functions $\xi'(t,\Delta f)$, containing only fluctuations with frequencies in the intervals Δf (in practice these components can be separated by means of filters with passbands Δf).

Practical spectral analysis is always based on the assumption that the random function $\xi'(t)$ is a **steady function**, i.e., that the probability distributions for its values over any finite set of time instants are independent of the time taken as the origin. On this assumption, functions $\xi'(t,\Delta f)$ corresponding to nonintersecting spectral intervals Δf are found to be noncorrelated and, for sufficiently narrow spectral intervals Δf, they can always be approximated over any specified finite time interval with any accuracy by the random harmonic oscillations

$$\xi'(t,\Delta f)\simeq A(\Delta f)\cos 2\pi ft+B(\Delta f)\sin 2\pi ft, \tag{1}$$

where f is any point from the interval Δf, while A and B are noncorrelated random quantities with zero mean magnitudes and with the same rms values. Here the rms value of the function $\xi'(t, \Delta f)$, which is approximated by the rms value of the amplitude of the random harmonic oscillation (1), can be written as

$$\overline{\xi'^2(t, \Delta f)} = s(f)\,\Delta f, \tag{2}$$

where $s(f)$ is a non-negative function, which is called the **spectral density** or simply the **spectrum** of oscillations $\xi'(t)$ (which may contain delta-type components of the form $a_n^2\delta(f - f_n)$, although not more than an even number of these). Equation (2) points to a practical way of determining the spectral density: $s(t)$ is obtained by dividing the rms signal at the output of a filter with passband Δf, by Δf. A second method is based on calculating the **correlation function** of oscillations $\xi(t)$, defined by the expression

$$b(\tau) = \overline{\xi'(t)\,\xi'(t+\tau)}, \tag{3}$$

and on using relationships

$$\begin{aligned} b(\tau) &= \int_0^\infty \cos 2\pi f\tau\, s(f)\, df \\ s(f) &= 4\int_0^\infty \cos 2\pi f\tau\, b(\tau)\, d\tau \end{aligned} \tag{4}$$

which are valid for stationary random functions.

A more detailed presentation of the idea of spectral analysis of stationary random functions can be found, for example, in the article by Kolmogorov /14/ and in the exhaustive work of Yaglom /27/. Practical methods for calculating spectra from empirical data are presented in the book by Blackman and Tukey /28/.

The first of equations (4) shows that the total variance $\overline{\xi'^2(t)} = b(0)$ of ξ can be represented in the form

$$\overline{\xi'^2} = \int_0^\infty s(f)\, df = \int_0^\infty S(f)\, d\ln f; \quad S(f) = fs(f). \tag{5}$$

In cases when the spectrum of oscillations under study extends over a wide frequency range within which the frequencies change by several orders (for example, within the limits of the atmospheric-turbulence spectrum the frequencies change by an order of 8—9), it is convenient to use $\ln f$, rather than f, as the spectral-curve abscissa. If the ordinate is then not $s(f)$, but $S(f) = fs(f)$, then the area beneath the curve of $S(f)$ over the interval $\Delta \ln f$ will be equal to the contribution of this frequency interval to the total variance $\overline{\xi'^2}$. Hence it will sometimes be more convenient to express spectra in the coordinate system $(S, \ln f)$, rather than (s, f).

It was probably noted that the average values $\bar{\xi}(t)$ of meteorological characteristics are periodic time functions with a daily or annual period (some investigators also assume that the 11.5-year solar activity period demonstrates itself in the fluctuations of a number of meteorological characteristics). In these cases the fluctuations are most frequently found to be not stationary, but **periodic random functions** of t. This means that the probability distributions for the values of $\xi'(t)$ over any finite set of time instants do not change when the time origins are shifted by amounts kT, $k = \pm 1, \pm 2, \ldots$, where T is some constant period, but they may vary for other shifts in the time origin. As an illustration we present the fluctuations in the planetary circulation index, which have been studied in /9, 15, 2, 10, 37/. Fluctuations of this index are perceptibly more intensive in winter than in summer. However, methods of spectral analysis of periodic random functions have not as yet been sufficiently developed (see /37/), and spectral analysis can be carried out in practice only on the assumption that the random function $\xi'(t)$ is at least approximately stationary.

The time spectra of meteorological characteristics should be limited from the direction of high frequencies. Thus, the maximum frequencies of **turbulent fluctuations** of meteorological fields have the magnitude (see equation (7) below), where u is the wind speed, $\lambda = (\nu^3/\varepsilon)^{1/4}$ is the microscale of turbulence (ε is the average rate of kinetic energy dissipation, ν is the kinematic coefficient of molecular viscosity). Assuming that $u = 10$ m/sec, $\varepsilon \sim 1$ cgs unit and $\nu \sim 10^{-1}$ cgs unit, we find that $f_{max} \sim 10^3 \text{sec}^{-1}$.

Conversely, spectra of meteorological characteristics are not limited from the direction of low frequencies (large periods). In fact, fluctuations of meteorological fields are observed in the micrometeorological spectral range with periods of from hundredths of a second to minutes, as well as in the synoptic region, with periods of from days to weeks (the interval between the micrometeorological and synoptic domains may be termed the mesometeorological region), etc., up to secular and more prolonged climate fluctuations (described, for example, by Willett /22/).

SPACE SPECTRA AND THEIR COMPARISON WITH TIME SPECTRA

We shall now describe oscillatory changes of some meteorological characteristic in space. For the purposes of this paper it is sufficient to consider the fluctuations ξ of a meteorological element along some line L (for example, along a straight line parallel to the wind direction, or along a circle of latitude). Setting $\xi = \xi(x)$, where x is an arc length of line L, and assuming that fluctuations $\xi'(x) = \xi(x) - \overline{\xi(x)}$ are an at least approximately homogeneous random function of x (i.e., that the probability distributions for values of ξ over any finite set of points of line L are independent of the origin in measuring the arc length), we will be able to introduce the concept of a spectral density of oscillations $\xi'(x)$, similar to that above for $\xi'(t)$. The periods T will be replaced here by wavelengths Λ, and frequencies $f = 1/T$ will be replaced by **wave numbers** $k = 2\pi/\Lambda$. The spectral density $s(k)$ of fluctuations $\xi'(x)$ will thus be defined as

$$\overline{\xi'^2(x, \Delta k)} = s(k)\,\Delta k. \qquad (6)$$

The time spectrum $s(f)$ of fluctuations of some meteorological characteristic ξ at a given point in space is approximately comparable to the spatial spectrum $s(k)$ of fluctuations of the same characteristic along the straight line L passing through the point M parallel to the wind direction, if we assume that the inhomogeneities of the field of ξ at line L do not change shape with time (are as if "frozen"), and only move along this line at constant wind velocity u. This assumption, which is called the "freezing" hypothesis, was first put forward by Taylor for the case of small-scale turbulence /43/. On this assumption oscillations with wavelengths Λ on the straight line L will correspond to oscillations with periods $T = \Lambda/u$ at point M, which then yields

$$s(k) = \frac{u}{2\pi} s(f); \quad f = \frac{ku}{2\pi}. \tag{7}$$

Consequently, when $f = ku/2\pi$, $ks(k) = fs(f)$, i.e., $S(k) = S(f)$. It was shown by appropriate studies that equation (7) is quite satisfactorily satisfied for small-scale turbulent fluctuations ξ', and even better, the larger k (uf). This was established in 1938 by Taylor /43/. Of the later works one should mention that of Tsvang /24/, who compared values of $s(f)$ for temperature fluctuations measured directly from a meteorological tower with the values of $s(k)$ obtained by aircraft sounding along the line L. It was established that equation (7) is satisfactorily satisfied at least up to $k = 10^{-4} - 10^{-5}$ cm^{-1} (wavelengths of 0.63—6.30 km).

The statistical structure of the spatial inhomogeneities of the field of $\xi'(x)$ described by the space spectrum $s(k)$ is evidently independent of the coordinate system from which values of $\xi'(x)$ are measured. At the same time, the time spectrum $s(f)$ in the spectral region where equation (7) applies depends markedly on the wind speed u. Hence the $s(f)$ measured at different wind speeds can be compared only by converting them into spectra $s(k)$ using equation (7).

For low k, at which time Λ/u of transport past the point M of inhomogeneities with wavelength Λ is not small as compared with the characteristic time of natural evolution of these inhomogeneities, equation (7) is no longer necessarily satisfied. Thus, on west-east transport, under which, as is known, the middle troposphere of moderate latitudes performs a total rotation about the earth during approximately a month (which, for a circle of latitude of length 30,000 km, has corresponding to it a linear transport velocity of 30,000 km/month $\approx$ 10 m/sec), transport past a given point M in a baric region with dimensions of the order of the radius of interaction of the baric field $L = c/l$ ($c \sim 300$ m/sec is the speed of sound and l is the Coriolis parameter, equal to 10^{-4} sec, so that $L \sim 3000$ km) requires several days, i.e., a time commensurable with the "life" of synoptic baric formations. This also shows that equation (7) is inapplicable in the synoptic spectral range, so that spectra $S(k)$ and $S(f)$ in this region may be of different form.

Unlike time spectra, which are limited only from the direction of high frequencies, space spectra must be limited from the direction of large wave numbers (minimum scales of turbulent inhomogeneities are of the order of $\lambda \sim (\nu^3/\varepsilon)^{1/4} \sim 10^{-1}$ cm), as well as from the direction of small wave numbers (since the wavelengths of the inhomogeneities cannot exceed the length of the equator, which is approximately 40,000 km). In order to compare, even by order of magnitude, the maximum wavelength ($\lambda \sim 40,000$ km) of inhomogeneities of the meteorological fields of the general circulation of the atmosphere with some maximum period of meteorological-element fluctuations at a given point in space, we shall use the expression $T = \Lambda/u$, taking the velocity of western transport as u. We then obtain $T_{max} \sim 1$ month. Thus, while in the frequency range $f > 1$/month it may still be possible to compare the time and space spectra, this cannot be done for $f < 1$/month.

Unlike time spectra, space spectra are not only limited from the side of small wave numbers, but should also have a linear structure, i.e., the form

$$s(k) = \sum_n a_n^2 \delta(k - k_n). \qquad (8)$$

Let us examine, for example, the spectrum of the field of $\xi'(x)$ at some circle of latitude L with length Λ_1. Obviously, the inhomogeneities of the field of ξ' can consist only of such elementary waves which fit an integral number of times into the circle L, i.e., have lengths $\Lambda_n = \Lambda_1/n$ $(n = 1, 2, 3, \ldots)$. Consequently, the spectrum of the field of ξ' may include only the wave numbers $k = k_n = 2\pi n/\Lambda_1$. Its description reduces to pointing out the sequence $a_1^2, a_2^2, a_3^2, \ldots$ of rms amplitudes of elementary waves with wave numbers $k_1, k_2, k_3, \ldots$ (with numbers $n = 1, 2, 3, \ldots$). This description of the kinetic energy spectrum of the general circulation is given, for example, by Saltzman and Fleisher /41/, who suggested that synoptic waves with the lowest numbers $1 \leqslant n \leqslant 5$ be called long, those with $6 \leqslant n \leqslant 10$, **medium** (cyclonic), and those with $11 \leqslant n \leqslant 15$, **short**.

However, for sufficiently high (as compared with $2\pi/\Lambda_1$) wave numbers k it is more convenient to replace the line spectrum $s(k)$ of equation (8) by the continuous function $\tilde{s}(k)$, setting, for example, $\tilde{s}(k) = d\tilde{F}(k)/dk$, where $\tilde{F}(k)$ is a sufficiently smooth function, approximating the step function $F(k) = \int_k^\infty s(k)\,dk$. This replacement of a line by a continuous spectrum is fully justified from physical considerations, since at very short wavelengths $\Lambda \ll \Lambda_1$ the requirement that they fit an integral number of times into a latitudinal circle L is physically meaningless. This also makes it possible to consider the space spectra on the basis of continuous time spectra in the corresponding frequency intervals, which is precisely one of the purposes of this paper. With this in mind, we shall now consider the available actual data on the time spectra of meteorological characteristics.

WIND SPEED SPECTRA

Figure 1a, curve 1 depicts (in the coordinate system S, $\ln f$) the spectrum of the horizontal component of the wind speed in the frequency range of from $7 \cdot 10^{-4}$ to $9 \cdot 10^2$ cycles/hour, obtained by Van der Hoven from measurements made from a 125-meter meteorological tower in Brookhaven (U.S.A.) /44/. This spectrum was calculated from equations (3) and (4) for eight overlapping spectral intervals described in Table 1. The spectrum has a clearly expressed maximum in its synoptic part (whose ordinate changes markedly from one season to another), corresponding to a period of $T = 4$ days and attributable to the passing of large-scale baric systems, and a maximum in the micrometeorological region, corresponding to a period of about one minute and produced by small-scale turbulence of dynamic and convective origin (whose ordinate may change during the day and from one day to another several times). The particular maximum corresponding to a 12-hour period was regarded by Van der Hoven as insignificant (he subdivided the 42-day period from which the spectrum was calculated in range 3 of Table 1 into two halves and calculated the spectrum from each of these; a maximum for $T = 12$ hours was found only at one of these two spectra).

TABLE 1. Intervals of the horizontal wind speed spectra

Interval number	Altitude, m	Measurement interval	Averaging interval	Frequency, cycles/hour
1	108	10 months	5 days	$7 \cdot 10^{-4} - 4.2 \cdot 10^{-3}$
2	108	6.2 months	1 day	$3.5 \cdot 10^{-3} - 2.1 \cdot 10^{-2}$
3	108	42 days	5 hours	0.01 — 0.1
4	108	26 hours	30 min	0.1 — 1.0
5	108	50 hours	10 min	0.5 — 3.0
6	125	32 hours	75 sec	0.8 — 24
7	91	1 hour	30 sec	15 — 90
8	91	1 hour	2 sec	30 — 900

A remarkable property of the spectrum under study is its deep minimum in the mesometeorological region over a period range of from several minutes to several hours. This feature appears to be typical of spectra obtained in extremely different series of measurements (see, for example, Table 2, taken from /39/).

It is seen from Table 2 that the location of the minimum of the spectrum does not depend on the wind speed or wave number. It was suggested that it depends on the type of locality. The minimum spectral amplitude is small: it ranges from 0.1 to 0.2 m^2/sec^2. The minimum in all the cases is wide and flat, extending from a period of approximately 3.3 hours to 40 min.

TABLE 2. Characteristics of the mesometeorological spectral minimum

Observation point	Measuring interval, hours	Altitude, m	Wind speed, m/sec	Period, min	Wavelength, km	Ordinate, m^2/sec
Brookhaven	50	108	6.2	60	25.0	0.20
	43	125	7.8	20	9.1	0.15
Pennsylvania	48	30	1.8	30	3.2	0.15
Oak Ridge	43	100	4.1	10	2.4	0.10
	43	100	5.2	20	6.2	0.15
Idaho Falls	60	76	12.8	12	9.1	0.20
	60	76	8.9	7	3.6	0.10

The question as to the mesometeorological spectral minimum will be a recurring topic in this paper.

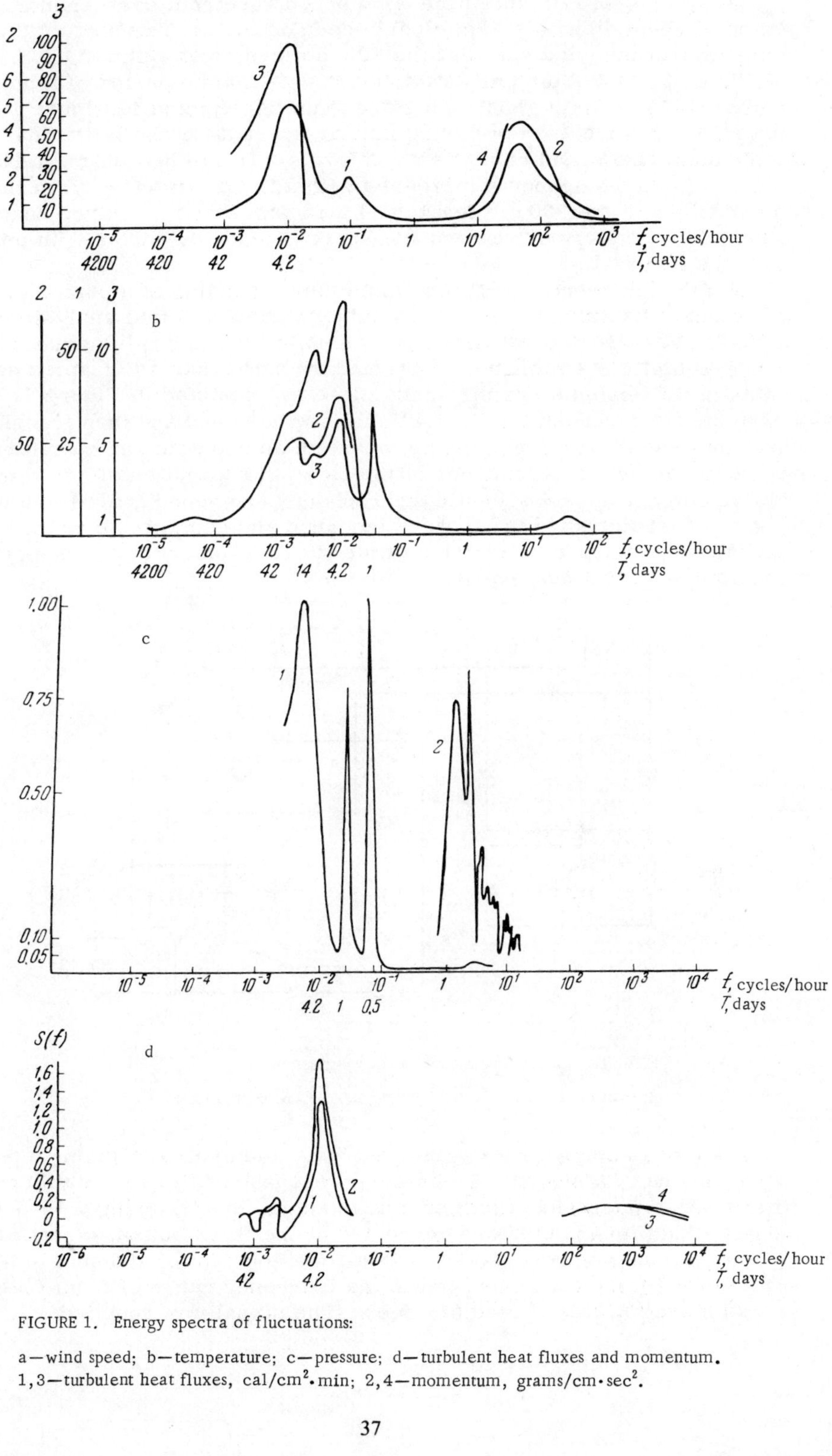

FIGURE 1. Energy spectra of fluctuations:

a—wind speed; b—temperature; c—pressure; d—turbulent heat fluxes and momentum. 1,3—turbulent heat fluxes, cal/cm^2·min; 2,4—momentum, grams/cm·sec^2.

The synoptic maximum of the wind speed spectrum corresponding to a period of about 4 days is also clearly seen on the spectrum of the latitudinal component of the wind speed at the 700 mb level over Belmar (U.S.A.), calculated in /45/ from daily radiosonde data for the period of from January 1953 to April 1954. This spectrum is depicted in Figure 1a by curve 3 (reference /45/ presents similar spectra for the latitudinal and meridional wind speed components at levels of from 700 to 50 mb). The energy content of both components increases with altitude, starting from the 700 mb and ending with the 200 mb level, and then decreases. A large contribution to the kinetic energy of western transport is made by eddies with periods of more than 30 days.

The micrometeorological maximum corresponding to a period of about one minute is satisfactorily confirmed by data of spectral analysis of longitudinal (i.e., directed parallel to the mean wind velocity) fluctuations of the wind speed at the 4 m altitude, obtained by Zubkovskii /12/ from measurements in the region of Tsimlyansk; these are depicted in Figure 1a by curve 2. He recorded the fluctuations of wind speed and then transformed them into electrical vibrations by means of an acoustic anemometer which was actuated by the dependence of the rate of propagation of sound on the flow velocity (compact acoustic microphones and sound generators for use in the ultrasonic range in air were designed and constructed at the Institute of Atmospheric Physics of the USSR Academy of Sciences under the direction of V.M.Bovsheverov).

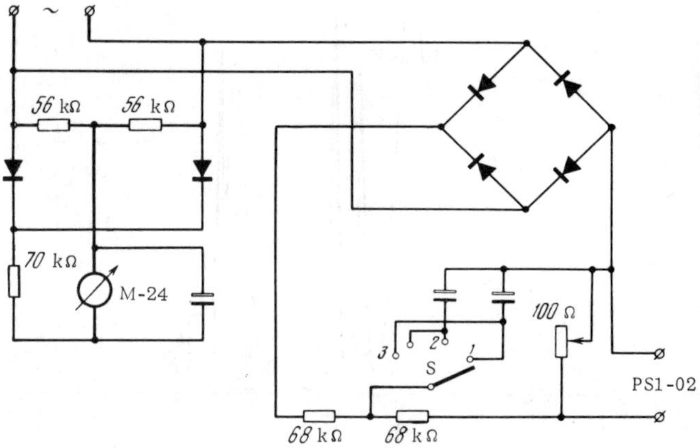

FIGURE 2. Block diagram of anemorhumbometer:

S — switch; 1, 2, 3 — C = 500, 1000, 1500 μF, respectively.

Zubkovskii carried out spectral analysis automatically (using equation (2)) by means of a system of electrical resonance filters (containing 30 filters with a passband of half an octave each, thus covering a 15-octave range). The linear detectors connected at the filter outputs were followed by a capacitance storage device with a time constant of 100 sec. The analyzer's filters were arranged in the frequency range of from 0.012 (T = 78.8 sec) to 52 Hz (T = 0.019 sec). The signal was supplied simultaneously to all the filters, while the storage devices were interrogated automatically by means of a step selector, which was connected in turn to each storage device for 3 secs each. The storage-device voltages were measured and recorded on an EPP-09 electronic recording potentiometer, and the frequency spectrum was calculated from the recordings by means of special templates. The spectral analyzer is described in detail in /23/.

We now present the results of calculations by the present authors of wind speed fluctuations measured during May 1963 at Zvenigorod (Figure 1a, curve 4). The fluctuations of wind speed were measured by an M-47 anemorhumbometer. The measuring unit of this instrument consisted of a sensor and an indicator, mounted on the adapter panel for monitoring the instrument's operation. A small ac generator powered by an eight-vane impeller served as the sensor. The alternating voltage generated by it, whose magnitude is proportional to the impeller's angular velocity, was supplied to two rectifiers. The signal from one of them was fed to reference indicator M-24, calibrated in m/sec, and that from the other was supplied, through a divider, to the PSl-02 recording instrument. To smooth out high-frequency fluctuations the adapter was provided with 500, 1000, and 1500μF filters (Figure 2). The recording was made on tape moving at 120 mm/hour. The wind-speed fluctuation measurements were prepared and carried out by D.F. Timanovskii. Spectral analysis was performed automatically by means of a spectral analyzer. The data-processing method will be described in the following section.

We shall also consider briefly the spectra of the vertical component of the wind speed. These spectra are limited from the side of low frequencies. The major portion of the fluctuation energy devolves upon the micrometeorological interval and is concentrated in the spectral region corresponding to scales of the order of altitudes at which observations are made /39, 7, 17/. The spectral maximum is shifted in the direction of higher frequencies upon passing from a less to a more stable stratification. The low-frequency boundary of the spectrum of the vertical wind-speed component is most probably close to 0.0005 Hz (period of 33 min). At frequencies of the order of 0.001 Hz (4 cycles/hour) the fluctuation energy becomes very small.

The fluctuations of the vertical component of the wind speed are most frequently measured by means of hot-wire anemometers. These operate on the basis of the dependence of the heat transfer from a thin heated wire on the flow velocity. Gurvich /7/ as well as Zubkovskii /12/ used an acoustic anemometer, and carried out spectral analysis by means of the previously described analyzer.

In earlier studies (for example, /19/) the distribution of the fluctuation energy of the vertical component of the wind speed was studied by changing the period of averaging in processing the fluctuation recordings. It was concluded that small eddies (with periods of less than 2 sec) supply one half of the turbulent heat flux. This result is in agreement with the data of MacCready /36/.

The period used in /19/ was also used by Panofsky /38/ who varied the averaging period from 1 to 490 sec.* He showed that the influence of large eddies increases with the altitude, since with increasing the altitude at which

* Averaging over θ actually results in multiplying the frequency spectrum by $[(\sin \omega\theta/2)/(\omega\theta/2)]^2$, which makes it possible to isolate the effect of the low-frequency part of the spectrum. In terminology of spectral analysis this is equivalent to passing the signal through a filter which passes only low frequencies $\omega < 2\pi/\theta$.

observations are made the spectra are shifted into the region of lower frequencies.

TEMPERATURE SPECTRA

Figure 1b shows time spectra of fluctuations of air temperature according to data of several investigators. Curve 1 depicts the spectrum in the period interval of from 2 to 7300 days, calculated from equations (3) and (4) in /33/. The spectrum was calculated separately over eight overlapping spectral intervals, whose characteristics are listed in Table 3. This spectrum has a clearly expressed synoptic maximum, corresponding to a period of $T = 4$ days. The secondary maximum, corresponding to a period of $T \approx 12$ days, is also quite well expressed; a similar maximum with $T \approx 12$ days was detected earlier in /15/ in the spectrum of fluctuations of the zonal circulation index. Figure 3 depicts this spectrum in the coordinates (s, T) (Figure 3 also shows a maximum with $T \approx 24$ days; maxima with even longer periods were detected in /10/ in the spectrum of fluctuation of the circulation index). The maximum at $T = 12$ days thus corresponds to planetary-scale fluctuations: it is sometimes attributed to the so-called index cycle (it was noted in /10/ that the half periods of about 6 days duration during which the circulation index retains relatively high or relatively low values possibly correspond to Mul'tanovskii's "synoptic periods").

TABLE 3. Intervals of the temperature spectrum (curve 1, Figure 1a)

Measurement period, years	Averaging period, days	Periods, days	Measurement period, years	Averaging period, days	Periods, days
1887—1954	365 or 366	370—7300	1934—1936	3	6—54
1934—1953	42	84—504	1934	2	4—48
1934—1953	14	28—252	1945	2	4—48
1934—1936	6	12—72	1934	1	2—24

Curve 2 in Figure 1b depicts the spectrum of temperature fluctuations at the 700 mb level over Belmar (U.S.A.), constructed on the basis of daily radiosonde data /45/. It also has a clearly expressed synoptic maximum with a period of approximately 4 days.

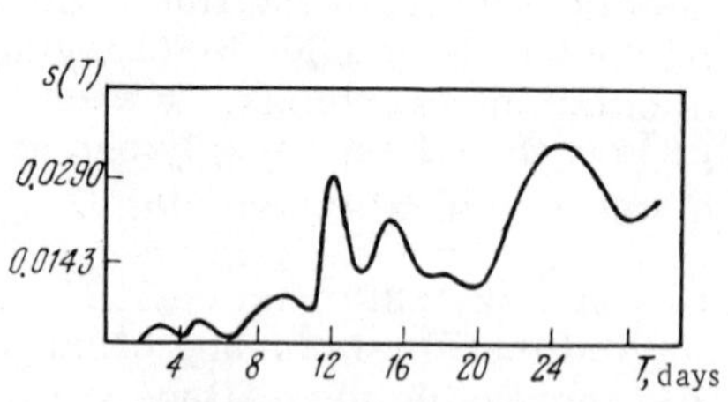

FIGURE 3. Spectrum of fluctuations of the zonal circulation index.

Curve 3 in Figure 1b depicts the spectrum of temperature fluctuations in the frequency range of from 10^2 to 10^{-3} cycles/hour, constructed by the present authors. This spectrum was determined separately at five frequency intervals (Table 4) on the basis of equation (2) by means of the electrical spectral analyzer mentioned previously in conjunction with curve 4 of Figure 1a. The temperature fluctuations over the 5-day period were recorded by M. I. Mordukhovich and A. G. Platov; the methodology and results of temperature-spectrum measurements in this frequency range are described by Tsvang in /23/. The spectra in intervals 1 and 2 were obtained on tapes of a standard meteorological thermograph. Four thermograms (30 July—4 August) were processed for the high-frequency end of an interval, and the average temperatures over time intervals of 15 min were recorded on the tape at 0.5 cm intervals (with a vertical scale of 1 cm = 2°C). To obtain the spectrum in the range of periods from several to several tens of hours, average temperatures for each hour were read off the thermograms and were then plotted on the tape with 2.5 mm steps between points (with the same vertical scale as in the preceding case). The spectrum in the range of large periods was obtained by plotting 6-hour average temperatures on the tape at 0.5 cm intervals (the scale on the ordinate was 2 cm = 1°C).

TABLE 4. Intervals of the temperature spectrum (curve 3, Figure 1b)

Interval number	Location of observations	Measurement altitude, m	Measurement period	Measuring instrument	Averaging period	Frequency range, cycles/hour
1	Kashira	2	January — December 1962	Thermograph	6 hours	0.002 — 0.05
2	Tsimlyansk	2	15 July — 31 August 1963	"	15 min 1 hour	0.05 — 0.1
3	"	2	31 June — 1 August, 5 — 6 and 19 — 21 August 1963	Aspiration resistance thermometer	—	0.1 — 1.3
4	"	4	1045 — 1730 hrs, 14 August 1963	"	—	1.3 — 15
5	"	4	1045 — 1127 hrs, 14 August 1963	Fluctuation micro-thermometer	—	15 — 4000

The temperature fluctuations recorded on tapes prepared as above were converted into current fluctuations supplied to the spectral analyzer input by means of a slave device, in which the tape with the recorded temperature fluctuations was pulled at a constant speed perpendicular to a slot with a pointer connected to a potentiometer slide. The block diagram of the device and the circuit of the slave device are shown in Figure 4 (the tape was pulled by the POB-12 oscillograph). An operator shifted the pointer manually while following the location of the curve on the recorder tape relative to the slot. The frequency of electric oscillations in the potentiometer circuit can be converted to the frequency interval for which the spectral analyzer was designed by changing the tape speed. The device for spectral analysis of thermogram data in the intervals 1 and 2 was constructed by A. K. Maiorov.

The method of spectral analysis makes possible the semiautomatic construction of spectra of a large variety of oscillations, independent of their natural time scale. In our opinion this method has a number of advantages over the methods based on the use of equations (3) and (4), which require a sufficiently complete determination of the correlation function of oscillations and then calculations of the values of a Fourier series, which virtually requires neglecting an infinite segment of the range of integration without any actual estimate of the contribution of this segment to the integral.

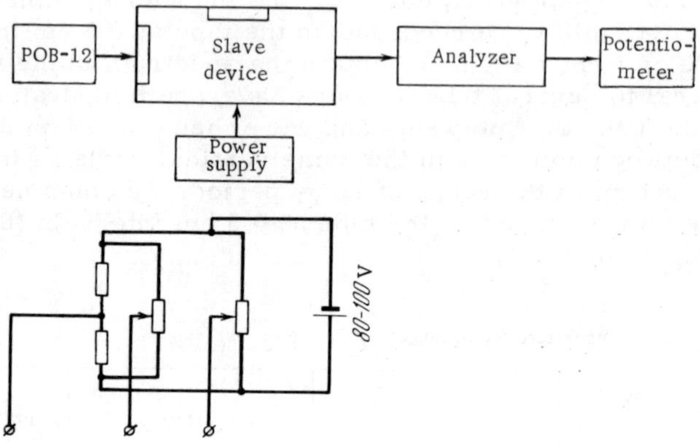

FIGURE 4. Block diagram of a device for obtaining statistical characteristics of fluctuations in T and U.

The circuit of the slave device is shown separately.

The temperature fluctuations in intervals 3 and 4 were measured by means of an aspiration resistance thermometer, designed and constructed at the Institute of Atmospheric Physics of the USSR Academy of Sciences.

The resistance thermometer was connected to a Wheatstone bridge (the bridge operated at 1.6 V dc). One of the legs of the bridge had a constant resistance of 500 Ω and a thermometer whose resistance is 500 Ω at 0°C. The other leg had a 500 Ω resistance and an R-33 bank of resistors (used for changing the instrument's scale; since the temperature has large diurnal variations, the resistance was adjusted to ± 12 Ω). Recording was with a PS1-01 automatic instrument (carriage travel time 2.5 sec). The graph paper was pulled at 80 mm/hour. The full-scale deflection of the instrument corresponded to 14.3 Ω or 6.3°, i.e., $dT/dR = 0.46$ deg/cm.

Continuous fluctuation records were taken for 10—40 hours. As an example, Figure 5 shows day and night-time recordings of temperature fluctuations. The temperature-fluctuation recordings on the graph paper were processed by the same principle as tapes with recordings of fluctuations in T in spectral intervals 1 and 2, except that the PS1-01 automatic recorder was used for pulling the tape and as the slave device.

Figure 6 shows the electrical connections between the PS1-01 and the analyzer. As can be seen from this figure, potentials in volts and in millivolts are supplied simultaneously to the analyzer and recorder, the latter being used for pulling the recording tape. Following was by means of the recorder pointer, which is pointed to the appropriate division of the recorder tape manually by varying the resistance R_1. A potential of from 0 to ±50 V was supplied to the analyzer input for continuous following of the record of T (or u).

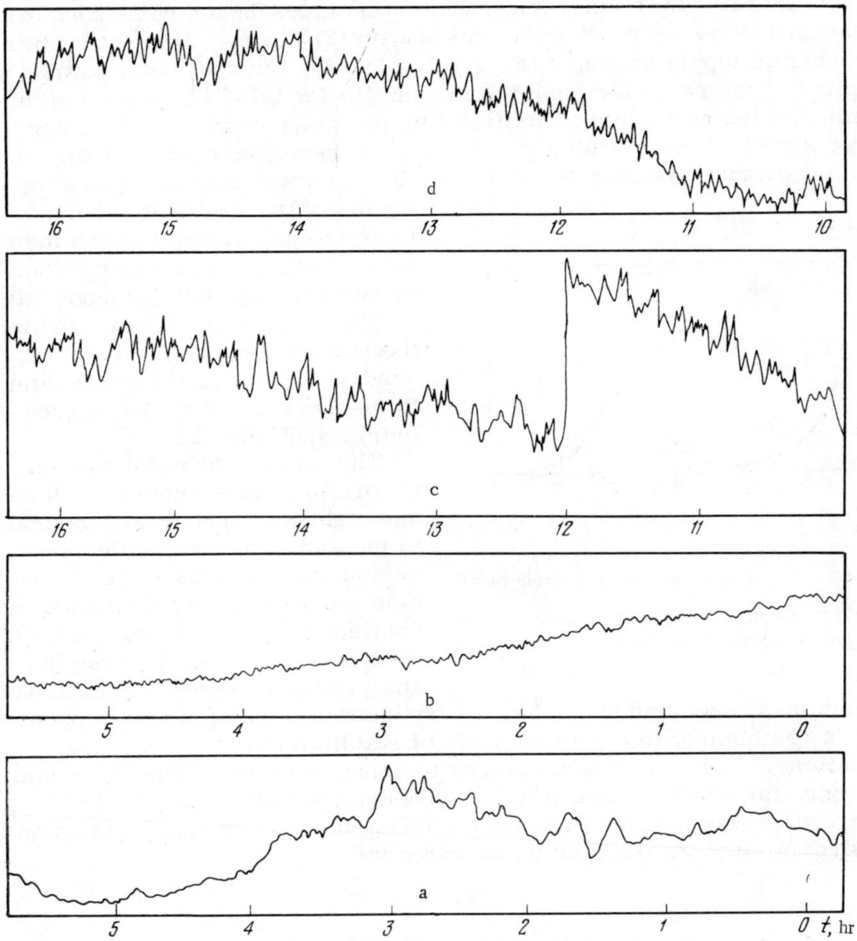

FIGURE 5. Temperature fluctuations:

a—6 August; b—8 August; c and d—30 June.

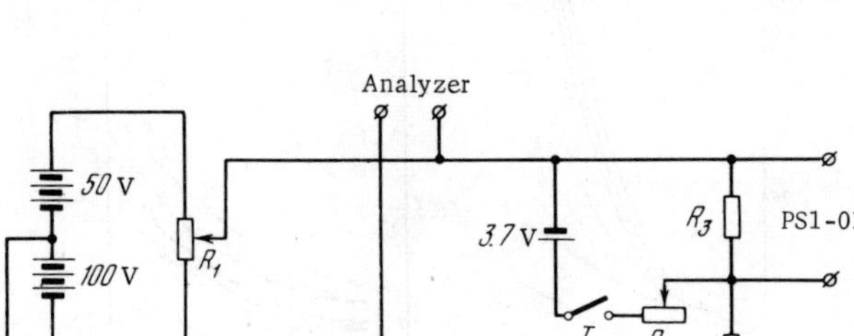

FIGURE 6. Electrical connections of the analyzer and PS1-01 recorder:

$R_1 = 33\ k\Omega$, $R_2 = 390\ k\Omega$, $R_3 = 100\ k\Omega$, $R_4 = 82\ k\Omega$.

In processing the temperature-fluctuation tapes the instrument scale has to be changed by resistance adjustments of up to 12Ω, i.e., "flips" had to be made by means of toggle switch *T* (Figures 5c and 6). An additional voltage was supplied to the recorder input to compensate for this "flipping" and thus eliminated the recording of artificial temperature changes. By using toggle switch *T* the operator could place the recorder pointer in any scale position without changing the signal to the analyzer (measurements by means of the aspiration thermometer were carried out and instruments for their processing were constructed by D.F. Timanovskii).

The same method was used by the authors for processing the wind-velocity spectrum according to measurements at Zvenigorod (curve 4, Figure 1a).

The temperature-fluctuation recordings made separately for day and night were processed basically in the same manner as the spectral regions in intervals 1 and 2. The tape was pasted into a ring and was continuously processed several times. This had to be done because the analyzer is designed for stationary random processes and hence has a very short recording whose reproduction time is commensurable with the time of stabilization of the analyzer indications. This made it necessary to convert the recording to a "stationary" one, for which reason it was reproduced several times. The interrogator was usually switched on during the second reproduction and the signals were transmitted to the recorder.

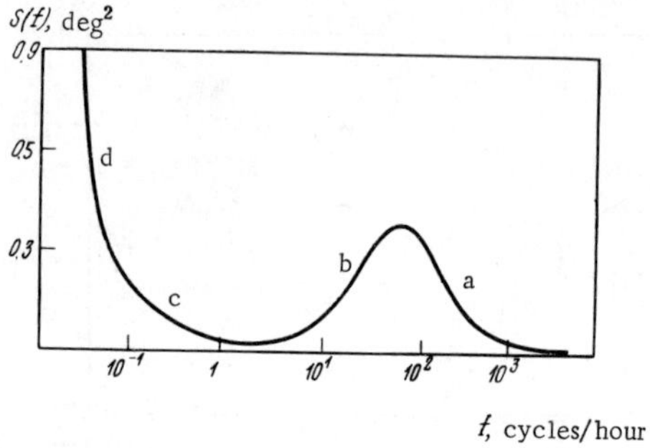

FIGURE 7. The temperature spectrum during 1963:

a — 1329 — 1359 hrs, 14 August; b — 1045 — 1730 hrs, 14 August; c — 5 — 6 August; d — 31 July — 1 August.

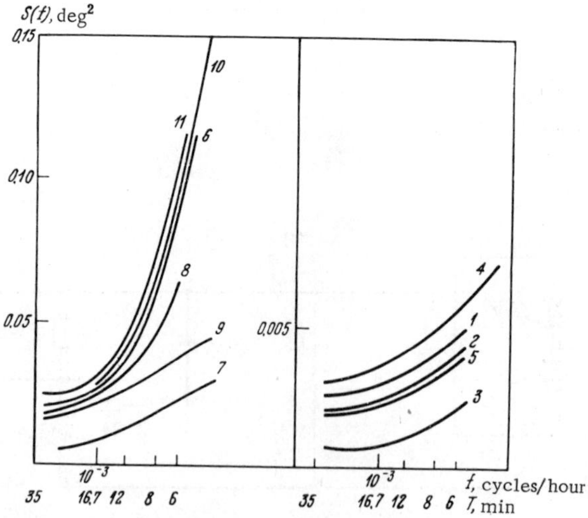

FIGURE 8. Temperature spectra 1 — 11 under different weather conditions (see Table 5).

Spectrum 3 has a clearly defined synoptic maximum with $T = 4$ days (this part of the spectrum was constructed from thermograms obtained at Kashira in 1962, see interval 1 in Table 4). The ordinate of the synoptic maximum may vary from season to season. Thus, according to data for interval 2 of Table 4 in the summer of 1963 at Tsimlyansk this ordinate was 50% larger than in 1962 at Kashira. The maximum corresponding to $T = 24$ hours (which should be represented in the spectrum by a line rather than a band, i.e., the expression for the spectral density s should contain a component such as $a^2\delta(f - 1/T_0)$, where $T_0 = 24$ hours) is also very sharply expressed; this is natural since the air temperature has a sharply delineated daily march. Then the curve shows clearly a wide and deep mesometeorological minimum and a relatively moderate micrometeorological maximum, which are fully analogous to the properties of the wind-velocity spectrum 1 in Figure 1a.

TABLE 5. Weather conditions for the spectra (see Figure 4)

Spectrum number	Date	Time of observation, hrs	Weather conditions
1	31 July	0000 — 0800	Slow-moving cold front, night
2	1 August	0015 — 0800	Cloudless, high-pressure ridge, night
3	8 August	0030 — 0645	Cloudless, after passage of cold front (weak wind), night
4	10 August	0000 — 0800	Cloudless, passage of cold front, night
5	19 — 21 August	2100 — 0800	Southeastern periphery of anticyclone, night
6	30 July	1100 — 1820	Slow-moving warm front, day
7	7 August	1100 — 1800	Secondary cold front associated with an Arctic invasion at 1500 hrs, day
8	9 August	1125 — 1850	Small-gradient high-pressure field, day
9	10 August	1100 — 1850	Occluded front located nearby, day
10	14 August	1100 — 1800	$^3/_3 - ^4/_4$. Cold front passed at 1500 hrs, day
11	20 August	0935 — 1855	Cloudless, southwestern periphery of anticyclone, day

The high-frequency region (frequency range from 10^{-5} to 1 cps) of the spectrum of temperature oscillations from data for the period 1329 — 1359 hrs on 14 August 1963 at Tsimlyansk is shown in Figure 7 on a much larger scale. It can be seen that the center of the mesometeorological minimum has corresponding to it a period of about 30 min and an ordinate of $S \approx 0.02$ deg^2, while the micrometeorological maximum has a period of about 1 min and an ordinate of $S \approx 0.4$ deg^2. These data, and particularly the ordinates of $S(f)$ in the high-frequency region, change markedly from one case to another, as is seen from Figure 8, which depicts eleven temperature spectra constructed from observations made in August at Tsimlyansk (the corresponding observation periods and weather conditions are given in Table 5).

THE PRESSURE SPECTRA

Figure 1c depicts the spectra of atmospheric-pressure fluctuations obtained by Gossard /32/. The low-frequency part of the spectrum was calculated from barograph records taken in San Diego during April—October 1956, while the high-frequency part of the spectrum was obtained from microphone recordings. This graph clearly shows a synoptic maximum with a period of about four days, and also clearly expressed 24 and 12 hour spectral lines and a mesometeorological minimum again appears. This spectrum is also depicted in the coordinate system ($\ln S$, $\ln f$) in Figure 9, from which it is seen that the spectral pressure density in the synoptic region decreases with a rise in f approximately in proportion to $f^{-7/3}$.

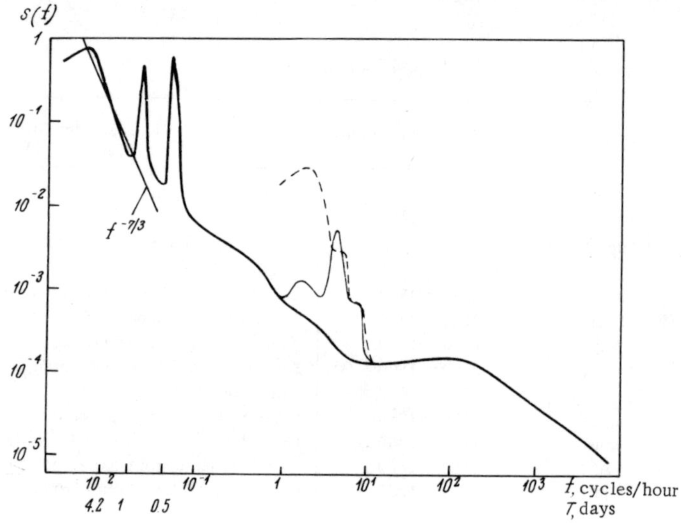

FIGURE 9. Spectral pressure density.

Gossard detected strong dependence of the spectral pressure fluctuations on the synoptic conditions in the range of 1—10 cycles/hour. On 24 June 1957 he detected strong pressure fluctuations in this range; the spectra are depicted in Figures 1c and 9 (the dashed curve is for the morning and the thin curve is for the evening). Gossard assumes that these fluctuations correspond to the generation of internal gravitational waves. Similar pressure fluctuations were detected by Clarke in Australia /30/ from spectra of recordings on a Dines microbarograph; two such spectra are shown in Figure 10. According to Clarke, these fluctuations are associated with the generation of unstable waves at inversion layers in the presence of a strong wind; these waves are indicated in particular by the formation of mammatus clouds.

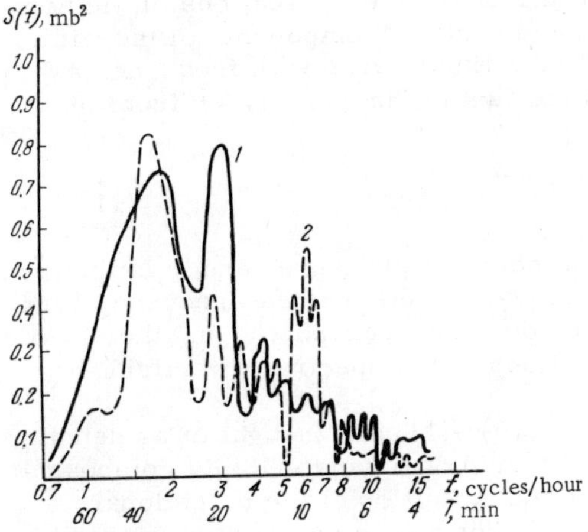

FIGURE 10. Pressure spectra on 26 January: 1—2200—0430 hrs; 2—1100—1730 hrs.

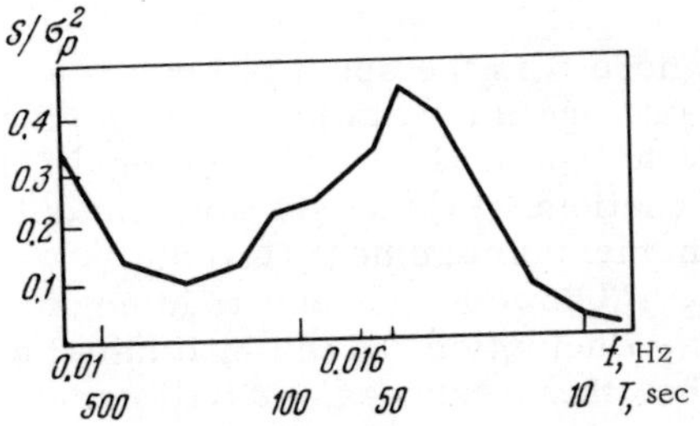

FIGURE 11. Average spectrum of microfluctuations of pressure.

Fluctuations of atmospheric pressure in the micrometeorological region of the spectrum can be produced by gravitational and acoustic waves, as well as by microturbulence. According to theoretical calculations by Monin and Obukhov /16/ and Dikii /11/, spectra of gravitational and acoustic waves hardly overlap, i.e., gravitational waves have periods in excess of 300 sec, while acoustic [sic] waves have periods of less than 300 sec. This prediction is satisfactorily confirmed by spectra of microfluctuations of pressure constructed by Golitsyn /4/ from microbarograms obtained by Pasechnik and Fedoseenko /18/. The average pressure spectrum in the 500—10 sec range obtained by Golitsyn is shown here in Figure 11 in the coordinate system $(S, \ln f)$ (the scale unit on the ordinate is the variance δ_p^2 of microfluctuations of pressure, with δ_p amounting to 10^{-2} millibars). The spectrum clearly shows a minimum corresponding to $T = 300$ sec and a maximum at $T \approx 50$ sec.

SPECTRA OF THE HEAT AND MOMENTUM FLUXES

In a manner similar to that used in equation (2) for expressing the spectral density of fluctuations of some meteorological characteristic $\xi(t)$, one can consider $\xi(t)$ and $\eta(t)$ by introducing the concept of their **cross** spectral density $s(f)$, setting

$$\overline{\xi'(t, \Delta f)\, \eta'(f, \Delta f)} = s(f)\, \Delta f \tag{9}$$

(the cross spectral density of two random functions $\xi(t)$ and $\eta(t)$ need not necessarily be positive everywhere). In particular, $s(f)$, which is defined by the expression

$$-\overline{\rho u'(t, \Delta f)\, w'(t, \Delta f)} = s(f)\, \Delta f, \tag{10}$$

where ρ is the air density, while u' and w' are the fluctuations of the horizontal (in the direction of the wind) and vertical components of the wind velocity, will describe the contribution of fluctuations with frequencies from the range Δf to the vertical momentum flux. The quantity $s(f)$ from the expression

$$\overline{c_p \rho w'(t, \Delta f) \cdot \theta'(t, \Delta f)} = s(f) \Delta f, \tag{11}$$

where c_p is the specific heat of air at constant pressure, and θ' is the temperature fluctuation, will describe the contribution of the frequency range Δf to the vertical turbulent heat flux [eddy heat transport]. In other words, equations (10) and (11) are the definitions of the **spectra of vertical momentum and heat fluxes.**

If, however, u' and w' in equations (10)—(11) are thought of as denoting the fluctuations of the latitudinal and meridional wind velocity components, then these expressions will serve as the definition of the **meridional** momentum and heat fluxes. Figure 1d shows such spectra calculated by Estoque /31/ from daily radiosonde data at the 850 mb level in the southeast of the U.S.A. (curve 1 is the spectrum of the meridional heat flux, curve 2 is the spectrum of the momentum flux). They again have a sharply expressed synoptic maximum (with a period of about four days) with a sharp drop in the mesometeorological region.

Figure 1d also displays spectra of vertical turbulent heat (curve 3) and momentum fluxes (curve 4) according to measurements by Gurvich and Tsvang /5, 6/ at an altitude of 1 m; these have a clearly expressed micrometeorological maximum (corresponding to a period of about 40 sec). No direct data on the spectra of horizontal or vertical heat and momentum fluxes in the mesometeorological region are available, but it can be inferred from the general shape of curves in Figure 1d that these spectra will also have a mesometeorological minimum.

EXPLANATION OF THE PRESENCE OF A MESOMETEOROLOGICAL MINIMUM IN THE SPECTRA

The most significant feature of all the above spectra is the fact that they have a synoptic maximum with a period of about four days, a micrometeorological maximum with a period of the order of a minute, and a wide mesometeorological minimum between them in the range of periods from several minutes to several hours (for example, in Figure 7 with a center at about a period of 30 min).

This separation of synoptic and micrometeorological fluctuations markedly reduces difficulties in determining the average values of micrometeorological characteristics associated with the appearance of "level evolution" of meteorological fields, i.e., with a dependence of the mean values on the averaging scale. It is possible to determine relatively stable average values, for example, average wind-velocity and temperature profiles in the surface air layer, by averaging over periods belonging to the range of mesometeorological spectral minima (over periods of 10—20 min, say, as done in practice). In addition, any perceptible fluctuations of meteorological fields with frequencies from the region of the mesometeorological minimum (and their corresponding spatial mesoinhomogeneities of meteorological fields) can be found only very rarely, so that the actual mesometeorology starts only at the high-frequency side of the synoptic spectrum with periods of several hours (space scales of the order of 100 km). We shall now attempt to give an explanation for the above feature of spectra of meteorological elements. We shall operate here with space spectra which should be simpler than time spectra (for example, space spectra should not contain features corresponding to the 24 and 12 hour lines of time spectra or to the maximum for periods of global circulation fluctuations of the "index cycle" type). The most important fact is that the atmosphere is relatively thin, i.e., its effective thickness (let us say a thickness of $H \sim 10$ km of the layer containing 80% of its mass) is small compared with its horizontal dimension (size of the planet). Hence the spectrum of the scales of space inhomogeneities of meteorological fields should be subdivided into three regions as follows:

synoptic region of scales; these are large compared with H, and apparently the inhomogeneities are **quasi-two-dimensional** (horizontal inhomogeneities);

micrometeorological region of scales; these are small compared with H, and the **three-dimensionality** of turbulent inhomogeneities is quite appreciable;

intermediate or **mesometeorological** region of scales, from several to many tens of kilometers. We wish to point out that according to the expression $T = \Lambda/u$ ("frozen turbulence") at $u = 10$ m/sec our "boundary" scale of inhomogeneities $\Lambda = H \sim 10$ km has a period of precisely $T \sim 20$ min, corresponding to the center of the mesometeorological minimum of time spectra.

As is known, the immediate source of air-movement energy is heat influx to the atmosphere from the surface by means of radiative and eddy heat transfer and transport of the latent heat by the evaporating moisture (the scientific study of the physics of these processes began only very recently; the rate at which they occur is as yet measured only at a small number of points over land, while virtually no measurements have as yet been made over oceans, and no scientific basis is available for climatic correlations). It may therefore be assumed that the maximum of space spectra $S(k)$ of the wind velocity (and most probably also of temperature and pressure) in the synoptic domain should correspond to scales $\Lambda_{max} = 2\pi/k_{min}$ of the most substantial inhomogeneities of the heat influx field, those associated with differences in solar heating of polar and tropic zones and in differences in the heat influx over land and over oceans, as well as those produced by synoptic cloud systems (which screen direct solar radiation as well as heat fluxes from the surface). These scales amount to thousands and tens of thousands of kilometers.

Motions on the above scales are found to be hydrodynamically unstable. Degenerating upon loss of stability, they generate motions at smaller scales and transfer to them a part of their energy. The quasi-two-dimensional, smaller-scale motions thus produced may in turn be found to be hydrodynamically unstable and generate motions with even smaller scales, etc. This gives rise to the cascade process of energy transfer over the spectrum from large to small scales, discovered by Richardson /40/ as early as 1926 and used by Kolmogorov as the basis for the theory of turbulent motions at very high Reynolds numbers /13/. A part of the energy from motions at synoptic scales of $\Lambda \sim 2000-3000$ km, which receive energy directly from external heat influxes, is apparently on the average transmitted also to large-scale ($\Lambda \sim 10^4$ km) motions, i.e., to flows of the general circulation of the atmosphere.

In addition to participating in the above cascade transfer of energy, all the synoptic-scale motions **directly** and continuously generate micrometeorological turbulences, transmitting to them a part of their energy. The microturbulence is generated here as a result of **vertical inhomogeneities** of the wind field, particularly near the earth surface (and also apparently in jet streams), where the vertical wind-velocity gradients are particularly high. In addition, there is a continuous exchange between the kinetic energy of the microturbulence and the potential energy of the vertical density stratification of the air which is brought about by the Archimedes forces attendant on vertical motions of air particles whose density is different from that of the surrounding medium; unstable stratification serves as a supplementary energy source for the microturbulence, while on stable stratification, conversely, microturbulence loses energy in opposing the Archimedes forces. This direct transfer of energy from synoptic motions to microturbulence takes place irrespective of the operation of the cascade energy transfer over the spectrum. For example, in a region with constant pressure gradient the plane-parallel air flow is hydrodynamically stable and cannot generate synoptic motions of smaller scale, but as a result of friction at the earth surface it does generate microturbulence.

Regarding microturbulence as a dissipative factor for synoptic motions, we can describe it by the effective kinematic coefficient of eddy viscosity ν_{eddy}. Then it can be assumed by analogy with the Kolmogorov hypotheses /14/, that in the short-wave part of the synoptic region $\Lambda < \Lambda_{max}$ the spectrum of quasi-two-dimensional atmospheric motions will be determined only by two parameters, viz., the rate ε of cascade energy transfer over the spectrum and the value of ν_{eddy} and that this value of ν_{eddy} will be substantial only at the extreme short-wave end of this region, at scales

$$\Lambda \sim \left(\frac{\nu_{eddy}^3}{\varepsilon}\right)^{1/4},$$ while in the scale range

$$\Lambda_{max} > \Lambda > \Lambda_1 = c\left(\frac{\nu_{eddy}^3}{\varepsilon}\right)^{1/4}, \qquad (12)$$

where c is some numerical constant, the form of the spectrum will be determined solely by the value of ε. This hypothesis for the range of scales given by equation (12) has been verified experimentally by a number of investigators. It is prompted by the early results of Richardson /40/, who found that the "four-thirds law" $K = \varepsilon^{1/3}\Lambda^{4/3}$ is valid for the eddy diffusion coefficient K up to scales Λ_{max} of the general circulation of the atmosphere. A similar relationship for time correlation functions of meteorological characteristics in the synoptic range of periods was detected by Hutchings /34/. Finally, for the spectral density of the pressure field this hypothesis yields $S(k) \sim \rho^2\varepsilon^{4/3}k^{-7/3}$, which is in rather satisfactory agreement with Gossard's spectrum (see Figure 9) in the synoptic region /32/.

The energy expended by quasi-two-dimensional synoptic formations to overcome the eddy viscosity (or, which is the same thing, to generate microturbulence) in the region of $\Lambda < \Lambda_1$ becomes appreciable and spectra $S(k)$ should decrease with increasing k much faster than in the region defined by equation (12). It can be concluded from this that if scale Λ_1 bounding region (12) from the short-wave side is greater than the characteristic scale $\Lambda_2 < H$ of vertical inhomogeneities of meteorological fields (corresponding to the maximum of the micrometeorological spectrum), then in region $\Lambda_1 < \Lambda < \Lambda_2$ spectra $S(k)$ of meteorological fields should have quite clear minima. Condition $\Lambda_1 > \Lambda_2$ can be rewritten in the form

$$\nu_{\text{eddy}} > \varepsilon^{1/3}\left(\frac{\Lambda_2}{c}\right)^{4/3}. \tag{13}$$

If it is assumed that $\varepsilon \sim 1$ cgs unit and that $\frac{\Lambda_2}{c} \sim 10^4$ cm, then the right-hand side of (13) will amount to about 10 m^2/sec, while estimates of ν_{eddy} from data on altitudinal wind changes in the planetary boundary layer of the atmosphere will usually yield values of about 10^2 m^2/sec. Thus, the above explanation of the mesometeorological minimum does not contradict available empirical data.

In the short-wave range of the micrometeorological region, with $\Lambda < \Lambda_2$, we will again have cascade energy transport from larger to smaller scales, determined by the parameter ε; the value of this parameter will depend on the synoptic circumstances, and on the average will be the same as for the range (12) of the synoptic region. In the scale interval of

$$\Lambda_2 > \Lambda \gg \lambda \sim \left(\frac{\nu^3}{\varepsilon}\right)^{1/4}, \tag{14}$$

where ν is now the **molecular** viscosity, the form of the turbulence spectrum for neutral thermal stratification of air will be determined, according to the Kolmogorov hypotheses, only by the parameter ε; this is verified by a large volume of experimental data (see, for example, the article by Gurvich and Tsvang /5/). Correlations associated with the effect of thermal stratification of the air will not be considered here.

THE UNDERSTANDING OF CLIMATE AND OF CLIMATIC FLUCTUATIONS

In our opinion climate can only be defined reliably by using concepts of time spectra of meteorological elements. One must imagine the total form of these spectra over the entire range of periods, starting with the minimum period of micrometeorological fluctuations up to those commensurable with scales of geological epochs and consider the fact that fluctuations of meteorological elements apparently occur (and consequently, the spectral densities of fluctuations are nonzero) in all the regions of the spectrum of periods.

So far, moving from small to large periods, we have established the existence of micrometeorological fluctuations, then of synoptic fluctuations with a maximum of spectra $S(f)$ about a period of $T \approx 4$ days, and mentioned "global" fluctuations of the "index cycle" type with a period of $T \approx 12$ days, and possibly with some greater periods. These "global" fluctuations with periods of from several weeks to several months have virtually not been considered as yet (if one does not count the poorly investigated annual fluctuations and higher-order harmonics of the annual variation); they would be of tremendous interest in problems of long-range weather forecasting.

Upon crossing the yearly "line" in time spectra, one will have to consider the year-to-year variation in meteorological elements. One of the first interesting subjects to arise here will be to compare the year-to-year variability with that within the year (synoptic micrometeorological) of a given meteorological element. Theoretically one should compare here the contributions made to the total variance $\sigma_\xi^2 = \overline{\xi'^2}$ by frequency intervals $0 \leqslant f < f_0$ and $f_0 \leqslant f < f_{max}$ (where $f_0 = 1/\text{year}$), i.e., estimate the value of the fraction

$$\mu^2 = \frac{\int\limits_0^{f_0-0} s_\xi(f)\, df}{\int\limits_{f_0-0}^{f_{max}} s_\xi(f)\, df}. \tag{15}$$

Had the year-to-year variability of the meteorological element under study been small compared with its variability within the year (i.e., were $\mu \ll 1$), the construction of the climatology of this element would have been much simpler; in solving some problems of climatology the year-to-year variation could have been simply neglected in this case. However, some data (part of which is presented below) force an opposite conclusion: **for the majority of meteorological elements usually considered, the values of μ are approximately the same and are not very small compared with unity.**

It is very difficult to obtain an accurate estimate of μ because the available data on meteorological elements are not available for sufficient time periods; it is thus impossible to estimate average values over very long time intervals and, in particular, to determine values of $s(f)$ at very small f. It is hence expedient to replace μ by some quantity which has approximately the same meaning and which can be estimated more easily. For example, if one has available an N-year series of observations of a given meteorological element, one can easily calculate

$$\mu_0 = \frac{a}{A}, \tag{16}$$

where a is the amplitude of fluctuations (peak-to-peak amplitude) of the annual average values of the element for the given N years, while A is the amplitude of the annual variation of element ξ, averaged for these N years. Replacement of μ_0 by the ordinary "variability factor," i.e., the ratio of the standard deviation of annual average values X_i to their average $\overline{X}$, can frequently be inconvenient, primarily for sign-variable quantities and particularly for those (like temperature) which have an arbitrary zero, since for them one may sometimes find that $\overline{X} = 0$.

As an illustration we used data obtained between 1951—1964 from 16 meteorological stations in the Soviet Union to calculate the values of μ_0 for the air temperature θ and the so-called "radiation balance" R, which denotes the algebraic sum of the radiative heat fluxes at the earth surface (the stations selected had data on R for at least 10 years). It was found that the values of $\mu_0(R)$ from different stations ranged from 13.5 to 27.0%, those of $\mu_0(\theta)$ varied from 19.0 to 27.8%, while the ratio $\mu_0(R)/\mu_0(\theta)$ ranged from 0.57 to 1.23, so that the values of μ_0 for θ and R were actually approximately equal. This shows the error and danger of assuming that the year-to-year variability of R is small, which some geographers postulate for verifying their substitution of unknown climatic average values of R by their averages for several years; this would also have required assuming that the year-to-year temperature variations, as well as the majority of meteorological elements under study, are small, which would have resulted in the negation of climatology as a whole.

Moving along the spectrum in the direction of longer periods, we recall the assumption made by some investigators that a relationship exists between fluctuations of meteorological elements and the solar activity. According to this assumption, solar activity periods, i.e., primarily the 11.5 year period (and possibly its 5.5 year harmonic) should appear in the weather fluctuations. From our point of view the majority of arguments in favor of this assumption were unconvincing and a more accurate analysis of actual data is needed here.

It can further be claimed that fluctuations of meteorological fields exist with periods of many tens of years. One may cite as an example the warming up in the first half of the 20th century which is apparently ending now. As regards secular fluctuations, we wish to point out a worsening of the climate during the 17th—19th centuries. Paleoclimatology also indicates the presence of fluctuations over the centuries: gradually increasing warmth upon termination of the glacial period (65th century BC), which resulted in the so-called **climatic optimum** between the 40th and 20th centuries BC, subsequent worsening of the climate during the so-called **subatlantic period** (between the 10th century BC and the 3rd century AD), its improvement between the 4th and 10th centuries AD, and worsening in the 13th and 14th centuries. Finally, the glacial periods as such can be regarded as fluctuations with periods of the order of tens and hundreds of thousands of years.

The above long-period fluctuations had quite impressive amplitudes. For example, during the "small climatic optimum" (8th—10th centuries), when Iceland and Greenland were being colonized by the Vikings, the Arctic, according to their chronicles, was apparently ice free. This, by the way, shows the error of claims based on the unjustified assumption as to the role of the local albedo for the Arctic regime (if the Arctic ice would melt, it would no longer be there). Since the Arctic within the limits of the 72°N circle amounts to less than 2.5% of the earth surface, it may be assumed that conditions in the Arctic are a result, rather than a cause, of the nature of the general circulation of the atmosphere.

It is clear from the above that it would have been incorrect to define climate as a **constant** "average background" of the weather. It is much more correct to define it as follows: **climate is the statistical regime of short-period fluctuations of meteorological fields undergoing long-term fluctuations.** The subdivision of the fluctuation spectrum into long- and short-period parts inherent in this definition is obviously somewhat arbitrary. It could become more definite only if a deep spectral minimum, which would have separate climatic fluctuations from the weather, had been detected in the range of periods in excess of a year. However, this range of periods has so far remained virtually unconsidered, and it can only be debated whether the boundary between "climate" and "weather" should be drawn at a period of more or less than 11 years. In our opinion the first version is correct.

SOLAR ACTIVITY AND TERRESTRIAL WEATHER

The relationship between variations in solar activity and certain geomagnetic events has been established quite reliably and has a graphic explanation as follows. Plasma clouds radiated by active regions of the solar surface "brush" against the terrestrial magnetosphere and thus disturb it. The question as to the possible relationship between solar activity oscillations and weather on the earth appears to be entirely different. Many arguments as to the existence of this relationship, collected for example in the books by Eigenson /25, 26/, Rubashev /20/, and Sazonov /21/, are not followed by any convincing support.

Since the relative fluctuations of the solar radiation incident on the earth are negligible, and even their absolute values are very small compared with the energy of the basic weather phenomena — cyclones and anticyclones — the effect of the solar activity can be imagined logically to be the result of the accumulation of some small effects during sufficiently long time periods. If one assumes that solar activity is a rapidly acting factor for changing the weather, one will have to agree that the terrestrial atmosphere is sufficiently frequently brought into a state of unstable equilibrium, "at the edge of instability," when small external disturbances can produce substantial changes. But this concept of the atmosphere as a system tending toward the peaks of "potential ridges" seems to us to be quite unnatural.

These considerations force us to have grave doubts as to the validity of arguments on the existence of a relationship between variations in solar activity and the terrestrial weather. These arguments were brought primarily by enthusiasts who had a priori assumed the existence of this relationship and then sought at all costs to find a basis for their views. The enthusiasm of these investigators, at times excessive, has comprised, in our opinion, the idea of the effect of solar activity on the weather.

When examining the relationship between solar activity and the weather a distinction should be made between three groups of weather fluctuations. The first of these are fluctuations on time scales of from several days to several weeks. Attempts are made to find the effects of the direct influence on the weather of each region of solar activity, the corpuscular radiation of which should, under the geometric conditions, reach the earth. From the point of view of time spectra of meteorological elements one should remember here that if the "life" of solar activity spots exceeds the period of the rotation of the sun about its axis (in the equatorial zone this is about 27 days) and if the radiations of these spots actually affect the weather, then the spectra of at least some meteorological elements should contain a 27-day period (and possibly its harmonics). At present no empirical data are available to prove this.

The second group of weather fluctuations of importance from the point of view of "heliophysics" are those with periods of from several to several tens of years, including the 11.5 years of the most reliably established solar-activity period and the higher-order harmonics of this period. The double cycle, which is basic from the point of view of the sign of magnetic-spot fields, is also of interest.

In our opinion, the most correct assertion on demonstrations of this cycle in the terrestrial weather is that the 11.5 year period does not demonstrate itself in the fluctuations of certain meteorological elements.

The third group of fluctuations which deserve separate consideration are secular, i.e., climatic fluctuations. According to the previously mentioned possibility of the accumulating small effect of a prolonged time the assumption that solar activity affects climatic fluctuations appears to be more probable than that it exerts its effect in the short-period spectral regions. At the same time one must contend with the nonavailability of long-period observations sufficient for statistical study, so that from the practical point of view only individual, and not statistical analysis of these series is possible. However, even this analysis can result in some useful conclusions. Thus, the most significant of all the information known to us on the effect of solar activity on the weather is the similarity, detected by Bezrukova /1/, between variations in the recurrences of atmospheric-circulation types according to Dzerdzeevskii, smoothed over decades, and the so-called Wolf numbers describing the solar activity during the first half of the 20th century; during this time both the recurrence of zonal circulations and the Wolf numbers increased. If a genetic relationship does indeed exist, then it would be of great interest to explain from physical considerations why an increase in solar activity should be instrumental in increasing the recurrence of zonal atmospheric circulations.

In the present paper we consider in more detail only the second group. The main question is: does the 11.5 year solar activity cycle demonstrate itself in weather fluctuations? It should first be noted that the state of the solar activity can be described by a number of indicators, the most popular of which is the Wolf number W, defined by

$$W = a + 10b, \tag{17}$$

where a is the number of spots on the visible side of the sun, while b is the number of spot clusters /47/. Wolf determined the values of W for the time interval from 1749 and found that they undergo perennial fluctuations with a period of about 11.5 years (epochs of maxima and minima of W were detected as early as a century before, from 1610). The monthly average Wolf numbers in the minima of this cycle amount to only several units, while the maxima are as high as 80—140 units.

The nature of fluctuations of monthly average values of W is clearly seen in Figure 12, where the curve of W is constructed for the years 1820—1950. It is seen from this graph that the variations of W within the 11.5 year cycle are not sinusoidal: W increases more rapidly (on the average over four years) than it decreases (on the average in seven years). It follows from this that the spectrum of fluctuations of W should, in addition to the 11.5 year period, also exhibit higher-order harmonics of this period.

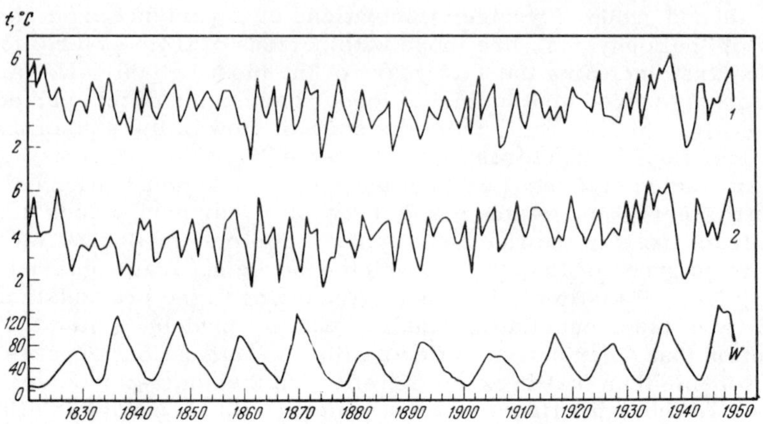

FIGURE 12. Monthly average temperatures and Wolf numbers:

1—Moscow; 2—Leningrad.

Using the previously described spectral analyzer we constructed the spectrum of W in a period range of from 2 to 25 years from data for 1749—1947. The band filters with half an octave width used in our analyzer for this interval are excessively wide (we recall that the system of band filters makes it possible to construct $S(f)$ only in the form of a stepped curve in which the width of each step is equal to the bandwidth of the corresponding filter). In order to improve the determination of the spectrum using this type of analyzer, we carried out spectral analysis twice with different ratios of the tape-travel speed to the recording speed; this made it possible to increase the number of points from which the spectral curve is constructed (a change in this ratio results in a shift of the filter passbands along the frequency axis). The spectrum $S(f)$ of Wolf numbers thus obtained is shown in Figure 13. A sharp maximum is seen at the 11.5 year period and a clear maximum with a smaller amplitude for a period of about 5 years. A similar spectrum of Wolf numbers was obtained numerically in /46/.

It should, however, be noted that Wolf numbers are apparently not the best index of solar activity for purposes of "heliophysics." According to some data they describe to some extent the oscillations in the ultraviolet rather than corpuscular solar radiation. The latter is described better, although also not too well, by the total area of solar spots σ (on the average $\sigma \approx 16.7W$; the correlation factor between σ and W is 0.85) /3/.

Many other solar activity indexes with different time variations are also considered. Which of these should, after all, be used for comparison with weather fluctuations ?

It is possible to avoid the difficulties involved in this question by first considering only the spectra of weather fluctuations and determining whether periods inherent to solar-activity fluctuations, primarily the 11.5 year cycle and its higher-order harmonics, are expressed in them. This approach not only makes it possible to avoid difficulties in selecting suitable solar-activity indexes, but also the subjectivity in estimating the similarity between these two random functions. It should also be noted that, while short-period fluctuations of meteorological elements may depend markedly

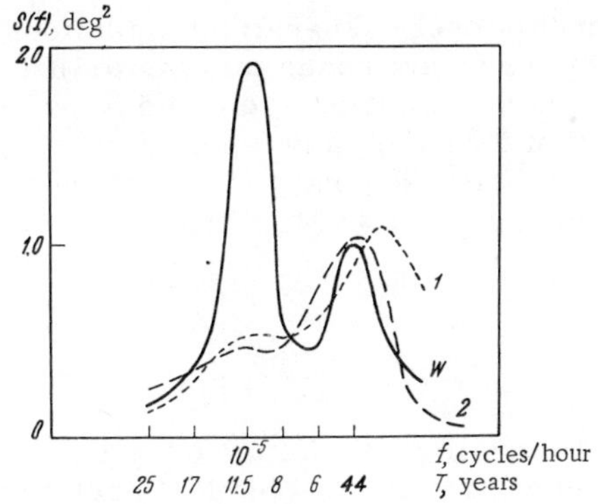

FIGURE 13. Spectra of temperatures and Wolf numbers:

1—Moscow; 2—Leningrad.

on local conditions and thus be nonrepresentative of the world's weather, the effect of local conditions should be weaker for long-period fluctuations, so that long-period fluctuations in the world's weather can be described not only by fluctuations of planetary indexes, but also by values of ordinary meteorological elements at individual points. In other words, it may be assumed that in the **long-period region the spectra of the majority of meteorological elements and of planetary indexes are at least approximately similar to one another.**

To verify this assumption and simultaneously to determine whether the 11.5 year solar activity cycle demonstrates itself in weather fluctuations, it is expedient to study the spectra of sufficiently long-term observation series which are available in meteorological records. We first determined the spectra of air-temperature variations in Leningrad (data collected for 198 years) and in Moscow (for 130 years). The graphs of these temperature variations constructed from monthly average values are shown in Figure 12. They show without additional calculations that the most pronounced temperature variations occur with a period approximately half that of the solar activity cycle.

In constructing temperature spectra in the long-period region one should first exclude fluctuations with $T = 1$ year, which has corresponding to it an extremely high maximum (theoretically it is a line). One should first construct a spectrum from a series of yearly average values, or analyze deviations of the temperature from the perennial average march (which appears to be preferable).

We constructed temperature variation spectra for Leningrad and Moscow using the previously described semiautomatic spectral analyzer (using the "ring method" described above); the Leningrad spectrum was also calculated from equation (3) on a computer (with the cooperation of V.F.Pisarenko). The spectra of Figure 13 fully confirm the conclusion drawn from examining the temperature curves of Figure 12: the spectrum has clearly expressed maxima of about half a period of the principal solar cycle (in the range between 2 and 5.5 years), while the 11.5 year cycle proper is hardly expressed. The spectra of temperature variations at some stations in Maryland (U.S.A.) constructed in /35/ have a similar shape. We should also recall spectra of 100—160 year annual precipitation totals at eight European stations obtained by Brier /29/. The spectrum of the series obtained by summation does not show any clear maxima at $T \approx 11.5$ or $T \approx 5.5$ years (nor at $T \approx 22$ years).

Do these data mean that the 11.5 year solar activity cycle is not expressed in terrestrial weather, while the maximum found in temperature spectra of Figure 13 in the range of 2 and 5.5 year periods is of "terrestrial" origin, or is it possible that this cycle generates fluctuations with a double period in the terrestrial weather in a manner similar to that in which, according to Simpson /42/, one solar radiation cycle produces two glaciation cycles on the earth ? We hope that future data on spectra of meteorological elements for periods ranging from several to several tens of years will help to resolve these problems.

References

1. Bezrukova, A. Ya. Kharakter tsirkulyatsii zemnoi atmosfery i solnechnaya aktivnost' (Nature of Terrestrial Atmospheric Circulation and Solar Activity). — Byulleten' Komissii po issledovaniyu Solntsa, Nos. 5 and 6. 1960.
2. Bykov, V. V. and S. A. Mashkovich. K kharakteristike zonal'noi tsirkulyatsii atmosfery (Characteristics of the Zonal Circulation of the Atmosphere). — Meteorologiya i gidrologiya, No. 2. 1956.
3. Vitinskii, Yu. I. Prognozy solnechnoi aktivnosti (Solar-Activity Forecasting). — Izdatel'stvo AN SSSR. 1963. [IPST translation Cat. No. 1400, TT 65-50115.]
4. Golitsyn, G. S. O vremennom spektre mikropul'satsii atmosfernogo davleniya (The Time Spectrum of Microfluctuations of Atmospheric Pressure). — Izvestiya AN SSSR, seriya geofizicheskaya, No. 8. 1964.
5. Gurvich, A. S. and L. R. Tsvang. O spektral'nom sostave turbulentnogo potoka tepla (The Spectral Composition of the Turbulent Heat Flux). — Izvestiya AN SSSR, seriya geofizicheskaya, No. 10. 1960.
6. Gurvich, A. S. O spektral'nom sostave turbulentnogo potoka kolichestva dvizheniya (The Spectral Composition of the Turbulent Momentum Flux). — Izvestiya AN SSSR, seriya geofizicheskaya, No. 10. 1961.
7. Gurvich, A. S. Spektry pul'satsii vertikal'noi komponenty skorosti vetra i ikh svyazi s mikrometeorologicheskimi usloviyami (Spectra of Variations of the Vertical Components of Wind Speed and their Relationships with Micrometeorological Conditions). — Trudy Instituta Fiziki Atmosfery AN SSSR, No. 4. 1962.
8. Dzerdzeevskii, B. L., V. M. Kurganskaya, and V. M. Vitvitskaya. Tipizatsiya tsirkulyatsionnykh mekhanizmov v severnom polusharii i kharakteristika sinopticheskikh sezonov (Classification of Circulation Mechanisms in the Northern Hemisphere and Description of Synoptic Seasons). — Trudy NIU GUGMS. 1946.
9. Dzerdzeevskii, B. L. and A. S. Monin. Tipovye skhemy obshchei tsirkulyatsii atmosfery v severnom polusharii i indeks tsirkulyatsii (Typical Models of the General Circulation of the Atmosphere in the Northern Hemisphere and the Circulation Index). — Izvestiya AN SSSR, seriya geograficheskaya, No. 6. 1954.
10. Dobryshman, E. M., S. A. Mashkovich, and Ya. M. Kheifets. Kharakteristiki zonal'noi tsirkulyatsii (Zonal-Circulation Characteristics). — Gidrometeoizdat. 1958.

11. Dikii, L.A. Chastoty svobodnykh kolebanii zemnoi atmosfery (Frequencies of Natural Oscillations of the Terrestrial Atmosphere). — Doklady AN SSSR, 157(3). 1954.
12. Zubkovskii, S.L. Chastotnye spektry pul'satsii gorizontal'noi komponenty skorosti vetra v prizemnom sloe vozdukha (Frequency Spectra of the Variations of the Horizontal Component of Wind Speed in the Surface Air Layer). — Izvestiya AN SSSR, seriya geofizicheskaya, No. 10. 1962.
13. Kolmogorov, A.N. Lokal'naya struktura turbulentnosti v neszhimaemoi zhidkosti pri ochen' bol'shikh chislakh Reinol'dsa (The Local Structure of Turbulence in an Incompressible Viscous Fluid for Very Large Reynolds Numbers). — Doklady AN SSSR, 30(4). 1941.
14. Kolmogorov, A.N. Statisticheskaya teoriya kolebanii s nepreryvnym spektrom (A Statistical Theory of Oscillations with a Continuous Spectrum). — Yubileinyi sbornik AN SSSR, Vol. 1. Izdatel'stvo AN SSSR. 1947.
15. Monin, A.S. O makroturbulentnom obmene v zemnoi atmosfere (The Macroturbulent Exchange in the Terrestrial Atmosphere). — Izvestiya AN SSSR, seriya geofizicheskaya, No. 4. 1956.
16. Monin, A.S. and A.M. Obukhov. Malye kolebaniya atmosfery i adaptatsiya meteorologicheskikh polei (Small Oscillations of the Atmosphere and Adaptation of Meteorological Fields). — Izvestiya AN SSSR, seriya geofizicheskaya, No. 11. 1958.
17. Panofsky, H.A. and R.J. Deland. One-Dimensional Spectra of Atmospheric Turbulence in the Lowest 100 Meters. — Advances in Geophysics, 6: 41. 1959.
18. Pasechnik, I.P. and N.E. Fedoseenko. Elektrodinamicheskii mikrobarograf s gal'vanometricheskoi registratsiei (An Electrodynamic Microbarograph with Galvanometer Recording). — Izvestiya AN SSSR, seriya geofizicheskaya, No. 1. 1958.
19. Perepelkina, A.V. Nekotorye rezul'taty issledovaniya turbulentnykh pul'satsii temperatury i vertikal'noi sostavlyayushchei skorosti vetra (Some Results of the Study of Turbulent Temperature Fluctuations and of the Vertical Component of Wind Velocity). — Izvestiya AN SSSR, seriya geofizicheskaya, No. 6. 1957.
20. Rubashev, B.M. Problemy solnechnoi aktivnosti (Problems of Solar Activity). — Izdatel'stvo Nauka. 1964.
21. Sazonov, B.I. Vysotnye baricheskie obrazovaniya i solnechnaya aktivnost' (High-Altitude Baric Formations and the Solar Activity). — Gidrometeoizdat. 1964.
22. Willett, H.C. Atmospheric and Oceanic Circulation as Factors in Glacial-Interglacial Changes of Climate. — In: Shapley, H. (editor), Climate Change, p. 51—71, Harvard Univ. Press. 1953.
23. Tsvang, L.R. Izmereniya chastotnykh spektrov temperaturnykh pul'satsii v prizemnom sloe atmosfery (Measurement of Frequency Spectra of Temperature Fluctuations in the Surface Air Layer). — Izvestiya AN SSSR, seriya geofizicheskaya, No. 8. 1960.
24. Tsvang, L.R. Izmerenie turbulentnykh potokov tepla i spektrov temperaturnykh pul'satsii (Measurement of Turbulent Heat Fluxes and of Spectra of Temperature Fluctuations). — Trudy Instituta Fiziki Atmosfery AN SSSR, No. 4. 1962.

25. Eigenson, M.S. Ocherki fiziko-geograficheskikh proyavlenii solnechnoi aktivnosti (An Outline of Physical-Geographical Manifestations of Solar Activity).—Izdatel'stvo L'vovskogo Gosudarstvennogo Instituta. 1957.
26. Eigenson, M.S. Solntse, pogoda i klimat (The Sun, Weather, and Climate).—Gidrometeoizdat. 1963.
27. Yaglom, A.M. Vvedenie v teorii statsionarnykh sluchainykh funktsii (Introduction to the Theory of Stationary Random Functions).—Uspekhi matematicheskikh nauk, 7(5(51)). 1952.
28. Blackman, K.B. and J.M. Tukey. The Measurement of Power Spectra.—New York, Dover Publications. 1958.
29. Brier, G.W. Some Statistical Aspects of Long-Time Fluctuations in Solar Atmosphere Phenomena.—Ann. N.Y. Acad. Sci., Vol. 95, Art. 1. 1961.
30. Clarke, R.E. Pressure Oscillations and Fall-out Down Draughts.—Quart. J. Roy. Meteorol. Soc., 88(378). 1962.
31. Estoque, M.A. The Spectrum of Large-Scale Turbulent Transfer of Momentum and Heat.—Tellus, 7(2). 1955.
32. Gossard, E.E. Spectra of Atmospheric Scalars.—J. Geophys. Res., 65(10). 1960.
33. Capt, H.L., H.L.Griffith, H.A.Panofsky, and J.Van der Hoven. Power Spectrum Analysis over Large Ranges of Frequency.—J. Meteorol., 13(3). 1956.
34. Hutchings, J.W. Turbulence Theory Applied to Large-Scale Atmospheric Phenomena.—J. Meteorol., 12(3). 1955.
35. Landsberg, H.E., J.M.Mitchell, and H.L.Gruntcher. Power Spectrum Analysis of Climatological Data for Woodstock College, Maryland.—Monthly Weather Rev., 87(8). 1959.
36. MacCready, P. Structure of Atmosphere Turbulence.—J. Meteorol. 10(6). 1953.
37. Monin, A.S. Statistical and Periodic Time Series in the General Circulation of the Atmosphere.—Time Series Analysis, N.Y. 1963.
38. Panofsky, H.A. The Variation of the Turbulence Spectrum with Height under Superadiabatic Conditions.—Quart. J. Roy. Meteorol. Soc., 79(339). 1953.
39. Panofsky, H.A. and J.Van der Hoven. Spectra and Cross Spectra of Velocity Components in the Mesometeorological Range.—Quart. J. Roy. Meteorol. Soc., 79(339). 1953.
40. Richardson, L.F. Atmospheric Diffusion Shown on a Distance Neighbour Graph.—Proc. Roy. Soc., A110(756). 1926.
41. Saltzman, B. and A.Fleisher. The Exchange of Kinetic Energy between Larger Scalars of Atmospheric Motion.—Tellus, 12(4). 1960.
42. Simpson, G.S. Ice Ages.—Nature, Vol. 141. 1938.
43. Taylor, G.I. The Spectrum of Turbulence.—Proc. Roy. Soc., Vol. 164. 1938.
44. Van der Hoven, J. Power Spectrum of Horizontal Wind Speed in the Frequency Range from 0.00007 to 900 Cycle per Hour.—J. Meteorol. 14(2). 1957.
45. Wan-Cheng-Chiu. The Wind and Temperature Spectra of the Upper Troposphere and Lower Stratosphere over North America.—J. Meteorol., 17(1). 1960.

46. Ward, F. and R. Shapiro. Solar Geomagnetic and Meteorological Periodicities.—Ann. N. Y. Acad. Sci., Vol. 95, Art. 1. 1961.
47. Wolf, R. Kortweg Sitzungsberichte. Wien. 1883.

G. V. Gruza

INTEGRAL CHARACTERISTICS OF THE GENERAL CIRCULATION OF THE ATMOSPHERE

Various characteristics of the general circulation of the atmosphere are investigated. Mean monthly zonal characteristics calculated for January and July 1956 from data for the 1000, 850, 700, 300, 200, and 100 mb levels are presented. The total kinetic energy, kinetic energy of the average zonal motion, energy of large scale eddies, meridional motion, and the meridional momentum transfer are considered. Qualitative data on the kinetic energy potentials of the geostrophic circulation in the northern hemisphere for 1958 are discussed and the relationships between their different components are examined. The annual variation of this process is considered. An estimate is made of the extent to which the data calculated on the basis of geostrophic movement correspond to the real circulation.

Meteorological studies of the dynamics of atmospheric processes over large areas and over sufficiently long time intervals make extensive use of various integral characteristics or "indexes" of atmospheric circulation. However, the need has been felt for quite some time for an objective method to estimate air flows which would make a quantitative description of the nature of those motions possible. A large number of localized and integral circulation indexes, which reflect the intensity of zonal and meridional motion, cyclonic activity, etc., have been suggested /1, 5, 8, 9, 11, etc. /. The zonal circulation index as suggested by Rossby and Haurwitz and the system of circulation indexes introduced by Kats have been generally recognized and are now extensively used.

Various parameters of the physical state of the atmosphere may be selected as integral characteristics of the circulation. Sufficiently general methods for their selection were pointed out by Monin /6, 7/ and applied by the present author /3/.

Suppose a certain physical quantity A describes the state of the atmosphere. It is natural to select as the corresponding integral characteristic the average (mathematical expectation) $\bar{A}$ which, subject to the selected time or space scale, describes some regular background or macrocomponent. The variations $A' = A - \bar{A}$ can be regarded as irregular fluctuations describing smaller-scale processes, or the microcomponent. These fluctuations should be examined statistically as random deviations from the background or macroprocess.

To determine which integral characteristics should be studied together with $\bar{A}$, one must examine dynamic equations describing the evolution of the macrocomponent. It is useful to include the terms of these equations in the system of integral circulation characteristics. One may find among these averages of other physical quantities (of the macrocomponent), as well as statistical characteristics of fluctuations (of the microcomponent).

As an illustration, we examine a quite simple dynamic model of the atmosphere. Let us consider the equations of the horizontal adiabatic atmospheric motion disregarding divergence and friction. These equations may be written in the form

$$\frac{\partial u}{\partial t}+\frac{1}{R\sin\vartheta}\frac{\partial u^2}{\partial\lambda}+\frac{1}{R\sin\vartheta}\frac{\partial}{\partial\vartheta}u\cdot v\cdot\sin\vartheta+\frac{uv\operatorname{ctg}\vartheta}{R}= \\ =-\frac{1}{R\sin\vartheta}\frac{\partial p}{\partial\lambda}-2\omega v\cos\vartheta; \qquad (1)$$

$$\frac{\partial v}{\partial t}+\frac{1}{R\sin\vartheta}\frac{\partial u\cdot v}{\partial\lambda}+\frac{1}{R\sin\vartheta}v^2\sin\vartheta-\frac{u^2\operatorname{ctg}\vartheta}{R}=-\frac{1}{R}\frac{\partial p}{\partial\vartheta}+2\omega u\cos\vartheta; \qquad (2)$$

$$\frac{\partial u}{\partial\lambda}+\frac{\partial}{\partial\vartheta}v\sin\vartheta=0; \qquad (3)$$

$$\frac{\partial Э}{\partial t}=-\frac{1}{R\sin\vartheta}\frac{\partial Э\cdot u}{\partial\lambda}-\frac{1}{R\sin\vartheta}\frac{\partial}{\partial\vartheta}Э\cdot v\cdot\sin\vartheta, \qquad (4)$$

where t is the time; λ and ϑ are the longitude and the colatitude; u and v are the velocity components along the latitude circle and the meridian, directed eastward and southward, respectively; R is the radius of the earth; ω is the angular velocity of the earth; p is the potential energy (geopotential of the isobaric surface); $Э=\rho c_p T$ is the enthalpy (heat content) of unit volume of air (ρ is the density, c_p is the specific heat at constant pressure and T is the temperature).

The state of the atmosphere in this case is described by the four quantities u, v, p, and $Э$ (or T). Their average values $\bar{u}$, $\bar{v}$, $\bar{p}$ and $\bar{Э}$ can therefore be used as integral characteristics of the macrocomponent. The deviations $u'=u-\bar{u}$; $v'=v-\bar{v}$; $p'=p-\bar{p}$, and $Э'=Э-\bar{Э}$ will define the field of fluctuations, i.e., the individual state of the microcomponent.

Averaging the initial system of equations and grouping terms containing the integral characteristics of the macroprocess and the statistical characteristics of the fluctuations, one easily obtains the expressions

$$\frac{\partial\bar{u}}{\partial t}+\frac{1}{R\sin\vartheta}\frac{\partial\bar{u}^2}{\partial\lambda}+\frac{1}{R\sin\vartheta}\frac{\partial}{\partial\vartheta}\sin\vartheta\bar{u}\cdot\bar{v}+\frac{\bar{u}\cdot\bar{v}\operatorname{ctg}\vartheta}{R}+\frac{1}{R\sin\vartheta}\frac{\partial\overline{u'^2}}{\partial\lambda}+ \\ +\frac{1}{R\sin\vartheta}\frac{\partial}{\partial\vartheta}\sin\vartheta\overline{u'v'}+\frac{\overline{u'v'}\operatorname{ctg}\vartheta}{R}=-\frac{1}{R\sin\vartheta}\frac{\partial\bar{p}}{\partial\lambda}-2\omega\bar{v}\cos\vartheta; \qquad (5)$$

$$\frac{\partial\bar{v}}{\partial t}+\frac{1}{R\sin\vartheta}\frac{\partial\bar{u}\bar{v}}{\partial\lambda}+\frac{1}{R\sin\vartheta}\frac{\partial}{\partial\vartheta}\sin\vartheta\bar{v}^2-\frac{\bar{u}^2\operatorname{ctg}\vartheta}{R}+\frac{1}{R\sin\vartheta}\frac{\partial}{\partial\lambda}\overline{u'v'}+ \\ +\frac{1}{R\sin\vartheta}\frac{\partial}{\partial\vartheta}\sin\vartheta\overline{v'^2}-\frac{\overline{u'^2}\operatorname{ctg}\vartheta}{R}-\frac{1}{R}\frac{\partial\bar{p}}{\partial\vartheta}+2\omega\bar{u}\cos\vartheta; \qquad (6)$$

$$\frac{\partial Э}{\partial t}=-\frac{1}{R\sin\vartheta}\frac{\partial\bar{Э}\bar{u}}{\partial\lambda}-\frac{1}{R\sin\vartheta}\frac{\partial\sin\vartheta\bar{Э}\bar{v}}{\partial\vartheta}-\frac{1}{R\sin\vartheta}\frac{\partial\overline{Э'u'}}{\partial\lambda}-\frac{1}{R\sin\vartheta}\frac{\partial\sin\vartheta\overline{Э'v'}}{\partial\vartheta}. \qquad (7)$$

One can see from these equations that the evolution of the macrocomponent, i.e., of the average parameters of the physical state, is also determined by the following statistical characteristics: $\overline{u'^2}$, $\overline{v'^2}$, $\overline{u'v'}$, $\overline{u'Э'}$, and $\overline{v'Э'}$.

These characteristics may be regarded as components of the momentum flux density and enthalpy flux density vectors. Here the given part of the flow is associated with pulsating motion. The divergences of these fluxes enter equations (5)—(7) in a manner similar to the Reynolds eddy stresses and eddy heat influx. However, by using the wind speeds, which are averaged in the process of measurement or calculation, as the initial data, we exclude small-scale turbulence. Averaging in equations (5)—(7) is carried out on scales commensurable with the radius of the earth, for example, along the entire latitude circle, and hence it is natural to associate these fluctuations with the macroturbulent state of the atmosphere. The main macroturbulence elements will be migratory cyclonic and anticyclonic eddies, as well as other quite large irregularities superimposed on the basic westerly transport. On the basis of these considerations, all the integral (statistical) characteristics of the fluctuation field shall subsequently be called large-scale eddy characteristics.

The contribution of these large-scale eddies to the atmospheric energy balance and to processes of heat, momentum, moisture, etc. transport is very large. In addition, large-scale eddies are one of the essential dynamic factors in the formation of many features of the general circulation.

To estimate the contribution of large-scale eddies to transport phenomena, we shall consider the flux density vector of some quantity A, with components $A \cdot u$ and $A \cdot v$. The average flux components will determine the integral latitudinal and meridional transport at a given point in space:

$$\left.\begin{aligned} \overline{A \cdot u} &= \bar{A} \cdot \bar{u} + \overline{A' \cdot u'}; \\ \overline{A \cdot v} &= \bar{A} \cdot \bar{v} + \overline{A' \cdot v'}. \end{aligned}\right\} \tag{8}$$

If averaging is carried out over time, the first term on the right-hand side describes the role of the time-averaged circulation, which is given by the resultant wind-velocity vectors, or, if the geostrophic wind is used, by the average baric-topography chart. In particular, v describes the permanent zonal-flow disturbances, which appear in the form of troughs and ridges on the average chart. The second term in these equations expresses the transport due to velocity fluctuations with time.

The local time variations of A are equal to the flux divergence taken with opposite sign, plus the dynamic changes. For conservative quantities $(dA/dt) = 0)$ the flux divergence is the sole time-varying factor, in which case one may isolate the divergence of the zonal specific momentum flux $(\overline{u \cdot u}, \overline{u \cdot v})$ in equation (5), the meridional specific momentum flux $(\overline{v \cdot u}, \overline{v \cdot v})$ in equation (6), and the enthalpy flux $(\overline{\vartheta \cdot u}, \overline{\vartheta \cdot v})$ in equation (7).

Resolution of these quantities into transport by the average motion and by fluctuations (as in equations (5)—(7)) is convenient for detailing the mechanism of large-scale transport, for determining the relative contribution of the average circulation and of the pulsating motion. When discussing local features of the general circulation, it is expedient to analyze these equations in the form of equations (8).

The average flux through the latitude circle is obtained when averaging over the latitude, and it has certain features of its own. There turns out to be no point in considering the zonal transport component, since its divergence disappears when averaging over the latitude. Meridional transport is determined by the average meridional flux density

$$\overline{A \cdot v} = \bar{A} \cdot \bar{v} + \overline{A' \cdot v'}.$$

In this case $\bar{v}$ describes the average meridional circulation. Transport by means of the average circulation, i.e., the first term on the right-hand side of equation (8), is sometimes called the "drift" effect. We note that it is quite difficult to determine $\bar{v}$ reliably. Latest studies in this direction were made by Palmen /13/ and Tucker /14/.

In the case of the simplified model (without divergence), which is used for illustration, as well as in the case of the geostrophic approximation $\bar{v} \equiv 0$, equation (8) reduces to transport by large-scale eddies only:

$$\overline{A \cdot v} = \overline{A' \cdot v'}. \qquad (8')$$

It is therefore natural to select the time-averaged quantities $\bar{u}$, $\bar{v}$, $\bar{p}$, and $\bar{\text{э}}$ (or $\bar{T}$) as integral characteristics of the general circulation, and corresponding characteristics of both the average motion: $\overline{u^2} = \overline{u}.\overline{u}$, $\overline{v^2} = \bar{v}.\bar{v}$, $\overline{u}.\bar{v} = \bar{v}.\bar{u}$, $\bar{\text{э}}.\bar{u}$, $\bar{\text{э}}.\bar{v}$ and the pulsating motion: $\overline{u''^2} = \overline{u'' \cdot u''}$, $\overline{v''^2} = \overline{v'' \cdot v''}$, $\overline{u'' \cdot v''}$, $\overline{\text{э}'' u''}$, $\overline{\text{э}'' v''}$. All these quantities vary in space and describe, at each point, the local features of the general circulation. A chart of the geographic distribution of these average quantities can be constructed. These quantities pertain to the entire time interval, and their variations from one time interval to another can determine large-scale "slow" changes in the general circulation.

It is expedient to select quantities defining the zonal structure of the general circulation, namely, the latitude-averaged values $\bar{u}$, $\bar{v}$, $\bar{p}$, and $\bar{\text{э}}$ (or $\bar{T}$) and the large-scale eddy characteristics $\overline{u}''$, $\overline{v}'^2$, $u'v'$, and $\text{э}'v'$, as the integral characteristics of the general circulation pertaining to large regions in space at a given time.

The above integral characteristics of the general circulation were taken from equations of motion expressing momentum and enthalpy changes. When considering changes in other characteristics, one may also consider (depending on the purpose of the study) new integral characteristics of the general circulation. Exhaustive studies have been made lately of the momentum, kinetic energy, and moisture-content balances. Analysis of applicable equations results in integral characteristics of these fluxes similar to equations (8) or (8'). We shall now consider several characteristics associated with the above quantities.

As is known, the absolute specific angular momentum is given by the quantity

$$M = (u + \omega R \sin \vartheta) R \sin \vartheta = u \cdot R \sin \vartheta + \omega R^2 \sin^2 \vartheta. \qquad (9)$$

The first term on the right-hand side defines the relative angular momentum, and the second expresses the translational angular momentum associated with the daily rotation of the earth.

Strictly speaking, we have already considered in equation (5) the angular momentum balance, since we used a spherical coordinate system associated with a rotating earth. However, the momentum characteristics were written without the applicable multipliers ($\sin \vartheta$), as if we were using Cartesian coordinates. The characteristics of angular momentum transport are proportional at each latitude to the characteristics of the zonal momentum transport, but their latitudinal variation is different.

We now give a more exact definition of the characteristics of the meridional transport of angular momentum

$$\overline{Mv} = R \sin \vartheta \, (\omega \cdot \overline{v} R \sin \vartheta + \overline{u} \cdot \overline{v} + \overline{u'v'}). \qquad (10)$$

The first two terms on the right-hand side, which are called the transport and drift terms, respectively, are associated with meridional circulation.

The third term denotes macroturbulent transport. It is easy to show that it is the divergence of the angular momentum flux contained in equation (5).

Empirical data on the angular-momentum balance and on the balance of other quantities can be found in /10/ or /4/.

The kinetic energy of horizontal motion of a unit mass of air is given by

$$\varepsilon = \varepsilon(u, v) = \frac{1}{2}(u^2 + v^2), \tag{11}$$

whence the average kinetic energy is

$$\bar{\varepsilon} = \frac{1}{2}(\bar{u}^2 + \bar{v}^2) + \frac{1}{2}(u'^2 + v'^2) = \bar{\varepsilon}(\bar{u}, \bar{v}) + b'(u, v) = E + b'. \tag{12}$$

Thus, the time-averaged kinetic energy is composed of the mean-velocity energy $\bar{\varepsilon}(\bar{u}, \bar{v}) = E$ and the kinetic energy of the fluctuations b'.

The terms in equation (12) may be grouped differently, combining separately the zonal flow kinetic energy

$$\bar{\varepsilon}_\lambda = \frac{1}{2}(\bar{u}^2 + \overline{u''^2}) \tag{13}$$

and the meridional flow kinetic energy

$$\bar{\varepsilon}_\vartheta = \frac{1}{2}(\bar{v}^2 + \overline{v''^2}). \tag{14}$$

A balance equation may also be constructed for the quantities $\bar{\varepsilon}$, E, and b' resulting in the kinetic energy fluxes and their components, as well as some other quantities which affect energy changes, being considered. All this can be obtained by known methods from equations of dynamics /2, 7/. We now consider the magnitude of the mean-velocity kinetic energy converted into the kinetic energy T_r of fluctuations. This quantity, with different signs, is contained in the expressions for dE/dt, the rate of change of mean-velocity kinetic energy, and for the rate of change of the kinetic energy of fluctuations, and shows the growth of the latter at the expense of the former. In the notation used, the expression for T_r is

$$T_r = -\frac{1}{R}\left\{\frac{\partial\bar{v}}{\partial\vartheta}\overline{v'^2} + \left(\frac{\partial\bar{u}}{\partial\vartheta} - u\operatorname{ctg}\vartheta + \frac{1}{\sin\vartheta}\cdot\frac{\partial\bar{v}}{\partial\lambda}\right)\overline{u'v'} + \right.$$
$$\left. + \left(\bar{v}\operatorname{ctg}\vartheta + \frac{1}{\sin\vartheta}\frac{\partial\bar{u}}{\partial\lambda}\right)\overline{u'^2}\right\}. \tag{15}$$

When averaging over a latitude circle and in the absence of the mean meridional circulation, we obtain the simpler expression

$$T_r = -\overline{u'v'}\frac{\sin\vartheta}{R}\frac{\partial}{\partial\vartheta}\frac{\bar{u}}{\sin\vartheta}, \tag{15'}$$

which expresses the rate of transformation of the mean zonal circulation in the energy of large-scale eddies.

To illustrate some regularities of the zonal structure of the general circulation by means of the above characteristics, we now describe several mean meridional sections. These were constructed from the average monthly zonal (mean latitude) characteristics of the mean circulation and of large-scale eddies calculated for the 1000, 850, 700, 500, 300, 200, and 100 mb levels for January and June, 1956.

The total kinetic energy of atmospheric fluxes in the average section (Figure 1a) exhibits a clearly expressed maximum, corresponding to the location of the subtropic jet stream. In summer this maximum is located at the 200 mb level at 45°N (the changes in altitudes between standard isobaric surfaces are given approximately). In winter the maximum shifts to the southern boundary of the section (35—30°N) or even farther south, and most likely rises somewhat.

A second maximum, associated with the polar-front jet stream, is clearly shown in summer at the northern (75°N) boundary of the section at the 300 mb level. In winter it is expressed weakly and is located at the 300 mb level at 55°N.

The subtropic total kinetic energy maximum occurs at velocities of 45 m/sec in winter and 25 m/sec in summer. For the polar-front kinetic energy maximum these values are 40 and 18 m/sec, respectively. (The kinetic energy is expressed here in terms of velocity obtained as the square root of double the kinetic energy.)

The kinetic energy of the mean zonal circulation (Figure 2a) has a distribution similar to that of the total kinetic energy, but it has a more sharply expressed maximum in the latitudinal variation. The latter is due to the energy transport in large eddies, as will be shown later.

The distribution of the mean circulation kinetic energy presented here is basically similar to that obtained by other authors, for example, by Mintz /12/. The only difference is the averaging method. Thus, instead of the mean latitude values of the zonal velocity, we used its rms values. It is possible that because of this, and also due to features peculiar to the time intervals under study, our sections show the polar-front maximum of the mean kinetic energy field quite clearly, viz., 10 m/sec in summer and 20 m/sec in winter. The subtropical maximum at the center has corresponding velocities of 22 m/sec in summer and 35 m/sec in winter. In both the above sections, as well as in subsequent ones, the maximum energy occurs at the 200 mb level in subtropics and descends to 300 mb in the circumpolar zone.

The energy distribution of large-scale eddies is shown in Figure 2b. The three maxima domains in this distribution lie about the maxima in the mean zonal flux. These maxima lie higher in the south and lower in the north. It should be noted that the northern maximum is stronger; this is characteristic of large-scale eddies. Three maxima are clearly shown to exist in summer. The other two maxima of large-scale eddy kinetic energy relative to the mean flow elements are retained in winter and are shifted southward relative to the latter. It is possible that regions with high-strength large-scale eddies exist in the mean circulation on both sides—to the south above and to the north below the jet streams; some of these may lie outside the section. It is seen from the figures that in the upper layers the energy of large-scale eddies is almost equal to, and in the lower layers it markedly exceeds, the energy of the mean zonal circulation.

The energy of zonal circulation basically varies in the same manner as the energy of mean circulation (Figure 3a). The polar-front maximum is expressed in summer somewhat better than that of the mean circulation.

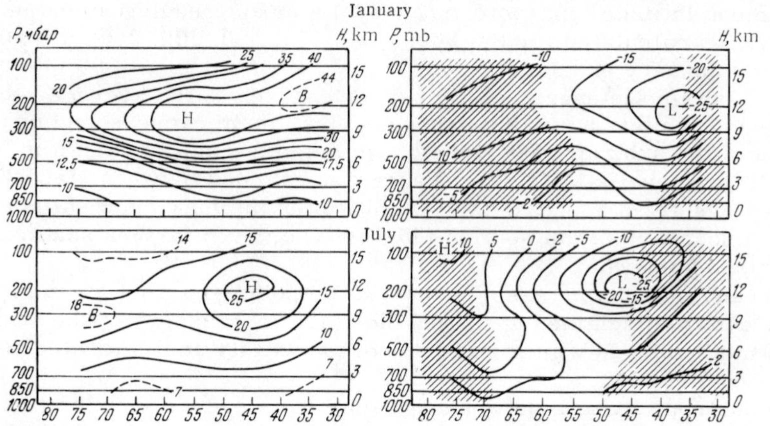

FIGURE 1. Mean-latitude kinetic energy (a) and meridional momentum flux (b) in January and July 1956.

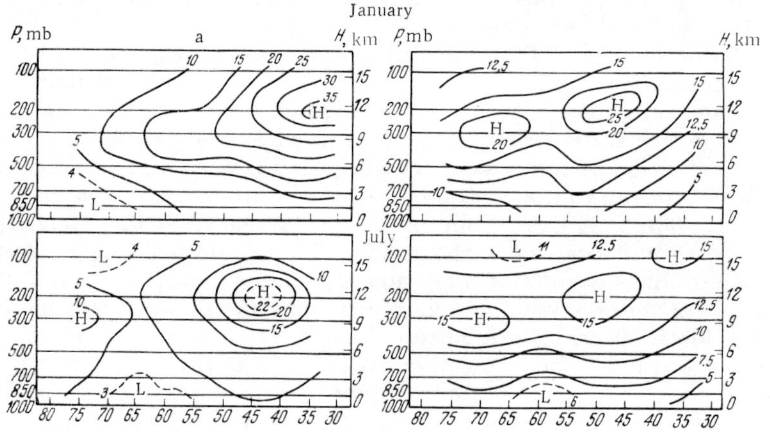

FIGURE 2. Mean-latitude kinetic energy of the mean circulation (a) and of large-scale eddies (b) in January and July 1956.

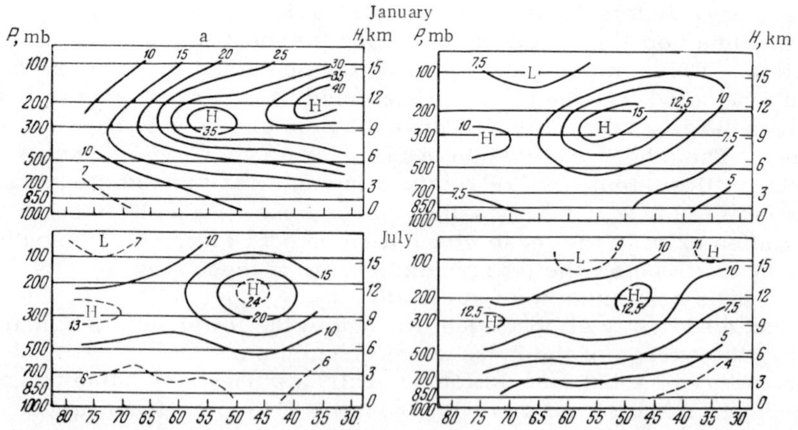

FIGURE 3. Mean-latitude energy of zonal (a) and meridional (b) atmospheric circulations in January and July 1956.

The distribution of meridional circulation energy (Figure 3b), on the other hand, is to a large extent similar to the energy distribution of large-scale eddies. A feature peculiar to the distribution of the meridional circulation energy is the greater horizontal length of maxima along the meridian and in general a small meridional variability in the moderate latitudes, although a sharp southward reduction in meridional circulation occurs in the subtropical zone. This is seen particularly well in the lower atmosphere.

The meridional momentum transport per unit mass is shown in Figure 1b. The negative values correspond to transport of positive (westerly) momentum northward, while the positive values correspond to southward transport. The reduction in transport from south to north indicates the divergence in the momentum (this reduction is crosshatched), while the rise indicates its convergence (accumulation).

The maximum northward momentum transport is located in winter and in summer to the north of the subtropic velocity maximum at the 200 mb level. A region with southward momentum transport exists in summer in the north at high altitudes, while everywhere else the fluxes are directed northward in both summer and winter. Most of the momentum convergence region is located between the subtropic and polar-front energy maximum of the mean and zonal circulations. This is probably the manner in which processes involving large-scale eddy diffusion produce momentum divergence in maximum-momentum zones and on the average are instrumental in weakening the latter. Western transport in moderate latitudes is maintained by convergence of the momentum.

Similar analyses show that the energy of large-scale eddies is converted into the mean circulation energy in the region of the subtropical maximum of the western transport. Consequently, large-scale eddies maintain this maximum. In moderate latitudes, on the other hand, the mean circulation energy is converted into energy of fluctuations and, consequently, the energy of large-scale eddies is maintained by the breaking-up of the mean westerly flux.

We now consider some quantitative data on the kinetic energy potentials of the geostrophic wind in the northern hemisphere in 1958. In averaging over the area the zonal circulation characteristics were assigned weights proportional to the areas of the corresponding latitude zones. In vertical averaging the data on the principal isobaric surfaces were assigned weights proportional to the layer thicknesses (in mb).

By comparing the results for 1958 with some climatic data one sees that anomalies of the energy characteristics do not exceed 20% of the average. Apparently this is the magnitude which should be taken as the possible error in our estimates of the kinetic energy.

Table 1 lists the total specific kinetic energy for the entire 32.5—77.5°N latitude belt averaged for each month. The maximum of the vertical-averaged kinetic energy was observed in December and the minimum in July. The extrema are observed at the 850 mb level later than at higher altitudes, i.e., the maximum is observed in November. This means that the extrema of the annual kinetic energy variations lag at the surface. The kinetic energy potentials in the atmosphere vary very markedly during the year: the amplitude of the annual variation exceeds the summer (minimum) values. The maximum of the specific kinetic energy lies on the 200 mb surface.

The available kinetic energy can be estimated from the values of the kinetic energy averaged over the entire depth of the atmosphere. For this one should multiply the average specific kinetic energy values by the mass of the atmosphere, which may be regarded as constant and equal to $115.6 \cdot 10^{19}$ g in the 32.5—77.5°N belt. The maximum available kinetic energy (in December, 1958) thus amounted to $3.14 \cdot 10^{27}$ erg, with the minimum (July, 1958) amounting to $1.34 \cdot 10^{27}$ erg.

TABLE 1. Average monthly specific kinetic energy of the geostrophic wind, m^2/sec^2

Isobaric surface	Jan.	Feb.	Mar.	Apr.	May	June	July	Aug.	Sept.	Oct.	Nov.	Dec.
850	60.8	54.6	46.8	46.4	35.4	29.8	26.8	24.2	35.7	45.4	52.5	58.8
500	205.4	189.0	185.2	171.3	125.8	102.1	78.8	81.2	122.0	163.2	196.8	219.7
300	439.6	399.0	413.2	376.2	286.0	248.1	197.8	197.8	297.2	378.0	478.0	465.4
200	487.3	403.8	408.3	356.6	310.5	269.8	226.9	245.0	339.6	430.0	482.0	491.0
Average	264.0	203.4	230.2	207.8	116.6	142.4	116.8	121.4	174.6	223.8	263.4	272.1

We now consider the kinetic energy components. First we examine the energy of the zonal and meridional circulations and the interrelationship between them. The data pertaining to the entire 32.5—77.5°N belt at different isobaric surfaces and for the entire troposphere are presented in Tables 2—7.

TABLE 2. Average monthly specific kinetic energy of the zonal circulation, m^2/sec^2

Isobaric surface	Jan.	Feb.	Mar.	Apr.	May	June	July	Aug.	Sept.	Oct.	Nov.	Dec.
850	32.7	31.1	27.6	27.0	19.4	16.6	15.4	14.6	20.4	26.8	30.7	36.1
500	139.9	127.0	134.4	123.4	84.0	69.3	56.0	58.2	87.5	114.6	144.4	160.6
300	314.6	282.6	311.1	275.7	198.6	174.4	145.8	146.6	220.0	270.6	361.2	354.4
200	375.5	321.2	323.3	280.2	230.6	203.0	175.8	188.6	263.0	325.5	381.6	391.7
Average	190.2	167.7	173.4	154.1	166.9	101.4	6.2	89.8	129.4	162.2	198.4	207.0

TABLE 3. Average monthly specific kinetic energy of the meridional circulation, m^2/sec^2

Isobaric surface	Jan.	Feb.	Mar.	Apr.	May	June	July	Aug.	Sept.	Oct.	Nov.	Dec.
850	28.1	23.4	19.3	19.4	16.0	13.2	11.4	9.6	15.3	18.6	21.8	22.8
500	65.6	62.0	50.8	47.6	41.8	37.8	22.8	23.0	34.4	48.6	55.4	59.1
300	125.0	116.4	102.2	100.6	87.4	73.7	52.0	52.6	77.3	107.4	116.8	111.1
200	111.8	82.5	85.0	76.5	79.8	66.8	51.2	56.4	76.6	104.6	100.4	99.4
Average	73.9	62.8	56.8	53.7	49.8	41.1	30.5	31.0	45.2	61.6	65.0	65.2

TABLE 4. Average monthly specific kinetic energy of the mean zonal circulation, m^2/sec^2

Isobaric surface	Jan.	Feb.	Mar.	Apr.	May	June	July	Aug.	Sept.	Oct.	Nov.	Dec.
850	10.7	10.7	8.3	7.6	5.6	3.6	4.6	4.2	7.0	11.4	13.2	14.8
500	86.2	82.1	85.2	77.9	51.0	40.2	35.2	37.6	61.4	75.8	97.0	107.9
300	200.4	183.3	228.2	195.7	130.4	110.2	97.1	101.8	107.2	188.2	231.0	271.5
200	274.8	254.6	279.6	212.0	155.6	146.0	137.3	142.1	210.0	248.4	285.4	312.8
Average	126.0	117.2	130.2	106.2	75.6	65.3	60.6	62.5	95.2	114.8	137.0	153.4

TABLE 5. Average monthly specific kinetic energy of large-scale eddies, m^2/sec^2

Isobaric surface	Jan.	Feb.	Mar.	Apr.	May	June	July	Aug.	Sept.	Oct.	Nov.	Dec.
850	50.1	48.8	38.6	38.9	29.8	26.2	22.2	20.2	28.7	34.0	39.6	44.0
500	119.3	106.8	100.0	93.4	74.8	61.8	43.6	43.6	60.6	87.4	99.8	111.8
300	239.2	215.8	185.0	180.6	155.7	138.0	100.7	97.4	140.1	189.9	246.9	194.0
200	212.5	149.1	128.8	144.7	155.0	123.8	89.6	102.9	129.6	181.6	196.6	178.2
Average	138.1	113.2	100.0	101.6	92.4	77.2	56.6	58.8	79.4	109.0	126.4	118.7

TABLE 6. Kinetic energy of zonal circulation as percentages of the total kinetic energy

Isobaric surface	Jan.	Feb.	Mar.	Apr.	May	June	July	Aug.	Sept.	Oct.	Nov.	Dec.
850	53.8	57.0	58.9	58.2	54.7	55.8	57.4	60.2	57.2	59.0	58.4	61.4
500	67.8	67.2	72.6	72.0	66.9	67.8	71.1	71.6	71.9	70.1	72.0	73.2
300	71.6	71.0	75.2	73.5	69.4	70.4	73.5	73.6	73.8	71.6	75.4	75.5
200	77.1	79.5	79.1	78.5	74.4	75.3	77.4	77.0	77.5	75.7	79.2	79.7
Average	72.0	72.8	75.3	74.2	70.1	71.2	74.0	74.0	74.0	72.6	75.4	76.1

TABLE 7. Kinetic energy of meridional circulation as percentages of the total kinetic energy

Isobaric surface	Jan.	Feb.	Mar.	Apr.	May	June	July	Aug.	Sept.	Oct.	Nov.	Dec.
850	46.2	43.0	41.1	41.8	45.3	44.2	42.6	39.8	42.8	41.0	41.6	38.6
500	32.2	32.8	27.4	28.0	33.1	32.2	28.9	28.4	28.1	29.9	28.0	26.9
300	28.4	29.0	24.8	26.5	30.6	29.6	26.5	26.4	26.2	28.4	24.6	23.5
200	22.9	20.5	20.9	21.5	25.6	24.7	22.6	23.0	22.5	24.3	20.8	20.3
Average	28.0	27.2	24.7	25.8	29.9	28.8	26.0	26.0	26.0	27.4	24.6	23.9

The specific kinetic energy in Tables 2 and 3 is given in absolute units, and in Tables 6 and 7 in relative units (as a percentage of the total kinetic energy).

The annual variation of the energy of zonal circulation is similar to that of the total kinetic energy. The extrema of the annual variation of the energy of meridional circulation occur in January and July without undergoing perceptible shifts with height.

Tables 6 and 7 show that the troposphere contains on the average more than 70% of its kinetic energy in the zonal circulation. The contribution of the zonal circulation increases with altitude: at the 850 mb level the kinetic energy amounts to slightly more than 50%, and it approaches 80% of the total kinetic energy at the 200 mb level.

The relationship between the contributions of the meridional and zonal circulations changes very little during the year. The contribution of the kinetic energy of meridional circulation is at a maximum at all altitudes in May, while the kinetic energy of zonal circulation makes its maximum contribution in December.

The above comparison of the energies of meridional and zonal circulations by means of the corresponding components of kinetic energy is quite rigorous. These two characteristics were obtained in the same manner for both kinds of circulations, they are easy to compare and, when taken together, describe the total energy of atmospheric circulations. Comparison of the strengths of meridional and zonal circulations using the Kats indexes /5/ is much more arbitrary from this point of view. While our estimate will give an average ratio of the strength of meridional to zonal circulations of the order of 0.33, according to Kats this ratio is 0.75.

Another method of resolving the kinetic energy into components consists, as far as we know, of isolating the energy of the mean zonal circulation (E_m) and the energy of large-scale eddies (b').

Data on the mean zonal circulation energy and the energy of large-scale eddies are presented in absolute units in Tables 4 and 5 and in relative units in Tables 8 and 9.

TABLE 8. Kinetic energy of the mean zonal circulation as percentages of the total kinetic energy

Isobaric surface	Jan.	Feb.	Mar.	Apr.	May	June	July	Aug.	Sept.	Oct.	Nov.	Dec.
850	17.5	19.5	17.6	16.2	15.6	12.0	17.1	17.5	19.6	25.0	25.0	25.2
500	42.0	43.5	46.0	45.4	40.6	39.4	44.6	46.2	50.4	46.2	49.3	49.2
300	45.6	45.9	55.2	51.8	45.6	44.5	49.2	51.2	53.4	49.8	48.4	58.4
200	56.4	61.5	68.4	59.5	50.2	54.2	60.4	58.2	61.8	57.8	59.4	63.6
Average	47.7	51.0	56.6	51.1	45.0	45.8	51.4	51.5	54.7	51.5	51.4	56.4

The energy of the mean zonal circulation increases more rapidly with altitude than that of large-scale eddies. The maxima of both these types of energy occur in December, and the minima occur in July at all the isobaric surfaces with the exception of the 850 mb level, where the extrema occur in January and August, respectively.

The contribution of the mean zonal circulation to the kinetic energy of the atmosphere is, on the average, almost equal to that of large-scale eddies. Below 300 mb the contribution of large-scale eddies usually predominates and only at 200 mb does the contribution of the mean zonal circulation become more significant, although not by much. Here the energy of the mean zonal circulation at the 850 mb surface does not exceed 25%.

TABLE 9. Kinetic energy of large-scale eddies as percentages of the total kinetic energy

Isobaric surface	Jan.	Feb.	Mar.	Apr.	May	June	July	Aug.	Sept.	Oct.	Nov.	Dec.
850	82.5	80.5	82.4	83.8	84.4	88.0	82.9	82.5	80.4	75.0	75.0	74.8
500	58.0	56.5	54.0	54.6	59.4	60.6	55.4	53.8	49.6	53.8	50.7	50.8
300	54.4	54.1	44.8	48.2	54.4	55.5	50.8	48.8	46.6	50.2	51.6	41.6
200	43.6	38.5	31.6	40.5	49.8	45.8	39.6	41.8	38.2	42.2	40.6	36.4
Average	52.3	49.0	43.4	48.9	55.0	54.2	48.6	48.5	45.3	48.5	48.6	43.6

The annual variation of the ratio of the mean zonal circulation energy and the energy of large-scale eddies does not have a large range, and the latter decreases with altitude. The relative contribution of the mean zonal circulation is larger in winter and smaller in summer.

All the above regularities were obtained in the final count from the pressure field. The integral characteristics of the circulation were defined by terminology corresponding to circulation systems, since the geostrophic wind speed was calculated from the pressure field. Hence all the data used here depend on the accuracy of the initial pressure field and may be expressed in terms of the applicable differential characteristics of this field (zonal and meridional components of gradients, etc.). These quantities thus also pertain to the actual circulation to the extent to which the geostrophic wind corresponds to the actual wind. This, naturally, means that the question as to the conformity of the geostrophic to the real wind is important. Data on the real wind are much more incomplete and their processing is more complicated. We were able to compare the integral (mean-latitude) characteristics of the circulation as obtained from the real and from the geostrophic winds only at the 500 mb level and only for January 1958.

All the stations of the northern hemisphere with wind data were grouped into 5° latitude (±2.5°) belts about the principal latitude circles (multiples of 5°) from 30 to 75°N. The wind speed and direction were obtained from synoptic reports for each day of January 1958. A special table was then used for each station to determine u, v, $u \cdot v$, u^2 and v^2. These were then averaged along the latitude circle and used for calculating all the integral energetic characteristics.

The extent to which the characteristics of the real and geostrophic circulation are close to one another can be judged from Table 10, which lists the relative values of the kinetic energy of large-scale eddies. It is seen from this table that the geostrophic circulation carries more energy than the real circulation. However, the perturbance (contribution of the energy of meridional circulation and of large-scale eddies) of the real circulation is greater.

The latitude variations of the integral characteristics of the real and geostrophic circulations were found to be quite similar.

TABLE 10. Average integral characteristics of the real and geostrophic circulations in January, 1958 at the 500 mb surface, m/sec

Quantity	Actual circulation	Geostrophic circulation
$\bar{u}$	9.8	11.6
$\sqrt{2[\bar{\varepsilon}]}$	19.0	20.2
$\sqrt{2\bar{b}'}$	15.0	15.4
$\sqrt{[\bar{u}^2]}$	15.1	16.7
$\sqrt{[\bar{v}^2]}$	11.5	11.4

It may be concluded that the integral characteristics of the general circulation of the atmosphere calculated from the geostrophic wind give an overall idea of the real circulation and very close values of the kinetic energy and all its components.

References

1. Belinskii, N. A. Opyt ustanovleniya indeksa tsirkulyatsii atmosfery (Experience in Establishing an Atmospheric Circulation Index).— Trudy NIU GUGMS, seriya 5, No. 4. 1946.
2. Gandin, L. S., D. L. Laikhtman, L. T. Matveev, and M. I. Yudin. Osnovy dinamicheskoi meteorologii (Principles of Dynamic Meteorology).— Gidrometeoizdat. 1955.
3. Gruza, G. V. Makroturbulentnost' v obshchei tsirkulyatsii atmosfery (Macroturbulence in the General Circulation of the Atmosphere).— Gidrometeoizdat. 1961.
4. Yeh Tu-cheng and Chu Pao-chen. Nekotorye vazhneishie voprosy obshchei tsirkulyatsii atmosfery (Some of the More Important Problems of the General Circulation of the Atmosphere).— Gidrometeoizdat. 1961.
5. Kats, A. L. Sezonnye izmeneniya obshchei tsirkulyatsii atmosfery i dolgosrochnye prognozy (Seasonal Variations of the General Circulation of the Atmosphere and Long-Range Forecasts).—Gidrometeoizdat. 1960.
6. Monin, A. S. O makroturbulentnom obmene v zemnoi atmosfere (Exchange by Large-Scale Eddies in the Terrestrial Atmosphere).— Izvestiya AN SSSR, seriya geofizicheskaya, No. 4. 1956.
7. Monin, A. S. O predvychislenii kharakteristik zonal'noi tsirkulyatsii atmosfery (Precalculation of the Characteristics of the Zonal Circulation of the Atmosphere).— Izvestiya AN SSSR, seriya geofizicheskaya, No. 10. 1958.

8. Pogosyan, Kh. P. and E. I. Savchenkova. O chislovom vyrazhenii vida atmosfernoi tsirkulyatsii (Numerical Expression of the Type of Atmospheric Circulation). — Meteorologiya i gidrologiya, No. 3. 1950.
9. Sadovnikov, A. V. Nekotorye novye kharakteristiki obshchei tsirkulyatsii atmosfery vo vnetropicheskikh shirotakh (Some New Characteristics of the General Circulation in Extratropical Latitudes). — In: Sbornik materialov mezhduvedomstvennogo seminara po dolgosrochnym prognozam pogody, Leningrad. 1960.
10. Haltiner, G. J. and F. L. Martin. Dynamical and Physical Meteorology. — McGraw-Hill. 1957.
11. Haurwitz, B. The Motion of Atmospheric Disturbances. — J. Mar. Res., 3(1). 1940.
12. Mintz, E. The Observed Zonal Circulation of the Atmosphere. — Bull. Amer. Meteorol. Soc., No. 5. 1954.
13. Palmen, E. The Mean Vertical-Meridional Circulation in Low Latitudes of the Northern Hemisphere. — Sci. Proc. Internat. Assoc. Meteorol. X Gen. Assembly, I. U. G. G., Rome. 1954.
14. Tucker, G. B. Mean Meridional Circulation in the Atmosphere. — Quart. J. Roy. Meteorol. Soc., 85(365). 1959.

E. P. Borisenkov

ENERGETICS OF THE GENERAL CIRCULATION OF THE ATMOSPHERE

The present article examines problems of the general circulation of the atmosphere from the viewpoint of energy principles. The question of using energy invariants for a more thorough analysis of forecasting schemes is considered. A qualitative analysis is presented of one of the possible mechanisms of interaction between the processes in the upper and lower atmosphere. The thermal regime of the upper atmosphere is also examined.

Systems of various atmospheric circulations of different scales, which comprise the general circulation of the atmosphere, are determined by the available kinetic energy. This circulation would be impossible if energy sources which continuously replenish the kinetic energy in the atmosphere were not available. Any system of atmospheric circulation, be it small-scale convection or a system of large-scale eddies and planetary circulation, is either maintained by transformation of some kind of energy into kinetic energy, or by conversion of the kinetic energy of circulations of some scale into that of circulations of a different scale.

From the point of view of atmospheric energetics the general circulation is a continuous process of production of kinetic energy, which maintains the atmospheric circulation. The study of the mechanism of transformation of one form of energy into another is thus definitely one of the principal links in the solution of this problem.

The study of the general circulation involves the solution of a large number of specific problems. Below are listed those problems which in the author's opinion are of fundamental importance.

1. The study of the physical regularities of atmospheric processes and of the factors responsible for them, with extensive utilization of numerical simulation of these processes.

2. The construction, on this basis, of physically justified numerical schemes of short-range and particularly long-range weather forecasting with allowance for all the interacting factors of terrestrial, cosmic, or astronomic origin.

3. The control of atmospheric processes with clear prediction of any effects of this control. The latter problem cannot be solved without solving the first two.

None of the above problems can really be solved successfully without studying the mechanism of transformation of one kind of energy into another.

Despite the obvious importance of studies of energy processes in the terrestrial atmosphere and hydrosphere, insufficient attention was and, unfortunately, still is being paid to this problem.

After the initial studies of Margules /19/, Murgetroyd /20/, and others, who established the basis for the theory of atmospheric energetics, there was quite a long break in the development of the theory of this problem as well as in the accumulation of experimental data. During the last few years, publications appeared by Borisenkov /1, 2, 3, 5, 6/, Gruza /7/, Lorenz /16, 17/, Palmen /22, 23/, Saltzman /24/, and Starr /26/. Nevertheless, the attention being paid to the energetics of atmospheric processes is still insufficient.

Although the different forms of energy are derivatives of the directly measured parameters of the state of the atmosphere, the application of energy principles to the study of both tropospheric and upper atmospheric processes has many positive aspects. In addition to the possibility of comparing such apparently different elements as wind, temperature, geopotential, radiation-balance components, corpuscular fluxes, energy of the earth magnetic field, etc. in the same units of measurement, the principal advantage of the energy approach consists in the feasibility of a quantitative estimation and physical interpretation of given processes for detecting the cause-and-effect relationships in the complex circulation mechanism of the terrestrial atmosphere and hydrosphere. The energy approach is virtually the only one in the study of individual circulation mechanisms, for example, in the upper atmosphere, where the thermodynamic relationships generally used do not apply.

The present paper is a generalized presentation of the principal trends of studies in the field of atmospheric energetics in conjunction with the study of the general circulation of the terrestrial atmosphere, on the basis of studies carried out in the USSR and in the West.

Quite naturally, it is impossible to consider every detail of the problem at hand, and the reader is referred to the bibliography. In addition, almost no reference is made to earlier works in the field of atmospheric energetics; these are discussed in /1/.

ENERGETICS AND NUMERICAL EXPERIMENTS

The statistical study of the general circulation was and is based on the statistical correlation of directly measured characteristics of the state of the atmosphere (pressure, temperature, geopotential, wind, etc.). In exactly the same way, the study of atmospheric dynamics, whose purpose is to construct numerical schemes of weather forecasting and simulate numerically individual components of the general circulation, is based on the solution of a system of hydrodynamic and thermodynamic equations and on their numerical integration with respect to time. At the same time one of the basic and most difficult problems of numerical experiments is to precalculate the evolution of thermobaric fields.* Since from the point of view of atmospheric energetics consideration of this evolution involves clarifying the sources responsible for changes in the atmospheric kinetic energy, it is very important that the dynamics equations (especially the simplified ones) used in a given numerical scheme have corresponding

* The term numerical experiment will be used to denote the construction of forecasting schemes, as well as the simulation of individual components of the general circulation, based on numerical integration of the equations of atmospheric dynamics.

energy invariants. It is then possible to check the various schemes from a physical point of view and prevent violation of the law of conservation of energy.

One such approach within the framework of the adiabatic approximation which is extensively used in short-range forecasting was suggested by Lorenz /16/.

As is known, the labile energy of some mass of the atmosphere contained in a given volume may, in the adiabatic approximation, be expressed in terms of the potential temperature by

$$L = I + \Pi = c_p \cdot p_0^{-\lambda} \int\limits_{(M)} p^{\lambda} \theta \, dm, \tag{1}$$

where

$$\lambda = \frac{c_p - c_v}{c_p};$$

θ is the potential temperature; p_0 and p are the pressure at the surface and at any altitude, respectively; c_p and c_v are the specific heats at constant pressure and volume; dm is an element of mass; L, I, and Π are the labile, internal, and potential energies. Integration is carried out over the atmospheric mass M.

If one imagines a state under which the isentropic surfaces are horizontal, then the pressure at them will equalize; this will be the state of static stability, the quiescent state. In this case

$$(I + \Pi)_{\Pi} = c_p \cdot p_0^{-\lambda} \int\limits_{(M)} \tilde{p}^{\lambda} \theta \, dm, \tag{2}$$

where $\tilde{p}$ is the average pressure at the given isentropic surface.

The quantity

$$A = c_p \cdot p_0^{-\lambda} \int\limits_{(M)} (p^{\lambda} - \tilde{p}^{\lambda}) \theta \, dm \tag{3}$$

defines the portion of labile energy which must be expended in order to bring the atmosphere from the quiescent to the given state, or that portion of the energy which is released when the atmosphere is brought from the given to the quiescent state. Lorenz called this energy the available potential energy.

As was shown in /17/,

$$A \approx \frac{1}{2} \lambda c_p \cdot p_0^{-\lambda} \int\limits_{(M)} p^{\lambda - 1} \left(\frac{\overline{\partial \theta}}{\partial p} \right)^{-1} (\theta - \bar{\theta})^2 \, dm \tag{4}$$

(the bar over the quantities denotes averaging over a given isobaric surface).

The available potential energy is not some new kind of energy. It is equal to the excess of that labile energy which would have theoretically existed had the atmospheric mass been transformed under adiabatic conditions to a state when the isentropic surface occupied a horizontal position.

It is possible to imagine yet another hypothetical restructuring in the atmosphere, associated, for example, with processes of influx of hot onto cold air, also accompanied by changes in the static stability of the atmosphere. For this purpose one introduces the concept of the energy of macrostatic stability which, in the same manner as the available potential energy, is not a new form of energy.

The term macrostatic stability denotes that part of the labile energy which is released (expended) when the mass of the atmosphere passes adiabatically from a given state to one under which the isentropic surfaces become vertical. This will result in a state of neutral stratification and θ will no longer be a function of p. The labile energy of a system in this state can be calculated from the expression

$$(I+\Pi)_{S}=\frac{c_p}{1+\lambda}\int\limits_{(M)}\theta\,dm. \tag{5}$$

The energy of the macrostatic stability is defined as the difference between equations (5) and (1), i. e.,

$$S=c_p\cdot p_0^{-\lambda}\int\limits_{(M)}\left(\frac{p_0^{\lambda}}{1+\lambda}-p^{\lambda}\right)\cdot\theta\cdot dm. \tag{6}$$

Integration by parts yields

$$S=(1+\lambda)^{-1}c_p\cdot p_0^{-\lambda}\int\limits_{(M)}(p_0^{\lambda}\cdot p-p^{1+\lambda})\cdot\left(-\frac{\partial\theta}{\partial p}\right)dm. \tag{7}$$

Thus, if it is assumed that the passing of the atmosphere from one state to another takes place under adiabatic conditions, then either the system's kinetic energy is modified by its labile energy, or the change in the kinetic energy of a system with one scale takes place due to changes in the kinetic energy of a system of another scale (without external sources). This means that the simplified equations of dynamics used for numerical experiments must satisfy the following requirements.

1. The total kinetic energy of the entire system should be constant.
2. The sum $I+\Pi$ and S should be constant; this means that $(I+\Pi)_{\Pi}$ is constant.
3. The difference between $I+\Pi$ and A should remain constant; this means that $(I+\Pi)_{\Pi}$ is constant.
4. The sum of $I+\Pi$ and K should remain constant, which presupposes equality of the difference of S and K and of the sum of A and K.

In order that any numerical forecasting scheme using a given simplification be substantiated from a physical point of view, the energy invariants must be such as to satisfy the above requirements.

We represent the components of wind velocity as the sum of the potential and solenoidal parts, i. e.,

$$u=u_2+u_3;\qquad v=v_2+v_3, \tag{8}$$

where

$$u_2 = -\frac{\partial \psi_2}{\partial y}, \qquad v_2 = \frac{\partial \psi_2}{\partial x},$$
$$u_3 = \frac{\partial \varphi_3}{\partial x}, \qquad v_3 = \frac{\partial \varphi_3}{\partial y};$$

ψ_2 is the stream function and φ_3 is the velocity potential. Under these conditions the vertical component of the vorticity is $\Omega_{z_2} = \nabla^2\psi_2$, while the horizontal velocity divergence is $D_3 = \nabla^2\varphi_3$. It shall henceforth be assumed that terms with subscript 2 denote the vorticity group of terms defined by the stream function, while those with subscript 3 denote the divergent group, defined by the velocity potential. Subscript 1 is assigned to terms containing the potential temperature or characteristics which are expressed in terms of it (temperature, specific volume). In this case the dry adiabatic equation has the form

$$\frac{\partial \theta_1}{\partial t} = -(\psi_2, \theta_1) - \frac{\partial \varphi_3}{\partial x} \cdot \frac{\partial \theta_1}{\partial x} + \frac{\partial \varphi_3}{\partial y} \cdot \frac{\partial \theta_1}{\partial y} - \tau_3 \frac{\partial \theta_1}{\partial p}, \tag{9}$$

where τ_3 is the vertical velocity.

The vorticity and velocity divergence equations can be written in the form

$$\frac{\partial \Omega_{z_2}}{dt} = -(\psi_2, \Omega_{z_2}) - (\psi_2, l) - \nabla l v_3 - v_3 \nabla \Omega_{z_2} - \Omega_{z_2} D_3 - \tau_3 \frac{\partial \Omega_{z_2}}{\partial p} - $$
$$- \nabla \tau_3 \cdot \nabla \frac{\partial \psi_2}{\partial p} - \left(\tau_3, \frac{\partial \varphi_3}{\partial p}\right); \tag{10}$$

$$\frac{\partial D_3}{\partial t} = -g\nabla^2 z + \nabla(l\nabla\psi_2) - (l, \varphi_3) - \nabla(v_2 \cdot \nabla v_2) - \nabla(v_2 \cdot \nabla v_3) -$$
$$\nabla(v_3 \cdot \nabla v_2) - \nabla ;_3 \frac{\partial v_3}{\partial p} - \nabla(v_3 \cdot \nabla v_3) - \nabla \tau_3 \frac{\partial v_3}{\partial p}. \tag{11}$$

The total kinetic energy of a unit mass is defined by

$$K = \frac{1}{2}c^2 = \frac{1}{2}\left[\left(\frac{\partial \psi_2}{\partial x}\right)^2 + \left(\frac{\partial \psi_2}{\partial y}\right)^2\right] + \frac{1}{2}\left[\left(\frac{\partial \varphi_3}{\partial x}\right)^2 + \left(\frac{\partial \varphi_3}{\partial y}\right)^2\right] + (\psi_2, \varphi_3). \tag{12}$$

On integrating over the entire mass of the atmosphere the integral of the last term in equation (12) vanishes. Then,

$$K = K_2 + K_3 = \frac{1}{2}\int\limits_{(M)} \left[\left(\frac{\partial \psi_2}{\partial x}\right)^2 + \left(\frac{\partial \psi_2}{\partial y}\right)^2\right] dm +$$
$$+ \frac{1}{2}\int\limits_{(M)} \left[\left(\frac{\partial \varphi_3}{dx}\right)^2 + \left(\frac{\partial \varphi_3}{\partial y}\right)^2\right] dm. \tag{13}$$

Combining in equations (9)—(11) terms with subscripts 2 and 3 and constructing the applicable energy invariants, we can estimate the kind of departure from the energy balance which will occur upon dropping a given term or terms. According to Lorenz, the following conclusions may, in particular, be drawn.

1. The requirement of quasisteadiness is not approximate. This condition has its corresponding energy invariants.

2. There is no point in retaining in the dynamics equations terms containing the vertical velocity and consequently also the ageostrophic

components of the wind velocity while at the same time dropping terms including the ageostrophic advection of the vorticity. In precisely the same manner, the simplified scheme, using the balance equation for transition from the field of the geopotential to the field of ψ_2, conversely has no energy invariant when the ageostrophic advection of the vorticity is neglected. Similar energy invariants were also obtained by Lorenz for the two-layer model of the atmosphere /16/.

The above examples do not exhaust the list of problems where energy principles will help substantially in a more thorough physical analysis of given forecasting schemes. However, one should note here a number of quite important problems which require further study, in particular, with reference to the general circulation.

1. If the calculations are made for a limited area, then energy fluxes at the boundaries of this area must be specified. In addition, some indeterminacy may appear due to fictitious boundary conditions. The method of constructing energy invariants is therefore more suitable for closed volumes. One can thus conclude that when constructing numerical models of the general circulation for a hemisphere or for the entire terrestrial sphere, one must check them from a physical point of view by constructing the applicable energy invariants. In these problems such checking will be sufficiently effective.

2. It may be claimed a priori with reference to processes occurring over several days or more that the adiabatic requirement will not be satisfied in the study of the general circulation. The atmosphere, even over a large area, cannot be regarded as a closed or even as an adiabatic system. In particular, consideration must be given to changes in the labile energy due to heat influx /1, 2, 3/. This means that when constructing numerical models of the general circulation taking into consideration heat influxes one must construct the applicable energy invariants with allowance for the heat influx also. Energy principles in the construction of the theory of the general circulation were clarified to some extent in /11, 14/.

The applicable energy invariants can be constructed when considering wet adiabatic processes by replacing λ by $\lambda' = d\lambda$, where

$$d = \frac{\gamma_a'}{\gamma_a};$$

γ_a and γ'_a are the dry and wet adiabatic lapse rates.

3. Until now advances in numerical weather forecasting were made primarily by developing new methods for solving hydrodynamic and thermodynamic equations. However, a physical interpretation of the given forecasting schemes is now needed. In the opinion of the present author this cannot be done without constructing energy invariants for any numerical forecasting scheme, and especially for any hydrodynamic model of the general circulation. It appears that this principle could become a good basis for the physical classification of the many forecasting schemes and models of the general circulation.

ENERGY PRINCIPLES FOR THE STATISTICAL DESCRIPTION OF ATMOSPHERIC PROCESSES

Energy relationships can be used for obtaining a large number of characteristics of the general circulation which are of great importance for understanding the physical regularities underlying the processes being studied.

Estimates are given in /2, 3/ of seasonal transformations of various kinds of energy in the atmospheres of the northern and southern hemispheres in conjunction with the study of the interaction between processes in both hemispheres. Table 1 lists composite data on the energy balance for different seasons.

TABLE 1. The energy balance of the mean atmosphere of the northern and southern hemisphere during different seasons, cgs units

Form of estimate	Northern hemisphere			Southern hemisphere		
	winter	summer	winter minus summer	winter	summer	winter minus summer
I	$430.4\cdot10^{28}$	$447.7\cdot10^{28}$	$-17.3\cdot10^{28}$	$423.0\cdot10^{28}$	$437.0\cdot10^{28}$	$-7.7\cdot10^{28}$
Π	$176.9\cdot10^{28}$	$183.3\cdot10^{28}$	$-6.4\cdot10^{28}$	$175.9\cdot10^{28}$	$179.2\cdot10^{28}$	$-3.3\cdot10^{28}$
$I+\Pi$	$607.3\cdot10^{28}$	$631.0\cdot10^{28}$	$-23.7\cdot10^{28}$	$605.2\cdot10^{28}$	$616.2\cdot10^{28}$	$-11{,}0\cdot10^{28}$
K	$405.63\cdot10^{25}$	$191.5\cdot10^{25}$	$213.8\cdot10^{25}$	$706.5\cdot10^{25}$	$385.5\cdot10^{25}$	$305.9\cdot10^{25}$
$\Delta K/\Delta\times$ $\times(I+\Pi)$	—	—	0.009	—	—	0.028
M	$265\cdot10^{19}$	$264\cdot10^{19}$	$0.9\cdot10^{19}$	$264.5\cdot10^{19}$	$263.6\cdot10^{19}$	$0.9\cdot10^{19}$
$(I+\Pi)/M$	$225\cdot10^{7}$	$238\cdot10^{7}$	$-13\cdot10^{7}$	$228\cdot10^{7}$	$233\cdot10^{7}$	$-5\cdot10^{7}$
K/M	$1.53\cdot10^{6}$	$0.73\cdot10^{6}$	$0.89\cdot10^{6}$	$2.64\cdot10^{6}$	$1.48\cdot10^{6}$	$1.16\cdot10^{6}$

From the data in the table it follows in particular that the general level of potential energy in the southern hemisphere is lower than that in the northern hemisphere during the seasons being compared, while the level of the kinetic energy is markedly higher.

It was shown in /2, 3/ that there are sharply expressed seasonal differences in the transformation of energy in conjunction with the nature of the underlying surface. The last circumstance leads to the conclusion that one must investigate the processes of interaction between the atmosphere and hydrosphere, and without this it is impossible to construct a rigorous theory of the general circulation.

It follows from Table 1 that the sum of L and K does not remain constant throughout the year over a given hemisphere. This means that the labile energy in the atmosphere is continuously replenished by heat influx. Consequently, consideration of heat influxes is essential for studies of the general circulation and not a single model of the latter can have energy invariants without consideration of the heat influx.

It is seen from Table 1 that there is a very large contrast between the available kinetic energy of the two hemispheres for the summer of the northern hemisphere. During this time the kinetic energy of the southern hemisphere is 3.5 times that of the northern hemisphere. One can therefore assume that there is a very intensive exchange not only between different zones within the two hemispheres, but also between the hemispheres themselves, shown by the increased recurrence of meridional circulations precisely during the summer /3/. Similar exchange is brought about by large-scale eddies. During the past few years the interest in this problem has increased greatly /4, 6, 7, 12, 13, 15, 24, 25/.

In the majority of studies, most attention is paid to studying the kinetic energy spectrum of eddies of different scales, and to estimating the energy transfer from eddies of one scale to those of another scale. These studies resulted in a number of new and important conclusions regarding the general circulation. Thus Palmen et al. /23/ examines the meridional kinetic energy flux whose maximum in the northern hemisphere is observed at 30—40°N. The energy flux through this latitude circle, calculated for a layer of thickness 1 mb, is defined by the expression

$$T_k(\Phi, p, t) = \frac{2\pi a \cdot \cos\Phi}{g}\left[v\left(\frac{u^2+v^2}{2}\right)\right], \qquad (14)$$

where a is the radius of the earth; Φ is the latitude; u and v are the components of wind velocity; g is the acceleration of gravity. Representing u and v as the average values and deviations from the mean at a given latitude circle, we have

$$u = [u] + u'; \quad v = [v] + v', \qquad (15)$$

in connection with which

$$T_k = T_{k_0} + T_{k_1} + T_{k_2}, \qquad (16)$$

where

$$T_{k_0} = \frac{2\pi a \cdot \cos\Phi}{g}[v]\left\{\frac{[u^2]+[v^2]}{2} + [v'^2]\right\};$$

$$T_{k_1} = \frac{2\pi a \cdot \cos\Phi}{g}[u][u' \cdot v'];$$

$$T_{k_2} = \frac{2\pi a \cdot \cos\Phi}{g}\left[v'\left(\frac{u'^2+v'^2}{2}\right)\right].$$

The eddy kinetic energy flux, defined by the terms T_{k_1} and T_{k_2}, has maxima in the 35—40°N belt and exhibits a clearly defined seasonal periodicity.

Reference /24/ presents the calculated energy fluxes from eddies of different scales. The geopotential field at the 500 mb level was expanded in a Fourier series. This expansion was used to determine the mean zonal velocity and the eddy components of the zonal and meridional velocity components for wave numbers $1 \leqslant n \leqslant 15$.

All the waves were subdivided into three groups:

$$\overline{L}_1 = \sum_{i=1}^{5} L_i(n) \text{— long waves,}$$

$$\overline{L}_2 = \sum_{i=6}^{10} L_i(n) \text{— average (cyclonic) waves,}$$

$$\overline{L}_3 = \sum_{i=11}^{15} L_i(n) \text{— short waves.}$$

The total kinetic energy flux for the closed system is $\sum_{j=1}^{3} \overline{L}_j = 0$. In the belt for which calculations are made,

$$\sum_{i=1}^{15} L_i = \frac{2\pi a \cos \Phi}{g} \left[v' \left(\frac{u'^2 + v'^2}{2} \right) \right] \Bigg|_{\Phi=72.5^\circ}^{\Phi=22.5^\circ}, \qquad (17)$$

where u' and v' are the zonal and meridional components of the wind velocity deviations (the square brackets denote averaging along a given latitude circle).

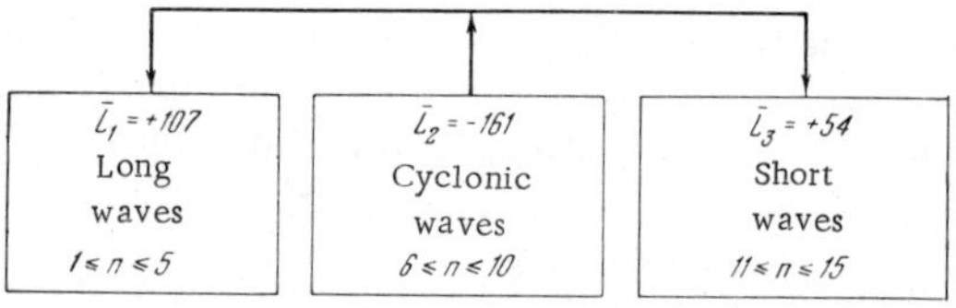

FIGURE 1. Fluxes of eddy kinetic energy, 10^{-3} erg/sec·cm^2·mb.

The calculated results, in units of 10^{-3} erg/sec·cm^2·mb, are:

$$\overline{L}_1 = +124 \pm 81; \; \overline{L}_2 = -145 \pm 96; \; \overline{L}_3 = +71 \pm 54; \; \sum_{j=1}^{3} \overline{L}_j = +50 \pm 56.$$

The average values and their rms errors are given.

Since due to computational errors $\sum_{j=1}^{3} \overline{L}_j = 0$, the 50-unit disagreement is spread out among the three wave groups. As a result, the kinetic energy transfer from eddies of different scales is described by the numbers in Figure 1.

A more general picture is now presented, taking energy transfer by zonal fluxes into consideration (Figure 2).

These results modify substantially the previously held ideas as to kinetic energy transfer in the atmosphere, according to which the energy of small eddies is obtained from that of larger eddies. The kinetic energy of the largest eddies, in the final count, is replenished by the kinetic energy of the mean zonal circulation. It is found that the kinetic energy of cyclonic eddies is the kinetic-energy source for large- as well as small-scale eddies. In its turn, the kinetic energy of the zonal circulation is replenished by energy supplied by eddies of all scales. This requires a new approach to the role of cyclonic activity in the general circulation. It is shown in /1, 24/ that medium-scale eddies, while damping the kinetic energy at some instant of time, create conditions which maintain the kinetic energy of small and large eddies as well as the energy of the zonal circulation.

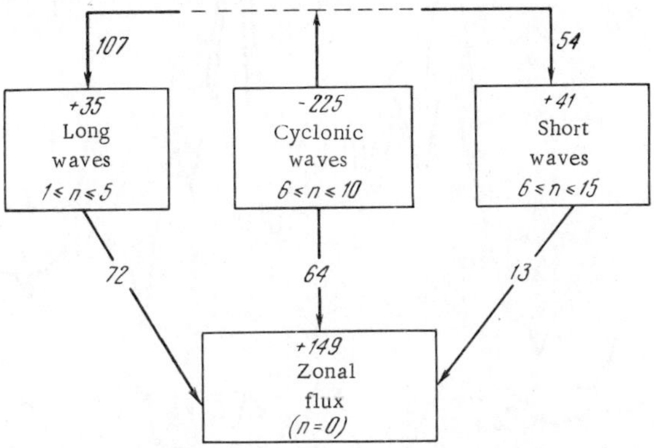

FIGURE 2. Average exchange of kinetic energy between the zonal circulation, long waves, and medium and short waves, 10^{-3} erg/sec·cm^2·mb.

The kinetic energy is formed as a result of the baric gradient forces, which appear by virtue of ageostrophic components of wind velocity and vertical motion, produced by the unsteadiness of processes, surface friction, and the orography, as well as by the heat flux. This approach forces a new estimate of the role of ageostrophic atmospheric processes when estimating the evolution of baric formations. However, these kinetic energy transformations are not the only source which maintains the energy of eddies of a given scale at the expense of that of eddies of another scale.

The present author calculated the mean zonal flux and the kinetic energy spectrum of eddies of different scales (with wave numbers $0 \leqslant n \leqslant 15$) for a period of about two months at different latitudes of the northern hemisphere at the 500 mb level. The calculations were carried out for latitude circles $35° \leqslant \Phi \leqslant 75°$N with a 5° latitude interval by the method presented in /4/ and using the geopotential field as the initial data.

The kinetic energy of the eddies was averaged over three groups: long waves with wave numbers $1 \leqslant n \leqslant 15$, medium waves with $6 \leqslant n \leqslant 10$, and short waves with wave numbers $11 \leqslant n \leqslant 15$. Figure 3 shows the time variation of the kinetic energy of eddies for these three groups for three latitude circles. The following conclusions may be drawn from this and other information.

1. About 80% of the kinetic energy of the atmosphere is contained in eddies with wave numbers of from 1 to 15. The kinetic energy of mean circulation amounts to only about 20% of the total.

2. The kinetic energy of small-scale eddies ($11 \leqslant n \leqslant 15$) is small compared with that of medium- and large-scale eddies and in a number of problems it may be neglected.

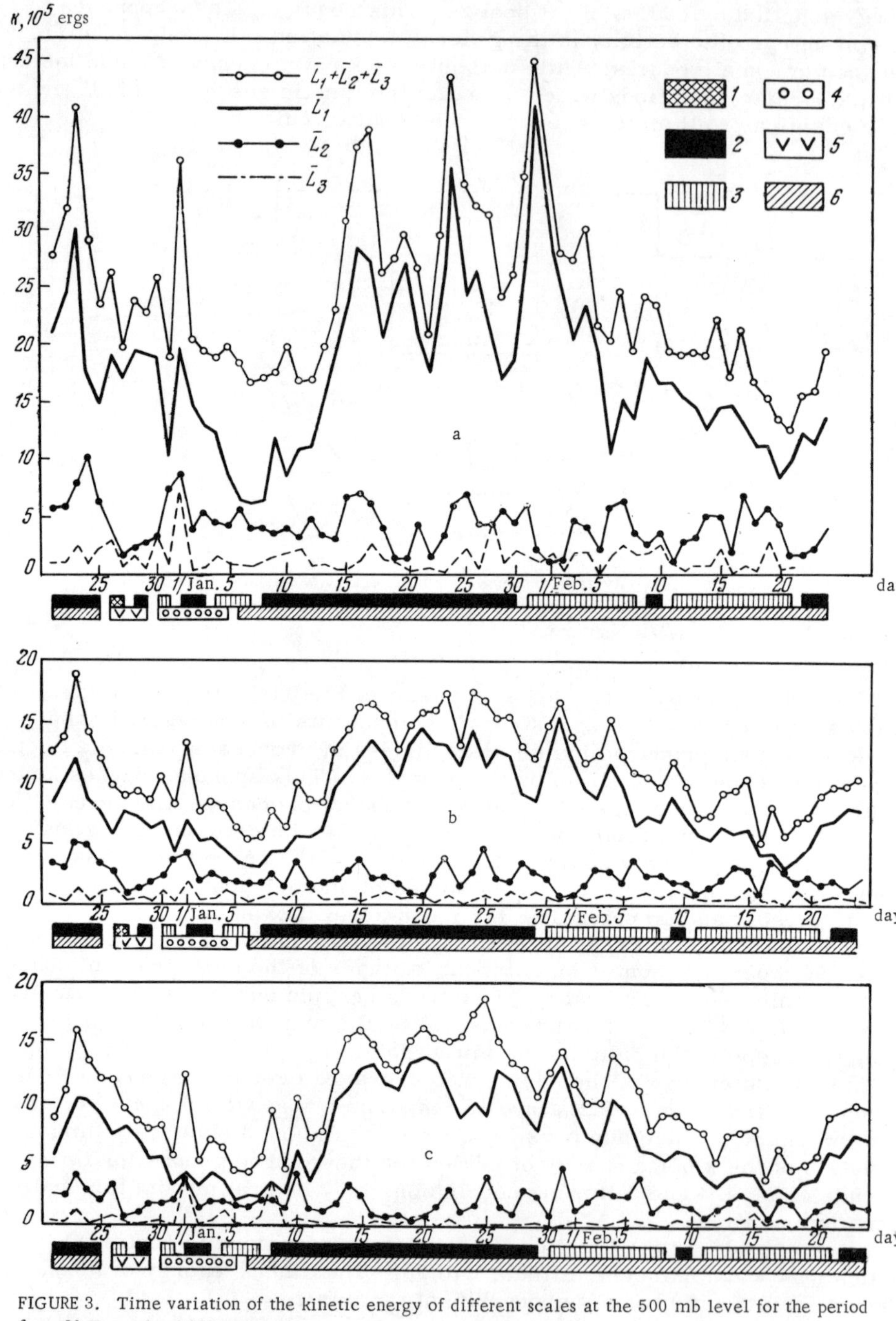

FIGURE 3. Time variation of the kinetic energy of different scales at the 500 mb level for the period from 20 December 1962 to 23 February 1963.

At latitudes: a — 35°N; b — 45°N; c — 55°N. The circulation modes are according to G. Ya. Vangengeim and A. A. Girs. Top scale: 1 — western (W); 2 — meridional (C); 3 — eastern (E); bottom scale: 4 — western (W); 5 — meridional (M_1); 6 — meridional (M_2).

3. The greatest amounts of kinetic energy are contained in eddies with $1 \leqslant n \leqslant 5$. These amounts are approximately double those in medium-scale eddies and 6—8 times the amount of kinetic energy contained in small-scale eddies.

4. The time variation of the kinetic energy of large-scale eddies has a clearly expressed periodicity with periods larger than in the time variation of the kinetic energy of medium-scale and especially of small-scale eddies. This is of importance, for example, when constructing statistical schemes for long-range weather forecasting. These data lead one to doubt the feasibility of long-range forecasting of the evolution of medium- and small-scale eddies, although this is more feasible for large-scale eddies.

5. Time periods exist when the kinetic energy of the mean circulation as well as that of eddies of all scales increases (decreases) simultaneously at all latitudes of the northern hemisphere. In precisely the same manner periods with increasing (decreasing) kinetic energy may be observed at all latitudes at the same time. This indicates the transformation of other kinds of energy into kinetic energy, and again forces one's attention to the interaction between processes of the two hemispheres. It becomes important to estimate the transformation of the kinetic and of the available potential energy simultaneously.

From the point of view of energetics, one may not, in particular, neglect the consideration of phase transitions in the moisture in the middle latitudes and particularly in the tropical belt. According to calculations, the energy of phase transitions in the tropical zone calculated on the basis of precipitation suffices to maintain the kinetic energy of cyclonic eddies in the entire hemisphere.

Energy calculations thus make possible a new approach to the study of individual components of the general circulation and guide studies in the most important directions. In the present author's opinion, numerical methods of synoptic studies using electronic computers should use this approach.

ENERGETICS OF THE UPPER ATMOSPHERE

The energetics of the circulation of the upper atmosphere is an important link in the problem of the general circulation by reason of a number of circumstances.

1. Knowledge of the energetics of the upper atmosphere is important in the study of the dynamics of these layers, particularly of the stratosphere and mesosphere.

2. The upper atmosphere is a classic example of a region where atmospheric dynamics are basically determined by radiative eddy, and a number of other heat influxes, and when not a single hydrodynamic scheme of the real circulation can have energy invariants unless the heat influx is considered.

3. The laws governing local thermodynamic equilibrium are not satisfied in the upper atmosphere (above 80—100 km). In this case the energy approach is virtually the only method which can be used for estimating the factors affecting the nature of the processes occurring.

4. A great deal of interest has lately been shown in the study of cosmophysical factors and in estimating the effect of processes in the upper atmosphere associated with solar activity on processes in the lower atmosphere; the mechanism of interaction of upper and lower atmospheric processes is still not entirely clear.

Below we shall examine one of the possible mechanisms of this interaction, developing to some extent the ideas put forward by Starr /26/.

Circulations in the upper atmosphere may be subdivided into two groups. The first contains motions whose energy sources are located in the uppermost atmosphere (for example, ionospheric turbulence). The second contains motions with energy sources in the lower atmosphere. The main problem for the first group of motions is to determine the energy sources and the mechanism of their transformation into kinetic energy. The main problem for the second group is to determine the mechanism by means of which energy exchange between the upper and lower atmosphere occurs. This mechanism is also of interest for solving the converse problem of determining the effect of processes in the upper atmosphere on those in the lower.

On the basis of the energy equation obtained by considering the three equations of motion, we have

$$\int\limits_{(V)} \frac{dK}{dt} \cdot dV = -\int\limits_{(V)} \frac{1}{\rho} c \cdot \operatorname{grad} p \, dV - D, \tag{18}$$

where D is the dissipation of kinetic energy; K is the kinetic energy of unit mass; p is the pressure; ρ is the density; c is the wind speed; dV is an element of volume.

Integration is carried out over volume V, whose lower boundary is the upper boundary of the troposphere.

Setting $\int\limits_{(V)} \operatorname{div} pc dV = 0$, which is correct on the scale of the general circulation, we have

$$\int\limits_{(V)} \frac{1}{\rho} c \operatorname{grad} p \, dV = -\int\limits_{(V)} \frac{1}{\rho} p \operatorname{div} c dV.$$

Then

$$\frac{\partial}{\partial t} \int\limits_{(V)} \rho K \, dV = \int\limits_{(S)} \rho \cdot K \cdot c_n \cdot dS + \int\limits_{(V)} p \operatorname{div} c dV - D, \tag{19}$$

where c_n is the normal component of the wind velocity of surface S bounding the given volume.

We denote the first term on the right-hand side of equation (19), which describes the kinetic energy flux at the boundary of the region, by A. The second term, which describes the generation of kinetic energy due to work of the baric-gradient forces (work of expansion), by s. Since the kinetic energy of the system remains unchanged over large time intervals, the left-hand side of equation (19) may be set equal to zero. In this case we have

$$D = A + s. \tag{20}$$

Under the selected conditions, when the area of the lower base of the volume represents the upper boundary of the atmosphere, the normal velocity component c_n will be identified with the vertical velocity at the lower boundary of the layer (there are no horizontal energy fluxes). In connection with this

$$A = \int_{(S)} K \cdot \rho \cdot w \cdot dS, \tag{21}$$

where w is the vertical velocity in the Cartesian coordinate system.

Over a sufficiently long time interval $\int_{(S)} w\rho dS = 0$ (the upward and downward mass transports are equal to one another). This means that $A \neq 0$ only under one physically acceptable condition, according to which a correlation must exist between w and K. $A > 0$ only when large values of K have ascending motions corresponding to them; $A < 0$ in the opposite case. Experiments carried out when studying radioactive contamination confirm this hypothesis regarding the role of horizontal currents in the interaction between the upper atmosphere and the troposphere.

We estimate the magnitude of s by

$$s = \int_{(V)} p \operatorname{div} c \, dV = \int_{(V)} (p \operatorname{div} c) \cdot dS \cdot dz. \tag{22}$$

Since

$$\tau = \frac{\partial p}{\partial t} + c \operatorname{grad} p$$

(where τ is the vertical velocity in the isobaric coordinate system), we have

$$p \operatorname{div} c = -\tau + \frac{\partial p}{\partial t}. \tag{23}$$

Upon substitution of equation (23), equation (22) yields

$$s = -\int_{(V)} \tau dV + \frac{\partial}{\partial t} \int_{(V)} p \cdot dV. \tag{24}$$

Over a large time interval the last integral on the right-hand side of equation (24) is zero. Hence

$$s = -\int_{(V)} \tau dV = -\int_{(S)} \int_0^{p_0} \tau \cdot \alpha \, dS \cdot dp. \tag{25}$$

Since the sum of vertical motions in the sphere or hemisphere is zero, this means that $s \neq 0$ only in the case when there is a correlation between τ and α, $s > 0$ when the ascending motions are in a warm region and $s < 0$ when these are in a cold region. A number of investigators noted that vertical motions and the work of the baric-gradient field make a substantial contribution to kinetic-energy genesis both in the upper and lower atmosphere. Thus, according to Palmen's calculations for the northern hemisphere to the north of $\Phi = 30°N$, dissipation (D) of the kinetic energy amounts to $60 \cdot 10^{10}$ kJ/sec /22/. The kinetic energy generated by the work of the horizontal baric-gradient forces is $28 \cdot 10^{10}$ kJ/sec. If the kinetic energy of the hemisphere is constant, the energy budget can be balanced by taking into account the kinetic energy flux through the 30°N latitude; it is estimated at $20-22 \cdot 10^{10}$ kJ/sec. But even this does not suffice for balancing. It hence follows that genesis of kinetic energy should be estimated at $80-100 \cdot 10^{10}$ kJ/sec and that a large contribution is made here by the work of baric-gradient forces due to vertical motions.

The above qualitative analysis provides a basis for the following conclusions.

1. Vertical motions are one of the principal mechanisms of interaction between the upper and lower layers of the atmosphere.

2. Since the heat sources in the upper atmosphere are determined primarily by processes within it and not in the troposphere, the kinetic-energy sources of the upper atmosphere are located in its uppermost layers and are related to heat-influx conditions.

3. According to available estimates, the effect of tropospheric processes on those in the upper atmosphere cannot be of substantial importance. It is demonstrated only at the troposphere-stratosphere boundary /22/.

4. Studies of the energetics of the general circulation of the upper atmosphere should be based on an examination of factors determining the thermal regime of the upper atmosphere, the dynamics of these layers, and in the first place on the combined study of the kinetic-energy and vertical-velocity fields in the upper atmosphere.

Here we come to the main problem of upper-atmosphere energetics, viz., its thermal regime. This problem is thoroughly examined in /8/, where a large number of studies in this field are discussed, as well as in /5, 20, 21/.

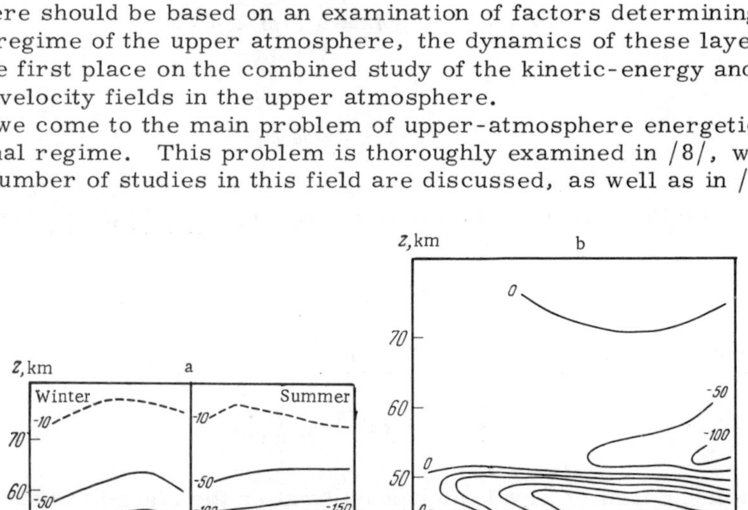

FIGURE 4. Seasonal distribution of the total radiation balance in the stratosphere and mesosphere (a) and the radiation-energy balance variation from winter to summer (b) in a two-kilometer layer, 10^{-5} cal/cm^2·min.

Unfortunately, studies of the thermal regime of the upper atmosphere are usually carried out without reference to the dynamics of these layers. In precisely the same manner, quantitative estimates of the dynamics of processes in the upper atmosphere are usually made without estimating the radiation balance, which is extremely necessary and useful. Many factors are responsible for this.

1. No network for the vertical sounding of the upper atmosphere has been organized, and this region is therefore studied solely on the basis of randomly collected data.

2. The vertical distribution of absorbing gases has not been measured. No data on the vertical distribution of absorbing substances such as water vapor, carbon dioxide, and particularly ozone are available for the stratosphere and mesosphere. It is therefore impossible to make theoretical calculations of the energy balance of the stratosphere and mesosphere on the basis of regular daily observations.

3. Virtually nothing is known about the composition of absorbing gases above 50—60 km, and in particular on the concentration and distribution of atomic and molecular oxygen, oxides of nitrogen, etc., which make the greatest contribution to the thermal regime at these altitudes.

4. Very little reliable experimental data are available on the transmission (absorption) functions for various spectral intervals and absorbing substances. These data are extremely necessary for all kinds of theoretical calculations.

Estimates of the radiation-balance components were therefore obtained from some average and not entirely reliable data. They are preliminary, but can still be used for drawing conclusions of importance in the study of the energetics of the general circulation of the upper atmosphere. Thus, reference /5/ presents an estimate of the seasonal features of the energy balance of the upper atmosphere of the northern hemisphere. Average data were used for calculating the short-wave and long-wave radiation balance components due to absorption by ozone, water vapor, and carbon dioxide over different spectral intervals. These data were used in turn to construct charts of the vertical distribution of the radiation balance at different latitudes during the warm and cold parts of the year with respect to each of the absorbing substances. In addition, charts were constructed of the vertical and latitudinal distribution of the total radiation balance (Figure 4), which were compared with the wind-velocity and kinetic-energy fields. The calculations were carried out from the tropopause level up to 80 km.

RESULTS

1. Zones with maximum horizontal radiation-balance gradients coincide quite well with zones of stratospheric jet streams and zones with maximum vertical wind-velocity gradient.

2. Zones with maximum radiation-balance gradients coincide quite well with zones of elevated turbulence in the upper atmosphere.

3. Maximum seasonal variations of the radiation balance occur at altitudes of 35—40 km, while maximum variations of kinetic energy occur at altitudes of 50—60 km. This is in agreement with conclusions obtained from the thermal wind model.

4. By comparing the temperature and kinetic energy fields with the field of radiation balance above 60—70 km and considering the above three absorbing gases, it was shown that maximum changes in the temperature and kinetic energy fields do not coincide with zones of radiation-balance variation.

This shows that one must have data on the concentration and distribution of absorbing surfaces and also on transmission (absorption) functions to study the energetics of the upper atmosphere (above 50—60 km).

A great deal still remains unknown regarding the effect of cosmic factors on the dynamics of the upper atmosphere. It appears that energy principles would be very useful and fruitful in this field.

CONCLUSION

The study of the energetics of the general circulation involves the entire volume of observations of the temperature, pressure, and wind fields at different levels, at least over a hemisphere. Data for the entire terrestrial sphere are needed for more accurate studies. More approximate estimates can be obtained from the concept of the average energy level, introduced by the present author in 1956 /1, 2, 4, 9/.

This concept refers to a level located on the average at the 410—420 mb surface, for which the ratio of the internal to the potential energy is the same as for the entire atmosphere. The average energy level has the following features: a) the ratio of the internal to the potential energy is the same as for the entire atmosphere, i. e., 2.5; b) the level coincides with the isopycnic level, i. e., with the constant density level; c) the function $\bar{\psi} = pz$ reaches its maximum at the average energy level; d) the ageostrophic wind-velocity components are uniquely defined here by the temperature advection; e) the geostrophic wind velocity is equivalent to the average geostrophic wind for the entire column of the atmosphere; f) the processes at this level occur polytropically with an average polytropic specific heat of 0.224 cal/g·deg; g) the temperature of the average energy level is equivalent to the mean temperature of an atmospheric column of unit cross section and its height corresponds to the location of the center of gravity of this column; h) according to calculations using sounding data, as well as from measurements of the integral departing radiation obtained with meteorological satellites, the departing radiation of the earth-atmosphere system is closest to the radiation of a perfect black body at the temperature of the average energy level.

The above properties of the average energy level make it possible to replace the investigation of energy transformation over the entire height of the atmosphere by the study of these transformations at this level. Naturally, this is not permissible for all purposes, but it appears to the author to be quite promising for the study of the main regularities in the general circulation of the atmosphere.

References

1. Borisenkov, E. P. Voprosy energetiki atmosfernykh protsessov (Problems of Energetics of Atmospheric Processes). — Gidrometeoizdat. 1960.
2. Borisenkov, E. P. Sezonnye preobrazovaniya energii v atmosfere severnogo i yuzhnogo polusharii (Seasonal Energy Transformation in the Atmosphere of the Northern and Southern Hemispheres). — Trudy AANII, Vol. 253. 1963.
3. Borisenkov, E. P. Sezonnye preobrazovaniya energii v atmosfere severnogo i yuzhnogo polusharii i voprosy vzaimodeistviya protsessov v oboikh polushariyakh (Seasonal Energy Transformation in the Atmosphere of the Northern and Southern Hemispheres and Problems of Interaction Between Processes in Both Hemispheres). — In: Sbornik "Meteorologicheskie issledovaniya," No. 9, Izdatel'stvo Nauka. 1965.
4. Borisenkov, E. P. Fiziko-statisticheskie metody analiza i predvychisleniya meteorologicheskikh polei (Physical and Statistical Methods of Analysis and Precalculation of Meteorological Fields). — Trudy AANII, Vol. 263. 1963.
5. Borisenkov, E. P. and B. A. Osipov. Otsenka sezonnykh osobennostei energeticheskogo balansa verkhnei atmosfery severnogo polushariya (An Estimate of the Seasonal Features of the Energy Balance of the Upper Atmosphere of the Northern Hemisphere). — Trudy AANII, Vol. 271. 1964.
6. Borisenkov, E. P. Vzaimodeistvie protsessov v severnom i yuzhnom polusharii v probleme obshchei tsirkulyatsii atmosfery (Interaction of Processes in the Northern and Southern Hemispheres as Applicable to the General Circulation of the Atmosphere). — Problemy Arktiki i Antarktiki, No. 20, Gidrometeoizdat. 1965.
7. Gruza, G. V. Makroturbulentnost' v obshchei tsirkulyatsii atmosfery (Macroturbulence in the General Circulation of the Atmosphere). — Gidrometeoizdat. 1961.
8. Kondrat'ev, K. Ya. and O. P. Fillipovich. Teplovoi rezhim verkhnikh sloev atmosfery (The Thermal Regime of the Upper Layers of the Atmosphere). — Gidrometeoizdat. 1960.
9. Borisenkov, E. P., K. Ya. Kondratiev, and Ya. P. Doronin. Structural Characteristics of the Radiative Field of the Earth as a Planet. — COSPAR Sympos. Florence, Italy. May 1964.
10. Benton, G. S. Spectra of Large-Scale Atmospheric Flow at 300 Millibars. — J. Meteorol., 15(2). 1958.
11. Charney, J. G. and P. C. Drazin. Propagation of Planetary Scale Disturbances From the Lower Into the Upper Atmosphere. — J. Geophys. Res., 66(1). 1961.
12. Eliassen, A. and P. Enok. On the Transfer of Energy in Stationary Mountain Waves. — Geophys. Publ., 22(3). 1961.
13. Eliassen, A. A Study of the Long Atmospheric Waves on the Basis of Zonal Harmonic Analysis. — Tellus, 10(2). 1958.
14. Fjortoft, R. On the Control of Kinetic Energy of the Atmosphere by External Heat Sources and Surface Friction. — Quart. J. Roy. Meteorol. Soc., 86(370). 1960.

15. Harper, B. P. Energy Spectra of 500 mb. Meridional Circulations Index.— J. Meteorol. 18(4). 1961.
16. Lorenz, E. N. Energy and Numerical Weather Prediction.— Tellus, 12(4). 1960.
17. Lorenz, E. N. Available Potential Energy and the Maintenance of the General Circulation.— Tellus, 7(2). 1955.
18. Litwin, W. Energy of Conditional Instability.— Monthly Weather Rev., Vol. 64. 1936.
19. Margules, M. Über die Energie der Stürme.— Jahrb. k. k. Zentral-anst. Meteorol. Annog. 1905.
20. Murgatroyd, R. J. and R. M. Goody. Sources and Sinks of Radiative Energy From 30 to 90 km.— Quart. J. Roy. Meteorol. Soc., 84(361). 1958.
21. Ohring, G. The Radiation Budget of the Stratosphere.— J. Meteorol., 15(5). 1958.
22. Palmen, E. On the Generation and Frictional Dissipation of Kinetic Energy of the Atmosphere.— Comment. phys.-math. Soc. scient. Fennica, 24(11). 1960.
23. Palmen, E., H. Riehl, and L. Vuorela. On the Meridional Circulation and Release of Kinetic Energy of the Tropics.— J. Meteorol., 15(3). 1958.
24. Saltzman, B. and A. Fleisher. The Exchange of Kinetic Energy Between Larger Scales of Atmospheric Motion.— Tellus, 12(4). 1960.
25. Shapiro, R. and F. Ward. The Time-Space Spectrum of the Geostrophic Meridional Kinetic Energy.— J. Meteorol., 17(6). 1960.
26. Starr, V. P. Questions Concerning the Energy of Stratospheric Motions.— Arch. Meteorol., Geophys. und Bioklimatol., 12(1). 1960.

E. M. Feigel'son

CERTAIN RADIATION PROBLEMS RELATED TO THE PROBLEM OF THE GENERAL CIRCULATION

Problems are examined which depend on making allowance for nonadiabatic factors in the study of the general circulation of the atmosphere; viz., the state of radiative equilibrium, interaction of radiation and developing cloudiness, and the radiative interaction between the atmosphere and the underlying surface. An expression is obtained for the effective surface temperature, as well as for the radiative heat influx in the atmosphere.

The main source of energy for the general circulation is solar radiation. Nevertheless, until recently the theory of the general circulation was constructed without satisfactory allowance for radiative transfer and results of interesting studies in this field were published only lately. In particular, Rakipova /11/ calculated the yearly-average and seasonal-average zonal circulation with allowance for the distribution of incoming solar radiation.

First results of a numerical experiment by Manabe, Smagorinsky, and Holloway /23/ pertaining to the relationship between radiation and dynamics in the atmosphere, as well as to the interaction between processes in the troposphere and stratosphere, were reported to the International Symposium on Radiation held in Leningrad in 1964. In this work the dynamics is represented by general equations, while radiation is taken into account in the same manner as in /24, 25/ for a yearly-average distribution of absorbing gases and solar radiation. Inhomogeneity of the surface (oceans, continents) and cloudiness were not considered.

The study of the interaction between radiation and dynamics over the oceans attendant to developing cloudiness is doubtlessly one of the most interesting questions. The sun-heated ocean serves as the main heat source of the atmosphere. One of the forms of transfer of this heat is long-wave radiation from the ocean surface, and the clouds are the main regulating mechanism in heating the atmosphere and the ocean /9/.

Allowance for radiation when developing the theory of the general circulation is extremely difficult. This is not because it is necessary to solve the transport equations, as some investigators think, but because an entire group of interacting processes are present: motion, radiative transfer, moisture exchange and cloud genesis, heat exchange with the ocean, and changes in the turbidity of the atmosphere. Here are several examples of changes in the radiative heat flux depending on specific conditions.

1. When the zenith distance of the sun changes from 30 to 70°, the albedo of the cloud layer changes by 10—20%. Depending on the optical thickness of the cloud (within the limits of observed optical thicknesses amounting to 10—60) the albedo may vary by a factor of 2 /14/.

2. If a cloud layer of thickness 1 km moves from the 2 km (upper-boundary altitude) to the 8 km level, the radiative cooling in the entire troposphere is halved /5/. When this same cloud layer is located at the 2 km level it increases the solar radiation influx to the atmosphere by 50% as compared with the cloudless case /14/. Radiative cooling of the entire atmosphere then increases by 20%. At the same time the influx of solar energy to the earth surface is reduced by approximately a half. As a result, according to Manabe /25/, the cloudiness of the lower layer causes the temperature in the thermal equilibrium state (see below) to drop by approximately 45° at each level in the 0—8 km range.

3. When good transmission conditions ($P = 0.8$) are replaced by a turbid cloudless atmosphere ($P = 0.6$), the insolation at the earth surface decreases by approximately 10—20%, depending on the zenith distance of the sun /15/.

The problem of the general circulation includes not only describing planetary circulation over large time intervals such as a year or season, but also over short time periods such as a day, week, or month. In the first case equilibrium is attained, while in the other cases the process is basically unsteady.

However, even when considering the equilibrium state, the presence of nonlinear effects, such as cloud genesis, makes direct solution of the equilibrium problem unreliable. It is shown in /22/ that the solution of the problem of the thermal and photochemical equilibrium of the mesosphere does not coincide with the time-limiting solution in the corresponding unstable problem. Lately the latter approach is increasingly used to determine equilibrium states /22, 25/ and is also useful when developing methods for solving long-range forecasting problems.

The theory of the general circulation thus assumes a solution of the system of unsteady equations of motion, heat and moisture exchange, interaction between the atmosphere and the ocean, as well as equations of radiative transfer. Allowance must be made for changes in the optical state of the atmosphere and microphysical properties of the clouds.

Together with the general problem, consideration is given to particular problems related to the study of individual aspects of the above-mentioned complex group of phenomena. Three of these problems are: 1) study of the state of radiative equilibrium; 2) study of the interaction between radiation and developing cloudiness; 3) interaction between the atmosphere and the underlying surface.

Of greatest interest among the many studies to determine the radiative equilibrium are /24, 25/. In these works sufficiently complete and accurate allowances are made for the basic absorbing substances (H_2O, CO_2, O_3) up to an altitude of about 40 km. The following points are also clarified: a) what features of the steady thermal state of the atmosphere can be attributed to purely radiative processes; b) the role of the individual absorbing substances; c) the process of establishing the radiative and thermal equilibrium (see below); d) the effect of the average cloudiness of the different levels. The principal results of /24, 25/ are presented in Figures 1—3.

One notices the duration of the process of establishing the equilibrium state, which is about a year (Figure 1). Figure 2 shows that, contrary to the widely-held opinion, it is impossible to neglect the direct heating of the atmosphere by solar radiation within the limits of the troposphere (compare curves describing the heating rate as a result of absorption by water vapor of the long-wave ($L\,H_2O$) and short-wave ($S\,H_2O$) radiation). Neglect of absorption by CO_2 is well justified.

A weak point in /25/ is the use of the artificial "convective adjustment" approach, which replaces the consideration of nonradiative forms of heat transfer and consists of imposing a lapse rate of 6.5 deg/km for the troposphere, instead of the superadiabatic lapse rate which is characteristic of radiative-equilibrium conditions. The equilibrium state which is thus obtained is termed the state of thermal equilibrium.

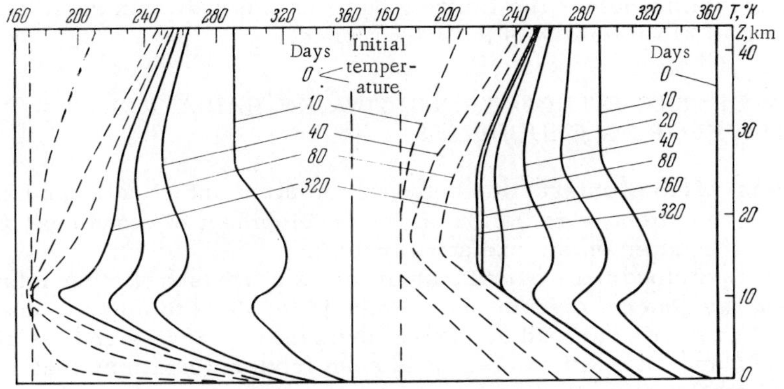

FIGURE 1. Establishing the temperature in the case of radiative (dashed lines) and thermal (solid lines) equilibrium:

curves plotted for $T_0 = 360$ and 170°K, where T_0 is the initial, vertically constant temperature.

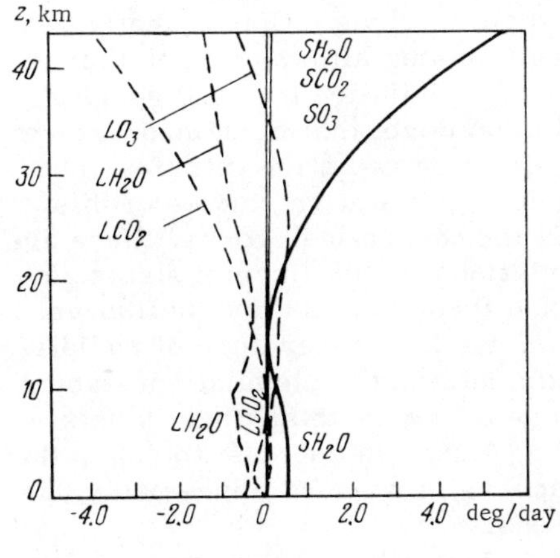

FIGURE 2. Heating rate due to absorption of long-wave and short-wave radiation by different gases.

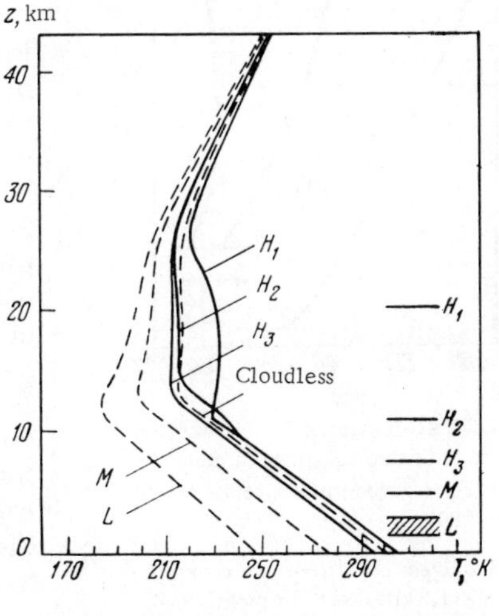

FIGURE 3. Equilibrium temperature distribution for a cloudless atmosphere with lower-, medium-, and upper-layer clouds (H_1, H_2, H_3):

The cloud levels are listed. High cloudiness is ascribed to all levels.

It seems strange that the atmosphere cools sharply with lower and middle level clouds, as found in /25/ (see Figure 3). A combination of two heating effects and one cooling effect of cloudiness may, under certain conditions, produce such a result, but it appears to be very unlikely.

References /24, 25/ present a sufficiently complete and accurate description of the absorption of radiation by atmospheric gases for a given average content of the latter. At the same time, the consideration of cloudiness cannot be regarded as satisfactory, since a steady mean cloudiness of given strength is used with poorly substantiated data on the optical properties of clouds. No consideration is given to moisture exchange, which makes it very uncertain whether the limiting equilibrium radiation regime in the presence of cloudiness was in fact obtained.

STUDY OF THE INTERACTION BETWEEN RADIATION AND DEVELOPING CLOUDINESS

The problem of interaction between radiation and cloudiness has two aspects: the effect of radiation on the development of cloudiness and the effect of the latter on the radiative transfer.

Stratified cloudiness over large areas is obviously of most interest when studying the general circulation. Methods for determining the characteristics of this type of cloud are being developed by Shvets and his co-workers /1, 8, 19/ and Matveev /6, 7/. It is seen from calculations that ordered vertical motions are the main factor in the formation of clouds. Vertical eddy heat and moisture exchange play a secondary role. The first calculations of cloud genesis taking long-wave radiation into consideration showed a quite perceptible (of the order of 1 km) rise in the upper boundary of the cloud layer due to this factor /10/.

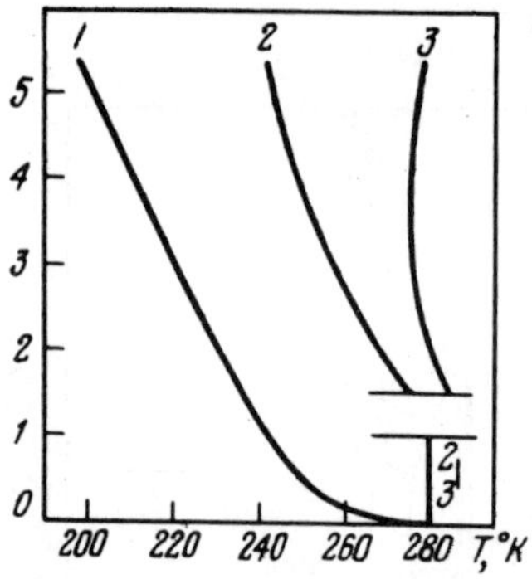

FIGURE 4. Temperature distribution under radiative equilibrium conditions for a given surface temperature $T_S = 280°$:

1—cloudless conditions; 2—cloud in the 1—1.5 km layer, at night; 3—same cloud by day.

A detailed vertical resolution is unattainable in the majority of problems on the dynamics of large-scale motions. Consequently, one must forego making allowances for the contribution of radiation to cloud genesis.

There is no doubt that cloudiness exerts the main effect on radiative transfer, but there are few quantitative studies of this effect. Of the theoretical works, there are only calculations of equilibrium states /5, 14, 25/, and these are mainly qualitative. In /14/, where the temperature of radiative equilibrium outside the cloud at constant temperature of the earth surface is determined, it is found, in contrast to /25/, that the atmosphere is warmed appreciably due to cloudiness.

Figure 4 shows the vertical temperature distribution in a cloudless atmosphere and in a cloud layer for a "gray" model of the atmosphere /14/. Infrared solar radiation hardly penetrates into the subcloud layer; the night and day temperature is the same, provided that the surface temperature remains constant.

It is shown in /5/ how the radiation of the entire thickness of the atmosphere outside the clouds cools when the level of the continuous cloud layer rises. Below are presented the results of these calculations using the expression

$$R = [F_1(o) - F_2(o)] - [F_1(h_1) - F_2(h_1)] + [F_1(h_2) - F_2(h_2)] - F_1(H),$$

where F_1 and F_2 are the fluxes of the ascending and descending long-wave radiation, respectively; h_1 is the level of the lower cloud-layer boundary; h_2 is the level of the upper boundary; $h_2 = 0$ is the case of a cloudless sky; H is the altitude of the tropopause which here and subsequently is regarded as the upper boundary of the atmosphere.

H_2, km	0	3	5	7	9
$-R$, cal/cm^2·min	0.138	0.086	0.070	0.060	0.054

The main difficulty in calculating the radiation regime of a cloudy atmosphere is to describe short-wave radiation. There are no mathematical methods for taking broken clouds into account, and sufficient data on absorption and transfer of solar radiation by clouds are unavailable.

Manabe /25/, in his calculations, used data of Korb /21/ obtained from an approximate theory and did not estimate the error. At the Institute of Atmospheric Physics of the USSR Academy of Sciences, calculations were made of almost exact values of the albedo of monodisperse clouds in the visible spectral region /5, 14/; in addition, a very limited amount of data were obtained (with an error of 20%) on the spectral albedo in the infrared region. These data are presented in Figures 5 and 6. However, even if sufficient optical data were available, the possibilities for calculating the solar radiation flux remain limited, since the hydrodynamic theory of cloud genesis will not produce the droplet spectrum on which the optical properties of clouds depend. In addition, this theory describes only horizontally stratified cloud layers. Hence it is very important to establish relationships between the measured solar energy flux at the surface and the state of the clouds. This problem is traditional in actinometry /2, 4, 13, 20/, but quantitative consideration of radiation in problems of atmospheric dynamics requires solving them at a new level. The cloudiness measure or the duration of sunshine do not satisfactorily describe the state of the cloudiness. One must know the thickness of the clouds, their altitude, moisture content, and droplet spectrum, i. e., actinometric measurements at the surface must be combined with aircraft cloudiness observations.

Additional valuable information can be obtained by investigating the relationship between fluxes of departing solar radiation and cloud pictures obtained from meteorological satellites. The quantitative description of these relationships should be based on statistical methods, and not be reduced to obtaining empirical formulas, as was mainly done until now.

Calculation of the transfer of long-wave radiation in a cloudy atmosphere does not differ basically from calculations under cloudless conditions, if one does not consider scattering at droplets. One more absorbing substance is added, viz., water droplets, and the problem consists only in preparing the transmission function with consideration of the latter. Neglect of scattering, as well as of changes in the droplet spectrum, does not apparently introduce any appreciable error. In any version one can draw two basic conclusions: a) the radiative flux in the inner layers of the cloud is close to zero; b) radiation from the cloud boundaries is close to black body radiation; the deviations amount to about 10% /14/.

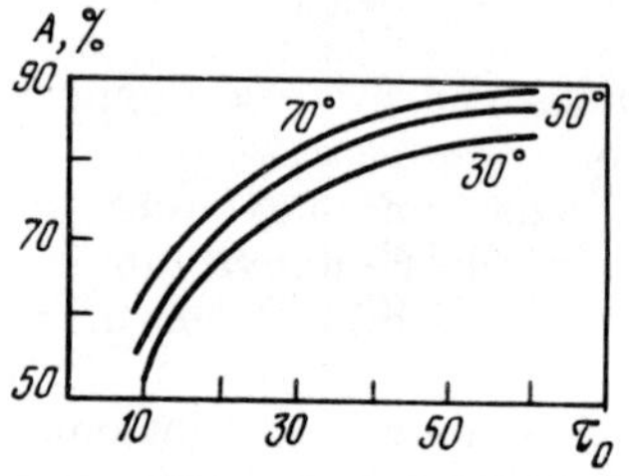

FIGURE 5. Albedo of clouds in the visible part of the spectrum as a function of τ_0, the optical thickness of clouds, for three zenith distances of the sun.

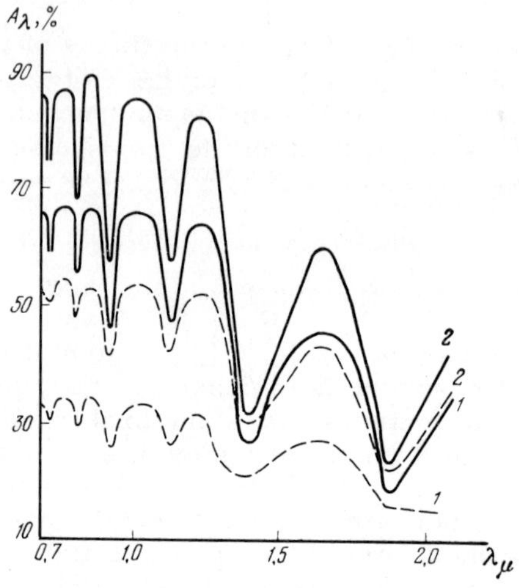

FIGURE 6. Spectral albedo of clouds in the infra-red region:

1—thick cloud; 2—thin cloud. The continuous curves represent $\xi = 60$; the dashed curves represent $\xi = 30$.

Since processes occurring inside the cloud layer are of no interest to problems of the general circulation, it may be found expedient when calculating heat transfer in the cloud atmosphere to regard the cloud as a surface of discontinuity between the above-cloud and subcloud regions. In this case consideration of scattering at droplets is replaced by consideration of reflection from the cloud boundary with precalculated values of the albedo /14/.

It is possible to formulate the following condition at the surface of the discontinuity, relating the heat transfer in the subcloud and above-cloud layers /17/:

$$\frac{dB}{dT_0}\left\{\hat{T}^{(2)}(z_0)+\hat{T}^{(1)}(z_0)+\int\limits_{\hat{T}^{(1)}(z_0)}^{\hat{T}(H)} D(m'-m_0)\,dT'-\int\limits_{\hat{T}(0)}^{\hat{T}^{(1)}(z_0)} D(m_0-m')\,dT'\right\}+$$
$$+(1-A_0)F(z_0)-(1-A_0-P)(1-A)F(z_0)+c_p\rho k^{(2)}(z_0)\left.\frac{d\hat{T}^{(2)}}{dz}\right|_{z_0}-$$
$$-c_p\rho k^{(1)}(z_0)\left.\frac{\partial\hat{T}^{(1)}}{\partial z}\right|_{z_0}=\Phi_1, \qquad (1)$$

where z_0 is the altitude of the discontinuity surface; $\hat{T}$ is the deviation of the temperature from the specified average T_0; $D(m)$ is the integral transmission function of the cloudless atmosphere; $m=\int_0^z \rho_w(z)\,dz$; $m_0=\int_0^{z_0}\rho_w(z)dz$;

$\rho_w(z)$ is the density of the absorbing substance (for simplicity we consider the case of the single absorbing substance, water vapor); subscript 1 denotes quantities pertaining to the subcloud layer; subscript 2 applies to the above-cloud layer; $F(z_0)$ is the solar radiation flux at level z_0; A_0 and A are the albedo of the cloud and the surface, respectively; P is the relative amount of solar radiation absorbed in the cloud; quantities A_0, A, and P pertain to the short-wave spectral region. For simplicity, the albedo of the cloud in the long-wave region is set equal to zero. The quantity $k(z)$ is the eddy exchange coefficient; ρ is the air density; c_p is the specific heat at constant pressure; H is the altitude of the upper cloud boundary.

Quantity Φ_1 includes data on the thermal regime of the cloud layer (these are regarded as known, since the cloudiness has been precalculated):

$$\Phi_1=H_0\left(c_p\rho\frac{\partial\theta^{(0)}}{\partial t}-L\overline{m}\right)+\{B\,[T^{(0)}(z_2)]+B\,[T^{(0)}(z_1)]-2B\,(T_0)\}-\\ -c_p\rho\gamma_a\,[k^{(1)}(z_1)+k^{(2)}(z_2)], \tag{2}$$

where $\theta^{(0)}$ and $T^{(0)}$ are the potential and absolute temperature in the cloud layer, whose boundaries are z_1 and z_2 (obviously, $z_1\leqslant z_0\leqslant z_2$). The bar denotes vertical averaging over (z_1, z_2); $L\overline{m}$ is the total amount of heat released in the cloud layer as a result of condensation; γ_a is the adiabatic lapse rate; $B(T)=\sigma T^4$ is the black body radiative energy.

INTERACTION OF THE ATMOSPHERE WITH THE ACTIVE LAYER OF LAND OR OCEANS

The thermal interaction with the underlying surface occurs theoretically through the boundary condition at $z=0$ for the heat transfer equation. However, an approach used extensively in solving problems of dynamic meteorology is to prepare a general system of equations such that the heat transfer equation is excluded (see, for example, /3/). Here the heat influx appears in the transformed equations of dynamics, for example, in the vorticity equation. It is easy to see that the thermal interaction of the underlying surface is in this case demonstrated by the long-wave radiative heat influx due to the boundary condition for the rate of ascending radiation.

We now consider in more detail the boundary condition for heat transfer at $z=0$. It is natural to specify continuity of the total heat fluxes or of their vertical components due to the greater horizontal homogeneity of the atmosphere as compared with the vertical homogeneity. The fluxes of long- and short-wave radiation at the surface are denoted by $F_i(0)$ and $\widetilde{F}_i(0)$ ($i=1$ denotes ascending and $i=2$ descending fluxes); $c_p\rho k\left.\frac{\partial\theta}{\partial z}\right|_0$ is the heat flux due to vertical eddy exchange; Lm_c is the heat released by evaporation from the surface. Using this notation, the continuity condition for the total heat flux can be written in the form

$$\Delta\left\{F_1(0)-F_2(0)+\widetilde{F}_1(0)-\widetilde{F}_2(0)-c_p\rho k\left.\frac{\partial T}{\partial z}\right|_0\right\}-Lm_c-c_p\rho k\gamma_a=0. \tag{3}$$

The braces in the above expression contain the values of all the above quantities on both sides of boundary $z=0$ (Δ is the difference sign), and hence it should be supplemented by the equation of heat transfer in the surface layer. Since this layer as such is not of interest to us, we can consider it, as a whole, by averaging the heat transfer equation over the layer thickness /17/. Using equation (3) we obtain an expression for the terrestrial-surface temperature $T_s(x, y, 0, t)$ (more precisely, the effective surface-layer temperature):

$$c_{v,s}\frac{\partial T_s'}{\partial t}+c_{v,s}\nabla T_s'=-c_{v,s}\nabla T_{s,0}-$$
$$-\left\{F_1(0)-F_2(0)-(1-A)\tilde{F}_2(0)-c_p\rho k\frac{\partial\theta}{\partial z}\Big|_0+\frac{c_{v,s}}{H_s}k_sT_s'-Lm_s\right\}. \qquad (4)$$

In equation (4) the quantity T_s is represented in the form

$$T_s(x, y, 0, t)=T_{s,0}(x, y)+T_s'(x, y, 0, t),$$

where $T_{s,0}$ is a known quantity; $T_s' \ll T_{s,0}$, $c_{v,s}=c_sH_s\rho_s$ is the specific heat per unit volume of the surface layer; k_s is the thermal conductivity coefficient in the layer;

$$\nabla T_s=u_s\frac{\partial T_s}{\partial x}+v_s\frac{\partial T_s}{\partial y},$$

where u_s and v_s are the horizontal velocity components. In the case of land, $\nabla T_s=0$.

According to observations, the orders of magnitude (in cal/cm$^2\cdot$ min) of heat fluxes contained in equation (4) are as follows: $F_1(0)-F_2(0)=0.1-0.2$ under cloudless skies and decreases sharply in cloudy weather; $\tilde{F}_2(0)=1$ at noon in summer and 0.1 in winter. In cloudy weather $\tilde{F}_2(0)$ decreases by a half and more. Over land, $c_p\rho\frac{\partial\theta}{\partial z}\Big|_0\approx 0.2$ by day; at night the sign of the turbulent flux changes and it decreases in magnitude. Over the sea, $c_p\rho\frac{\partial\theta}{\partial z}\Big|_0$ is much smaller than over land. The estimates of Lm_c are the poorest. According to Skvortsov /12/, $Lm_c=0.1-0.3$ for Central Asia. The above figures show that none of the fluxes in equation (2) can be neglected.

Using equation (4) as the boundary condition for the intensity of ascending radiation, we note that this expression contains two unknowns: T_s and $F_1(0)$. We note further that to solve the equation of radiative transfer one must know the intensity $I_{1,\lambda}(0,r)$ of ascending radiation at $z=0$, depending on the direction r and the wavelength λ. The value of $I_{1,\lambda}(0,r)$ is related to $F_1(0)$ by the expression

$$F_1(0)=\int_{\lambda_1}^{\lambda_2}\int_S I_{1,\lambda}(0, r)\cos(z, r)\,dw\,ds,$$

where the inner integral is taken over a solid angle of 2π.

Only when using the simplification

$$I_{1,\lambda}(0, r)=\text{const}$$

is it possible to determine the intensity

$$I_{1,\lambda}(0, r) = \frac{1}{\pi\Delta\lambda} F_1(0)$$

from the specified flux. In the general case an additional equation is needed.

We assume that the intensity of ascending radiation is continuous at $z=0$:

$$I_{1,\lambda}(0, r) = I_{1,\lambda,s}(0, r).$$

Assuming for simplicity that

$$I_{1,\lambda,s}(0, r) = B_\lambda(T_s),$$

where $B_\lambda(T)$ is the Planck function, we obtain

$$I_{1,\lambda}(0, r) = B_\lambda(T_s). \qquad (5)$$

From this we have $F_1(0) = \sigma T_s^4$ and equation (4) yields the surface temperature σ, more precisely, allows one to express it in terms of the temperature of the atmosphere and of other heat-transfer characteristics.

Using the specified values of T_s, one can determine $I_{1,\lambda}(0, r)$ from equation (5), after which one obtains the following expression for the radiative heat influx in the atmosphere /18/:

$$R(z) = \rho_v(z)\Big\{-[B(T_s) - B(T_0)]\frac{dD}{dm} + B(T^x)\frac{dD}{dm}(m^x - m + \\ + \frac{d}{dm}\int_{B(T_0)}^{B(T^x)} D(|m - m'|)\,dB'\Big\}, \qquad (6)$$

where $T_0 = T(0)$; $T^x = T(H)$; $B(T) = \sigma T^4$; $m^x = \int_0^H \rho_v(z)dz$. Heat exchange with the surface, allowing for all the forms of heat transfer, is contained in the expression for the radiative influx by the temperature jump at the surface. In the case of land one can describe graphically the manner in which heat transfer demonstrates itself.

Since the specific heat at constant volume for land is small compared with that of the atmosphere, the time derivative in equation (4) can be neglected /17/. If at the same time we set $\nabla T_s = 0$ and $k_s = 0$ (i.e., if one neglects the molecular heat conduction in the soil), then, representing $F_2(0)$ by the integral transmission function of /18/, we obtain

$$B(T_s) - B(T_0) = (1 - A)\widetilde{F}_2(0) + Lm_c + c_p \rho k_v \frac{\partial\theta}{\partial z}\Big|_0 - \\ - B(T^x)D(m^x) + \int_{B(T_0)}^{B(T^x)} D(m')\,dB'. \qquad (7)$$

For oceans T_s is eliminated from equation (6) by means of equation (4).

References

1. Arrago, L.R. and M.E. Shvets. K teorii obrazovaniya i evolyutsii nekonvektivnoi oblachnosti (The Theory of the Genesis and Evolution of Nonconvection Clouds).— Trudy GGO, No. 121. 1961.
2. Barashkova, E.P., V.L. Gaevskii, and Z.I. Pivovarova. Radiatsionnyi rezhim territorii SSSR (The Radiation Regime of the USSR).— Gidrometeoizdat. 1961.
3. Buleev, N.I. and G.I. Marchuk. Nekotorye voprosy teorii kratkosrochnogo prognoza polei meteorologicheskikh elementov (Some Problems of the Theory of Short-Range Forecasting of Fields of Meteorological Elements).— Trudy Vsesoyuznogo nauchnogo meteorologicheskogo soveshchaniya, Vol. 2, Gidrometeoizdat. 1963.
4. Gal'perin, B.M. Sutochnyi prikhod summarnoi solnechnoi radiatsii pri razlichnykh oblakakh (Daily Influx of Total Solar Radiation for Different Cloudiness Conditions).— Trudy GGO, No. 125. 1962.
5. Gradus, L.M. and E.M. Feigel'son. Vliyanie oblachnosti na radiatsionnyi pritok tepla v atmosfere (Effect of Cloudiness on the Radiative Heat Influx in the Atmosphere).— Izvestiya AN SSSR, seriya fiziki atmosfery i okeana, 1(7). 1965.
6. Matveev, L.T. Usloviya obrazovaniya i evolyutsii oblakov pod vliyaniem vertikal'nykh tokov i turbulentnogo obmena (Conditions for the Genesis and Evolution of Clouds Due to Vertical Current and Eddy Transport).— Izvestiya AN SSSR, seriya geofizicheskaya, No. 1. 1961.
7. Matveev, L.T. Trebovaniya k tochnosti opredeleniya potokov infrakrasnogo izlucheniya pri pomoshchi iskusstvennykh sputnikov zemli i metod rascheta vysoty verkhnei granitsy oblakov (Requirements for the Accuracy in Determining Infrared Radiative Fluxes by Satellites and a Method for Calculating the Altitude of the Upper Cloud Boundary).— Kosmicheskie issledovaniya, Part 2, No. 1. 1964.
8. Meleshko, V.P. K voprosu o chislennom prognoze oblachnosti i osadkov s uchetom transformatsii vozdushnoi massy (Numerical Prediction of Cloudiness and Precipitation with Allowance for Modification of the Air Mass).— Trudy GGO, No. 151. 1964.
9. Monin, A.S. O fizicheskom mekhanizme dolgosrochnykh izmenenii pogody (The Physical Mechanism of Long-Range Weather Changes).— Meteorologiya and gidrologiya, No. 8. 1963.
10. Petrova, L.V. and E.M. Feigel'son. Rol' radiatsii v razvitii oblachnosti (Role of Radiation in the Development of Cloudiness).— Izvestiya AN SSSR, seriya geofizicheskaya, No. 8. 1964.
11. Rakipova, L.R. and B.E. Shneerov. Uchet radiatsionnykh pritokov tepla v zadachakh dinamicheskoi meteorologii (Consideration of Radiative Heat Influx in Problems of Dynamic Meteorology).— Present collection.
12. Skvortsov, A.A. Oroshenie sel'skokhozyaistvennykh polei i mikroklimat (Irrigation of Agricultural Fields and the Microclimate).— Gidrometeoizdat. 1964.
13. Undla, I. O zakonomernostyakh izmenchivosti dnevnykh summ summarnoi radiatsii (Regularities in the Variation of Diurnal Total Radiation Sums).— Issledovaniya po fizike atmosfery, Institut fiziki i astronomii Akademii Nauk Estonskoii SSR, No. 4. 1963.

14. Feigel'son, E. M. Radiatsionnye protsessy v sloistoobraznykh oblakakh (Radiative Processes in Stratified Clouds). — Izdatel'stvo Nauka. 1964.

15. Feigel'son, E. M., M. S. Malkevich, et al. Raschet yarkosti sveta v atmosfere pri anizotropnom rasseyaniyu (Calculation of the Light Intensity in the Atmosphere for Anisotropic Scattering). — Trudy Instituta Fiziki Atmosfery, Vol. 1. 1958.

16. Feigel'son, E. M. Opticheskie svoistva oblakov (Optical Properties of Clouds). —Kosmicheskie issledovaniya, Part 2, No. 3. 1964.

17. Feigel'son, E. M. O nekotorykh vozmozhnostyakh issledovaniya teploobmena i vlagoobmena v atmosfere (Some Possibilities of the Study of Heat and Moisture Exchange in the Atmosphere). — Izvestiya AN SSSR, seriya fiziki atmosfery i okeana, 1(1). 1965.

18. Feigel'son, E. M. Radiatsionnyi pritok tepla v atmosfere (Radiative Heat Influx in the Atmosphere). — Izvestiya AN SSSR, seriya geofizicheskaya, No. 10. 1964.

19. Shvets, M. E. O kondensatsii vodyanogo para v atmosfere (The Condensation of Water Vapor in the Atmosphere). — Izvestiya AN SSSR, seriya geofizicheskaya, No. 6. 1955.

20. Haurwitz, P. Insolation in Relation to Cloud Type. — J. Meteorol., 5(3). 1948.

21. Korb, G. Absorption von Sonnenstrahlung in Wolken. — Univ. München, Meteorol. Inst. Wiss. Mitt., No. 6. 1961.

22. Leovy, C. Radiative Equilibrium of the Mesosphere. — J. Atmos. Sci., 21(3). 1964.

23. Manabe, S., J. Smagorinsky, and J. L. Holloway. A Numerical Integration of the General Circulation of the Atmosphere with the Explicit Calculation of Radiative Transfer. —Report at the International Symposium on the Study of Radiative Processes, Leningrad. 1964.

24. Manabe, S. and F. Möller. On the Radiative Equilibrium and Heat Balance of the Atmosphere. — Monthly Weather Rev., 89(12). 1961.

25. Manabe, S. and R. Strikler. Thermal Equilibrium of the Atmosphere with a Convective Adjustment. — J. Atmos. Sci., 21(4). 1964.

L. R. Rakipova and B. E. Shneerov

CONSIDERATION OF RADIATIVE HEAT INFLUX IN PROBLEMS OF DYNAMIC METEOROLOGY

The first part of this article surveys studies of the theory of climate in which allowance is made for radiative heat influx. It is shown that the "gray" atmosphere model is unsatisfactory. A more complete incorporation of physical factors determining the thermal regime of the atmosphere makes it possible to obtain not only the temperature distribution close to actual values, but also satisfactory values for the components of the radiation and thermal balance of the atmosphere and the underlying surface.

The second part of the article considers a method for constructing a nonadiabatic model of atmospheric circulation. The heat influx throughout the atmosphere is calculated with allowance for radiation. It is shown that incorporation of nonadiabatic factors improves the forecasting quality.

Radiative transfer as one form of heat influx is considered in problems of dynamic meteorology when one is concerned with the thermal state of the atmosphere, in the theory of climate, and in nonadiabatic weather forecasting.

A completely quantitative description of thermal and dynamic processes in the atmosphere can be obtained by solving hydro- and thermodynamic equations together with equations of radiative transfer. In dynamic meteorology this complete system of equations is subjected to many kinds of simplifications, depending on the approach to the problem.

We begin our survey with studies of the theory of climate. One of the principal problems of this theory is the vertical temperature distribution in the atmosphere, which involves considering only the system of equations of heat and radiative transfer. The simplifications made in this system reduce to selecting only several from among all the factors determining the temperature of the atmosphere. The results which are obtained should be regarded as an estimate of the effect of the particular factors selected on the temperature regime of the atmosphere.

The first works (Humphreys, Gold, Emden) considered only radiative heat influx. As a result it became possible to clarify the thermal regime of the principal layers of the atmosphere. Emden obtained an isothermal temperature profile for the stratosphere and a value of the yearly average temperature in the moderate latitudes (−68.5°C) close to the actual one. It was concluded from this that the thermal regime of the stratosphere is determined in the first approximation by the condition of radiative equilibrium. However, calculations using the Emden formula do not yield a correct curve of the latitudinal temperature variation in the stratosphere, i. e., do not show that the temperature increases toward the pole. This is because his postulation of "gray" absorption of radiation in the atmosphere and, in addition, his assumption that the only absorbing medium is water vapor, is too approximate a simplification of the actual radiative processes in the atmosphere. Attempts to make allowance for selectivity of radiation absorption by water vapor showed that not only the horizontal, but also the vertical temperature distribution in the stratosphere do not correspond to radiative equilibrium relative to water vapor. This forced Linke and Möller /30/ to the conclusion that radiation absorption by ozone and carbon dioxide must be taken into account in determining the thermal regime of the atmosphere. Goody /28/ did this and obtained as a result the correct latitudinal temperature variation in the stratosphere.

Kibel' /7/ included in his study of the yearly average temperature distribution not only radiative, but also eddy heat transfer. Apart from this he stated the problem in the same manner as Emden. Before Kibel', an attempt to improve Emden's results by incorporating vertical flows was made by Fridman /23/.

Kibel' was able to obtain a vertical temperature distribution in the troposphere close to the observed one (Table 1—values of T). With his results it was possible to determine the effect of various factors on the temperature. For example, if Γ, the albedo of the "earth-atmosphere" system is taken as 0.4 rather than 0.7, and the water-vapor density at the earth surface ρ_0 is estimated at 13 g/m^3 instead of 6.2 g/m^3, we obtain the values of T'. These temperatures were calculated for the 42° latitude, since the effect of horizontal eddy transport, which is not considered in this problem, is small in middle latitudes.

TABLE 1. Vertical temperature distribution (after Kibel')

Quantity	z, km								
	0.3	1.0	2.0	4	6	8	10	12	14
T	8	9	4.5	−6.5	−19.5	−33.5	−45.0	−52.0	−55
T_{av}	8.1	8.9	5.2	−6.1	−19.5	−33.7	−47.15	−55.7	—
T'	52	52	52	49	42	35	30	13	—
T''	−27	−21	−22	−28	−40	−48	—	—	—

If the thermal effect of water-vapor condensation in the atmosphere and of evaporation from the surface is taken into account, and the temperature is calculated for the middle latitudes (without changing Γ and ρ_0), then we obtain T''. The Schwarzschild approximation makes no allowance for diffusive propagation of long-wave radiation. When this effect is taken into account, as was done by Feigel'son, the temperature at all altitudes is found to be 6—10° higher than the distribution obtained when considering parallel radiation fluxes /21/.

The following should be stated here about the "gray" atmosphere model.

1. It is easy to show that the long-wave radiation fluxes in the "gray" atmosphere model are far from their actual values in the terrestrial atmosphere. This can be done by simply integrating the transfer equations for "gray" radiation and then calculating the departing radiation $B(M)$ and the back radiation of the atmosphere $A(0)$ for the actual vertical temperature distribution and water-vapor density, together with the usually utilized "gray" atmosphere parameters: absorption coefficient of water vapor for

long-wave radiation $\alpha = 7.25$ cm/g, the Hulbert parameter $f = 0.6$, accounting approximately for the selectivity of radiation absorption. By definition

$$f = \frac{\int_0^{8\mu} E_\lambda \, d\lambda + \int_{12\mu}^{\infty} E_\lambda \, d\lambda}{E},$$

where $E_\lambda(T)$ is Planck's constant, and $E = \sigma T^4$. We obtain $B(M) = 69$ kcal/cm$^2\cdot$year and $A(0) = 444$ kcal/cm$^2\cdot$year, while climatological calculations using the most refined radiation diagrams yield $B(M) = 166$ kcal/cm$^2\cdot$year; according to /6/ $A(0)$ is of the order of 230 kcal/cm$^2\cdot$year. If $\rho_0 = 6.2$ g/m^3 in the "gray" model we obtain $B(M) = 80$ kcal/cm$^2\cdot$year and $A(0) = 289$ kcal/cm$^2\cdot$year. Consequently, this "gray" atmosphere model yields underrated values of the departing radiation and overrated values of the back radiation.

2. The superadiabatic lapse rates obtained by Emden for a troposphere in radiative equilibrium are not necessarily inherent to the "gray" atmosphere. In fact, Milankovich /7/ obtained a vertical temperature profile under radiative equilibrium conditions for average annual conditions close to actual ones. This is because for a "gray" atmosphere with two absorption coefficients (α_1 for long-wave and α_2 for short-wave radiation) the radiative equilibrium at the terrestrial surface $T(0)$ and at the upper boundary $T(M)$ is determined by the ratio of these coefficients:

$$\sigma T^4(0) = \frac{1}{2} S(M) \left[1 + \frac{\alpha_1}{\alpha_2} + P_a \left(\frac{\alpha_2}{\alpha^1} - \frac{\alpha_1}{\alpha_2}\right)\right]; \quad (1)$$

$$\sigma T^4(M) = \frac{1}{2} S(M) \left[1 + \frac{\alpha_2}{\alpha_1}\right]. \quad (2)$$

Here $S(M) = \frac{I_0}{4}(1 - \mathrm{T})$; I_0 is the solar constant; P_a is the transparency of the atmosphere to short-wave radiation. According to Emden, $\alpha_1/\alpha_2 = 21.8$, while according to Milankovich it is equal to 7.5. The values of P_a and T used by Milankovich were close to actual ones ($P_a = 0.67$, $\mathrm{T} = 0.4$). The temperature $T(0)$ and $T(M)$ obtained from equations (1) and (2) with these values of P_a and Γ are listed in Table 2. This table also presents temperatures obtained by Shekhter for a selectively absorbing atmosphere /25/.

TABLE 2. Temperature (°C) for different parameters

Quantity	After Emden	After Milankovich	After Kibel'	After Shekhter
$T(0)$	69	10	−6 32	70
$T(M)$	−65	−61	−91 −67	−79

If the same calculations are carried out using the values of parameters selected by Kibel' ($\mathrm{T} = 0.7$; $\alpha_1/\alpha_2 = 5.0$; $P_a = l - \alpha_2/\alpha_1\tau_0$; $\tau_0 = 12.6$), then we obtain different results; these are also listed in Table 2. The use of the factor $f = 0.6$ results in higher temperatures (32 and − 67, respectively).

In order to obtain temperatures close to actual ones in the troposphere when using the "gray" atmosphere parameters selected by Emden and Kibel', one must assume vertical eddy heat transport; this is not necessary when using the parameters assumed by Milankovich. In this case satisfactory agreement with the actual temperature distribution can be obtained by slight adjustment of α_1 and α_2 about the selected values, although Milankovich himself chose a different way: he reduced the Stefan-Boltzmann constant by 8% as compared with its presently assumed value. This made it possible to obtain a value close to the actual one of the average annual surface-air temperature (15.2°C).

It is thus seen that the "gray" atmosphere model can yield satisfactory temperature distributions not only without allowing for evaporation and condensation, but also without considering the vertical eddy heat transport, i.e., under conditions far removed from the actual. This example again shows that the obtaining of correct quantitative data should be regarded as a necessary, but by far not sufficient condition for the correctness of a given physical model. A sufficient condition for its correctness should be incorporation of all the physical factors of importance for the given problem or, which is the same thing, obtaining correct values of both the quantities required and those directly related to them. In problems of the thermal regime of the atmosphere these will primarily be radiative and eddy heat fluxes.

In the paper of Kibel' the solar radiation fluxes and eddy fluxes have the following values: at the surface $S(0) = 6$ kcal/cm²·year (for $\alpha_2/\alpha_1 = 0.2$ and $\tau_0 = 12.6$) and $P(0) = 79$ kcal/cm²·year, while according to climatic calculations of absorption by the earth surfaces $S(0) = 113$ kcal/cm²·year and eddy transport accounts for $P(0) = 14$ kcal/cm²·year. According to Milankovich, the question of long-wave radiation fluxes has been very badly studied, since the parameters selected by him for the "gray" atmosphere give values of fluxes obtained from the transfer equation which do not agree with their values given by the boundary conditions. Thus, from equations (47), (57), and (62) (Chapter 2 of /17/) we have

$$B(0) - A(0) = (1 - \Gamma)\frac{I_0}{4}P_a = 104 \text{ kcal/cm}^2\cdot\text{year} \qquad (3)$$

for $\Gamma = 0.4$ and $I_0 = 1.98$. Since $B(0) = 274$ kcal/cm²·year for $t = 15$°C (obtained with σ reduced by 8%), then $A(0) = 170$ kcal/cm²·year. From condition (80) /17/ $B(M) = (1 - \Gamma)\,I_0/4 = 156$ kcal/cm²·year. This value of $B(M)$ is close to its climatological value, while the value of $A(0)$ is on the low side, since actually equation (3) should have $A(0) - LE - P(0)$ instead of $A(0)$. If, however, the same fluxes are determined from transfer equations, we obtain $B(M) = 123$ and $A(0) = 233$ kcal/cm²·year. This lack of agreement is due to the fact that the parameters defining the long-wave radiation fluxes in the Milankovich model are not selected from quantitative agreement between the values obtained from boundary conditions and those obtained from transfer equations, but are rather "external" characteristics.

The problem of the thermal regime of the atmosphere as stated by Kibel' was markedly advanced by Blinova /2/, who succeeded in incorporating such an important factor as the horizontal eddy heat transport. On the basis of this, Mashkovich calculated the zonal temperatures at different latitudes in the northern and southern hemispheres /12/, but he did not succeed in obtaining a correct latitudinal temperature variation in the stratosphere. Later Blinova worked together with Marchuk and extended her results to constructing a theory of seasonal zonal-circulation variations /4/.

Kutznetsov /9/ did particularly important work on the thermal regime of the atmosphere. He paid particular attention to the theoretical consideration of radiative heat transfer features characteristic of the terrestrial atmosphere, but only for conditions of radiative equilibrium. He examined the effect on the temperature distribution of the albedo of the surface and clouds for long- and short-wave radiation, horizontal discontinuity surfaces, and scattering of long-wave radiation. For example, he showed that only in the case of radiative equilibrium is the radiation flux continuous when passing through the interfaces. Solar energy scattering has a marked effect on the vertical temperature distribution, increasing it somewhat in the lower layers and decreasing it in the upper layers. This is a result of an increase in the optical thickness of the atmosphere by the scattering. Unfortunately, due to the cumbersomeness of the method developed in /9/ for integrating the radiative transfer, the solution is only completed for the simplest model of the "gray" atmosphere.

The work of Kuznetsov was continued by Feigel'son. In addition to the diffusivity effect of long-wave radiation propagation, he studied the thermal regime of an atmosphere containing a cloud layer /22/. Here, as before, the radiative equilibrium condition was assumed for the subcloud and above-cloud layer. In the cloud layer, allowance is made for the vertical eddy heat transport in addition to the radiative heat transfer. It was shown that for a sufficiently thick cloud the temperature in the subcloud layer is independent of the altitude and is equal to the temperature of the terrestrial surface (Figure 1, curve 3, lower part). Hence, in the presence of mechanisms transporting heat from the lower to the upper cloud boundary (for example, eddy transport) the temperature above the cloud will be higher than at the same level without a cloud (upper part of curve 3). When scattering of radiation in the cloud is taken into account, the temperature in the atmosphere above the cloud becomes less than in the case of pure absorption, since then a part of the radiation arriving from the top is as if reflected from the upper boundary of the latter (curve 4). Curve 1 shows the temperature distribution in a cloudless atmosphere at night (as do curves 3 and 4). The temperature distribution in a cloudless atmosphere for daytime conditions is given by curve 2 and that above clouds by curve 5. It was shown that the temperature profile above the clouds depends on the moisture distribution and on the nature of the absorption spectrum. All the models (curves 1—5) were constructed with the condition of "gray" absorption of long-wave radiation. Segments with very weak absorption in the true atmospheric absorption spectrum are taken into account (approximately) by using two absorption coefficients ($\alpha_1 = 0.2$ and $\alpha_2 = 4\ \mathrm{cm}^2/\mathrm{g}$) and the method suggested by Kuznetsov. Obviously, when taking into account the transparent interval ($8-12\mu$) the temperature in the subcloud atmosphere is found to be lower than the "gray" atmosphere (curve 6). However, the method developed by Kuznetsov for approximating transfer [transmission] functions of the individual absorption bands can only be used for constructing a rough model of the long-wave absorption spectrum.

When studying the thermal regime of the atmosphere, selective radiation absorption can be most completely taken into account by the Ambartsumyan—Lebedinskii method, which can be regarded as satisfactory /1/. Ambartsumyan integrates over all the wavelengths λ in two steps: 1) integration over the λ for which the absorption coefficient lies in the range $\alpha < \alpha_\lambda < \alpha + d\alpha$; 2) integration over all the $d\alpha$.

This method was developed further by Lebedinskii and used to study the radiative equilibrium of the terrestrial atmosphere /10/. Lebedinskii introduced the function $\omega(\alpha, T)$, which represents that fraction of the total black body radiation in the wave range λ for which $\alpha < \alpha_\lambda < \alpha + d\alpha$. This fraction depends on the absorption coefficient α as well as on T, the radiator temperature. Lebedinskii used approximate equations of radiative energy transfer: the diffusivity was taken into account by a 60% increase in the absorption coefficients for parallel beam radiation. In addition, he considered only one absorbing substance—water vapor.

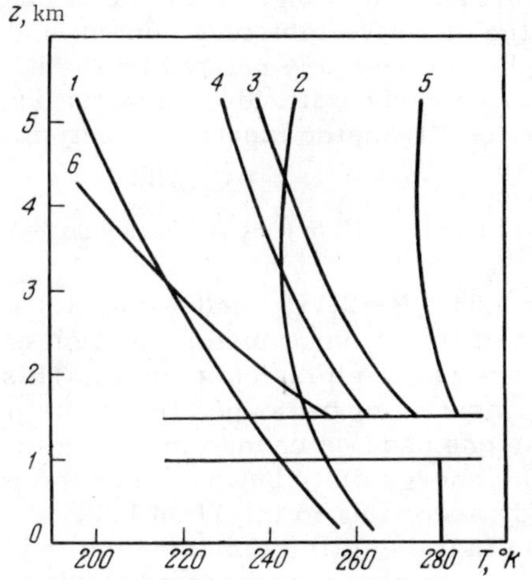

FIGURE 1. Vertical temperature distribution in different models:

1 and 2—cloudless atmosphere, night and day; 3 and 4—cloud at night, with and without allowance for scattering; 5—cloud, by day; 6—cloud, at night, selective absorption (unlike curves 1—5, where the water vapor absorption spectrum is regarded as "gray").

Shekhter /25/ extended the Ambartsumyan-Lebedinskii method to the case of exact transfer equations and made allowance in radiative processes for the participation of two absorbing substances: water vapor and carbon dioxide. Upon integrating over all the wavelengths (allowance for selectivity), over all directions in space (allowance for diffusion), and over all the elementary layers comprising the total layer M (the radiation of which was being considered), she obtained the following expressions for the fluxes of descending and ascending radiation:

$$A(m) = E(m) - E(M)D(M-m) + \int_m^M D(\tau - m)\frac{dE}{d\tau}d\tau; \tag{4}$$

$$B(m) = E(m) - (1-\delta)E(M)D(M-m) + $$
$$+ (1+\delta)\int_0^M D(\tau + m)\frac{dE}{d\tau} - \int_0^m D(m-\tau)\frac{dE}{d\tau}d\tau, \tag{5}$$

where $E(m)$ is the perfect black body radiation flux; $D(m)$ is the transparency function for diffuse long-wave radiation; δ is the absorptivity of the terrestrial surface; m is the effective moisture content ($m = \int_0^z \rho \sqrt{p/p_x}\, dz$; the quantity $\sqrt{p/p_x}$ takes into account the vertical variation in the absorption coefficient and p_x is the pressure at which the absorption coefficient was determined).

The transfer functions for water vapor and carbon dioxide were constructed from experimental data. In order to be able to express the total transfer function in terms of the single variable m, a relationship was established between the effective moisture content and the quantity of carbon dioxide in the atmosphere (these are related by the baric pressure if it is assumed that the relative carbon dioxide concentration does not depend on the altitude). The total transfer function is satisfactorily represented by the expression

$$D(m) = QH_S(q\sqrt{m}) + PH_S(p\sqrt{m}), \tag{6}$$

where $Q = 1.884$, $q = 0.54$, $P = 2.116$, and $p = 6.94$.

A known limitation of the above transfer function is that it neglects the temperature dependence of the emission source. This can be justified by some degree of compensation of two opposite effects produced by temperature changes: on the one hand, a change in the emitter temperature produces a shift on the energy distribution curve in the radiation spectrum of a perfect black body according to the Wien law, which decreases absorption with a rise in the emitter temperature, while on the other hand, absorption increases with a rise in temperature due to an increase in the intensity of absorption bands. The dependence of absorption coefficients on pressure used in the above expression is also not too convincing. However, available experimental data show that the Shekhter transfer function is satisfactory at least for the troposphere. Thus, the rms difference between the aircraft-measured effective radiation of the atmosphere and that calculated by the Shekhter radiation diagram amounted to about 4 % for 16 cases /26/.

The above survey of studies of the thermal regime of the atmosphere shows that a physically justified theory of the temperature distribution in the atmosphere cannot be constructed without making allowance for all the principal factors determining the thermal state of the atmosphere. The Main Geophysical Observatory decided that it was necessary to formulate and solve the problem of the steady-state zonal temperature field as completely as possible /20/. The physical model underlying this theory reflects the characteristic features of the general circulation of the atmosphere: ordered zonal transfer and large-eddy meridional exchange. The following factors were considered.

1. The radiative heat flux with allowance for latitudinal distribution of the solar energy incident on the upper-atmosphere boundary and the albedo of the "earth-atmosphere" system; the selectivity of absorption by water vapor and carbon dioxide in their latitudinal distribution and diffusivity in the distribution of long-wave radiation in the atmosphere. The Shekhter and Möller transfer functions for long- and short-wave radiation were used.

The latter contained corrections for attenuation by dust and clouds.

2. Turbulent heat fluxes, both vertical and horizontal, with altitude-dependent eddy transport coefficients.

3. Heat fluxes associated with condensation of water vapor in the atmosphere and evaporation of water from the earth surface, with allowance for their latitudinal distribution.

4. Heat exchange between the ocean surface and the atmosphere and the ocean subsurface layers, as a function of latitude.

5. Refinements were also made in some relationships which are usually used in calculating the thermal regime of the atmosphere:

a. In climatological calculations of the albedo of the "earth-atmosphere" system the portion of radiation reflected by the earth is defined by

$$\Gamma'''_{cl} = (1 - n)\,\alpha, \tag{7}$$

where α is the experimentally determined value of the surface albedo, and n is the degree of cloudiness.

The above form of the equation implies that α% of solar radiation is reflected from the upper boundary of the atmosphere. It is more correct, with allowance for attenuation of solar radiation as it moves from the upper boundary of the atmosphere to the terrestrial surface and back, to calculate the planetary albedo of the terrestrial surface from the expression

$$\Gamma''' = \alpha\,(1 - \Gamma' - \Gamma'')^2 D_S^2(M) \sum_{i=1}^{l}\left[\left(\frac{D_S(M)}{P}\right)^2 \alpha\,(\Gamma' + \Gamma'')\right]^{i-1} = K\alpha, \tag{8}$$

where $D_S(M)$ is the magnitude of the transfer function for solar radiation which traveled through the entire thickness of the atmosphere; P is the coefficient of cloud transparency for solar radiation; l is the number of successive reflections of the solar ray from the earth surface; Γ'' is the planetary albedo of the atmosphere; Γ' is the planetary albedo of the clouds.

The values of Γ', Γ'', and P were taken from data of London /31/. Table 3 presents values of coefficient K as a function of n and α for average atmosphere transparency conditions $D_S^2(M)/P^2 = 0.62$, $\alpha' \simeq 0.50$ (albedo of clouds), and $\alpha'' \simeq 0.14$ (albedo of the atmosphere).

If the clouds are broken, the climatological calculations yield higher than actual planetary albedos of the terrestrial surface; in the case of continuous cloudiness the climatological value is zero, while actually the clouds are transparent to ascending radiation.

b. It is usually assumed that

$$S(m) = W\,(1 - \Gamma)\,D_S(M - m), \tag{9}$$

where W is the solar radiation flux incident on the upper boundary of the atmospheric layer under consideration (approximately 20 km).

Hence

$$S(0) = W\left(1 - \Gamma' - \Gamma'' - \frac{z(M)}{W}\right) D_S(M) = Q - r(M)\,D_S(M),$$

where Q is the short-wave radiation flux arriving at the terrestrial surface, and $r(M)$ is the magnitude of the radiation reflected from the surface at level M.

Climatological calculations employ a different expression:

$$S_{cl}(0) = Q(1-\alpha) = Q - r(0). \qquad (10)$$

TABLE 3. Values of coefficient *K*

α	n				
	0	0.25	0.50	0.75	1.0
0.10	0.58	0.46	0.36	0.22	0.20
0.30	0.59	0.47	0.39	0.23	0.21
0.70	0.61	0.50	0.42	0.25	0.24
$1-n$	1	0.75	0.50	0.25	0

Obviously $r(M)D_S(M) < r(0)$, since $r(M) < r(0)$ and $D_S(M) < 1$. Equation (10) yields values of the radiation flux reflected from the earth which are a factor of 1.5—2 below actual values (this follows from $r(0) = \frac{1-\Gamma'-\Gamma''}{\Gamma'''}\alpha r(M)D_S(M)$). To avoid this, $S(m)$ was expressed in the form

$$S(m) = W(1-\Gamma'-\Gamma'')[D_S(M-m) - \alpha D_S(M)D_S(m)],$$

which coincides with the climatological form of this expression when $m = 0$.

The solution of this problem provided the zonal temperature distribution in the atmosphere and values of radiation and eddy fluxes.

Particular attention was paid to an experimental and physical substantiation of the numerical values of the parameters used.

The vertical temperature profile (yearly average for the entire earth) is:

z, km	t, °C	z, km	t, °C
0	15	4	—4
1	12	6	—15
2	8	7	—24
3	3	20	—64

We now give the heat-balance components for the earth as a whole; their climatological values according to Budyko (in kcal/cm^2·year) are given in parentheses.

Radiation balance of the "earth-atmosphere" system

$S(M) = W_0(1-\Gamma) = 162$ (solar radiation absorbed by the "earth-atmosphere" system).

$B(M) = 162$ (166) (departing radiation of the "earth-atmosphere" system).

$W_0 = 263$ (solar radiation flux incident on the upper boundary of the atmosphere).

$\Gamma = 0.38$.

Heat balance of the terrestrial surface

$S = Q(1-\alpha) = 113$ ($Q = 130$) (solar radiation flux absorbed by the earth).

$I = B - A = 51$ (effective radiation of the terrestrial surface).

$R = 62$ (72) (radiation balance of the terrestrial surface).

$LE = 54$ (59) (heat expended for evaporation).

$P = 8$ (13) (eddy heat flux at the terrestrial surface).

Heat balance of the atmosphere

$S_a = 49$ (solar radiation absorbed in the atmosphere).
$\Phi = 54$ (heat of condensation).
$P = 8$.
$(B - A)_a = 111$ (long-wave radiation absorbed in the atmosphere).

Table 4 shows the average annual zonal temperature distribution at different latitudes, while Table 5 lists the meridional profile of the departing and effective radiation. The corresponding climatological values are also listed there /31/. It is seen from these tables that there is quite satisfactory agreement between the predicted and actual temperatures and heat-balance components.

TABLE 4. Zonal temperatures of the northern hemisphere, °C (yearly average)

z, km	φ									
	0	10	20	30	40	50	60	70	80	90°
0	28	26	24	18	12	5	—3	—10	—16	—18
1	24	22	19	14	8	2	—4	—10	—13	—14
2	18	18	14	10	4	—2	—8	—12	—16	—17
4	6	4	2	—2	—7	—12	—17	—22	—24	—26
6		—6	—8	—12	—19	—25	—31	—35	—40	—41
10	—35	—34	—37	—41	—47	—50	—51	—52	—50	—50
16	—74	—75	—73	—69	—63	—56	—51	—48	—51	—52

TABLE 5. Departing ***B*** (***M***) and effective ***I*** radiation, $kcal/cm^2 \cdot year$

Radiation	φ									
	90	80	70	60	50	40	30	20	10	0°
B (M)	146	148	153	159 148	163 155	166 164	165 175	164 181	162 180	161 178
I	39	39	42	45 34	47 34	50 45	52 50	54 52	54 37	54 37

The studies of the thermal regime of the atmosphere carried out in the West are quite formal from the point of view of physical content. Defant /27/ determined the meridional temperature profile in the atmosphere from a given surface temperature with allowance for vertical and horizontal eddy transport and the heat of condensation. Hess and Frank /29/ determined the zonal distribution and the annual march of the surface temperature from considerations of the radiation balance of the terrestrial surface and the horizontal eddy heat conduction.

The rapid development of numerous methods of short-range weather forecasting resulted in the appearance of operative forecasting schemes which are used for predicting the geopotential field with a higher accuracy than can be done with synoptic methods. The initial success of numerical forecasting is doubtless attributable to the fact that large-scale atmospheric processes requiring several days to develop can be regarded as adiabatic in the first approximation. However, any further development of forecasting methods requires making allowance for heat influx.

At least two difficulties arise when attempting to use nonadiabatic models of the atmosphere. First, the mechanism of the genesis of heat sources and sinks in the atmosphere is still not too clear. The established concepts of the dependence of eddy and radiative heat transfer on meteorological characteristics of the atmosphere require further refinement. Second, accounting for heat fluxes markedly increases the number of elements which are forecast (moisture content, cloudiness, precipitation, etc.). The number of initial equations accordingly increases and the problem becomes more complicated and cumbersome.

A number of nonadiabatic forecasting schemes have lately been developed in the Soviet Union. Of these one should mention first the work by Blinova /3, 5/ which served as a basis for quite a number of studies. In 1961 she suggested a spatial circulation scheme with allowance for eddy and radiative heat influx. The five functions, viz., velocity potential, flux, vertical velocity, and deviations of the air temperature and geopotential from their standard values, are determined from eddy transport equations, the generalized balance equation at the sphere, heat influx, statics, and continuity. Reference /5/, in addition to formulating the problem proper, also gives a method for the numerical integration of the initial system of equations. Unfortunately, the radiation part of the Blinova model is limited by the assumption of nonselective absorption in the atmosphere. In addition, processes associated with the atmospheric transfer of moisture and the formation of cloudiness are not considered, although there is no doubt that they contribute to the formation of radiative heat fluxes. This same approach to the description of radiative processes was used in Purganskii's forecasting scheme for the lower layers of the atmosphere /19/. His calculations show that the radiative heat influx contributes 1 — 1.5 mb/day to the surface pressure forecast. Allowance for nonadiabatic factors usually changes the pressure forecast by 15 — 20%. A more refined scheme for taking radiative factors into account was suggested by Mashkovich /14, 15/. He solved the nonlinear problem of developing large-scale processes for an arbitrary initial distribution of meteorological elements and time variable influx of solar energy by incorporating the equation of radiative energy transfer and the moisture redistribution equation in the system of forecasting equations. In addition to radiative and eddy heat influx, allowance is made for the influx of the heat of condensation, with the quantity of condensed moisture determined in the course of the forecast. One of the versions of this scheme allows for selection of long-wave radiation absorption using the method of Shekhter. Numerical experiments using the above scheme show that for an initially steady zonal circulation the inception of perturbances in the temperature field is due to nonadiabatic factors (eddy and radiative heat influxes) with circulation having an effect only after several hours.

The most complete of all the existing schemes, in the sense of variety of energy forms incorporated and the generality of equations describing the dynamics of atmospheric processes, is the theoretical model suggested recently by Marchuk /11/. He attempts to incorporate radiative processes occurring in an atmosphere containing a cloud system. Although this scheme also makes no allowance for selectivity of absorption, the effect of cloudiness may be incorporated to some extent by selecting different absorption and scattering coefficients for water, ice, and water vapor. The radiative transfer equations are integrated with boundary conditions which depend on the nature of the underlying surface. The boundary conditions are based on the physically correct assumption that in the case of land all the radiation incident at the underlying surface is returned to the atmosphere after modification. For the sea surface the departing long-wave radiation is completely determined by the self-radiation of the water surface, whose temperature is regarded as constant during the entire forecasting period.

The above theoretical models of the atmosphere require knowledge of the vertical heat-influx distribution. However, attempts to calculate this distribution for a real atmosphere meet a number of difficulties associated with the lack of systematic data on the vertical profile of absorption coefficients for long- and short-wave radiation, water content, and other features of the cloudiness, its boundaries, etc., as well as difficulties due to the fact that the influx of heat at a thin layer of the atmosphere consists of a small difference of large quantities and the error in calculating it is quite high. Methods which would circumvent the above difficulties are obviously needed.

Kibel' /8/ suggested using the so-called thermal boundary layers, where the effect of nonadiabatic factors is most strongly felt. According to his calculations these layers are the surface air layer several hundred meters high and the boundary layer in the stratosphere. By isolating the thermal boundary layers it is possible to solve the adiabatic problem as before for the entire atmosphere, writing the boundary conditions at the atmospheric boundaries in such a form that they incorporate corrections for nonadiabatic processes inside the boundary layer. Studying the Green's functions in the solution obtained in the above manner, Nemchinov /18/ showed that it is most important to make allowance for air mass modification when forecasting the surface baric field. The idea of Kibel' was also used by Mashkovich /15/ to construct a forecasting scheme making allowance for the effect of the heat influx on the development of large-scale atmospheric processes.

Another method, suggested by Shvets /24/, consists in using the magnitude of heat influx averaged over the entire thickness of the atmosphere when constructing the forecasting scheme. This quantity is calculated more exactly and, what is most important, with consideration of cloudiness. The initial system of equations is composed of:

1. Vorticity equation with the turbulence friction forces included:

$$\frac{d(\Omega+l)}{dt}-\frac{g}{p_0^2}\frac{\partial}{\partial\xi}\left(k\rho^2\frac{\partial\Omega}{\partial\xi}\right)=l\frac{\partial\omega}{\partial\xi}, \tag{11}$$

where Ω is the vertical component of the vorticity; l is the Coriolis parameter; g is the acceleration of gravity; p_0 is the standard pressure at sea level; p is the pressure; $\xi = p/p_0$; ρ is the air density; k is the eddy viscosity coefficient; $\omega = d\xi/dt$; t is the time.

2. Heat influx equation:

$$\frac{dT}{dt}-\frac{RT(\gamma_a-\gamma)}{g\xi}\omega=-\frac{g}{p_0}\frac{\partial\theta}{\partial\xi}+\frac{Lr}{c_p}, \tag{12}$$

where T is the air temperature; R is the gas constant; γ and γ_a are the actual and dry adiabatic lapse rates, respectively; c_p is the specific heat of air at constant pressure; L is the latent heat of condensation; r is the amount of vapor condensing per unit air mass per unit time. The first term on the right-hand side of this equation describes the heat influx to unit air mass by eddy and radiative heat exchange.

3. Equation for the dew point τ:

$$\frac{d\tau}{dt} - \frac{RT(\gamma_{a\tau} - \gamma_\tau)}{g\xi}\omega = \frac{g}{p_0}k_2\frac{\partial}{\partial\xi}\left(\rho D\frac{\partial q}{\partial z}\right) - k_2 r,$$

$$r = \begin{cases} -\dfrac{64}{p_0\left(\dfrac{L}{c_p} + k_2\right)}\dfrac{\omega}{\xi} & \text{for } T - \tau \neq 0 \\ & \text{for } T - \tau = 0, \end{cases} \tag{13}$$

where γ_τ and $\gamma_{a\tau}$ are the vertical dew-point and the wet adiabatic lapse rates; q is the specific moisture content; D is the eddy exchange coefficient; k_2 is a temperature-dependent factor. The above equation assumes that condensation occurs only when the air temperature is equal to the dew point. The problem is solved in the quasistatic and quasigeostrophic approximations.

If the initial equations are integrated vertically over some atmospheric layer, then instead of the heat influx the equations will contain the difference between the fluxes at the boundaries of the layer. On the other hand, by statistically analyzing results of aerological sounding it is seen that the fields of the horizontal temperature gradients at different atmospheric levels are approximately similar. This empirically established fact of approximate "thermotropicity" of the atmosphere makes it possible to express the vertically-integrated functions by values of meteorological elements, usually ascribed to some isobaric surface inside the layer, averaged over the layer. If the heat fluxes at the boundaries of the atmosphere are now related to these average values of the temperature $\overline{T}$, dew point $\overline{\tau}$, altitude $\overline{H}$, and circulation velocity components $\overline{u}$, $\overline{v}$, and $\overline{\omega}$ (which is much easier than determining the vertical heat-influx distribution), then the system of forecasting equations is complete.

The above method was tested for two particular problems, which are part of the more general problem. The diagnostic problem consisted in determining the evolution of the temperature, geopotential, and velocity fields, produced by the nonadiabatic properties of the processes being considered, by solving the system of equations

$$\Delta\frac{\partial\overline{H}}{\partial t} + \left(\overline{H}, \frac{g}{l}\Delta\overline{H} + l\right) + m_1\left(\overline{T}, \frac{g}{l}\Delta\overline{T}\right) + k^*\left(\Delta\overline{H} - \frac{R}{g}\Delta\overline{T}\right) = 0;$$

$$(\Delta - \mu^2)\frac{\partial T}{\partial t} - \frac{g\mu^2}{l}(\overline{H}, \overline{T}) - m_2\left(\overline{T}, \frac{g}{l}\Delta\overline{T}\right) + \left(\overline{T}, \frac{g}{l}\Delta\overline{H} + l\right) +$$
$$+ \left(\overline{H}, \frac{g}{l}\Delta\overline{T}\right) - m_3 k^*\left(\Delta\overline{H} - \frac{R}{g}\Delta\overline{T}\right) = \frac{\mu^2 g}{c_p p_0}(Q_0 - Q_1 + Lr);$$

$$(a, b) = \frac{\partial a}{\partial x}\frac{\partial b}{\partial y} - \frac{\partial a}{\partial y}\frac{\partial b}{\partial x};$$

$$\Delta = \frac{\partial^2}{\partial x^2} + \frac{\partial^2}{\partial y^2}; \quad (\overline{\ \ }) = \int_0^1 (\ )\, d\xi$$

(m_1, k^*, m_2, m_3 and μ are constant coefficients). Then $Q_0 - Q + Lr$, the heat influx to the atmosphere, is calculated as a function of coordinates and time from observations over synoptic intervals.

In the other problem, which was solved by Meleshko /16/, the temperature, moisture content, and cloudiness in the lower layer of the atmosphere are forecast using the system of equations

$$\begin{gathered}\overline{\frac{\partial T}{\partial t}} + \frac{g}{l}(\overline{H}, \overline{T}) = W_2 + \frac{g}{p_0(\xi_1 - \xi')}(Q' - Q_1) + \frac{L}{c_p}\bar{r};\\ \frac{\partial \tau}{\partial t} + \frac{g}{l}(\overline{H}, \bar{\tau}) - k_1(\overline{T}, \bar{\tau}) = W_1 - \frac{k_2 g}{p_0(\xi_1 - \xi')}E_1 - k_2\bar{r};\\ W_i = \frac{1}{\xi_1 - \xi'}\int_{\xi'}^{\xi} \lambda_i \frac{\omega}{\xi}\, d\xi;\ (\bar{\ }) = \frac{1}{\xi_1 - \xi}\int_{\xi'}^{\xi_1} (\ )\, d\xi;\\ E_1 = \rho D \frac{\partial q}{\partial z}\Big|_{\xi=1}.\end{gathered} \tag{14}$$

Here the components of wind velocity and the vertical velocity are specified for the entire forecasting interval, while the magnitude of the heat influx to the layer and the cloudiness are calculated during the forecast using data obtained from the preceding time interval. The cloudiness is forecast using a correlation function between the cloudiness, dew-point deficit, and vertical velocity.

As a first approximation the values of u, v, and ω in this scheme can be taken from any given adiabatic scheme, since the effect of nonadiabatic factors on the circulation elements is relatively small.

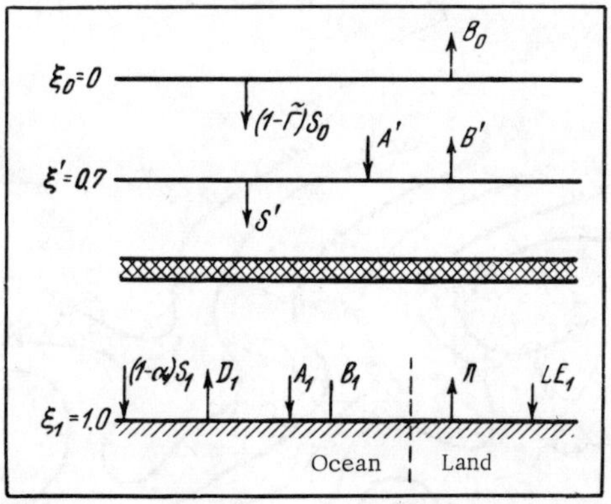

FIGURE 2. Scheme of the atmospheric heat balance.

In both problems the amount of heat supplied to the atmosphere is determined from the heat fluxes at the layer boundaries which are schematically depicted in Figure 2. Here the heat fluxes at the upper boundary of the layer reduce to determining the radiation-balance components (A' and B' are long-wave radiation fluxes, S' is the short-wave radiation flux), since eddy heat fluxes are small here. The radiation balance at the upper boundary of the atmosphere is, as known, expressed by the formula

$$Q_0 = S_0(1 - \Gamma) - B_0. \tag{15}$$

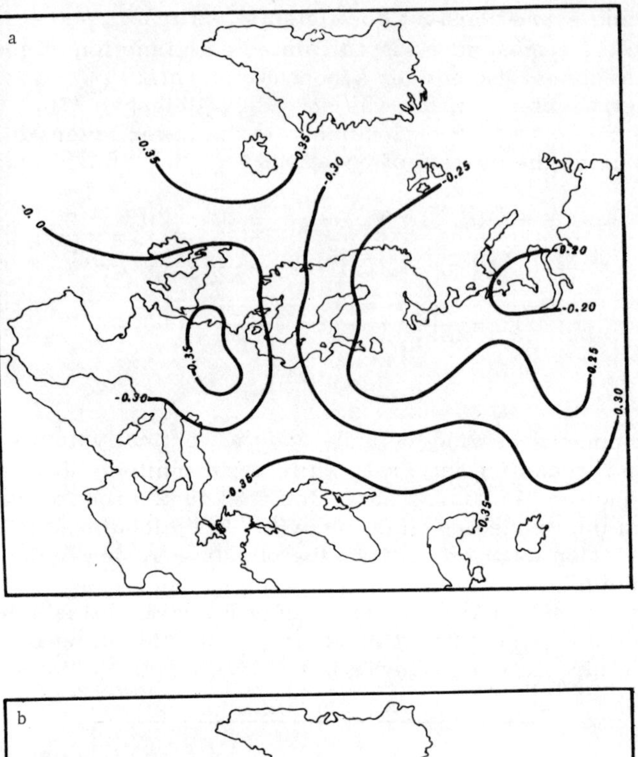

FIGURE 3. Radiation balance of the "earth-atmosphere" system in 1960: a—0300 hrs, Jan. 15; b—1500 hrs, Jan. 15; c—0300 hrs, Jan. 16.

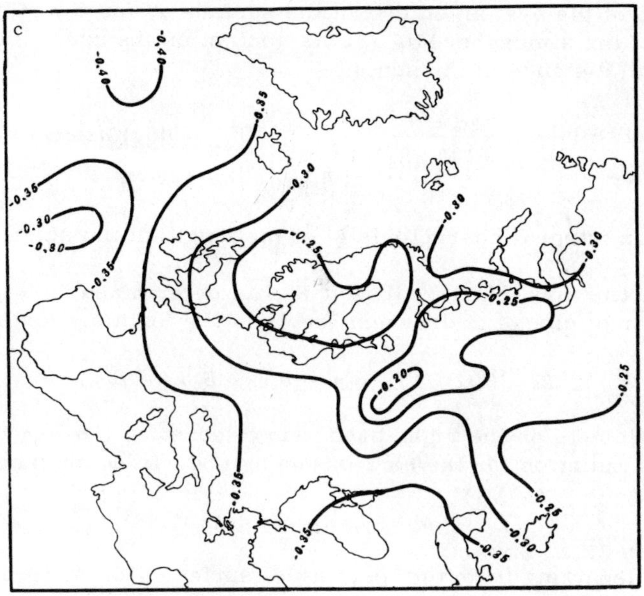

FIGURE 3, c.

The solar radiation flux S_0 passing through unit horizontal area at the upper boundary of the atmosphere depends only on astronomic factors and is related to the solar constant I_0 and the zenith distance v of the sun by the expression

$$S_0 = I_0 \cos v.$$

The albedo Γ of the "earth-atmosphere" system was calculated with allowance for anisotropic scattering and absorption of short-wave radiation in the atmosphere as a linear function of the amount of clouds at different levels $(\tilde{n}_1, \tilde{n}_d, \tilde{n}_{cp}, \tilde{n}_u)$. The cloud albedo was regarded as a specified quantity (use was made of average estimates of albedo for different cloud forms), where n_1 is the amount of lower-level clouds as a fraction of unity, visible from the top; n_d is the amount of vertical-development clouds, as a fraction of unity, visible from the top; n_m is the amount of middle-level clouds, as a fraction of unity, visible from the top; and n_u is the amount of upper-level clouds, as a fraction of unity, visible from the top:

$$\Gamma = \Gamma_0 (1 - N) + \Gamma_1 \tilde{n}_1 + \Gamma_d \tilde{n}_d + \Gamma_m \tilde{n}_m + \Gamma_u \tilde{n}_u. \qquad (16)$$

In the case of clear skies (Γ_0) the fluxes of diffuse short-wave radiation at the upper boundary of the atmosphere were determined by the Sobolev method using corrections for the absorption of direct and scattered radiation in the close infrared spectral region. Allowance was made for the dependence of the albedo of the system on that of the surface α, for the mean optical properties of the atmosphere (τ_0, x_1), its content of absorbing substances (u), and the zenith distance of the sun v:

$$\Gamma_0 = 1 - \frac{R_0(\tau_0, v)(1-\alpha)}{2 + 0.5\,\tau_0(3 - x_1)(1-\alpha)} - \{(0.095 + 0.015\alpha)(u \sec v)^{0.303} + \\ + 0.125\,[1 - (0.95)^{\sec v}]\}(\alpha + a), \tag{17}$$

where x_1 is an integral characteristic of the length of the scattering indicatrix.

The departing long-wave radiation B_0 was determined as a linear function of the number of clouds of different levels in the form suggested by Berlyand:

$$B_0 = B_0^0(1 - N) + B_0^{1}\tilde{n}_1 + B_0^{d}\tilde{n}_d + B_0^{m}\tilde{n}_m + B_0^{u}\tilde{n}_u. \tag{18}$$

The coefficients on the right-hand side of equation (18) represent fluxes of departing radiation for the corresponding underlying surfaces:

$$N = \tilde{n}_1 + \tilde{n}_d + \tilde{n}_m + \tilde{n}_u.$$

The flux departing from the terrestrial surface, for example, is

$$B_0^0 = \delta E(T_1)\,p(v) + \int_v^0 E(T)\,dp(u) - (1-\delta)\int_v^0 E(T)\,dp(2v-u), \tag{19}$$

where $u = 1/g_0 \int_0^\xi q(\xi)\sqrt{\xi}\,d\xi$; δ is the relative emissivity of the underlying surface.

Formula (19) uses the transfer function due to Niilisk, calculated for a "gray" atmosphere, where the total ozone content is constant and the vertical distribution of water vapor and carbon dioxide is exponential.

The remaining heat fluxes in equation (18) were calculated similarly.

Figure 3 shows the variation, in 12-hour intervals, of the radiation balance of the "earth-atmosphere" system calculated from the above formulas on the basis of aerological sounding data. The isolines were drawn at intervals of 0.05 kcal/cm$^2\cdot$ min. Similar formulas were used by Meleshko for calculating long- and short-wave radiation fluxes at the upper boundary of the 700—1000 mb layer, the only difference being that he made an allowance for the difference between the ascending and descending radiation fluxes, while going into less detail about cloud forms.

The heat fluxes from the terrestrial surface to the atmosphere were most satisfactorily calculated from the formulas for the land and ocean:

$$Q_1 = \begin{cases} (1-\alpha)S_1 - F_1 + D_1 & \text{for the ocean,} \\ \Pi - LE_1 & \text{for the land.} \end{cases} \tag{20}$$

The total radiation S_1, effective radiation F_1, eddy heat fluxes D_1, and moisture E_1 were calculated by the tried and extensively used methods suggested by Shekhter, Budyko, Monin, and Obukhov:

$$\begin{aligned} S_1 &= \frac{I_0 \cos v}{1+f\cos v}(1-c_1 n_1 - c_m n_m - c_в \Pi_u); \\ F_1 &= (B-A)_1 = F_0(1-N) + F_1 n_1 + F_m n_m + F_в n_u; \\ F_0 &= \delta\left[E(T_1) + \int_0^M E(T)\,dP(m)\right]; \\ m &= \frac{1}{g}\int_{\xi}^{\xi_1} q(\xi)\,d\xi; \\ D_1 &= a_1 V_1 (T_w - T_1); \\ E &= \begin{cases} a_2 V_1 (q_w - q_1) & \text{for } q_w - q_1 > 0; \\ 0 & \text{for } q_w - q_1 \leqslant 0, \end{cases} \end{aligned} \tag{21}$$

where a_1 and a_2 are constants; T_w and q_w are, respectively, the ocean-surface temperature and the saturated-air moisture content at this temperature. The value of Π, heat fluxes to the soil, can be found from the integral equation

$$\Pi - D_1\left[\left(T_1 - \frac{1}{\sqrt{\pi c\rho\lambda}}\int_{-\infty}^{t}\frac{\pi\,d\xi}{\sqrt{t-\xi}}\right)V_1\right] = R_1 \tag{22}$$

for a specified radiation balance R_1 of the surface, temperature T_1, wind speed V_1 at the surface, and thermophysical properties of the soil c, ρ, and λ.

Both the above schemes were solved on a computer. The temperatures and altitudes of the 500 mb surface were calculated according to the first scheme (6 cases) and the average temperature and dew point for the 700—1000 mb layers and the cloudiness were calculated using the second scheme (4 cases). By analyzing the results one sees that incorporation of heat influxes introduces substantial corrections to the temperature, moisture content, and cloudiness forecasts. Thus, the relative error of middle-level temperature forecasting decreases by 13% (in individual cases by 21%). Allowance for nonadiabatic factors improves the reliability of temperature and dew-point forecasting for the lower atmosphere level. On the average the correlation factor is improved by 3—7%, while in individual cases this improvement may be as high as 15%. Figure 4 shows the distribution of correction to the 24-hour forecasting of temperature $\bar{T}$ and of the dew point $\bar{\tau}$ for the bottom layer of the atmosphere due to heat influxes. The isolines were drawn for each 0.5°. It is found that above the ocean in winter the lower layer of the atmosphere is always heated by 1—2°, while above land this layer is cooled by approximately the same amount. This cooling and heating depends to a large extent on the cloudiness. For example, above the ocean this layer of the atmosphere is heated most markedly in regions with appreciable cloudiness during the entire forecasting interval. Most pronounced heating of the atmosphere is observed near the European continent and at the ice limit as a result of greater contrasts between air and ocean temperatures.

Figure 5 shows the 24-hour cloudiness forecast made from a non-adiabatic scheme and the synoptic chart for the same period. Estimates show that when heat and moisture influx are considered, the "cloudy" forecast for an

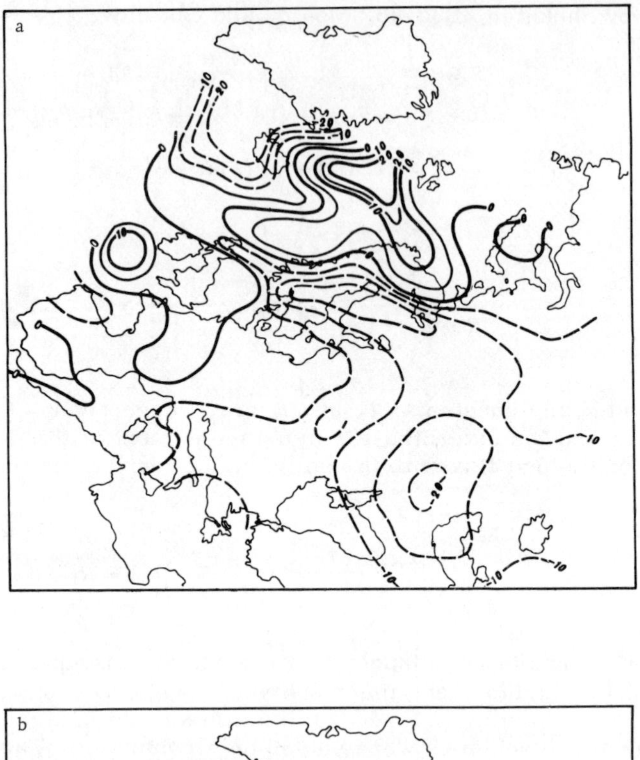

FIGURE 4. Corrections to the 24-hour forecast due to nonadiabatic properties of processes:

a—temperature T; b—dew point τ.

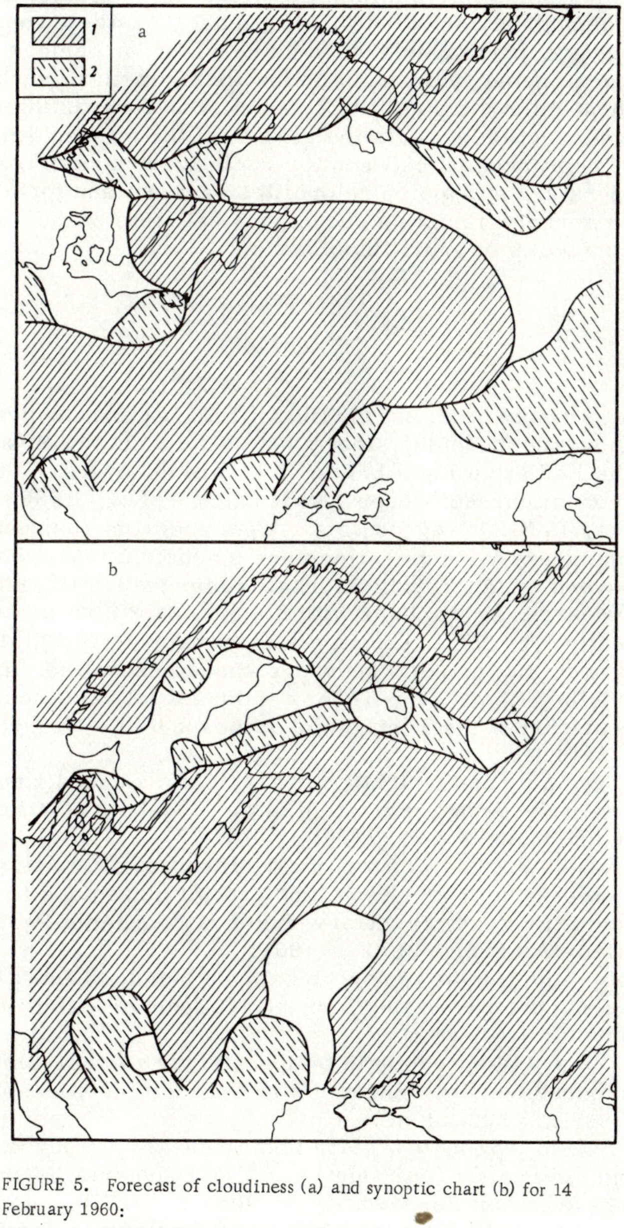

FIGURE 5. Forecast of cloudiness (a) and synoptic chart (b) for 14 February 1960:

1—continuous clouds; 2—partly cloudy.

area is improved by 5—12%, while the "precipitation" forecast is improved by, on the average, 12%.

Calculations show that nonadiabatic processes in the atmosphere (including radiative processes) have a marked effect on the temperature, moisture content, cloudiness, and precipitation forecasts and their consideration is most necessary, particularly when forecasting for periods of more than one day. Modern numerical weather forecasting methods make it possible to allow for heat influx as a result of radiation in forecasting schemes. Hence, the construction of radiation models of the atmosphere which would yield reliable relationships between the radiative heat influxes and the vertical temperature and moisture profiles, amount and form of cloudiness, as well as other characteristics of the physical state of the atmosphere and terrestrial surface is of great importance.

References

1. Ambartsumyan, V. A. O vliyanii linii pogloshcheniya na luchevoe ravnovesie vneshnikh sloev zvezd (The Effect of Absorption Lines on the Radiative Equilibrium of External Layers of Stars). — Trudy Astronomicheskoi Observatorii LGU, Vol. 6. 1936.
2. Blinova, E. N. K voprosu o srednem godovom raspredelenii temperatury v zemnoi atmosfere s uchetom materikov i okeanov (Concerning the Yearly Average Temperature Distribution in the Terrestrial Atmosphere with Allowance for Continents and Oceans). — Izvestiya AN SSSR, seriya geofizicheskaya, 11(1). 1947.
3. Blinova, E. N. K teorii godovogo khoda nezonal'noi tsirkulyatsii atmosfery Zemli (The Theory of the Annual Variation in the Nonzonal Circulation of the Terrestrial Atmosphere). — Trudy IFA AN SSSR, No. 2. 1958.
4. Blinova, E. N. and G. I. Marchuk. Godovoi khod zonal'noi tsirkulyatsii (The Annual Variation in the Zonal Circulation). — Trudy IFA AN SSSR, No. 2. 1958.
5. Blinova, E. N. Gidrodinamicheskoe issledovanie nestatsionarnykh atmosfernykh protsessov planetarnogo masshtaba (A Hydrodynamic Study of Unsteady Planetary-Scale Atmospheric Processes). — Doklady AN SSSR, 140(2). 1961.
6. Budyko, M. I. Atlas teplovogo balansa zemnogo shara (An Atlas of the Heat Balance of the Terrestrial Sphere). — Gidrometeoizdat. 1963.
7. Kibel', I. A. Raspredelénie temperatury v zemnoi atmosfere (The Temperature Distribution in the Terrestrial Atmosphere). — Doklady AN SSSR, 39(1). 1943.
8. Kibel', I. A. Vvedenie v gidrodinamicheskie metody kratkosrochnogo prognoza pogody (Introduction to Hydrodynamic Methods of Short-Range Weather Forecasting). — Moskva, GITTL. 1957.
9. Kuznetsov, E. S. Raspredelenie temperatury po vertikali pri luchistom ravnovesii (Vertical Temperature Distribution Under Radiative Equilibrium Conditions). — Trudy Instituta Teoreticheskoi Geofiziki, Vol. 1. 1946.

10. Lebedinskii, A. I. Luchevoe ravnovesie zemnoi atmosfery (Radiative Equilibrium of the Terrestrial Atmosphere). — Uchenye Zapiski LGU, Vol. 31, seriya matematicheskaya, No. 63. 1939.
11. Marchuk, G. I. Teoreticheskaya model' prognoza pogody maloi zablagovremennosti (A Theoretical Model for Short-Range Weather Forecasting). — Izvestiya AN SSSR, seriya geofizicheskaya, No. 5. 1964.
12. Mashkovich, S. A. Raspredelenie temperatury po vysote na raznykh shirotakh severnogo i yuzhnogo polusharii (The Vertical Temperature Distribution at Different Latitudes of the Northern and Southern Hemispheres). — Trudy TsIP, No. 21. 1950.
13. Mashkovich, S. A. O sezonnykh izmeneniyakh struinykh sechenii (Seasonal Variations of Jet Streams). — Meteorologiya i gidrologiya, No. 6. 1956.
14. Mashkovich, S. A. Teoreticheskaya model' dlya izucheniya formirovaniya obshchei tsirkulyatsii atmosfery i klimaticheskikh polei meteoelementov (A Theoretical Model for the Study of the Formation of the General Circulation of the Atmosphere and of Climatic Fields of Meteorological Elements). — Trudy TsIP, No. 78. 1958.
15. Mashkovich, S. A. K uchetu neadiabaticheskikh vozdeistvii pri chislennom kratkosrochnom prognoze davleniya (The Incorporation of Nonadiabatic Effects in Short-Range Numerical Pressure Forecasting). — Trudy TsIP, No. 126. 1963.
16. Meleshko, V. P. K voprosu o chislennom prognoze oblachnosti i osadkov s uchetom transformatsii vozdushnykh mass (Numerical Forecasting of Cloudiness and Precipitation with Allowance for Modification of Air Masses). — Trudy GGO, No. 151. 1964.
17. Milankovich, M. Matematicheskaya klimatologicheskaya i astronomicheskaya teoriya kolebanii klimata (The Mathematical, Climatological, and Astronomical Theory of Climate Variations). — GONTI. 1939.
18. Nemchinov, S. V. Ob uchete transformatsii vozdushnykh mass pri kratkosrochnom prognoze nezonal'nykh otklonenii vysot absolyutnoi topografii izobaricheskoi poverkhnosti (Allowing for Modification of Air Masses in Short-Range Forecasting of Nonzonal Deviations of the Altitudes of the Absolute Topography of Isobaric Surfaces). — Izvestiya AN SSSR, seriya geofizicheskaya, No. 5. 1959.
19. Purganskii, N. S. K uchetu vliyaniya neadiabaticheskikh faktorov pri kratkosrochnom prognoze polei geopotentsiala na nizhnikh urovnyakh atmosfery (Allowing for the Effect of Nonadiabatic Factors in Short-Range Forecasting of Geopotential Fields at Lower Atmosphere Levels). — Trudy TsIP, No. 126. 1963.
20. Rakipova, L. R. Teplovoi rezhim atmosfery (The Thermal Regime of the Atmosphere). — Gidrometeoizdat. 1957.
21. Feigel'son, E. M. Raspredelenie temperatury zemnoi atmosfery po vysote (Vertical Distribution of the Terrestrial Atmosphere Temperature). — Izvestiya AN SSSR, seriya geograficheskaya i geofizicheskaya, 14(4). 1950.
22. Feigel'son, E. M. Radiatsionnye protsessy v sloistoobraznykh oblakakh (Light and Heat Radiation in Stratus Clouds). — Izdatel'stvo Nauka. 1964.

23. Fridman, A. A. Raspredelenie temperatury vozdukha s vysotoi (Vertical Air-Temperature Distribution). — Geofizicheskii sbornik, No. 1. 1914.
24. Shvets, M. E. and B. E. Shneerov. K uchetu rezul'tatov radiatsionnykh izmerenii so sputnikov v neadiabaticheskoi modeli atmosfernykh dvizhenii (The Incorporation of Radiation Measurements from Satellites in the Nonadiabatic Model of Atmospheric Motions). — Doklady AN SSSR, 152(3). 1963.
25. Shekhter, F. N. K vichisleniyu luchistykh potokov tepla v atmosfere (Calculation of Radiative Heat Fluxes in the Atmosphere). — Trudy GGO, Vol. 22. 1950.
26. Shlyakhov, V. I. Issledovanie balansa dlinnovolnovoi radiatsii v troposfere (A Study of the Long-Wave Radiation Balance in the Troposphere). — Gidrometeoizdat. 1956.
27. Defant, F. Das mittlere meridionale Temperaturprofil in der Troposphäre als Effekt von vertikalen und horizontalen Austauschvorgängen und Kondensationswärme. — Arch. Meteorol., Geophys. und Bioklimatol, Ser. A, 2(2—3). 1950.
28. Goody, R. The Thermal Equilibrium at the Tropopause and the Temperature of the Lower Stratosphere. — Proc. Roy. Soc. A, 197(487). 1949.
29. Hess, S. and R. M. A. Frank. Theory of the Temporal and Latitudinal Distribution of Temperature. — J. Meteorol., 10(2). 1953.
30. Linke, F. and F. Möller. Die dunklen Strahlungsströme in der Atmosphäre und die Strahlungsbilanz. — Handbuch der Geophysik. 8(3):12. 1943.
31. London, J. Study of Atmospheric Heat Balance. — New York Univ., Final Rept. Contract, N AF 19. 1957.
32. Normal Weather Maps, Northern Hemisphere Upper Level. Washington. 1945.

M. S. Malkevich

DETERMINATION OF METEOROLOGICAL ELEMENTS FROM SATELLITES

Methods are considered for determining the variation in meteorological elements from measured terrestrial radiation in different spectral regions. Statistical characteristics of the temperature and moisture fields are used as supplementary information, allowing one to solve inverse problems. Examples of the determination of temperature and moisture profiles are presented.

Meteorological satellites supply information on the global distribution of the cloud cover, radiation balance, and certain meteorological elements (temperature, humidity, pressure). All this information is contained in the characteristics of the radiation field of the earth measured directly by satellites with different spectral, angular, and spatial resolution. Hence one of the major tasks of satellite meteorology is the interpretation of radiation measurements, i. e., the extraction from them of the required meteorological parameters.

The main difficulties in this task are due to the following.

1. All the parameters of interest vary in time and space, the variations of interest in atmospheric physics having a random nature. This requires the application of statistical methods in order to obtain objective characteristics of fields of meteorological elements and of cloudiness.

2. Interpretation of radiation measurements reduces in a number of cases to solving inverse problems, the solutions of which are incorrect from a mathematical point of view. This means that even small measurement errors may result in large errors when determining the required parameters.

3. The determination of the radiation balance from satellites is made difficult by the nonuniform nature of the underlying surface and cloud cover, which appear in the field of view of radiometers, as well as the anisotropicity of the radiation field of the earth. This reduces the possibility of obtaining reliable time and space variations of the radiation balance with the accuracy needed for atmospheric circulation problems.

However, the terrestrial atmosphere is studied from many aspects and a large number of its parameters are being measured, somewhat compensating for the above difficulty.

The present paper considers some methods for the determination of the variation of meteorological parameters from satellite radiation measurements. Specific examples are also presented on the utilization of supplementary information on fields of meteorological elements, based on statistical analysis of radiosonde data.

DETERMINATION OF THE UNDERLYING SURFACE TEMPERATURE

The determination of the underlying surface temperature $\boldsymbol{T}(\boldsymbol{p_0})$ from measurements of terrestrial radiation intensity in the $8-12\ \mu$ transparent window (examined in /1, 6, 17, 19/) requires the use of additional information on the vertical temperature $T(p)$, humidity $W(p)$ and ozone $m(p)$ distributions, since the atmosphere in this range may appreciably modify the surface emission. The intensity I of the departing radiation is related to $T(p_0)$ and the above atmospheric parameters by the expression

$$I = B[T(p_0)]\,\tau(w^*, m^*) - \int_0^1 B[T(p)]\,d\tau[W(p), m(p)], \tag{1}$$

where τ is the atmosphere transfer function in the spectral region at hand; B is the integral of the Planck function in this region; w^* and m^* are the water-vapor and ozone masses in the entire column of the atmosphere.

It is suggested in /19/ that the absorption and emission of the atmosphere be taken into account on the basis of I calculated for many models of the atmosphere. This allows one to calculate the correction for modification of the radiation by the surface. Since this correction is highly variable, the use of its average value may yield large errors in $T(p_0)$. This correction cannot be incorporated in each measurement, since its variations depend on the variability in $T(p)$ and $W(p)$ which are not monitored.

The present author wishes to suggest the following method for incorporating the correction for atmospheric absorption and emission. We introduce the following transfer function for the atmosphere:

$$P = \frac{I}{B[T(p_0)]}, \tag{2}$$

which, as before, depends on the values of $T(p)$, $W(p)$ and $m(p)$; these latter functions vary much in time and over the globe. These variations can be incorporated in determining the transfer function by using information on the vertical structure of $T(p)$ and of the specific humidity $q(p)$, i.e., the average vertical profiles ("norms") $\overline{T}(p)$ and $\overline{q}(p)$, as well as correlation functions (or matrices for variations of T and q at discrete levels)*

$$B_T(p_k, p_l) = \overline{T'(p_k)\,T'(p_l)}; \tag{3}$$

$$B_q(p_k, p_l) = \overline{q'(p_k)\,q'(p_l)}; \tag{4}$$

$$B_{qT}(p_k, p_l) = \overline{q'(p_k)\,T'(p_l)}. \tag{5}$$

Here T' and q' denote deviations from the norm; the bar denotes averaging over all the obtained vertical profiles. Examples of correlation matrices (3)–(5) are listed in Table 1.

Using average profiles for the region where I was measured one easily obtains the transfer function $\overline{P}$ for these profiles, and, consequently, the

* As shown in /19/, the value of I is hardly sensitive to variations in $m(p)$.

first approximation $T_1(p_0)$ of the surface temperature and of its deviation from the norm $T_1'(p_0) = T_1(p_0) - \overline{T}(p_0)$. Using then an optimum extrapolation method (see, for example, /3/) one can reconstruct, in the first approximation, the deviation T' from the norm at any level p_k, for example

$$T_1'(p_k) = C_0^{(k)} T_1'(p_0), \tag{6}$$

where

$$C_0^{(k)} = B_T(p_0, p_k)/B_T(p_0, p_0). \tag{7}$$

The absolute error of this extrapolation is

$$\delta_T(p_k) = \left[B_T(p_k, p_k) - \frac{B_T^2(p_0, p_k)}{B_T(p_0, p_0)}\right]^{1/2}. \tag{8}$$

TABLE 1. Correlation matrices (3)—(5) for land (Bismarck [North Dakota] 46°50' N, 100°35' W)

p, mb \ p, mb	1000	850	700	500	400	300	200	100
				$B_T(p_k, p_l)$				
1000	27.7	23.4	18.4	10.0	10.5	9.0	—9.4	—8.3
850	23.4	22.2	17.9	9.4	10.3	9.0	—8.2	—6.5
700	18.4	17.9	21.2	9.1	10.7	10.1	—7.0	—7.3
500	10.0	9.4	9.1	7.5	7.8	6.4	—5.4	—3.9
400	10.5	10.3	10.7	7.8	9.4	8.0	—5.2	—4.5
300	9.0	9.0	10.1	6.4	8.0	8.2	—3.8	—4.5
200	—9.4	—8.2	—7.0	—5.4	—5.2	—3.8	17.7	0.6
100	—8.3	—6.5	—7.3	—3.9	—4.5	—4.5	0.6	8.5
				$B_q(p_k, p_l)$				
1000	4.0	2.5	1.0	0.4	0.1			
850	2.9	3.0	1.2	0.3	0.1			
700	1.0	1.2	2.4	0.3	0.1			
500	0.4	0.3	0.3	0.4	0.1			
400	0.1	0.1	0.1	0.1	0.1			
				$B_{qT}(p_k, p_l)$				
1000	4.9	1.0	2.1	0.4	0.1			
850	0.3	3.0	2.8	0.9	0.3			
700	1.3	2.1	3.0	1.0	0.3			
500	—0.5	1.1	1.1	1.7	0.2			
400	—0.4	1.4	1.0	0.7	0.3			
300	0.2	1.7	0.9	0.7	0.3			
200	0.2	—1.3	—2.1	—0.6	—0.2			
100	—0.4	—0.9	—0.2	—0.4	—0.1			

Deviation $q_1'(p_0)$ from the norm can be determined similarly by

$$q_1'(p_0) = \sum_{i=1}^{n} d_i^{(0)} T_1'(p_i), \tag{9}$$

where coefficient $d_i^{(0)}$ is determined from the system of equations

$$\sum_{j=0}^{n} B_T(p_i, p_j)\, d_j^{(0)} = B_{qT}(p_0, p_i) \qquad (10)$$

and

$$q_1'(p_k) = \tilde{d}_0^{(k)} q_1'(p_0); \qquad (11)$$

here

$$\tilde{d}_0^{(k)} = B_q(p_0, p_k)/B_q(p_0, p_0). \qquad (12)$$

The extrapolation errors are given by expressions similar to (8). Using the above profiles of $T_1(p)$ and $q_1(p)$ one may refine the transfer function and find $T_2(p)$, the second approximation. This makes it possible by using supplementary information on the vertical structure of the $T(p)$ and $q(p)$ fields not only to determine the surface temperature and to calculate the rms error, but also to reconstruct the optimum vertical profiles of $T(p)$ and $q(p)$. Examples of such calculations using structural features are shown in Tables 2 and 3.

TABLE 2. Determination of the surface temperature and optimum reconstruction of profiles of $T(p)$, in °K, and $q(p)$, in g/kg, for two soundings at Bismarck [North Dakota]

p, mb	$\bar{T}$	$\bar{q}$	7 July 1958				11 July 1960			
			T	q	T_1	q_1	T	q	T_1	q_1
1000	302.2	7.7	292.0	6.5	297.2	6.8	309.3	11.1	311.2	9.2
850	291.6	6.8	280.8	5.1	287.4	6.2	297.1	6.7	298.7	7.9
700	278.9	3.9	271.9	2.4	275.8	3.7	284.1	3.1	284.5	4.3
500	262.1	0.96	257.1	0.63	260.3	0.88	263.9	1.8	265.1	1.1
400	249.3	0.52	245		247.4		252.0		252.5	
300	234.7	0.10	229.2		233.1		236.0		237.4	
250	226.6		218.7		225.7		233.5		228.0	
200	220.7		220.4		222.4		225.7		217	
150	217.9		223.0		219.8		216.2		214	
100	216.2		221.2		217.7		210.6		214.4	

TABLE 3. Determination of surface temperature and optimum reconstruction of profiles of $T(p)$, in °K, and $q(p)$, in g/kg, from measurements of TIROS-II Satellite over Medford [Oregon] (42°23' N, 122°52' W) on 12 July 1961

p, mb	$\bar{T}$	$\bar{q}$	T	q	T_1	q_1
1000	296.5	6.7	290.2	8.4	300	7.5
850	291.2	6.0	297	6.2	296	6.8
700	280.7	3.7	285.8	4.5	284.5	4.0
500	263.8	1.1	264.5	1.6	265.8	1.2
300	235.6		237.7	0.2	237.4	
200	219.6		221.6		217.7	
100	212.5		208.7		210.8	

The first of these examples illustrates a method of optimum extrapolation as used to determine the vertical profiles of $T(p)$ and $q(p)$ over Bismarck, while the second illustrates the determination of $T(p_0)$ for real measurement of I from TIROS-II under deep inversion conditions. The latter case is the more difficult, since the temperature calculated in the above manner has to be referred to some level between 1000 and 700 mb. It is easy to see that similar difficulties will arise when the radiometer has clouds in its field of vision. From this follows the necessity of separating cases of vertical structure of fields of $T(p)$ and $q(p)$ at least for cloudy and cloudless skies. An item of importance is the latitudinal variation of correlation matrices observed by Boldyrev /1/, in particular the very weak correlation between q' and T' at low latitudes. As was pointed out in /1/, in this case one must use the spatial structure of the temperature and moisture fields, remembering that radio sounding at the meteorological and ocean station network is carried out simultaneously with satellite measurements.

The use of supplementary information does not entirely resolve the difficulties in determining the surface temperature, due both to the statistical nature of this information, and because reliable information is unavailable on the transfer function in the transparent window for a real atmosphere, as well as on the emissivity of the natural surface.

DETERMINATION OF THE VERTICAL TEMPERATURE PROFILE

Kaplan /15/ suggested that the vertical distribution $T(p)$ of the temperature of the atmosphere be determined from satellite measurements of the intensity I_ν of departing radiation in the $15\,\mu$ CO_2 pass band, related to $T(p)$ by the expression

$$I_\nu = B_\nu[T(p_0)]\,\tau_\nu(p_0) - \int_0^1 B_\nu[T(p)]\,d\tau_\nu(p), \qquad (13)$$

where $\tau_\nu(p)$ is the transfer function of the atmosphere, averaged over some frequency range $\Delta\nu$.

An appreciable difficulty in solving this problem, even when sufficiently reliable data are available on the spectral variation of I_ν as well as on the transfer in the atmosphere and on the emissivity of the terrestrial surface, is the error in the inverse problem for equations such as (13), i.e., the fact that it is possible to obtain solutions for small changes in I_ν and for differences as large as desired. This problem has as yet not been afforded sufficient attention in /20, 21/, where some methods for determining $T(p)$ for the stratosphere are put forward. In particular, Yamamoto /21/ suggested representing solutions of problems such as (13) as power series expansions or as expansions of orthogonal polynomials (Legendre and Chebyshev). At the Symposium on Radiative Processes in Leningrad (in August 1964) Wark suggested the use of trigonometric functions for this purpose. Examples of the determination of $T(p)$ from I_ν for four spectral regions /21/ show that first, second, and third order polynomials cannot be used for approximating real variations of $T(p)$ with sufficient accuracy at all the levels under study. On the other hand, the use of additional terms of the expansion (even a fourth term) makes the approximation poorer, since the error in the inverse problem is then exhibited (Figure 1). Methods for adjusting operators involving Fredholm equations of the first kind (see, for example, /12/) require as yet a large volume of initial information about the specified I_ν.

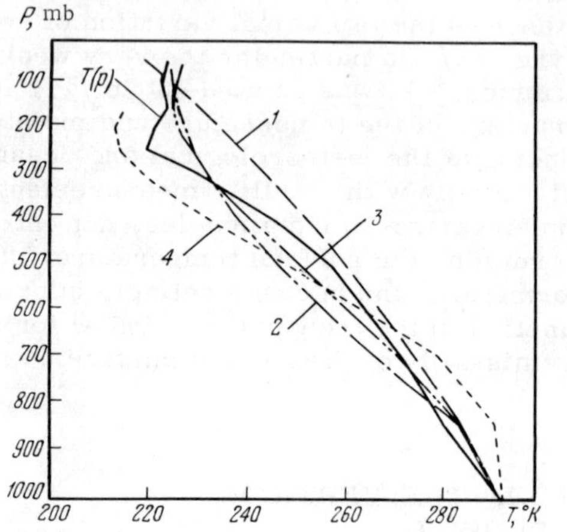

FIGURE 1. Reconstruction of the vertical temperature distribution by the Yamamoto method /21/ using polynomials of different orders

$$T_n(p) = \sum_{k=0}^{n} t_k p^k$$

1— $n=1$; 2— $n=2$; 3— $n=3$; 4— $n=4$.

In order to obtain a reliable solution of equation (13) one must carry out optimum approximation of $T(p)$. This is done by means of a system of statistical orthogonal functions, the rigorous theory of which was developed by Obukhov /9/. This theory can be applied to the determination of $T(p)$.* If $T(p)$ is expressed in the form

$$T(p) = \overline{T}(p) + T'(p), \tag{14}$$

where $T'(p)$ are random deviations from the norm $\overline{T}(p)$ $(T' \ll \overline{T})$, which is known for any region on the globe and for any time interval, or may be determined from aerological data, then equations (13) and (14) will yield an expression from which to determine $T'(p)$:

$$f_\nu = \frac{\partial B_\nu[\overline{T}(p_0)]}{\partial T}\tau_\nu(p_0)T'(p_0) - \int_0^1 \frac{\partial B_\nu[\overline{T}(p)]}{\partial T} T'(p)\,d\tau_\nu(p); \tag{15}$$

* The suggestion that an optimum parameter method be applied to the problem at hand was made by Tatarskii /7/.

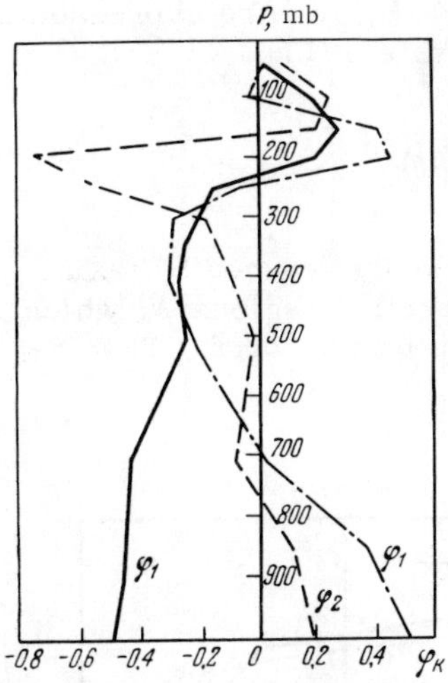

FIGURE 2. Examples of eigenvectors $\varphi_k\,(p_i)$ calculated from the correlation matrix $B_T\,(p_k,\,p_l)$ for Bismarck [North Dakota].

this latter expression is linear in $T'(p)$. Here $f_\nu = I_\nu - \bar{I}_\nu$, where $\bar{I}_\nu$ is the intensity of departing radiation for the average profile $\bar{T}(p)$. As was shown in /9/, an optimum approximation of a random variable, for example $T'(p)$, is obtained by a system of orthogonal functions, consisting of eigenfunctions corresponding to the correlation function, in the given case B_T; for a given T' at a discrete number of levels this will be a system of base vectors which are the eigenvectors of correlation matrix $B_T(p_k,\ p_l)$.

Examples of three eigenvectors $\varphi_k\,(p_i)$ of matrix $B_T(p_k,\ p_l)$ are shown in Figure 2. It is found that any application of $T'\,(p_i)$ can be approximated by 3 or 4 eigenvectors with an rms error not exceeding 10—15%. This error can be expressed in terms of the eigenvalues of a matrix, the former having the meaning of variances of the expansion coefficients. Thus, representing T' as the expansion

$$T'_n\,(p) = \sum_{k=1}^{n} t_k \varphi_k\,(p) \tag{16}$$

and substituting into equation (15) we derive a system of algebraic equations governing t_k:

$$\sum_{k=1}^{n} D_{\nu k} t_k = f_\nu \quad (\nu = 1, 2, \ldots, N \geqslant n), \tag{17}$$

whence

$$t_k = \sum_{\nu=k}^{n} D_{\nu k}^{-1} f_\nu,$$

where $D_{\nu k}^{-1}$ are elements of the inverse matrix of system (17). As was shown in /7/, the use of optimum systems does not eliminate the instability of the solution, due to errors in the inverse problem for equation (15). However, since these systems ensure the best approximation using a small number of terms of series (16), it becomes possible to solve equation (15) with sufficient accuracy (Figure 3). Here it is found that such a satisfactory reconstruction of vertical profiles for each of the cases under study (land, ocean) can be obtained only by means of an "extraneous" system of eigenvectors (curve 4 of Figure 3b).

A problem of importance is the effect of errors in measuring I_ν or f_ν on the error in determining $T'(p)$. It is easy to show /7/ that if the measurement errors are uncorrelated, then the amplification factor of the rms error (given by

$$K_T\,(p) = \delta T / \delta f,$$

where δT is the rms error in determining T'; $\delta f = \sqrt{\overline{f_\nu^2}}$) can be expressed in terms of elements $D_{\nu k}^{-1}$ of the inverse matrix of system (17):

$$K_T(p) = \left[\sum_{k=1}^{n}\sum_{l=1}^{n}\sum_{\nu=1}^{n} D_{\nu k}^{-1} D_{\nu l}^{-1} \varphi_k(p)\,\varphi_l(p)\right]^{1/2}. \tag{18}$$

The amplification factor of the error can be markedly reduced by taking initial information on I_ν for a large number of spectral regions N, which exceeds the number n of base vectors being taken into account. Then K_T is reduced by approximately a factor of $\sqrt{N/n}$.

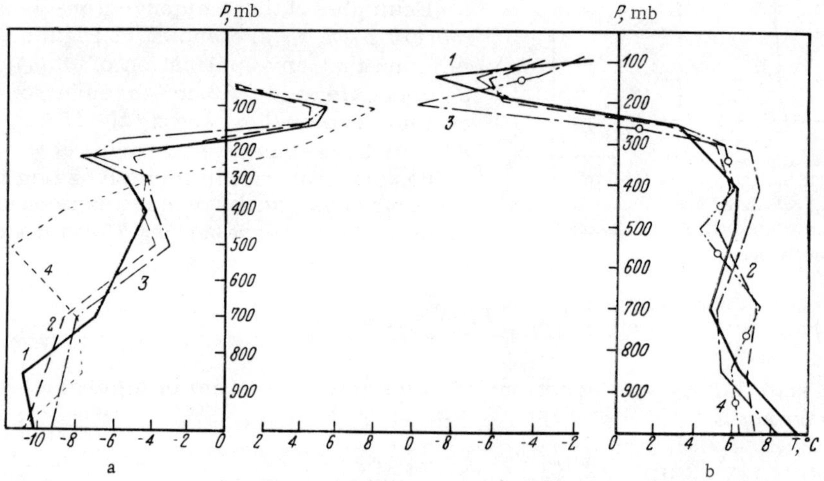

FIGURE 3. Reconstruction of the vertical temperature distribution:

a—by an optimum approximation method over land (Bismarck); b—optimum parameter method over ocean;

1—$T'(p)$; 2—$T_3'(p_i) = \sum_{k=1}^{3} \tilde{t}_k \varphi_k(p_i)$; $\tilde{t}_k = \sum_{l=1}^{10} T'(p_i)\varphi_k(p_i)$; 3 — $T_3'(p) = \sum_{k=1}^{3} t_k \varphi_k(p)$, t_k are determined from system (17); 4—solution of equation (13) by the Yamamoto method /9/. Curve 4 in Figure 3b is obtained from $\varphi_k(p_i)$ determined for land.

The above results pertain to the case when the radiation of water vapor, whose bands overlap the 15 μ CO_2 band, is neglected. At the same time, the incorporation of water vapor may be essential when determining $T(p)$. Hence the vertical temperature profile should be determined at the same time as the vertical moisture distribution from the departing radiation in the H_2O band (for example, 6.3 μ).

DETERMINATION OF THE VERTICAL DISTRIBUTION OF MOISTURE

The convenience in simultaneously determining $T(p)$ and $q(p)$ (as well as the O_3 concentration) from simultaneous measurements of I_ν in the 15 μ CO_2

and 6.3 μ H_2O bands (as well as in the 9.6 μ O_3 band) consists in the fact that determination of one of these parameters requires knowing the others (this approach was suggested by Kaplan /16/). In other words, having determined the first approximation of $T_1(p)$ from I_ν in the 15 μ CO_2 band without allowance for overlapping with the H_2O bands, it can then be used for determining $q(p)$ from the I_ν in the 6.3 μ H_2O band. When the latter is obtained, $T(p)$ can be refined with consideration of band overlapping, and then iteration can be applied.

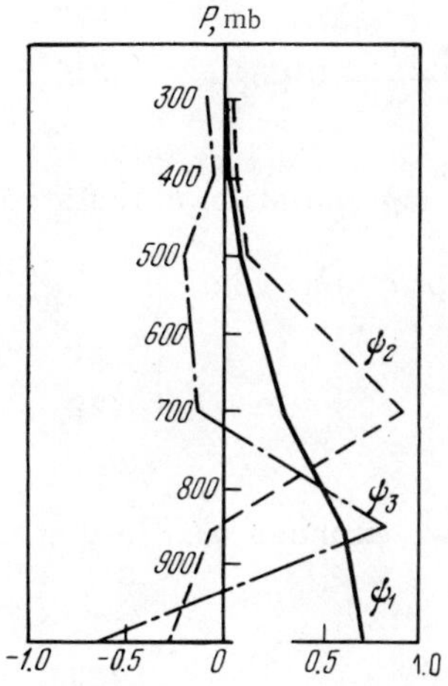

FIGURE 4. Examples of eigenvectors $\psi_k(p_i)$ calculated from the correlation matrix $B_q(p_k, p_l)$ for Bismarck.

The problem of determining $q(p)$ is similar to that of determining $T(p)$ with only one difference: equation (13), which can be used for this purpose, though with different ν, will be substantially nonlinear ($\tau_\nu(p)$ is a complicated function of $q(p)$ and has no analytical solution). In solving equation (13) by a mathematical method one can use the approximate expression for function $\tau_\nu(p)$ based on the Elsasser line shape:

$$\tau_\nu(x) = 1 - \Phi\left[\sqrt{\frac{l_\nu}{2} W_0 Q(x)}\right], \qquad (19)$$

where $x = p/p_0$; $W_0 = p_0 q_0/q$ is the water-vapor mass in an air column reduced to standard pressure p_0; l_ν is a generalized absorption coefficient;

$$\Phi(y) = \frac{2}{\sqrt{\pi}} \int_0^y e^{-t^2}\, dt \qquad (20)$$

is the error function;

$$Q(x) = \int_0^x t q(t)\, dt$$

is the reduced water-vapor mass.

Equation (13) is solved by an approximation method using empirical orthogonal functions obtained for the specific humidity. Suppose

$$q(x) = \bar{q}(x) + q'(x),$$

where $\bar{q}(x)$ is the average specific-humidity distribution over the given region and during the given time period; $q'(x)$ is the deviation of the profile from the average; and $q'(x)$ is regarded as a random variable. We can now construct a system of orthogonal functions (vectors), which ensure optimum parameters for any application of $q'(x)$. Examples of the first three vectors of such a system obtained from radiosonde data over Bismarck (in summer) are shown in Figure 4.

If it is assumed that the vertical temperature distribution $T(x)$ and the average distribution of the specific humidity $\bar{q}(x)$ are known, then it is possible to construct an equation for determining the value of $q'(x)$, which is small as compared with $\bar{q}(x)$:

$$f_\nu = -B_\nu[T(1)]\frac{\partial\Phi_\nu[\bar{Q}(1)]}{\partial Q}Q'(1) + \int_0^1 B_\nu[T(x)]\,d\left\{\frac{\partial\Phi_\nu[\bar{Q}(x)]}{\partial Q}Q'(x)\right\}, \tag{21}$$

where $f_\nu = I_\nu - \bar{I}_\nu$, $\bar{I}_\nu$ being the intensity of the departing radiation calculated from equation (13) for $\bar{q}(x)$; $\bar{Q}(x) = \int_0^x t\bar{q}(t)dt$, $Q'(x) = \int_0^x tq'(t)dt$ and

$$\frac{\partial\Phi_\nu[\bar{Q}(x)]}{\partial Q} = \frac{l_\nu W_0 \exp\left[-\frac{l_\nu}{2}W_0\bar{Q}(x)\right]}{\sqrt{2\pi l_\nu W_0\bar{Q}(x)}}. \tag{22}$$

To solve equation (21), which is linear in $q'(x)$, we express $q'(x)$ in the form

$$q'_n(x) = \sum_{k=1}^{n} q_k\psi_k(x). \tag{23}$$

Substituting equation (23) into (21), we obtain an algebraic system of equations for the coefficients:

$$\sum_{k=1}^{n}\Gamma_{\nu k}q_k = f_\nu \quad (\nu = 1, 2, \ldots, N \geqslant n), \tag{24}$$

where

$$\Gamma_{\nu k} = -B_\nu[T(1)]\frac{\partial\Phi_\nu[\bar{Q}(1)]}{\partial Q}g_k(1) + \int_0^1 B_\nu[T(x)]\,d\left\{\frac{\partial\Phi_\nu[\bar{Q}(x)]}{\partial Q}g_k(x)\right\},$$

$$g_k(x) = \int_0^x t\psi_k(t)\,dt. \tag{25}$$

The effectiveness of this method was checked as follows. For specified applications of $T(x), q(x)$ and $\bar{q}(x)$, values of I_ν and $\bar{I}_\nu$ were calculated from equation (13) for several spectral regions of width $\Delta\nu = 25\ \text{cm}^{-1}$ in the range $\nu = 1241 - 1512.5\ \text{cm}^{-1}$. Absorption coefficients l_ν were taken from /19/. Then function f_ν obtained from these calculations and the specified $T(x)$ and $\bar{q}(x)$ were utilized for determining the coefficients and the required function $q'(x)$ by the above method. The results of these calculations are plotted in Figure 5 for $n = 3$ and $N = 3$. Note the large difference between the true and calculated curves of $q'(x)$ at the surface. This is apparently due to the fact that here $q'(x)$ is commensurable with the average profile $\bar{q}(x)$, which means that one must retain also nonlinear terms in equation (21). We also note that, unlike the similar problem for the vertical temperature profile, the specific humidity variations have a smaller effect on the variability of the spectral variation of the intensity of departing radiation than temperature variations. In fact, as follows from equation (19), calculation of I_ν involves double smoothing of the specific humidity.

The solution of this problem can be refined by applying the Newton method, extended by Kantorovich /5/ to nonlinear operators, to nonlinear equation (13). Expressing the latter equation in the form

$$F_\nu(q_k) \equiv B_\nu[T(1)]\left\{1-\Phi\left(\sqrt{\frac{l_\nu}{2}W_0\left[\overline{Q}(1)+\sum_{k=1}^{n}q_k g_k(1)\right]}\right)\right\}+$$
$$+\int_0^1 B_\nu[T(x)]\,d\Phi\left\{\sqrt{\frac{l_\nu}{2}W_0\left[\overline{Q}(x)+\sum_{k=1}^{n}q_k g_k(x)\right]}\right\}-I_\nu=0 \qquad (26)$$

and regarding (26) as a nonlinear equation in coefficient q_k, whose first approximation $q_k^{(1)}$ is considered to be given by the solution of system of equations (24), we determine the following approximations:

$$\sum_{k=1}^{n}\frac{\partial F_\nu[q_k^{(i)}]}{\partial q_k}[q_k^{(i+1)}-q_k^{(i)}]+F_\nu[q_k^{(i)}]=0 \qquad (27)$$
$$(i=1,2,\ldots).$$

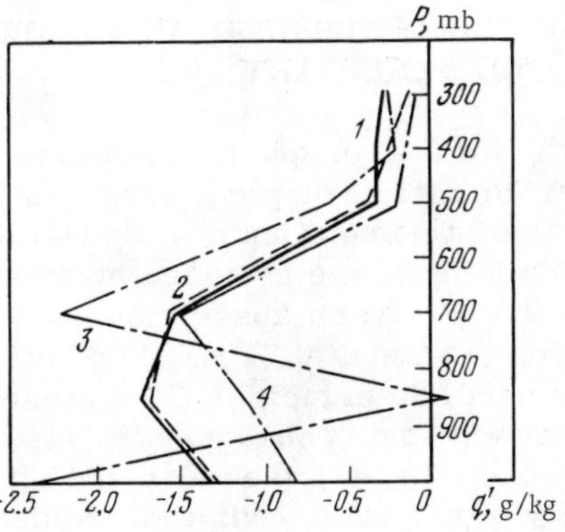

FIGURE 5. Reconstruction of the vertical specific humidity distribution over Bismarck:

1 — $q'(p)$; 2 — $q'_s(p_i)=\sum_{\kappa=1}^{3}\tilde{q}_\kappa\psi_\kappa(p_i)$,

$q'_k=\sum_{i=1}^{6}q'(p_i)\psi_\kappa(p_i)$; 3 — $q'(p)=\sum_{\kappa=1}^{3}q_k\psi_k(p)$,

q_κ are determined from system of equations (24); 4 — second approximation of $q'(p)$, obtained from system of equations (27).

It is easy to see that the coefficients of system of equations (27) are identical with $\Gamma_{\nu k}$ when $q_k^{(0)}=0$. Hence for simplicity in calculations one may use in equations (27) the same coefficients as in (24), changing only the free term. The solution of equations (27) for $i = 2$ is plotted in Figure 5, which shows that this method ensures satisfactory agreement between the calculated and actual $q'(x)$ profiles. The rms error of the solution does not exceed 10% on condition that the quantity is measured accurately. If, however, there is some error σ_f in measuring f_ν, then the coefficient

$$K_q(x) = \frac{\sigma_q}{\sigma_f},$$

which shows the factor by which the error in determining the specific humidity σ_q increases, can be expressed in terms of elements of the inverse matrix $\Gamma^{-1}_{\nu k}$ of system of equations (24) and base vectors ψ_k:

$$K_q(x) = \left[\sum_{k=1}^{n}\sum_{l=0}^{n}\sum_{\nu=1}^{n}\Gamma^{-1}_{\nu k}\Gamma^{-1}_{\nu l}\,\psi_k(x)\,\psi_l(x)\right]^{1/2}. \qquad (28)$$

Values of $K_q(x)$ calculated using the above solution for $q'(x)$ were listed above. As expected, the error amplification factor attains a maximum at the surface. If the number of spectral intervals N over which I_ν is measured exceeds the number of base vectors n, then, as in the case of temperature, the error decreases by approximately a factor of $\sqrt{N/n}$.

DETERMINATION OF THE PRESSURE AND ALTITUDE OF THE UPPER CLOUD BOUNDARY

Determination of the total atmospheric pressure p_0 from satellites with an accuracy sufficient for meteorological purposes is one of the most difficult problems of the interpretation of radiation measurements. Various possibilities for determining p_0 are provided by measurements of reflected solar radiation in the 0.76 μ oxygen absorption band and in the neighboring transparent window (for example, 0.74 μ). For reflection by clouds, these measurements can be used, in principle, for determining the pressure at the upper cloud boundary and, consequently, also the altitude of this boundary, as suggested in /14, 21/ (see also /8/).

If $I_i^{(0)}$ is the intensity of the solar radiation incident at angle ζ to the vertical at the upper boundary in the 0.76 μ O_2 absorption band ($i=1$) and in the transparent window ($i = 2$), b_i is the albedo of the reflecting surface in this spectral interval, $u(z_0)$ is the mass of oxygen in unit cross section of air column from the reflecting surface level z_0 to the upper boundary of the atmosphere, then the ratio of outgoing radiation intensities I_1 and I_2 from the upper boundary of the atmosphere at angle θ to the vertical will be related to $I_i^{(0)}$ and the parameters of the atmosphere by the following expression:

$$\frac{I_1}{I_2} = \frac{b_1 I_1^{(0)}}{b_2 I_2^{(0)}}\,\tau\left[u(z_0)(\sec\zeta + \sec\theta)\right]\frac{1+\varepsilon_1}{1+\varepsilon_2}, \qquad (29)$$

where τ is the transfer function of O_2 in the 0.76 μ band; ε_i is the ratio of the intensity of the entire multiply scattered part of outgoing radiation to the intensity of direct solar radiation reflected from the surface. It is assumed in deriving equation (29) that the scattering coefficient of the atmosphere and the spectral sensitivity of the instrument are identical at both spectral intervals. It is thus possible, in principle, by measuring the ratio I_1/I_2 to determine $u(z_0)$ and, consequently, the partial pressure of the atmosphere and the altitude z_0. However, this approach involves the following complications, which are primarily due to variations in the transfer function of the atmosphere.

1. Unavailability until recently of information on the transfer function of oxygen in the 0.76 μ band for the real atmosphere.* Some calculated values of this function obtained by Wark for different levels in the atmosphere were quoted by Saiedy at the Leningrad Symposium on Radiative Processes /18/.

2. Differences in the albedo b_i of some surfaces and clouds in the absorption band and in the window, calculated, for example, by Feigel'son /13/ and Rozenberg /10/. This also follows from aircraft measurements of the partial pressure of oxygen obtained by Saiedy /18/, who measured the radiation reflected by clouds immediately above their upper boundary. The difference between the true and "radiative" pressures at the cloud level was on the average 137 mb with a standard deviation of ± 30 mb, the "radiative" pressure always being on the high side. This is due to the fact that multiple scattering in the cloud increases the optical path and consequently also the mass of oxygen absorbing solar radiation.

3. The large variability of radiation multiply scattered in the atmosphere, the intensity of which can be appreciable at large ζ and θ, shows that quantities ε_i cannot be neglected.

The simplest problem is the determination of τ if we know the structural parameters of the absorption band: intensity S_j, location of centers ν_{0j} and half-width γ of lines. A study of this band was carried out by Dianov-Klokov /4/, who also determined some of the above parameters. These data were used to calculate the transfer function on the assumption that the contour of each line is described by the Lorentz line-shape equation:

$$K_\nu^{(j)}(p) = \frac{S_j}{\pi} \frac{\gamma}{(\nu - \nu_{0j})^2 + \gamma^2}, \tag{30}$$

where γ is a function of $p(\gamma = \gamma_0\, p/p_0)$ only. Here allowance is made for overlapping of all the lines of the O_2 absorption band. The calculations were carried out on the Minsk-2 computer.

The calculation of the albedo b_i is much more difficult. These quantities can be calculated for clouds by solving the transfer equation for optically dense media (11) and from data on the fine structure of absorption by O_2. The band-average albedo can be found by integrating b_ν over all the ν.

Making allowance for multiple scattering in the atmosphere is just as difficult, since the aerosol responsible for variations in scattering is highly variable. Hence it is possible to estimate ε_i only by solving the transfer equations for model atmospheres with different degrees of turbidity.

* These data were published as the present collection appeared.

References

1. Boldyrev, V.G. Raschet peredatochnykh funktsii atmosfery v intervale spektra 8—12 mk dlya territorii severnogo polushariya (Calculation of Transfer Functions of the Atmosphere in the 8—12 micron Spectral Range for the Northern Hemisphere).— Izvestiya AN SSSR, seriya fizika atmosfery i okeana, 1(7). 1965.
2. Boldyrev, V.G., L.I.Koprova, and M.S.Malkevich. Ob uchete variatsii vertikal'nykh profilei temperatury i vlazhnosti pri opredelenii temperatury podstilayushchei poverkhnosti po ukhodyashchemu izlucheniyu (On Incorporating Variations in the Vertical Temperature and Humidity Profiles when Determining the Surface Temperature from Departing Radiation).— Izvestiya AN SSSR, seriya fizika atmosfery i okeana, 1(7). 1965.
3. Gandin, L.S. Ob optimal'noi interpolyatsii i ekstrapolyatsii meteorologicheskikh polei (On Optimum Interpolation and Extrapolation of Meteorological Fields).— Trudy GGO, Vol. 14. 1962.
4. Dianov-Klokov, V.I. K voprosu o proiskhozhdenii spektra zhidkogo i szhatogo kisloroda (12,600—3000 A) (On the Origin of the Spectrum of Liquid and Compressed Oxygen (12,600—3000 Å)).— Optika i spektroskopiya, 6(4). 1959.
5. Kantorovich, L.V. O metode N'yutona (On Newton's Method).— Trudy Matematicheskogo instituta im. V.A. Steklova, Vol. 28. 1949.
6. Malkevich, M.S. Nekotorye voprosy interpretatsii radiatsionnykh izmerenii so sputnikov (Certain Problems of the Interpretation of Satellite Radiation Measurements).—Kosmicheskie issledovaniya, 2(2). 1964.
7. Malkevich, M.S. and V.I.Tatarskii. Opredelenie vertikal'nogo raspredeleniya temperatury atmosfery po ukhodyashchemu izlucheniyu v polose pogloshcheniya CO_2 (Determination of the Vertical Distribution of the Air Temperature from the Departing Radiation in the CO_2 Absorption Band).— Kosmicheskie issledovaniya, 2(2). 1964.
8. Malkevich, M.S. Nekotorye voprosy interpretatsii polei ukhodyashchei radiatsii Zemli (Some Problems of the Interpretation of Departing Terrestrial Radiation Fields).— Trudy GGO, Vol. 166. 1964.
9. Obukhov, A.M. O statisticheski ortogonal'nykh razlozheniyakh empiricheskikh funktsii (On Statistically Orthogonal Expansions of Empirical Functions).— Izvestiya AN SSSR, seriya geofizicheskaya, No.3. 1960.
10. Rozenberg, G.V. Svetovye kharakteristiki tolstykh sloev rasseivayushchei sredy s malym udel'nym pogloshcheniem (Optical Characteristics of Thick Layers of a Scattering Medium with Low Specific Absorption).—Doklady AN SSSR, **145**(4). 1962.
11. Romanova, L.M. Pole izlucheniya v ploskikh sloyakh mutnoi sredy s sil'nym anizotropnym rasseyaniem (The Emission Field in the Plane Layers of a Turbid Medium with Strong Anisotropic Scattering).—Optika i spektroskopiya, 14(2). 1963.
12. Tikhonov, A.N. O reshenii nekorrektno postavlennykh zadach i metode regulyarizatsii (On Solving Incorrectly Formulated Problems and on a Method of Regularization).—Doklady AN SSSR, **151**(3). 1963.

13. F e i g e l ' s o n, E. M. Opticheskie svoistva oblakov (Optical Properties of Clouds). — Kosmicheskie issledovaniya, 2(3). 1964.
14. C h a p m a n, R. M. Cloud Distribution and Altitude Profiles from a Satellite. — Planetary and Space Sci., Vol. 9, p. 70. 1962.
15. K a p l a n, L. D. Inference of Atmospheric Structure from Remote Radiation Measurements. — J. Opt. Soc. Am., 49(10). 1959.
16. K a p l a n, L. D. The Spectroscope as a Tool for Atmospheric Sounding by Satellites. — J. Quant. Spectrosc. and Radiative Transfer, 1(2):86—95. 1961.
17. L ö n n q u i s t, O. A Window Radiation Chart for Estimating Surface Temperature from Satellite Observations. — Tellus, 15(4):382—386. 1963.
18. S a i e d y, F., D. T. H i l l e a r y, and W. A. M o r g e n. On Cloud-Top Altitude Measurements from a Satellite. — Paper submitted at the International Symposium on Radiative Processes, Leningrad, August 1964.
19. W a r k, D. Q., G. Y a m a m o t o, and J. H. L i e n e s c h. Methods of Estimating Infrared Flux and Surface Temperature from Meteorological Satellites. — J. Atmos. Sci., 19(5). 1962.
20. W a r k, D. Q. Indirect Temperature Sounding of the Stratosphere from Satellites. — J. Geophys. Res., 66(1). 1961.
21. Y a m a m o t o, G. Numerical Method for Estimating the Stratospheric Temperature Distributions from Satellite Measurements in the CO_2 Band. — J. Meteorol., 18(5). 1962.
22. Y a m a m o t o, G. and D. W a r k. Discussion of the Letter by R. A. Hanel "Determination of Cloud Altitude from a Satellite". — J. Geophys. Res., 66(10): 3596. 1961.

Kh. P. Pogosyan, A. A. Pavlovskaya, and M. V. Shabel'nikova

NATURE OF THE INTERACTION BETWEEN AIR CIRCULATION IN THE TROPOSPHERE AND STRATOSPHERE OF DIFFERENT REGIONS OF THE NORTHERN HEMISPHERE

The article presents results of studies of the zonal and meridional components of air circulation in the troposphere and lower stratosphere of three large areas of the northern hemisphere for five years (1958—1959 and 1961—1963).

Computed circulation indexes are used in an attempt to establish the nature of the interaction of atmospheric processes in the troposphere and lower stratosphere. This interaction is illustrated by examples.

The achievements of meteorology over the past 2 or 3 decades are quite impressive. New information was obtained on the structure of the atmosphere and on processes developing in the terrestrial troposphere and stratosphere. Complicated mathematical weather forecasting problems are being solved by high-speed electronic computers. Much new information was obtained on calculating the heat-balance components of the atmosphere and of the terrestrial surface. Vertical air velocities and other parameters are being computed, precipitation is being calculated, etc. However, the quality of long-range weather forecasts has not improved perceptibly.

This was to be expected. All the studies of atmospheric processes are based on observed data, since large-scale laboratory experiments are impossible in meteorology. Observations of meteorological elements are not technically refined, and some of these elements do not even lend themselves to measurement. Thus, the free-atmosphere temperature, air humidity and wind are measured, but not very accurately, while no methods are available for measuring the vertical air velocity.

During the past several decades synoptic methods of long-range weather forecasts were developed primarily on the basis of empirically established asynchronous relationships between atmospheric processes. However, since they were determined qualitatively, the results were markedly affected by the subjective approach of the investigators and the relationships thus obtained were frequently found to be invalid.

The hydrodynamic method of long-range weather forecasting developed and applied in 1943 in the USSR by Blinova /4/ is of a quality comparable with forecasts compiled by the synoptic method. Long-range hydrodynamic forecasting methods are gradually being refined, but there are still many difficulties ahead.

Numerous studies were carried out by Vangengeim /5/, Baidal /1/, Girs /7/, Dzerdzeevskii /10/, Kats /12/, Pagava /16/, Khrabrov /31/ and others in order to determine the relationship and interaction between processes developing over various large regions of the northern hemisphere. In particular Vangengeim suggested taking into account the meridional and zonal types of circulation (W, C, E); these terms were used by him and his followers to describe asynchronous relationships between macrosynoptic processes for long-range weather forecasting. A place of importance in the Vangengeim and Girs studies is occupied also by problems of synchronous transformations of the geopotential field in the troposphere of the northern hemisphere.

The classification of processes suggested by Dzerdzeevskii has as its aim the determination of their relationship and order of succession in different regions of the northern hemisphere.

The above authors were mainly concerned with qualitative determination of the circulation form.

The large-scale atmospheric circulation can be efficiently estimated by the method developed by Kats /12/ and was successfully used by him for calculating the recurrences of the zonal and meridional circulations over Europe from average 500 mb charts of synoptic periods in the belt contained between 35 and 70°N.

Two monographs containing results of a special study of the coupling of the atmospheric circulation in the northern hemisphere were published in 1962. Vorob'eva /6/ considered the effect of processes in the American sector of the northern hemisphere on processes occurring in the Atlantic-European sector. She clarified certain features in the atmospheric circulation in Europe and contingent regions which arise due to changes in the circulation intensity in the western hemisphere. She attempted, by making allowance for this intensity and for conditions of coupling between processes in these two regions, to obtain recommendations on long-range forecasting of temperature anomalies for the European USSR.

The book by Spiridonova /30/ also contains data on the relation between different processes in various regions of the northern hemisphere. She attempted to describe the coupling between processes in different regions of the hemisphere by means of qualitative features of the trajectories of large atmospheric eddies, and used stable air mass transport in the troposphere as the indicator. She defines regions with different circulation forms in the hemisphere and presents an analysis of synchronous relationships between processes in regions of smaller area.

The conclusion reached in /6/ and /30/ is the same: definite relationships exist between the circulation in different regions of the hemisphere, but these vary depending on the intensity of zonal or meridional transport. Hence when compiling weather forecasts one must take into account the intensity of the circulation and the relationship between atmospheric processes in different regions of the hemisphere. The cause for this coupling of atmospheric circulation over the hemisphere is seen by Spiridonova in solar spot fluctuations and in the self-oscillatory properties of the atmosphere proper.

With the same purposes in mind the present authors made a somewhat different approach to the study of the nature of the atmospheric circulation over the northern hemisphere. We intend to study features of the zonal and meridional circulation components in the troposphere and stratosphere and to resolve the problem of the extent to which the air circulation in the lower stratosphere up to altitudes of 15—16 km depends on tropospheric processes, i.e., to establish a relationship between processes in the vertical direction and in adjoining regions of the northern hemisphere.

Our study of the coupling of processes over the hemisphere was carried out for three regions, whose conventional boundaries coincide with those of the so-called synoptic regions. The first region contains: North Atlantic, Europe and West Siberia (45°W—90°E); the second region consists of Eastern Asia and its adjoining part of the Pacific (90°E—160°W); while the third region includes the Eastern Pacific and North America (160—45°W).

The atmospheric circulation was described quantitatively by the previously mentioned method /12/, which makes it possible to obtain by quite simple calculations the zonal (I_z), meridional (I_m and general (I') circulation indexes for the entire northern hemisphere and for its individual parts at the surface and at different altitudes.

In this paper these indexes were calculated from the daily 100 mb charts of the General Forecasting Institute [TsIP] for 1958—1959, 1961—1963, from the monthly average 100 mb charts of the Scientific Research Institute of Aeroclimatology, and the monthly average 30 mb and 10 mb charts of Berlin University. The indexes were computed for the 35—70 and 52—70°N belts. In addition, the present paper uses circulation indexes at 500 mb for the same five-year period, calculated in the Long-Range Weather Forecasting Department of TsIP under the direction of Kats, who kindly permitted us to use them /25/.

The study of circulation features in the aforementioned three belts was carried out from the 100 mb charts, pertaining to the lower stratosphere, as well as from 500 mb charts which describe the troposphere. Unfortunately, the relative paucity of observational data at higher altitudes made it impossible to obtain daily circulation characteristics above 100 mb for all the five years under study. Hence we restricted ourselves to the study of individual processes in the middle stratosphere, attempting to establish the effect of tropospheric processes in variations in the geopotential field and of air currents in the atmosphere.

The annual variation of circulation indexes in the troposphere and lower atmosphere. Belinskii /3/, Blinova /4/, Vorob'eva /6/, Gromova /9/, Kats /12/, Papinashvili /17/, Sadovnikov /28/ and Rossby /33/ showed that the value of the zonal circulation index at the ground and at different altitudes has a yearly variation, with a maximum during the cold season and a minimum during the warm season. The yearly variation of the meridional component is in general similar to the yearly variation of I_z, but unlike I_z, two minima (in spring and autumn) and a secondary maximum in summer are observed against the background of the general decrease in I_m from the cold to the warm season.

In analyzing the values of indexes obtained by us we have studied the following: 1) changes in the magnitudes of daily as well as monthly average indexes over the year; 2) their interaction in the different belts; 3) their interaction in the troposphere and lower stratosphere.

The data we obtained showed that both the zonal and meridional circulation components in the troposphere and lower stratosphere reach their maxima during the cold season, i.e., precisely during the time when the most significant air-temperature gradients are formed, namely, the South-North and the ocean-shore gradients. The highest values of the indexes occur in the third belt due to its physicogeographic features (cold, relatively small continent with a complex orography, bordering in the west and east on warm oceans).

The day-to-day changes in the value of indexes in the troposphere and lower stratosphere are in the majority of cases of the same kind: when the intensity of tropospheric circulation increases, the same also happens in the lower stratosphere. The only exceptions are individual small fluctuations. However, the dates when maximum values of I_z and I_m are observed in both spheres differ somewhat. It hence may be concluded that in spite of the generally satisfactory agreement between the variation in daily indexes, there is no exact quantitative correspondence in the variations of these quantities.

The correlation between the zonal circulation at the 500 mb level and of the same component at the 100 mb level, calculated from data for three winter months of the five years, was found to be 71 % with a tolerance of ±0.3.

The cold season not only has the highest values of both circulation components, but also the highest range of their daily values. During transition seasons, when the temperature differences between the continents and oceans are smoothed out, the values of the zonal and meridional circulation components decrease markedly.

The yearly variation in the zonal components has a quite clear seasonal nature. While the values of I_z on the 100 mb chart in winter exceed in the majority of cases the values of the same index on the 500 mb chart, the relationship between these quantities in summer is opposite. No such clear relationship is observed during the transition periods.

Comparison of the daily values of I_{z500} and I_{z100} thus makes it possible to establish quite objectively the following general boundaries of seasons over the hemisphere:

$$I_{z100} > I_{z500} \text{ — winter}$$
$$I_{z100} \approx I_{z500} \text{ — spring}$$
$$I_{z100} < I_{z500} \text{ — summer}$$
$$I_{z100} \approx I_{z500} \text{ — autumn}$$

The yearly variation of monthly average values of I_z and I_m at the 500 and 100 mb levels for each of the five years is shown in Figure 1, while that of the five-year average values is given in Figure 2.

An examination of the graphs shows quite substantial differences between them. Thus, in the first and second belts the meridional component in the troposphere and lower stratosphere changes little from one year to another; this is obviously due to the relatively small variation of zonal temperature differences between the land and the ocean from year to year. In the third belt, the physicogeographic conditions of which differ from the other two, the yearly range of fluctuations in I_m increases markedly in both spheres.

The zonal circulation component undergoes much greater changes during the year than the meridional one, which is particularly clear in the stratosphere, where the main contribution to changes in the circulation intensity devolves upon the South-North temperature gradient in winter and on the North-South gradient in summer.

The circulation intensity in summer in all the regions for different years hardly changes, and thus the curves of Figure 1 almost coincide. The

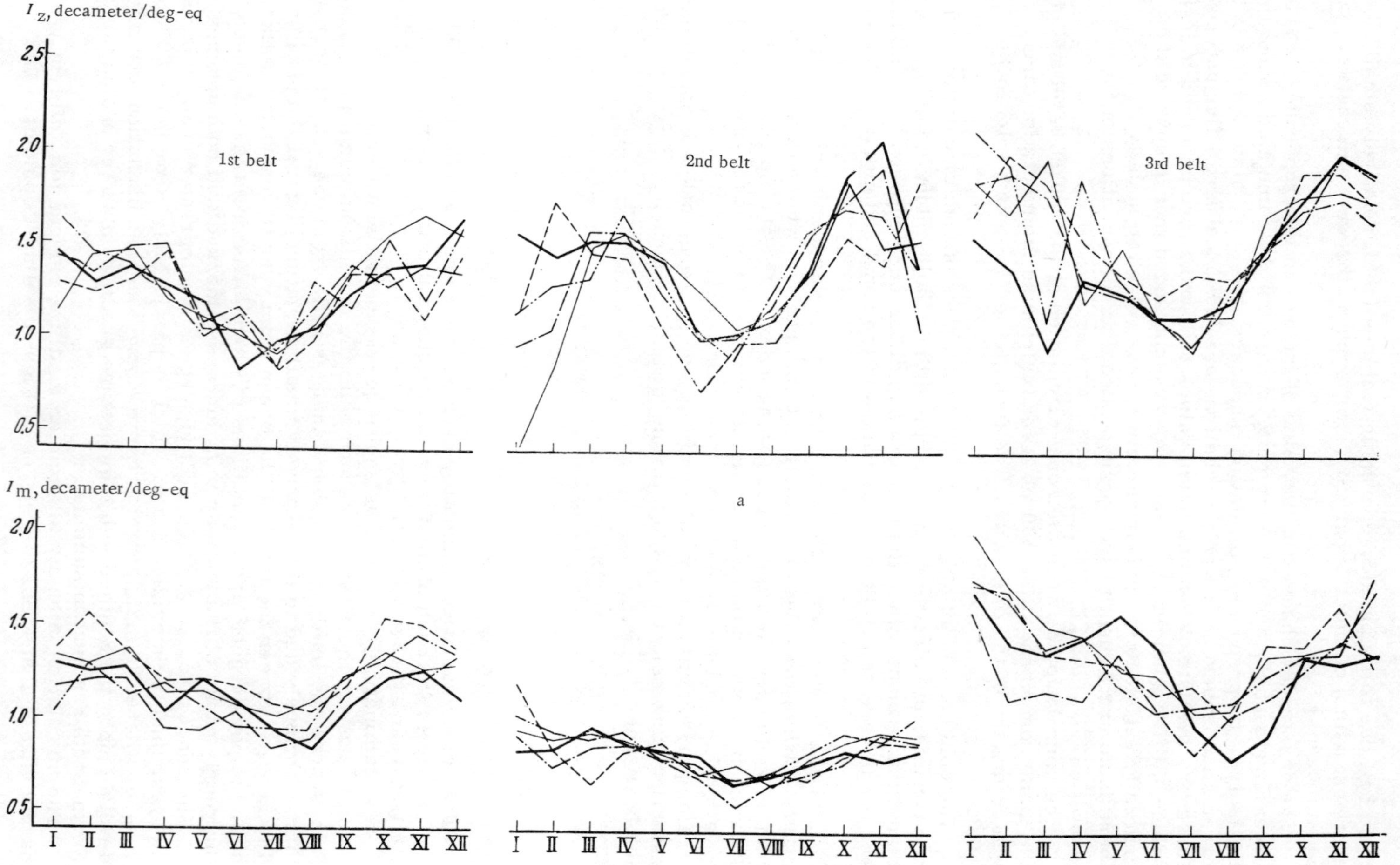
I_z, decameter/deg-eq
1st belt
2nd belt
3rd belt
2.5
2.0
1.5
1.0
0.5
I_m, decameter/deg-eq
2.0
1.5
1.0
0.5
I II III IV V VI VII VIII IX X XI XII
I II III IV V VI VII VIII IX X XI XII
I II III IV V VI VII VIII IX X XI XII

a

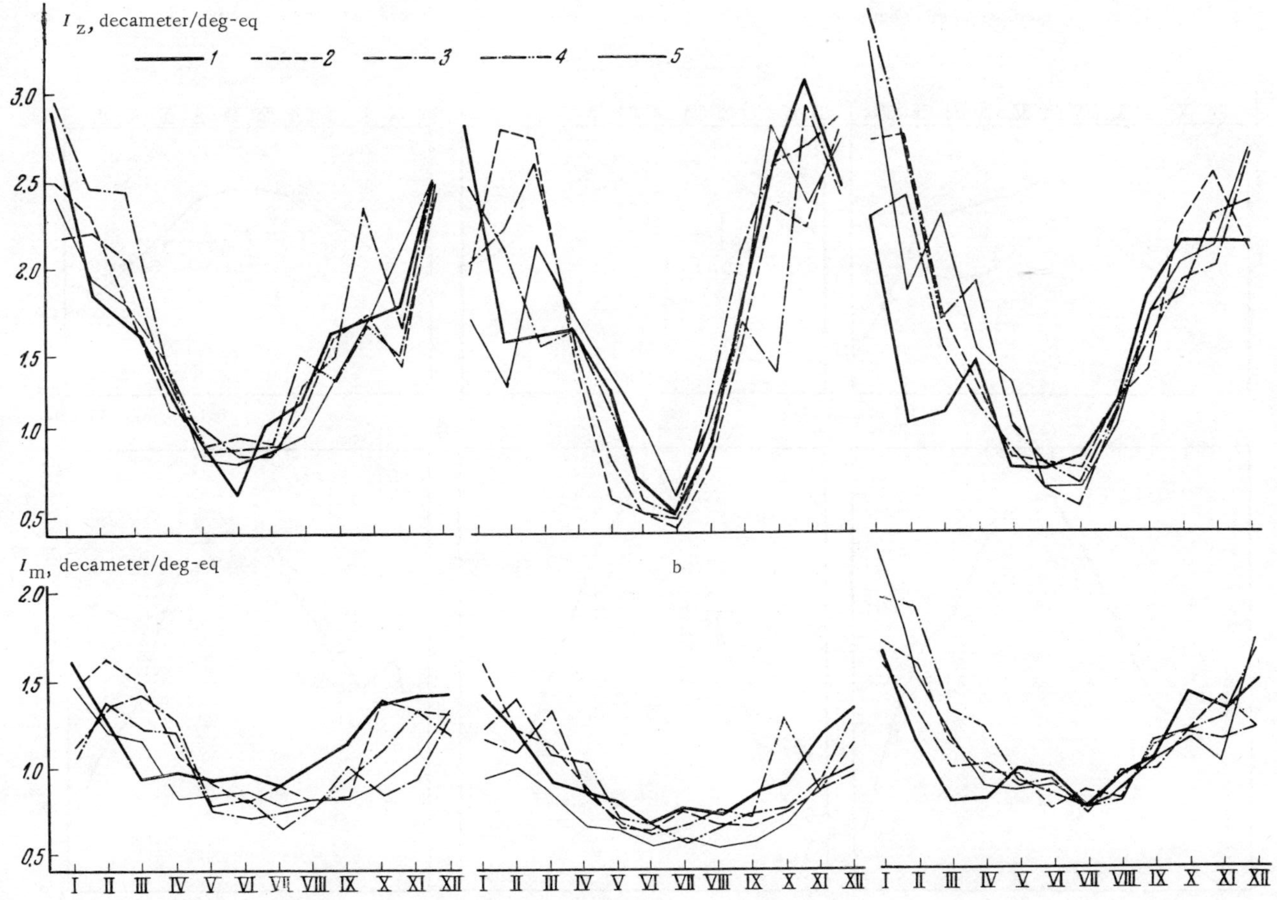

FIGURE 1. Yearly variation of zonal and meridional circulation indexes in the northern hemisphere:

a — 500 mb; b — 100 mb; 1) 1958; 2) 1959; 3) 1961; 4) 1962; 5) 1963.

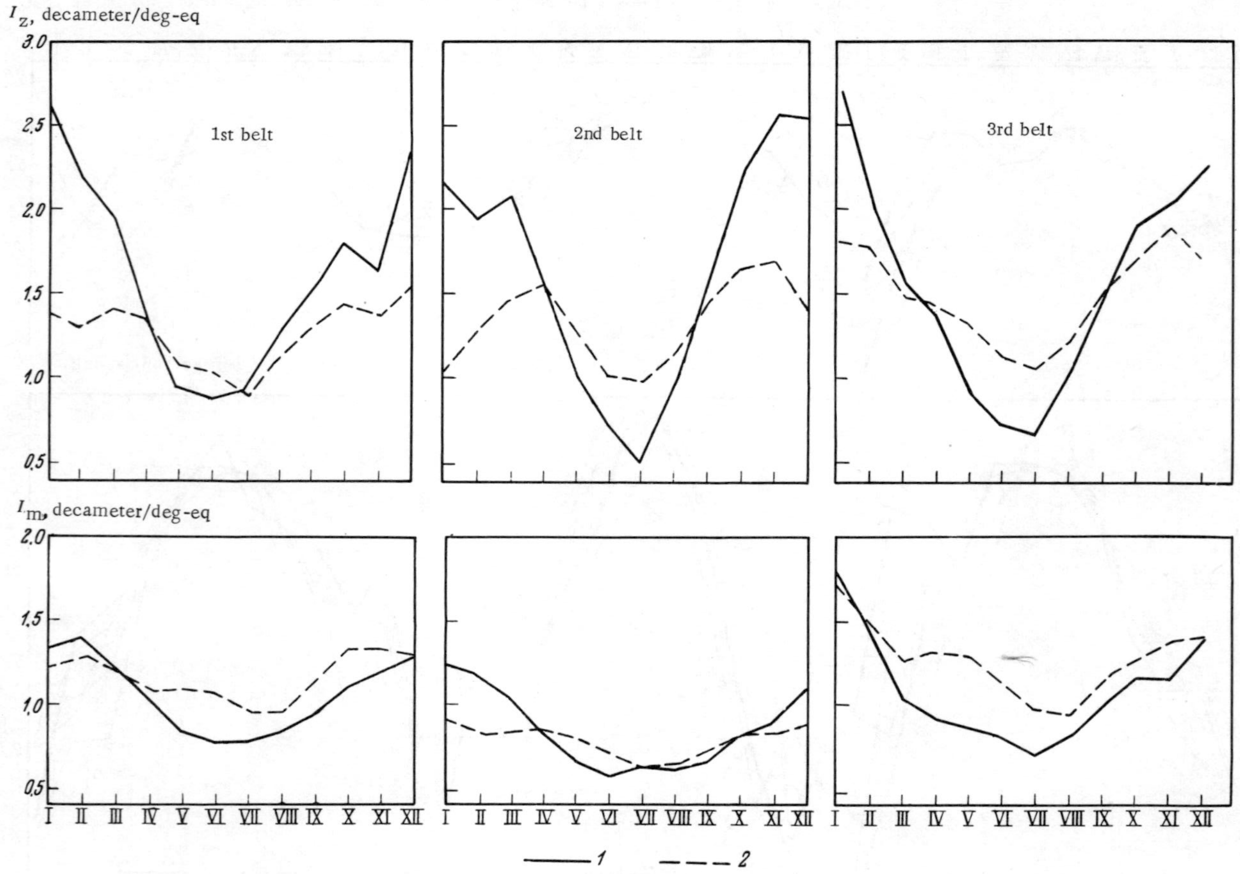

FIGURE 2. Yearly variation of the five-year average zonal and meridional circulation indexes in the northern hemisphere: 1—100 mb; 2—500 mb.

most significant differences between the values of I_m and particularly of I_z are observed in winter. Thus, the intensity of the zonal circulation in winter of some years can differ from that of other years by 100 and even 200 %.

Large differences in the intensity of circulation (particularly of zonal) between the troposphere and stratosphere are observed in winter. These differences are so great that they are reflected even in the values of average five-year indexes. It is seen from Figure 2 that I_z increases from the 500 to the 100 mb surface by 22 % in all the regions.

The yearly average characteristics of circulation calculated from the five-year data for the three latitude belts are presented in Table 1. It is seen that the rate of zonal air exchange increases with altitude by 5 % in the third and by 26 and 27 % in the first and second belts. At the same time the meridional circulation in the first and third belts is attenuated with altitude by 10 and 17 % respectively, and increases by 8 % in the second belt. This feature is attributable to the dynamics of processes helping to increase the meridional component in the North Pacific.

TABLE 1. Five-year average meridional ($\bar{I}_m$), zonal ($\bar{I}_z$) and general ($\bar{I}' = \bar{I}_m/\bar{I}_z$) circulation indexes (decameters/deg-eq)

Circulation index	1st belt			2nd belt			3rd belt		
	500 mb	100 mb	change in index, %	500 mb	100 mb	change in index, %	500 mb	100 mb	change in index, %
$\bar{I}_m$	1.15	1.04	−10	0.76	0.82	+8	1.25	1.05	−17
$\bar{I}_z$	1.26	1.60	+27	1.32	1.66	+26	1.49	1.57	+5
$\bar{I}'$	0.91	0.65	−29	0.58	0.49	−16	0.85	0.67	−21

The average values of the general index ($\bar{I}'$) describing the degree of turbulence of the zonal flow decrease with altitude by 16 − 29 %.

Comparison of the circulation indexes by seasons points to the well known fact that the intensity of both circulation components decreases toward the summer; this is characteristic of the troposphere and the stratosphere. The zonal component undergoes much more marked changes; hence I' in all the three belts in summer is greater than the yearly average. Index I' increases from winter to summer. The greatest changes in I_z between the 500 and 100 mb surfaces occur in winter and may be as high as 70 − 80 %.

Zonal and meridional processes. The average values of the general circulation indexes at the 100 mb level, reduced from five-year data of Table 1, were used as criteria for distinguishing between the zonal and meridional atmospheric processes developing in each of the belts under study. The process was regarded as zonal if the value of I' calculated from the daily constant pressure maps were below the value of $\bar{I}'$ given in Table 1, and meridional when the value of I' was greater than or equal to $\bar{I}'$.

For convenience in classifying different processes by circulation forms and for brevity in designation, we used the following method: zonal processes were designated by the numeral 1, meridional processes were assigned the numeral 2; homogeneous zonal processes, i. e., processes for which the value of I_z in the entire 35 − 70°N belt, including its northern part in the 52 − 70° strip, were below the critical value, were designated by (11);

homogeneous meridional processes, for which I_m in the entire belt and in its northern part separately were equal to or higher than the critical value, were designed by (22). The term "mixed" is applied to processes in which the deviation of the circulation indexes from the critical value in the entire latitude belt and in its northern part were of different sign. Here designation (12) was used in cases when the inequality $I' < \bar{I}'$ was satisfied in the entire 35—70°N latitude belt as a whole (zonal process), while in its northern half $I' \geqslant \bar{I}'$, i.e., the meridional process predominated. In the opposite case, when $I' \geqslant \bar{I}'$ in the entire 35—70°N belt while in its northern part $I' < \bar{I}'$, designation (21) was used.

The processes were thus classified as homogeneous zonal (11), homogeneous meridional (22) and mixed (12) and (21). These are the data which were used to determine the monthly-average recurrence of different circulation forms. The results of calculations are shown in the form of graphs in Figure 3, which also shows for comparison analogous 500 mb curves.

It is easy to see that the monthly average recurrence of zonal and meridional processes changes appreciably from one month to another and also differently for each of the three belts. At the same time the yearly recurrence curve shows a number of features common to all the belts.

The yearly variation of the meridional circulation recurrence is opposite to that of the zonal: the maximum number of days with meridional processes is observed in summer, and the minimum in winter; the variation in the recurrence of zonal and meridional processes in the lower stratosphere occurs on much larger scales than in the troposphere.

We now consider in more detail the features of the yearly variation of the recurrence of zonal and meridional processes in each belt.

First belt. The winter zonal planetary transport here is determined primarily by large temperature gradients between the cold Arctic and the relatively warm south of Europe and Hither Asia* and Soviet Central Asia, as well as between the cold Greenland and the tropics of the Atlantic. Since the temperature contrasts are greatest during the cold season, the yearly maximum of zonal process occurrence is observed here in winter. The number of days with zonal processes in December and January at the 100 mb level is by 35—40% greater as compared with the 500 mb level. A small minimum is observed in February in the yearly variation of zonal-circulation recurrence.

The transition from the winter to the summer radiation regime in the lower stratosphere starts from March and is accompanied by a marked reduction in temperature and geopotential gradients. However, the south-north temperature contrasts in the troposphere are still so large during this month that they determine not only the nature of processes in the troposphere, but also in the lower stratosphere. Hence zonal processes have greater recurrence in both spheres.

The disappearance of the snow cover and the rapid warming of the continent in March and April result in spring equalization of the temperature in the troposphere between the continent and ocean and in a moderate amplification of the zonal circulation. Toward summer a warm ridge is formed over the continent with the temperature gradient directed toward the ocean. This is responsible for the rise in the number of meridional processes and reduction in the number of zonal processes (see Figure 3). The smallest yearly recurrence of zonal circulation is observed in June and July.

* [Southwestern part of Asia, including the Bosporus, Mediterranean and Red Sea to the Plateau of Iran.]

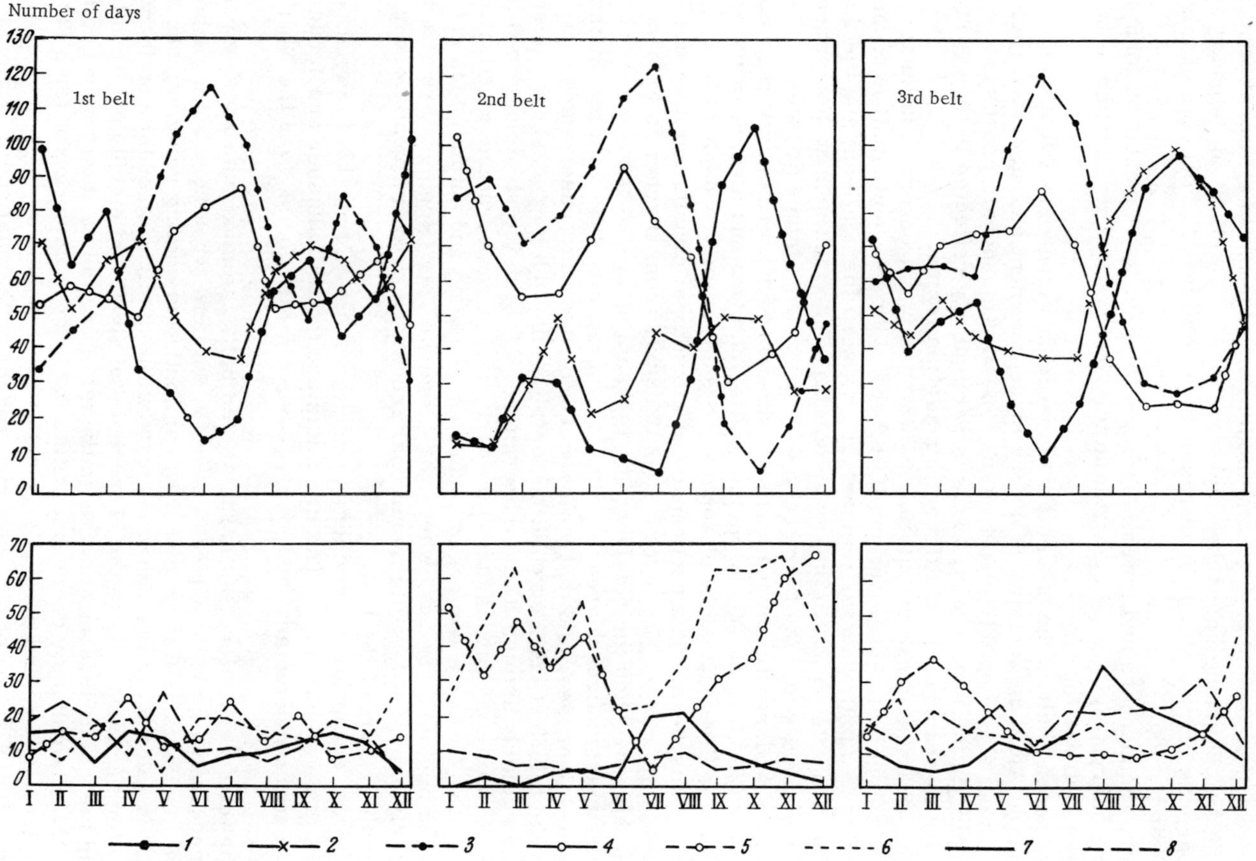

FIGURE 3. Monthly average recurrence of zonal and meridional processes calculated from daily data for 1958–1959 and 1961–1963: 1–11, 100 mb; 2–11, 500 mb; 3–22, 100 mb; 4–22, 500 mb; 5–12, 100 mb; 6–12, 500 mb; 7–21, 100 mb; 8–21, 500 mb.

In August and September the summer regime starts weakening, the temperature gradients decrease between the continent and ocean and increase between the lower and higher latitudes, and the number of zonal processes increases sharply. The moderate reduction in the zonal process recurrence occurring in October and November, as well as in February, is apparently attributable to features of the dynamics of processes of this five-year period.

In the troposphere and lower stratosphere the mixed circulation forms have no sharply delineated variation in the first belt. They are shown in the lower part of the figure.

The second belt. The extremely complex physicogeographical conditions of Eastern Asia, the vicinity of a very large continent and an even larger ocean, have a perceptible effect on the type of atmospheric processes developing in the second belt. This is particularly reflected in the yearly variation of recurrence of different circulation forms.

The maximum zonal-circulation recurrence is not observed in winter, as in the first belt, but in autumn (October), with a second small maximum occurring in April when the temperature between the continent and the ocean washing it levels out. The intensive cooling of the continent in November and December, while the temperature of the ocean surface decreases relatively moderately, increases the ocean-continent air temperature gradient, and results in development of meridional circulation and a corresponding reduction in the recurrence of zonal processes in the troposphere and lower stratosphere. The recurrence of meridional circulation increases even more in winter. It increases continuously from October to February; only in March, when the continent starts warming up and the ocean-continent temperature gradient decreases, does the number of days with meridional processes decrease, and that with zonal processes increases somewhat.

Unlike the first belt, circulation form (12) has a high recurrence in spring and particularly autumn in the troposphere, while in the stratosphere this occurs in winter. This is due to the fact that the temperature equalization between the continent and ocean occurring in spring and autumn does not extend to northern regions, where the ocean-continent temperature gradient is clearly expressed, while meridional processes most frequently occur in the northern part of the belt.

With the exception of summer, the recurrence of form (12) in the stratosphere is relatively high. The maximum occurs in December and January, when a stratospheric heat and high-pressure region forms over the North Pacific.

The third belt. Here, as in the second belt, the maximum zonal-circulation recurrence in both spheres occurs in autumn, when horizontal geopotential gradients between the still warm south and cooling polar regions of North America increase simultaneously with equalization of the continent-ocean temperature. As to the recurrence of zonal and meridional circulation during the warm season, here, as in the adjoining belts, the recurrence of form (11) also sharply decreases with increasing recurrence of form (22) in the troposphere and particularly in the stratosphere.

The classification of atmospheric processes as zonal and meridional yields only the most general concept of the direction of air currents during a given period of time and is insufficient for describing processes (particularly meridional) developing in some part of the belt. In order to somewhat take into account the details of processes under development, the recurrence was calculated of the location of high-altitude ridges in each of the belts. The recurrence of the different ridge locations at the 100 mb surface is given in Table 2.

TABLE 2. Recurrence of high-altitude ridges by seasons*

Season	1st belt				2nd belt				3rd belt			
	W	C	E	WE	W	C	E	WE	W	C	E	WE
Winter	186	117	41	106	27	29	155	239	207	49	11	183
	41	26	9	24	6	7	34	53	46	11	2	41
Spring	105	131	127	97	103	49	114	194	81	157	66	156
	23	28	28	21	22	11	25	42	18	34	14	34
Summer	23	202	181	54	177	148	44	91	85	221	62	92
	5	44	39	12	38	32	10	20	18	48	14	20
Autumn	164	202	22	67	61	95	154	145	85	144	84	142
	36	44	5	15	13	21	34	32	18	32	18	32

* W—western location; C—central; E—eastern; WE—high-altitude ridge in the west and east of the belt with a trough at the center. The first row gives the number of cases and the second expresses it as a percentage of the total.

Judging from the data of Table 2, the W processes, for which the high-altitude ridge is located over the British Isles or Western Europe while a high-altitude trough is situated over Eastern Europe, are most frequent in winter in the first belt. The frequency of these processes in winter is 41 %. Form C processes, for which the high-altitude ridge is situated at approximately the 30°E meridian, also have a relatively high frequency (26 %); form WE processes with ridges in the west and east of the region and a trough at the center also occur relatively frequently. The formation of a high-altitude ridge in winter is least probable over the eastern part of the continent which is then cold. The frequency of form E processes is only 9 %.

In spring, when the continent starts warming up, the recurrence of high-altitude ridges in the east of the belt increases and decreases in the west, with the result that all four process forms in this region have the same occurrence probability.

The most favorable conditions for the formation of a high-altitude ridge in summer are created over the heated continent; hence forms C and E have the highest occurrence. Ridges are very rare over the ocean, which is cold at this time (5 % of the total).

In autumn, when the continent cools rapidly and the ocean surface becomes warmer, the frequency of E processes decreases by a factor of 7 or 8 as compared with the summer while that of W processes increases by approximately the same amount. In general the type of process differs little from those in winter.

The diametrically opposite location of the continent and ocean in the second belt as compared with the first results also in an opposite type of process. In winter, high-altitude ridges form most frequently over the Pacific, with a frequency of E and WE processes here equal to 87 %. In summer, when high-altitude ridges are observed over the heated continent, the frequency of these forms decreases to 30 % and that of forms W and C accordingly increases to 70%. During transition periods the differences in the occurrence of the individual forms are markedly smoothed out. The form with greatest recurrence in spring is WE, for which the ridges are observed over the then warm ocean and western part of the continent, with a trough located over its still cold northeastern edge. The autumn processes are closest generally to those in winter, although the frequency of forms E and WE (66%) is still not as high as in winter.

The physicogeographical features of the third belt leave their imprint on the atmospheric processes of this belt and cause different processes than those occurring in neighboring belts. The frequency in winter of high-altitude ridges over the Pacific and Atlantic (forms W and WE) equals 87%. There are almost no ridges over the cold, eastern edge of this continent, where a high-altitude trough is usually located. The frequency of form E is only 2%. With the onset of spring and heating of the continent, the frequency of forms C and E increases and reaches its maximum in summer (62%).

In spring and autumn each of the four circulation forms has approximately the same recurrence. One's attention is attracted by the high occurrence of forms C and WE, the former of which is characterized by formation over the continent of a high-altitude ridge, while the second involves, on the other hand, a high-altitude trough. This is not only due to instability of processes arising in restructuring of the winter to the summer regime and conversely, but also to smoothing of differences in the summation of circulation forms by seasons. Analysis of the recurrence of these forms in each month of both seasons showed that processes C are most frequent during the first half of autumn and second half of spring, while during the second half of autumn and the first half of spring form WE processes predominate.

As we see, seasonal differences in the influx of heat over continents and oceans determine not only the total seasonal structure of the temperature and geopotential fields /18/, but also the predominance of some processes over the hemisphere and its individual parts.

Relationship between atmospheric processes in adjoining belts of the northern hemisphere. Large-scale processes developing in one region obviously should affect the reconstruction of the thermobaric field in adjoining regions. A great deal of attention was paid to this problem in the 1940's /19/, when the idea was developed of the relationship between meridional transformations of high-altitude deformation fields in adjoining regions. Primary attention in these studies was afforded to detection of the mechanism of thermobaric field transformations in adjoining regions.

Subsequently, some authors studying the coupling of different types of processes developing in adjoining belts of the hemisphere attempted to establish the frequency of the latter /5, 7, etc./. The results obtained were not too promising. It was even more difficult to obtain asynchronous relationships for the sequence of succession of processes, so necessary in the compilation of short- and long-range weather forecasts.

An attempt was made, on the basis of daily quantitative features of the zonal and meridional circulations of the atmosphere, to establish a relationship between the basic circulation forms in the troposphere and lower stratosphere in the three belts at hand.

The results of statistical processing of data are presented in Tables 3—5. Table 3 contains information on the number of cases with different circulation forms for the five-year period under study. As can be seen, the homogeneous meridional circulation (22), with a frequency exceeding 40—50% during all seasons in the majority of cases, is most frequently observed in the troposphere and lower stratosphere. The frequency of homogeneous zonal circulation (11) as a whole is lower, being most frequently observed during winter in the first belt and during autumn in the third belt. The mixed circulation forms (12) and (21), particularly the latter, are observed much less frequently.

TABLE 3. Recurrence of different circulation forms in the troposphere and lower stratosphere for the five-year period by seasons

Season	Circulation form	1st belt				2nd belt				3rd belt				Total number of cases per season
		500 mb level		100 mb level		500 mb level		100 mb level		500 mb level		100 mb level		
		number of cases	%	number of cases	%	number of cases	%	number of cases	%	number of cases	%	number of cases	%	
Winter	11	196	44	267	59	57	13	67	15	148	33	189	42	450
	22	159	35	110	25	251	56	227	50	182	41	171	38	
	12	49	11	37	8	118	26	152	34	78	17	67	15	
	21	46	10	36	8	24	5	4	1	42	9	23	5	
Spring	11	189	41	141	31	103	22	78	17	142	31	131	29	460
	22	178	39	231	50	191	42	248	54	222	48	229	50	
	12	40	9	51	11	151	33	125	27	36	8	80	17	
	21	53	11	37	8	15	3	9	2	60	13	20	4	
Summer	11	156	34	92	20	112	24	49	11	170	37	84	18	460
	22	222	48	286	62	247	54	325	71	197	43	292	64	
	12	54	12	57	12	79	17	43	9	41	9	25	5	
	21	28	6	25	6	22	5	43	9	52	11	59	13	
Autumn	11	193	42	168	37	131	29	263	58	280	61	275	60	455
	22	179	40	206	45	118	26	47	10	75	17	92	20	
	12	38	8	39	9	190	42	126	28	29	6	30	7	
	21	46	10	42	9	16	3	19	4	71	16	58	13	

The frequencies of the same circulation forms in the troposphere and lower stratosphere are close to one another. In rare cases they differ by more than 25% of the total number of cases at the 500 and 100 mb levels. These differences are most perceptible in summer, when the frequency of type (11) circulation decreases sharply and type (22) increases in all three belts. This may be attributed to the summer radiation regime, with the result that a cellular structure of the geopotential field appears as soon as the 100 mb level and the zonal circulation decreases perceptibly. Under these conditions the frequency of meridional circulation naturally increases as compared with the troposphere.

The recurrence of circulation forms in adjoining belts of the hemisphere was studied by taking each of the three belts as the basis, and by calculating the percentage recurrence of a given type of circulation in the adjoining belt by referring it to that in the base belt.

Tables 4 and 5 contain data, by seasons, on the recurrence of the circulation forms in the first belt, with the second and third belts regarded as adjoining. The recurrence is obtained here as a percentage of the number of cases of the given circulation form in the first region.

As is seen, the frequency of circulation form couplings in different belts in the troposphere (Table 4), as well as in the stratosphere (Table 5), in the majority of cases does not exceed 50—60 %, which points to the great complexity of interaction of atmospheric processes between adjoining regions. At the same time, the tables imply that some combinations are encountered more often than others. For example, corresponding to form (11) in the first belt in winter, form (22) is observed in 58 % of cases in the troposphere and in 45 % of cases in the stratosphere. The frequency is also high between combinations of the meridional circulation in the first and second belts: 60 % in the troposphere and 58 % in the stratosphere. Conversely, the frequency of combinations of form (11) in the second belt and of forms (11), (22) and (21) in the first is low (9—13 %), while the frequency of combinations of form (21) in the second belt and of all the four circulation forms in the first amounts to only 4—9 %. The frequency in winter in the first and third regions is usually below 50 %. Processes (22) recur most often in the first and third belts (48 % in the troposphere and 49 % in the stratosphere) and forms (21) in the first and (22) in the third belts (46 and 47 %).

The greatest recurrence in spring corresponds to processes (22) in the second belt attendant to zonal circulation in the first. They amount to 49 % in the troposphere and 55 % in the stratosphere. The frequency of combinations of all the circulation forms in the first and third belts does not in the main exceed 50 %. Form (22) in the first belt is an exception, since it is accompanied by the most frequent (71 %) occurrence in the third belt of forms (21) and (22).

The frequency increases in summer in both spheres of combinations (12) in the first and (22) in the second belt (61 and 72 %), of combinations (22) in both belts (54— 77 %) and of (21) in the first and (22) in the second belt (54 and 64 %). As in winter, the frequency of combinations of forms (21) in the second and third belts with all the circulation forms in the first belt is observed most infrequently. The remaining combination forms amount to 20—40 %.

In autumn one observes a relatively high recurrence of combinations of circulation forms (11) in the first and (11) and (12) in the second belts (75 % in the troposphere and 84 % in the stratosphere), of forms (22) in the first and (11) and (12) in the second belt (63—85 %), forms (21) in the first and (11) and (12) in the second belt (79—88 %).

The most frequently occurring combinations in the first and third belt are forms (11) in both belts (62 and 60 %), form (12) in the first and (11) in the third belt (74 and 62 %), and form (12) in the first and (11) and (12) in the third belt (79 and 67 %). While in summer the combinations of forms (22) and (21) recur more frequently than the others, the greatest recurrence in autumn devolves upon forms (11) and (12), due to thermal conditions at the surface /19/. As is known, in summer due to heating a warm ridge is produced over Eastern Asia, while in autumn as a result of equalization of the temperature over land and the ocean the isohypses conform more closely to parallels.

TABLE 4. Recurrence of different circulation forms in the second and third belts by seasons at the 500 mb level as a percentage of the number of cases in the first belt

Season	Circulation form in the 1st belt	2nd belt						3rd belt						Number of cases
		11	12	22	21	11 and 12	22 and 21	11	12	22	21	11 and 12	22 and 21	
Winter	11	13	23	58	6	36	64	40	16	33	11	56	42	196
	12	22	35	39	4	57	43	33	14	39	14	47	53	49
	22	9	27	60	4	36	64	25	21	48	6	46	54	159
	21	13	28	50	9	41	59	32	13	46	9	45	55	46
Spring	11	15	32	49	4	47	53	32	6	49	13	38	62	189
	12	18	38	40	4	56	44	33	7	50	10	40	60	40
	22	24	30	43	3	54	46	29	10	47	14	39	61	178
	21	13	40	47	0	53	47	30	8	49	13	38	62	53
Summer	11	26	20	50	4	46	54	49	10	31	10	58	42	156
	12	24	18	61	2	37	63	41	7	41	11	48	52	54
	22	24	16	54	6	40	60	29	8	52	11	47	63	222
	21	21	18	54	7	39	61	29	14	39	18	43	57	28
Autumn	11	31	44	24	1	75	25	62	7	15	16	69	31	103
	12	42	34	19	5	76	24	74	5	16	5	79	21	38
	22	22	41	31	6	63	37	60	7	16	17	67	33	178
	21	35	44	17	4	79	21	55	2	26	17	57	43	46

TABLE 5. Recurrence of different circulation forms in the second and third belts by seasons at the 100 mb level as a percentage of the number of cases in the first belt

Season	Circulation form in the 1st belt	2nd belt						3rd belt						Number of cases
		11	12	22	21	11 and 12	22 and 21	11	12	22	21	11 and 12	22 and 21	
Winter	11	18	36	45	1	54	46	48	15	30	7	63	37	267
	12	16	27	54	3	43	57	21	19	57	3	40	60	37
	22	9	32	58	1	41	59	35	12	49	4	47	53	110
	21	8	28	61	3	36	64	36	17	47	0	53	47	36
Spring	11	15	30	55	0	45	55	35	11	48	6	46	54	141
	12	18	31	47	4	49	51	25	25	45	5	50	50	51
	22	18	26	54	2	44	56	25	4	52	19	29	71	231
	21	19	16	60	5	35	65	27	19	51	3	46	54	37
Summer	11	19	13	55	13	32	68	35	3	45	17	38	62	92
	12	14	5	72	9	19	81	14	5	72	9	19	81	57
	22	8	8	77	7	16	84	15	6	68	11	21	79	286
	21	12	8	64	16	20	80	12	8	64	16	20	80	25
Autumn	11	59	25	13	3	84	16	60	8	20	12	68	32	168
	12	62	5	15	19	67	33	62	5	15	18	67	33	39
	22	56	29	10	5	85	15	58	7	24	11	65	35	206
	21	55	33	5	7	88	12	74	2	10	14	76	24	42

The recurrences of combinations of circulation forms in the second and third belts with the neighboring belts are not presented here but differ little from the data of Tables 4 and 5, where the percentages are presented relative to the first belt.

The tabulated data show that in the majority of cases the recurrences of combinations of various circulation forms in adjoining belts of the northern hemisphere are not clearcut enough to enable one to refer to a definite regularity of processes developing over the hemisphere. Obviously, to obtain any given specific conclusions, one must study the mechanism of transformations of high-altitude deformation fields in adjoining belts.

Variation in the temperature and wind fields in the stratosphere in connection with processes in the troposphere. It was shown above that the structures of the geopotential and air-current fields in the troposphere and lower stratosphere are approximately of the same type. Their modifications by atmospheric eddies and by thermobaric field transformations are reflected to almost the same extent at the 500, 300 and 100 mb surfaces. The difference consists primarily in the fact that horizontal geopotential and wind velocity gradients between middle and high latitudes change with altitude. As a rule these increase in winter and decrease in summer.

Since the lower stratosphere reacts to eddy genesis processes and to all the main transformations occurring in the troposphere, the results obtained up to now on the coupling or relationship of tropospheric processes over different regions of the northern hemisphere can be ascribed in their entirety also to the lower stratosphere. The question, however, arises whether this conclusion can be extended to the entire lower stratosphere or at least to its middle part.

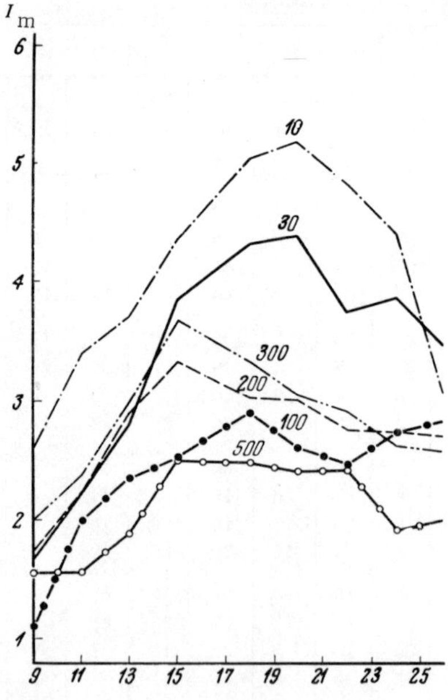

FIGURE 4. Meridional circulation indexes from 9 through 26 January 1963 at the 500—10 mb levels.

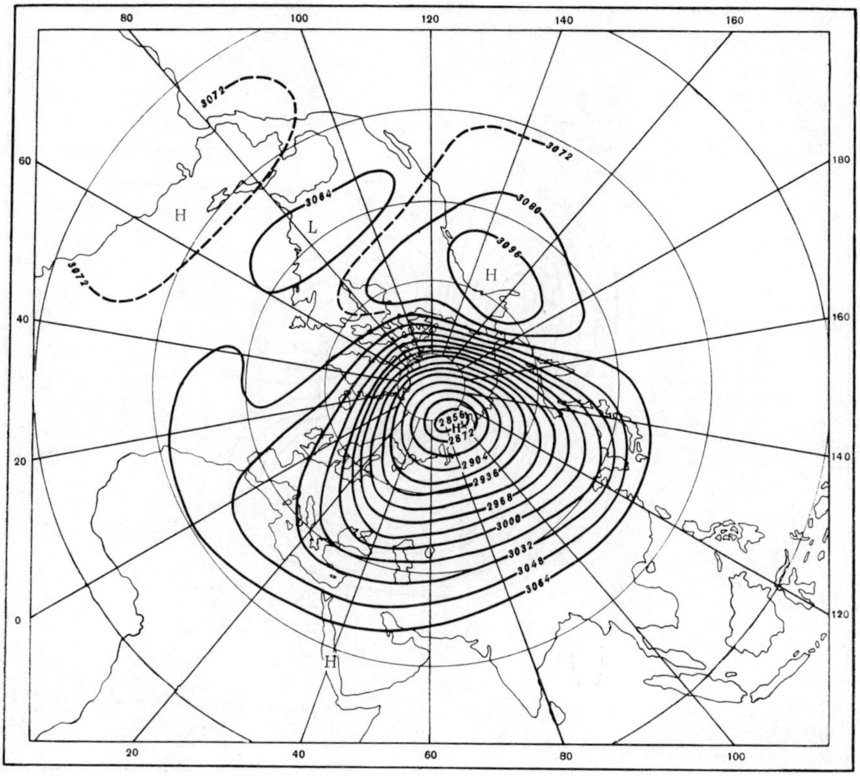

FIGURE 5a. 10 mb chart for 15 January 1961.

It is known that sharp changes of meteorological fields, accompanied by perceptible warming up, occur in the cold season in the stratosphere of the northern hemisphere. Earlier studies /14, 21, 22/ allowed the conclusion that aperiodic changes in atmospheric circulation, accompanied by sharp warming in the middle stratosphere of the Arctic in winter, are produced by advective dynamic factors. This is also the conclusion of many other meteorologists /11, 35—38/, the only exception being that some of them ascribe more importance to temperature advection from middle latitudes, while others regard adiabatic changes due to vertical air motions as more significant. At the same time some investigators ascribe anomalous processes in the stratosphere to changes in solar activity /27, 29, 34/.

Unlike many investigators, we studied the tropospheric and stratospheric processes and concluded that the troposphere makes the principal contribution to the development of processes in these two spheres.

As is known, the recurrence of atmospheric eddies at the terrestrial surface depends somewhat on the season. In the cold season cyclones occur frequently over oceans, with anticyclones predominating over continents /20/. This situation changes with altitude. As low as the 500 mb level cyclones become more frequent over land and anticylones over oceans /2, 13/. In summer the recurrence of cyclones at the earth surface

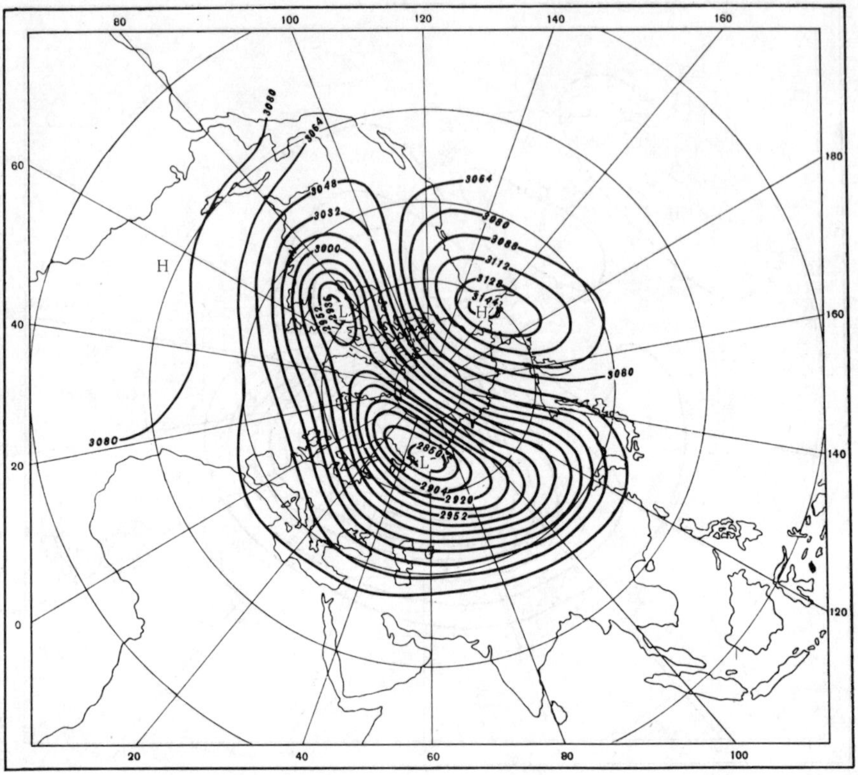

FIGURE 5b. 10 mb chart for 25 January 1961.

increases over land and that of anticyclones increases over oceans. Here their recurrence does not change with altitude to the same extent as in winter.

Not all the eddies generated in the troposphere participate in producing anomalous changes in the high-altitude atmospheric circulation. In addition, there is a difference whether these eddies are produced over land or over oceans. Many of them are difficult to find even in the upper troposphere, while medium-strength eddies are damped between the 100 and 500 mb levels.

Large meridional transformations of the thermobaric field, associated with developments of individual cyclones and anticyclones, occur in the troposphere. The meridionality usually decreases with altitude above the tropopause. Small disturbances of the geopotential field disappear and zonal circulation is established.

The circulation in the stratosphere behaves differently upon the development of cyclones which are accompanied by strong meridional transformations of the thermobaric field in the troposphere. In these cases the temperature and air current fields in the stratosphere adapt themselves to the meridional processes developing in the troposphere with meridional circulation also developing in the latter. These processes are more typical of the winter season, particularly of December-April.

Due to the summer heating of air under polar day conditions, the air temperature in the Arctic stratosphere above 12—15 km exceeds the temperature in middle latitudes. As a result the processes in the middle stratosphere are not as intensive as in winter.

During winter, in accordance with the radiation conditions and location of the low-temperature center in the polar region (above the 100 mb level), the temperature gradients increase and the winds become stronger. Under these conditions strong meridional transformations of meteorological fields in the troposphere have a direct effect on the similar reconstruction of circulation not only in the lower stratosphere, but in its entirety. This is understandable. When the horizontal temperature gradients and air-current velocities increase with altitude, both the horizontal heat transport and vertical air currents increase. Meridional reconstruction of the thermobaric field upon intensive eddy genesis takes place more intensively, when descending motions occur in heat-advection regions and ascending motions occur in cold-advection regions /14, 26/.

Studies of strong winter warmings in the middle stratosphere of the Arctic in 1957—1963 showed that all of them occurred as a result of eddy activity and strong meridional transformations of the thermobaric field, which started in the troposphere and after 2—4 days moved into the stratosphere. In particular, a sharply delineated warming up took place in January 1963 at altitudes of 25—30 km, and was not weaker than that in January 1958. The zonal circulation, which was well expressed both in the troposphere and in the stratosphere up to 9—10 January, was subsequently replaced by a meridional circulation. This was due to increasing tropospheric eddy activity over North America and Europe, and its associated meridional transformation of the thermobaric field, which was clearly expressed on 15—16 January at the 300 and 200 mb levels /15/. At this time first signs of increasing meridionality appeared in the middle stratosphere, while on 20—24 January apparently deformation fields with their corresponding circulation system formed in the entire stratosphere.

It is seen from the calculated zonal and meridional circulation indexes that the meridional circulation started gaining strength in the middle and upper troposphere and extended to the middle stratosphere after several days. The subsequent change in indexes of meridional circulation at different levels* is well seen in Figure 4. Since the largest transformations occurred in the first and third belts, the figure shows the variation in I_m only in the first belt.

We presented examples of meteorological-field transformations in the middle stratosphere from 15—25 January 1961. The geopotential and wind fields changed sharply during this time. A polar-cyclone center was located at the 10 mb level on 15 January over Northern Asia, with a high-pressure region situated over Canada (Figure 5a). As of 25 January one can see two cyclonic centers. The second appeared over the eastern seaboard of North America, where it replaced a high-pressure region. The geopotential at the 10 mb level over northeastern USA decreased during these ten days by 110—120 decameters (Figure 5b), with the reduction in the upper troposphere (300 mb) amounting to about 60 decameters.

* Tropospheric processes are most clearly represented at the 300 mb level. Lower-lying levels, including the 500 mb level, are less representative, particularly in summer. This is easily seen from the curves of daily values of I_m in Figure 4.

What is the cause of this substantial reconstruction of the geopotential and air-current fields in the upper stratosphere? Judging by the surface weather map, poorly expressed eddy activity without substantial meridional circulation existed at the surface on 15 January.

Western transport prevailed over the northern hemisphere. Subsequently, as a result of cyclonic activity developing over the eastern seaboard of North America and in the North Pacific, a strong meridional transformation took place in the troposphere as early as on 19 January; this is clearly seen at the surface and on the 500 mb chart. The same pattern was observed at the 300 and 200 mb levels. The eddy generation process at the eastern seaboard of North America continued until 20—22 January, in connection with which the meridionality increased, embracing higher layers, while on 25 January it was expressed in the entire troposphere and stratosphere.

Subsequently the eddy activity over the hemisphere gradually decreased. As early as during the first days of February no cyclonic developments were noted in the southern regions of the moderate belt, and the meridional circulation scales decreased. As of 2 February, western air transport was almost restored in the troposphere and lower stratosphere, while on 3—4 February it was restored also in the middle stratosphere (Figure 6).

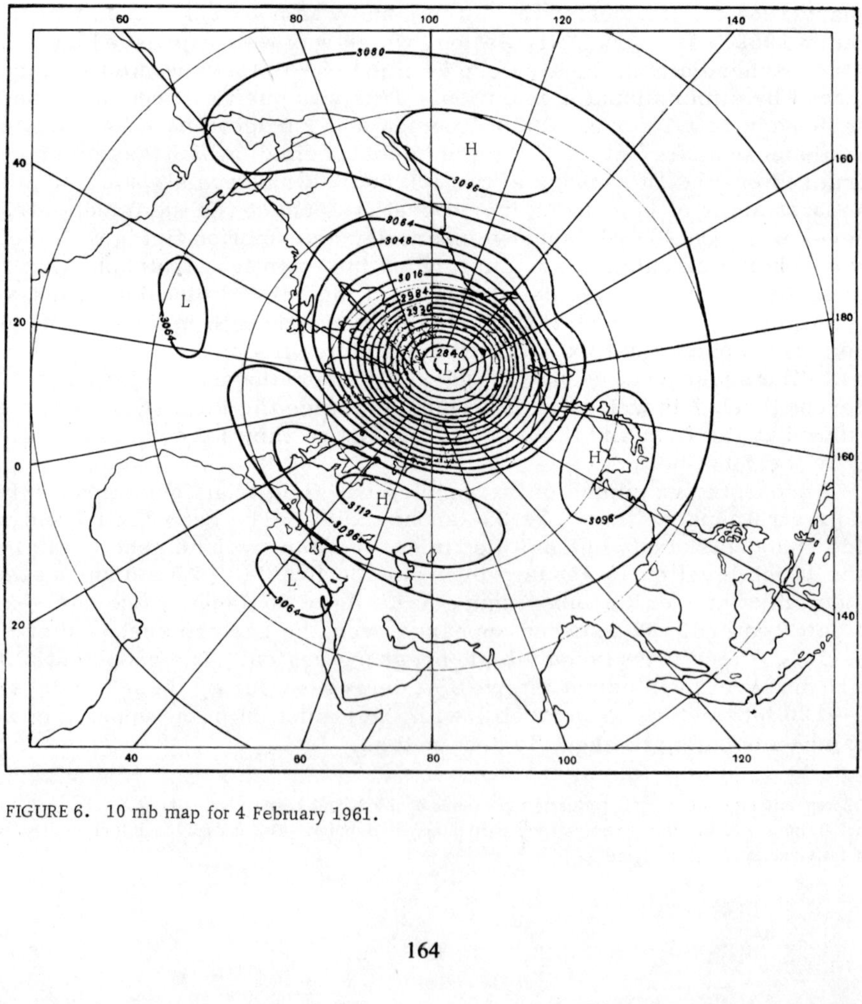

FIGURE 6. 10 mb map for 4 February 1961.

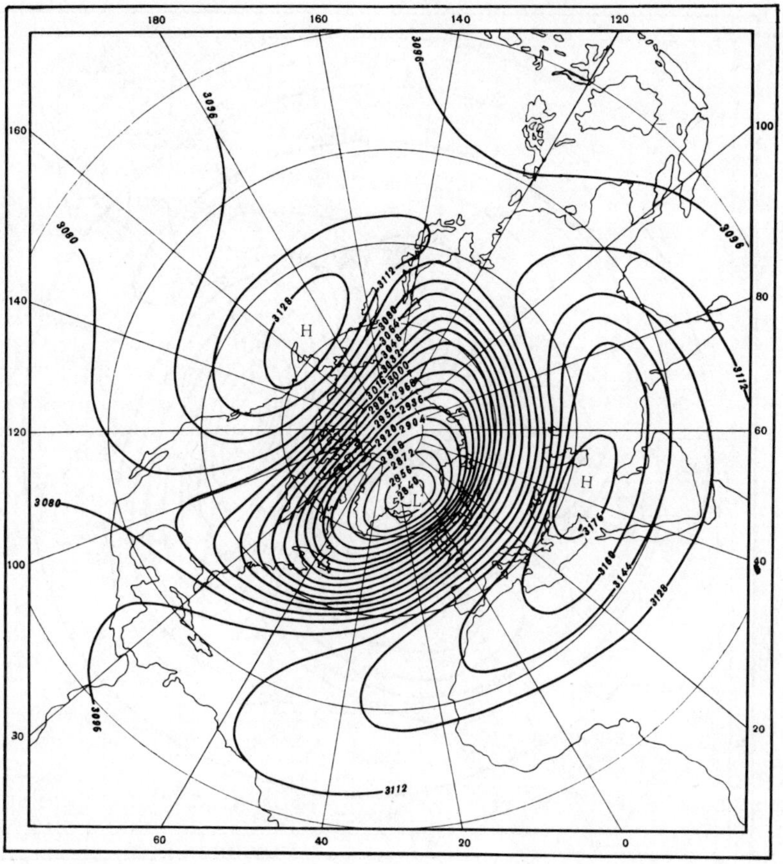

FIGURE 7a. 10 mb map for 23 January 1958.

The above examples of thermobaric field transformation in the troposphere and stratosphere are quite representative. More or less significant modifications of temperature and geopotential fields occur annually.

A similar reconstruction of the thermobaric field, but on a large scale, occurred at the end of January 1958. The scales of transformations, which embraced a large part of the northern hemisphere during this period, can be gleaned from examining the 10 mb maps for 23 and 30 January (Figures 7a and b).

We paid particular attention to the year 1958, which had large anomalies of temperature and of air currents in the troposphere. It was of particular interest from the viewpoint of the feasibility to represent quantitatively the relationship between atmospheric circulation and altitude.

Figure 8 shows curves of the annual variation of the zonal circulation index, calculated from maps of the monthly average values of the geopotential at the 500, 100, 50, 30 and 10 mb levels for the three belts of the northern hemisphere. It is easy to see the overall similarity of the annual variation

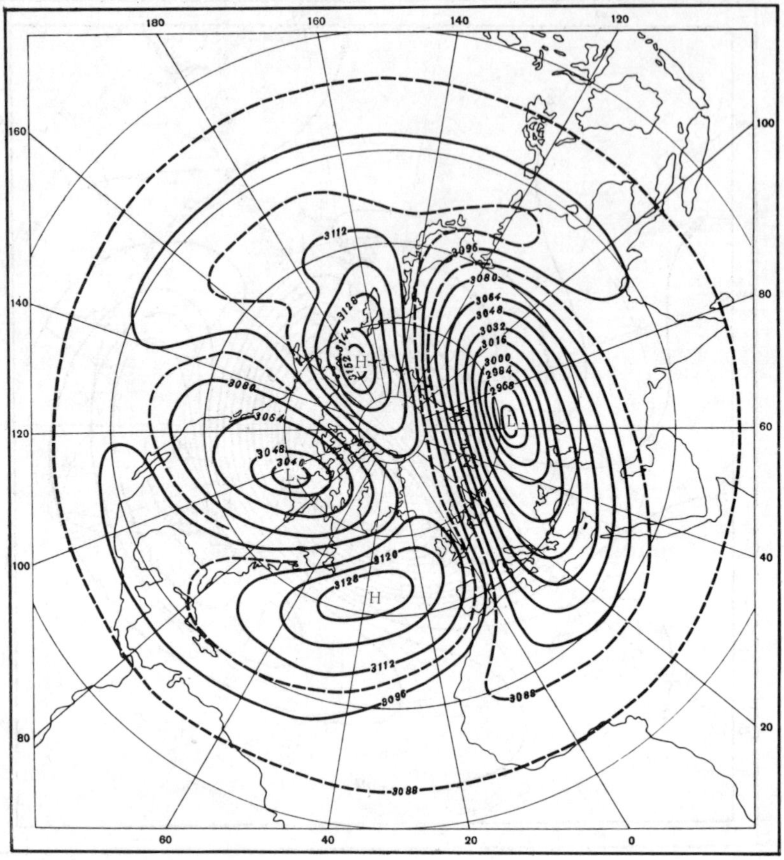

FIGURE 7b. 10 mb map for 30 January 1958.

in I_z at all levels for small annual variations in the troposphere and marked variations in the stratosphere.

In 1958, as usual, the winter western zonal circulation was, on the whole, intensive in the stratosphere of the northern hemisphere. It weakened toward the summer and, in keeping with the radiation regime in high latitudes of the stratosphere, was replaced in May-August by eastern circulation. Only the value of I_z decreased in the troposphere. However, as is indicated by the shape of curves in Figure 8, the month-to-month changes in I_z in 1958 were large, out of keeping with the yearly variation of this index. These were most significant in the first and third belt, i.e., over Europe and North America. The month-to-month changes in I_z at all the levels, with little exception, are quite similar, although the winter and spring of 1958 had exceptionally anomalous processes over the entire hemisphere. These were reflected also in the structure of the high-altitude baric field. In particular, February was among the most anomalous months; during February a strong high-pressure region formed in the region of the Pole and Northern Canada instead of the winter stratospheric cyclone.

The structure of the monthly average geopotential field at the 10 mb level in February, shown in Figure 9, is typical of great atmospheric depth. Cyclonic activity over Eurasia and a high pressure over Canada prevailed in February even at the surface. A negative pressure anomaly with a maximum of 8—12 mb between Scandinavia and the Urals existed along the 60° latitude over Europe and a part of Asia, including southern Yakutia, while a positive anomaly with a 8—10 mb maximum existed at the same latitude over North America. The monthly average value of H at the 500 mb level in Northern Canada was by 15—20 decameters higher than the long-term average, while over Scandinavia, on the other hand, it was lower by the same amount. The greatest deviations of the geopotential from the long-term average in February at the 50 mb level were: +72 decameters over Northern Canada; -32 to -40 decameters over Eurasia. Hence, unlike the strong January zonal circulation, a clearly expressed meridional circulation prevailed in February over the Atlantic and the oceans.

All the circulation features were reflected in the month-to-month variations in I_z and I_m. Thus, it is seen from Figure 8 that the large zonal circulation magnitudes in January at all levels in all belts of the northern hemisphere decreased sharply toward February. At the same time the magnitudes of meridional circulation increased in the first and decreased in the second and third belts, although it would appear that I_m should increase with decreasing I_z and conversely. The cause of this apparently unexplainable phenomenon is easy to explain by considering the fact that not only in the troposphere but also in the stratosphere the geopotential gradients decreased sharply from January to February, particularly over the western hemisphere. The zonal circulation increased again in March-April at all levels and in the entire hemisphere. However, the general yearly variation of zonal and meridional circulation did not differ from the long-term average variation in I_z and I_m. Their maximum devolves upon the winter and the minimum upon the summer months.

Thus, even when considering 1958 which was a year with complex processes, with little exception, the month-to-month changes in the zonal and meridional circulation components in the troposphere (500 mb) and in the stratosphere at the 30 and 10 mb levels, i.e., virtually up to the middle stratosphere, were of the same type.

A close relationship between processes in the troposphere and stratosphere was also detected in the study of a large series of sharply delineated anomalies during different periods of time.

Examples showing the significant effect of tropospheric processes on changes of the geopotential and air-current fields in the stratosphere are common. In our study of planetary-scale circulation processes we have not found a single case of reconstruction of the geopotential field and of tropospheric air currents, which did not occur under the influence of stratospheric processes.

The following conclusions can be drawn from the computed circulation indexes in the stratosphere of the northern hemisphere, comparing them with similar indexes in the troposphere (500 mb) and analyzing them.

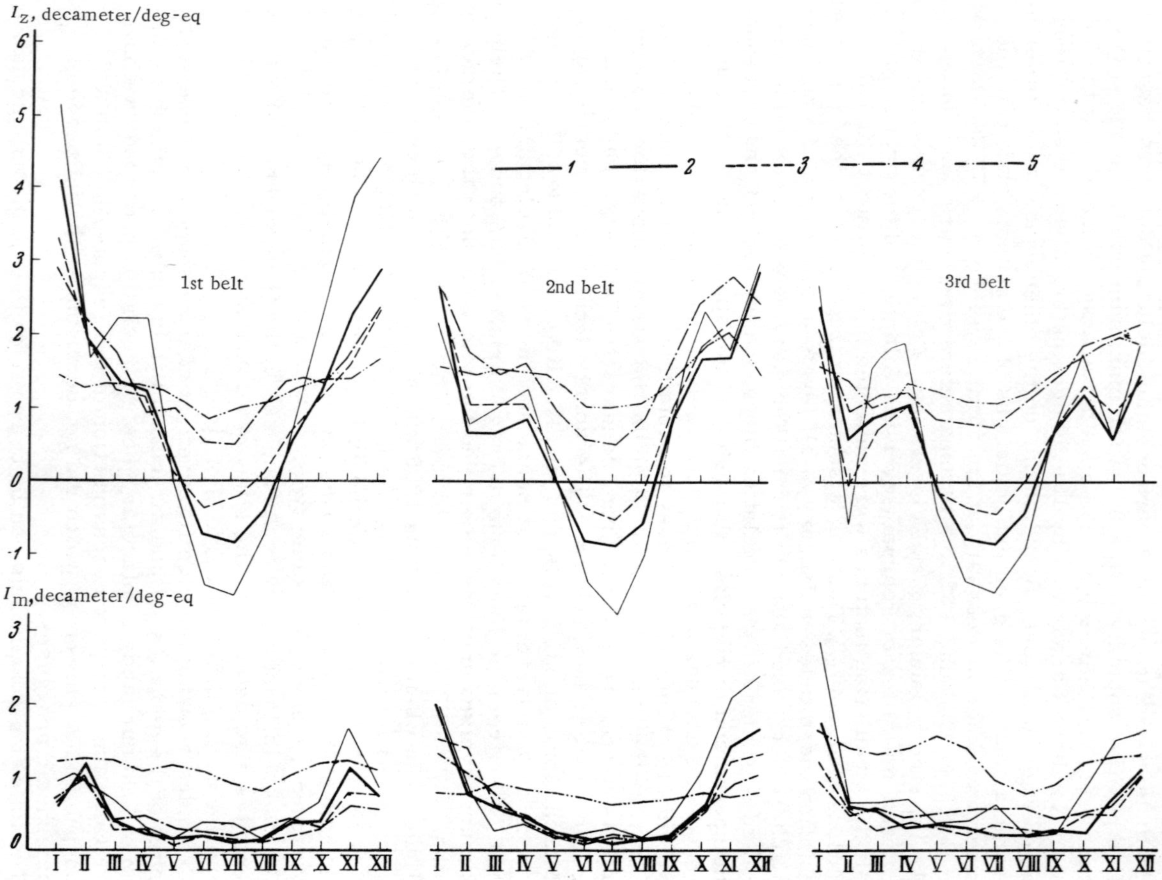

FIGURE 8. Yearly variation of zonal and meridional circulation indexes in the northern hemisphere in 1958:

1—10 mb; 2—30 mb; 3—50 mb; 4—100 mb; 5—500 mb.

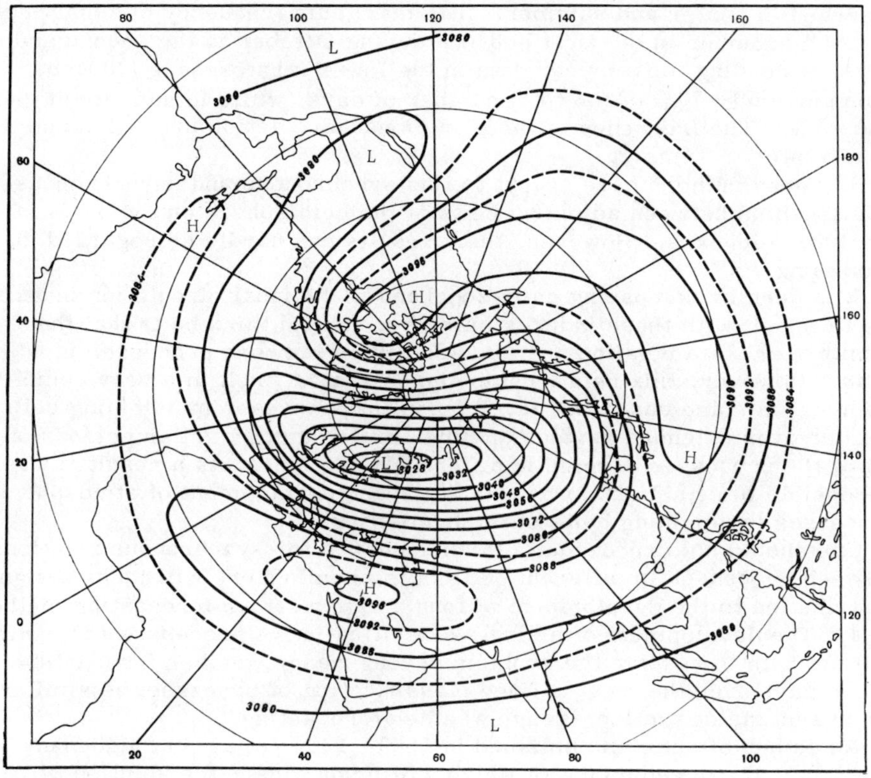

FIGURE 9. Average 10 mb map for February 1958.

CONCLUSIONS

1. Synoptic processes developing in the troposphere and stratosphere are closely related. This is verified by the annual variation in the daily zonal and meridional circulation indexes at the 500 and 100 mb levels for the five-year period, as well as by variations in I_z and I_m at the 500, 300, 200, 100, 50, 30 and 10 mb levels in individual cases of sharply anomalous processes during 1958—1963.

The change in the nature of the winter tropospheric circulation is accompanied by similar changes in the stratosphere. Sometimes these changes occur simultaneously, but usually they lag in the stratosphere by 1—2 or more days. It follows that processes in the stratosphere up to altitudes of 25—30 km are determined by tropospheric processes.

Radiative heating of air under polar day conditions opposes the effect of tropospheric processes in the reconstruction of stratospheric fields.

2. A quantitative estimate of tropospheric and stratospheric processes of three belts of the northern hemisphere using I_z and I_m as indicators shows that, depending on the season, the frequency of the meridional and zonal circulations markedly fluctuates. The most substantial differences are

observed in winter and summer. The maximum frequency of zonal processes during December in the first belt and during October in the second and third belts, according to five-year data in the lower stratosphere (100 mb), comprises 60—70% of the total number of days, while in summer it drops to 4—7%. The frequency of the meridional circulation increases appreciably from winter to summer.

3. Many authors have sought to find synchronous and asynchronous relationships between adjoining belts of the hemisphere on the basis of qualitative indexes. However, their results can hardly be regarded as promising.

It is seen by comparing daily zonal and meridional circulation indexes in the first belt with these indexes in the second and third belts that there deinitely exists a relation and a mutual effect between processes in adjoining belts. However, this relationship demonstrates itself in a very complicated manner: in some cases the development of processes in adjoining belts is similar over extended periods of time, while during similar periods of other years the processes develop in different directions. As a result it was impossible to establish any regularity in the development of atmospheric processes in adjoining belts of the hemisphere.

One should not expect that a further search for synchronous relationships between processes in different belts of the hemisphere will make a significant contribution to the development of long-range weather forecasting methods.

4. The development of a single classification of processes /8/ obviously will not help to resolve the problem of long-range weather forecasting. The use of any, even the most refined classification of processes is similar to the use of analogs in long-range weather forecasting.

A number of articles published in 1963—1964 /8, 23, 32/ list many problems, the development of which will bring closer the solution of problems of long-range weather forecasting, but these do not contain sufficiently substantiated information on the promise held by any of these approaches.

Great importance in the study of regularities in atmospheric circulation should be assigned to physicosynoptic studies. The aim of such studies is the detection of causes for transformation of high-altitude deformation fields in adjoining belts and the determination of the type of process transformation in the course of their development at the hemisphere, with allowance for radiation. This will make it possible to establish the nature of the stability of processes and their anomalies during individual time intervals over large areas.

Synopticostatistical studies should be expanded to obtain information on modifications of moving air, in a system of atmospheric eddies, with allowance for cloudiness and its changes and for establishing the probability of the given transformations of high-altitude deformation fields in different regions of the northern hemisphere and primarily of anomalous processes and the weather.

The development of any methods for long-range weather forecasting should involve making allowance for the heat influx on the basis of changes continuously introduced by cloudiness, the terrestrial-surface albedo, etc., rather than on the basis of average indexes. Assistance may be obtained from studies of the thermal balance of the earth, based on data collected by the extensive network of actinometric stations which can supply the required observations from the entire globe. Judging by the successes of satellite meteorology, these may make a great contribution to this goal.

It follows from the above study that the role of the stratosphere as well as of sunspot changes is not predominant in the development of synoptic processes in the troposphere.

References

1. Baidal, M.Kh. Kompleksnyi makrotsirkulyatsionnyi metod dolgosrochnykh prognozov pogody (Integrated Macrocirculation Methods for Long-Range Weather Forecasting). — Gidrometeoizdat. 1961.
2. Barabashkina, A.P. Povtoryaemost' baricheskikh obrazovanii v tsentral'nye mesyatsy sezonov za 1954—1958 v severnom polusharii (Recurrence of Baric Formations during the Central Months of Seasons for 1954—1958 in the Northern Hemisphere). — Trudy NIIAK, No. 19. 1962.
3. Belinskii, N.A. Opyt ustanovleniya indeksa tsirkulyatsii atmosfery (Experience in Establishing an Atmospheric Circulation Index). — Trudy NIU UGMS, Series 5, No. 14. 1946.
4. Blinova, E.N. Gidrodinamicheskaya teoriya voln davleniya i tsentrov deistviya atmosfery (The Hydrodynamic Theory of Pressure Waves and Atmospheric Activity Centers). — Doklady AN SSSR, 39(7). 1943.
5. Vangengeim, G.Ya. Smeshchenie tsentrov deistviya atmosfery i sredneshirotnogo zonal'nogo potoka v svyazi s preobrazovaniyami zapadnoi tsirkulyatsii (Shifting of Atmospheric Activity Centers and of the Latitudinal Average Zonal Flux in Connection with Western Circulation Transformations). — Problemy Arktiki, No. 5. 1958.
6. Vorob'eva, E.V. Sopryazhennost' atmosfernykh protsessov v severnom polusharii (Coupling of Atmospheric Processes in the Northern Hemisphere). — Gidrometeoizdat. 1962.
7. Girs, A.A. Nekotorye osobennosti sinopticheskikh protsessov v Arktike i ikh svyaz' s obshchei tsirkulyatsiei atmosfery (Some Features of Synoptic Processes in the Arctic and their Relationship to the General Circulation of the Atmosphere). — Trudy GGO, 56(118). 1956.
8. Girs, A.A. O sozdanii edinoi klassifikatsii makrosinopticheskikh protsessov severnogo polushariya (On Setting up a Uniform Classification of Macroscopic Processes in the Northern Hemisphere). — Meteorologiya i gidrologiya, No. 4. 1964.
9. Gromova, G.G. Kharakteristika zonal'nykh i meridional'nykh protsessov nad Vostochnoi Sibir'yu i prilegayushchei chast'yu Tikhogo okeana (Classification of Zonal and Meridional Processes over East Siberia and the Adjoining Part of the Pacific). — Trudy TsIP, 119. 1962.
10. Dzerdzeevskii, B.L. Tsirkulyatsionnye skhemy sezonov goda v severnom polusharii (Circulation Schemes for Seasons of the Year in the Northern Hemisphere). — Izvestiya AN SSSR, seriya geograficheskaya, No. 1. 1957.

11. Zubyan, G.D. O mezhshirotnom obmene teplykh i kholodnykh mass vozdukha v stratosfere zimoi (On the Interlatitude Exchange of Warm and Cold Air Masses during Winter in the Stratosphere).— Meteorologiya i gidrologiya, No. 1. 1959.
12. Kats, A.L. Sezonnye izmeneniya obshchei tsirkulyatsii atmosfery i dolgosrochnye prognozy (Seasonal Variations in the General Circulation of the Atmosphere and Long-Range Forecasts).— Gidrometeoizdat. 1960.
13. Morozova, M.I. and M.A. Petrosyants. Povtoryaemost' tsiklonov i antitsiklonov v troposfere nad severnym polushariem (Recurrence of Cyclones and Anticyclones in the Troposphere of the Northern Hemisphere).— Trudy TsIP, 106. 1960.
14. Pavlovskaya, A.A. Vnutrisezonnye izmeneniya atmosfernoi tsirkulyatsii v verkhnei troposfere i nizhnei stratosfere (Intraseasonal Changes of Atmospheric Circulation in the Upper Troposphere and Lower Stratosphere).— Trudy TsIP, 104. 1961.
15. Pavlovskaya, A.A. Svyaz' mezhdu protsessami v troposfere i nizhnei stratosfere (Relationship Between Processes in the Troposphere and Lower Stratosphere).— Trudy TsIP, 137. 1964.
16. Pagava, S.T. Estestvennyi sinopticheskii raion (A Natural Synoptic Region).— Trudy TsIP, 36(63). 1954.
17. Papinashvili, K.I. Atmosfernye protsessy v Zakavkaz'e i ikh svyaz' s makrotsirkulyatsionnymi protsessami nad Evraziei (Atmospheric Processes in Transcaucasia and their Relationship with Macrocirculation Processes over Eurasia).— Gidrometeoizdat. 1963.
18. Pogosyan, Kh.P. Sezonnye kolebaniya obshchei tsirkulyatsii atmosfery (Seasonal Fluctuations of the General Circulation of the Atmosphere).— Trudy TsIP, 1(28). 1947.
19. Pogosyan, Kh.P. Advektivno-dinamicheskie osnovy frontologicheskogo analiza (Advective-Dynamic Fundamentals of Frontolysis).— Trudy TsIP, 7(34). 1948.
20. Pogosyan, Kh.P. Obshchaya tsirkulyatsiya atmosfery (The General Circulation of the Atmosphere).— Gidrometeoizdat. 1959.
21. Pogosyan, Kh.P. Struinye techeniya v atmosfere (Atmospheric Jet Streams).— Gidrometeoizdat. 1960.
22. Pogosyan, Kh.P. Sezonnye i vnutrisezonnye izmeneniya temperatury i vozdushnykh techenii v stratosfere (Seasonal and Intraseasonal Changes of Temperature and Air Currents in the Stratosphere).— Trudy TsIP, 104. 1961.
23. Pogosyan, Kh.P. Sovremennye problemy obshchei tsirkulyatsii atmosfery (Modern Problems of the General Circulation of the Atmosphere).— Meteorologiya i gidrologiya, No. 3. 1963.
24. Pogosyan, Kh.P. and A.A. Pavlovskaya. Nekotorye osobennosti tsirkulyatsii vozdukha v stratosfere severnogo polushariya (Some Features of Air Circulation in the Stratosphere of the Northern Hemisphere).— Meteorologiya i gidrologiya, No. 8. 1964.
25. Pogosyan, Kh.P., A.A. Pavlovskaya, and M.V. Shabel'nikova. Vzaimosvyaz' protsessov v troposfere i stratosfere severnogo polushariya (Relationship Between Processes in the Troposphere and Stratosphere of the Northern Hemisphere).— Gidrometeoizdat. 1965.

26. Pogosyan, Kh. P. and M. V. Shabel'nikova. Evolyutsiya struinykh techenii pri meridional'nykh preobrazovaniyakh termobaricheskogo polya (Evolution of Jet Streams in Meridional Transformations of the Thermobaric Field). — Trudy TsIP, 104. 1961.
27. Rakipova, L. R. O vozdeistvii solnechnoi aktivnosti na obshchuyu tsirkulyatsiyu atmosfery (The Effect of Sunspots on the General Circulation of the Atmosphere). — Astronomicheskii sbornik. Izdatel'stvo L'vovskogo gosudarstvennogo universiteta. 1960.
28. Sadovnikov, A. V. Nekotorye novye kharakteristiki obshchei tsirkulyatsii atmosfery vo vnetropicheskikh shirotakh i ikh vremennoi khod (Some New Characteristics of the General Circulation of the Atmosphere in Extratropical Latitudes and Their Time Variation). — Trudy Pervoi nauchnoi konferentsii po obshchei tsirkulyatsii atmosfery. Gidrometeoizdat. 1962.
29. Sazonov, B. I. Vysotnye baricheskie obrazovaniya i solnechnaya aktivnost' (High-Altitude Baric Formations and Sunspots). — Gidrometeoizdat. 1964.
30. Spiridonova, Yu. V. Sopryazhennost' atmosfernoi tsirkulyatsii v raznykh chastyakh severnogo polushariya (Coupling of Atmospheric Circulation in Different Parts of the Northern Hemisphere). — Izdatel'stvo AN SSSR. 1962.
31. Khrabrov, Yu. B. Metodika sostavleniya prognozov pogody na 3—7 dnei (A Method of Preparing Weather Forecasts for 3 to 7 Days). — Gidrometeoizdat. 1959. [IPST catalog No. 216. OTS 60-51075. 1961.]
32. Khrabrov, Yu. B. Perspektivy razvitiya sinopticheskogo metoda dolgosrochnykh prognozov pogody (Prospects for the Development of a Synoptic Method of Long-Range Weather Forecasts). — Meteorologiya i gidrologiya, No. 3. 1963.
33. Rossby, C. G. Relation Between Variations in the Intensity of the Zonal Circulation of the Atmosphere and the Displacement of the Semipermanent Centres of Action. — J. Marine Res., 11(1). 1939.
34. Scherhag, R. Die explosionsartigen Stratosphärenerwärmungen des Spätwinters 1951/1952. — Ber. Dtsch. Wetterdienstes in der US-Zone, No. 38. 1952.
35. Teweles, S. and F. G. Finger. An Abrupt Change in Stratospheric Circulation Beginning in Mid-January 1958. — Monthly Weather Review, No. 1. 1958.
36. Teweles, S. Anomalous Warming of the Stratosphere over North America in early 1957. — Monthly Weather Review, No. 10. 1958.
37. Warnecke, G. Ein Beitrag zur Aerologie der Arktischen Stratosphäre. — Meteorol. Abhandl. Inst. Meteorol. und Geophys. Freien Univ. Berlin, 3(3). 1956.
38. Wexler, H. Seasonal and Other Temperature Changes in the Antarctic Atmosphere. — Quart. Roy. Meteorol. Soc., 85(365). 1959.

A. Kh. Khrgian, V. I. Bekoryukov, V. M. Berezin, and Yu. A. Shafrin

PROBLEMS OF OZONE CIRCULATION IN THE TERRESTRIAL ATMOSPHERE

The article examines variations in ozone in connection with the general circulation of the atmosphere and vertical air motions. It is shown that meridional circulation results in the accumulation of ozone in polar regions and the reduction of its content in tropical latitudes. Computation shows that the level of maximum ozone concentration increases for ascending motions while the total ozone content markedly decreases. The reduction in maximum ozone concentration and particularly the increase in the total quantity of ozone in descending motions are not evident. The vertical ozone distribution in the presence only of eddy mixing is examined. Eddy exchange does not result in extensive vertical redistribution of ozone and has little effect on its total content.

The compilation of synoptic maps of ozone content may serve as an important auxiliary tool in the study of the general circulation of the atmosphere. Ozone may serve as a tracer for atmospheric currents, its changes are undoubtedly related to the synoptic situation and, what is even more important, its vertical distribution (as will be shown below) is closely related to the type of stratospheric circulation.

On the other hand, ozone absorbs a substantial proportion of solar radiation and is hence responsible for the formation of a warm layer at altitudes of 30—50 km; consequently, differences in the amount of ozone can produce air current peculiarities at these altitudes. This means that ozone may also play an active role in the formation of the general circulation. However, this interaction is quite complicated. It is known, for example, from acoustical experiments by Vel'ken at Novaya Zemlya during the Second International Polar Year and from rocket observations at Heiss Island /1/, that the above warm layer also exists during the polar night, although ozone cannot then absorb solar radiation, since the latter is nonexistent. Thus, both the warm layer and most likely the ozone existing in winter in the circumpolar region are to a large extent transported there by general circulation currents.

In the study of interaction between ozone and circulation one notes four states, which have since developed into four separate studies:

1. study of the correlation between ozone and air temperature at different levels with pressure, tropopause height, etc.;
2. attempts to relate ozone to the surface synoptic situation and later on with the trajectories of ozone-conveying air masses;
3. study of statistical relationships between the total amount of ozone and features of the zonal and meridional circulations;
4. study of ozone transport.

A correlation (inverse relationship) between the total amount of ozone and the temperature of the upper troposphere was discovered by Dobson and Meetham in Oxford as long ago as 1934 /14/. These relationships were studied in great detail by Johansen from data collected in Tromsö in 1941—1945 /16/. He also found an inverse relationship between ozone and the temperature of the troposphere, the tropopause height and the pressure at altitudes of 3—12 km. He found a direct relationship between ozone and temperature at altitudes of 12—15 km. He also discovered that "ozone waves" over Tromsö are opposite in form to pressure waves at the 6 km altitude. It was found that ozone fluctuations are strongest in spring and autumn; Johansen treated this relationship as one of ozone with a specific synoptic formation—high-altitude ridges and troughs. Later these correlation studies were continued by many others /2, 6, 8, 10/, who arrived in general at the same conclusion of a direct relationship between ozone and the temperature of the lower stratosphere, as well as an inverse relationship between ozone and the temperature of the upper troposphere.

Such correlations, found from observations made at a single location, encourage investigators by verifying that they should exist, but have little to say about their physical mechanism. Nobody even attempted to correlate these relationships, or clarify their dependence on latitude and season. At present this stage of studies is only of historical interest.

There was also no lack of attempts to clarify the relationship between ozone and the synoptic situation, first with cyclones and anticyclones, which appeared to the synopticians of the old school as the main demonstrations of atmospheric circulation. As late as at the start of the 1930's quite detailed isoline charts were drawn of the average ozone distribution in cyclones and even anticyclones, usually compiled on the basis of observation at one point /14/. In 1946 Dobson /7/ reproduced such maps in his extensive and well known work. In 1954 Miyake and Kawamura /20/ studied the variation in ozone at Tateno [near Tokyo] as a function of the location of the observational point relative to a "moving anticyclonic center", and made an important step forward by studying ozone in conjunction with the location of the observational point relative to the location of the axis of a "typical trough" axis at the 500 mb level.

The most important of the above works are probably individual studies on determining the dependence of the ozone on given forms of motion. Thus, in 1944 Götz /15/ described a sharp rise in ozone in the Alpine foothills, thus proving that descending motions even in the troposphere may aid in downward ozone transport. At the same time Moser, using several examples, related changes in ozone to its advection at the 16 km level (about 100 mb) from polar or tropical regions. It is true that this left unresolved the question how far one must move along these trajectories in order to reach the region determining the amount of ozone at the end point of the trajectory. In 1959 Gushchin /5/ presented several examples, primarily for spring, showing that the ozone content is higher on the left and lower on the right side of a jet stream. Graphic examples of this kind are obviously very useful for future, more general studies.

Studies of statistical relationships between ozone and numerical or morphological characteristics of the general circulation were carried out primarily in connection with work done at the MGU /9/. Their main conclusion was as follows: an increase in zonal circulation (in the entire moderate-latitude belt of the hemisphere) creates as if a barrier, aiding the accumulation of ozone in high latitudes and reduction in its content in lower latitudes; conversely, a local amplification of meridional circulation, increasing the influx from the North, aids in increasing the local ozone concentration. Relationships of this kind are most pronounced in spring when the meridional ozone gradient is greatest, and disappears in summer and at the start of autumn when the ozone maximum about 75—80°N, typical of other seasons, disappears. These relationships are detected when analyzing the fluctuations of ozone and circulation day after day, as well as in the study of long-term fluctuations of both these variables.

From the above statistical and synoptic comparisons of circulation with ozone concentration, one wishes to isolate one small group, which is quite closely related to the main topic of the present report. As early as in 1934 Dobson and Meetham /14/ put forward the idea that if there exists a common descending current, extending over a large extent of the atmosphere, then the ozonized air will descend, while the air replacing it will rapidly become ozone-enriched; the ozone in the descending flow will be protected from the destructive solar radiation. No examples or calculations were presented in /14/ in support of this scheme and it was then forgotten for a long time. In 1954 Paetzold /23/ presented it again in the same hypothetical aspect as Dobson. The aforementioned observation of Götz concerning the background presented only a weak and isolated verification of this hypothesis. Martin and Brewer /18/ attempted to provide a statistical verification on the basis of relatively extensive data collected over northwestern Europe in 1952. They compared the total ozone content with the magnitude of the absolute eddy at the 100 mb level and found in the first case a correlation factor of +0.55 and in the second case one of even +0.64. Having reference to the known relationship

$$\frac{\Delta}{\Delta t}(\Omega + l) = (l + \Omega)\frac{\partial w}{\partial z},$$

it is clear that the elevated ozone content is associated with the stronger descending motions below the 100 mb level (as compared with motions above it). Martin and Brewer estimated the order of magnitude of the descending motion velocity responsible for the observed changes in ozone which, in their opinion, does not exceed 0.7 cm/sec. A year earlier Normand /21/ suggested that this requires velocities of the order of 1—2 cm/sec.

For a quantitative study of ozone transport phenomena in the general circulation one may write the continuity equation in the form

$$\frac{\partial \rho}{\partial t} + \frac{\partial(\rho w)}{\partial z} + \frac{\partial(\rho u)}{\partial x} + \frac{\partial(\rho v)}{\partial y} = \alpha(\rho_0 - \rho) + \frac{\partial}{\partial z}\left(D_z \frac{\partial \rho}{\partial z}\right) + \frac{\partial}{\partial x}\left(D_x \frac{\partial \rho}{\partial x}\right) + \frac{\partial}{\partial y}\left(D_y \frac{\partial \rho}{\partial y}\right), \quad (1)$$

where ρ is the partial density of ozone; u, v and w are the velocity components; D_x, D_y and D_z are diffusion (molecular or eddy) coefficients, which may be functions of coordinates along three directions; ρ_0 is the equilibrium ozone density (concentration), which would have been described solely by photochemical reactions; α is a factor inversely proportional to time τ for restoring one half of the ozone by means of the same reactions. The values of τ were calculated by Kondrat'eva /12/ (using the most recent data on the extraterrestrial solar spectrum and upper-atmosphere temperatures for the 47° latitude).

According to observations it may be assumed that $\rho = 0$ at some low altitude z, since in the troposphere the ozone is in fact rapidly decomposed due to high temperatures, and ρ is actually zero at some high altitude, where ozone formation on triple molecular collisions also becomes non-productive.

The first attempt to solve equation (1) with allowance only for vertical motions and neglecting diffusion was made by Kondrat'eva and Khrgian in 1962 /12/. Equation (1) was integrated numerically in such a manner that it would be possible to find changes in the vertical distribution 6, 12 and 24 hours after the start of an ascending or correspondingly descending motion, i. e., the unsteady problem was solved. Here the diffusion of ozone was neglected, but then an allowance was made for differences in the initial ozone concentration and its associated, not too great, difference in α.

This numerical integration of equation (1) showed that the ozone content of ascending motions decreases below the 22 km level and increases above it. The total amount X of ozone for 24 hours here decreases, in the entire layer, by 0.042 — 0.052 cm for a velocity (w) of 10 cm/sec, and by 0.014 — 0.019 cm for a velocity linearly decreasing downward. Such processes can therefore fully explain day to day changes in the amount of ozone, for example, over strong warm fronts, over which w possibly reaches such relatively high velocities as +10 cm/sec.

The results of calculations for descending motions ($w < 0$) are less definite and changes in X may be both positive and negative, depending on the form of motion and initial conditions. Thus, for a constant value of $w = -10$ cm/sec an increase in X in the entire layer is still perceptible after 12 hours (+0.010 cm), but it may be replaced by a reduction if the initial ozone layer was relatively thick. Velocities decreasing in the downward direction (for example, for a strong divergence in the lower stratosphere) are particularly unfavorable to increasing the ozone concentration in descending motions. Thus, the air descent (referred to by Dobson and Meetham) still does not suffice for accounting for sudden increases in ozone concentration, its relation with the cold front, the existence of polar ozone and of its maximum.

Subsequently Berezin and Shafrin /4/ examined the unsteady problem of ozone transport with consideration of turbulent diffusion and of the effect of the latter. Only the one-dimensional problem ($u = v = 0$, $D_x = D_y = 0$) was considered. The vertical air velocity was taken equal to $\pm cz$, where $c = 4.5 \cdot 10^{-6}$ sec^{-1}. It was assumed that either $D_z = 3$ m^2/sec (A), or $D_z \sim u^2$ (B), where u is the wind speed calculated from the average circulation index according to data of Mashkovich and Kheifets /11/. Figure 1 shows the distribution of ρ after 4, 12 and 24 hours for ascending and descending motions. An initial total amount of ozone in the 6 — 50 km layer amounting to 0.300 cm was reduced after 24 hours of ascending motion to 0.219 cm under version A (and to 0.218 in version B). The reduction first proceeded at a low rate, and then accelerated. After 24 hours it was most pronounced in the 6—22 km layer (by a factor of two). Diffusion changes the ozone content only slightly, and also at altitudes below 25 km; the difference between versions A and B in the 6—22 km layer after 24 hours does not exceed 2 %, and in the entire layer this difference is less than 0.5 %. The calculated ozone distribution clearly shows the manner in which ascending motion results, alongside with a total reduction in ozone concentration and a sharp rise of the layer in which it attains a maximum. Both these features recall the ozone distribution in the tropics, described, for example, by Ramanathan /24/.

For descending motions the ozone concentration maximum is shifted downward from 26 to 21.5 km in 25 hours, although slower than it was shifted upward in the preceding case. Here the ozone concentration increases sharply by 77—79 % in the lower layers (6—22 km) and decreases in the upper layer (22—50 km) by 26 %. The rise in the total amount of ozone, amounting to about 14 % in 24 hours, is less substantial than its reduction corresponding to ascending motions in the preceding case.

This case can be ascribed to polar conditions (in the same manner as the preceding case was attributed to tropical conditions), in which the increase in ozone concentration is very clearly expressed in winter and spring at the time of development of circumpolar anticyclones. Here, however, it becomes clear that vertical motions alone cannot account for the difference between the polar and tropical ozone distribution, as long as one assumes that the vertical motions have velocities of the same order of magnitude.

It was shown in /4/ that differences in the assumption regarding the turbulence intensity, even if the latter is quite strong, have little effect on the type of distribution, as well as on the total ozone concentration. This effect is perceptible only in the lower part of the ozone layer. Below we shall again verify that this factor is of secondary importance in ozone physics.

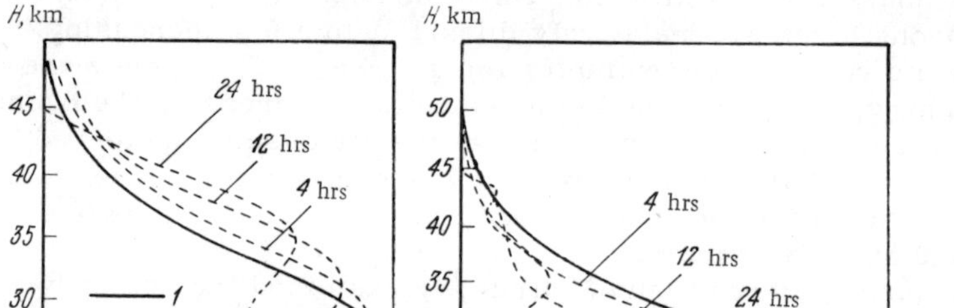

FIGURE 1. Modification of the vertical ozone-concentration profile:

a — in ascending currents; b — in descending currents; 1 — equilibrium curve; 2 — D_{const}; 3 — D_{var}.

Reference /3/ considers changes in ozone concentration when there exists a meridional ozone circulation, which is a closed circulation between the 20 and 70° latitudes. Taking into account the motion along the meridian and making allowance by means of a factor ctgθ for the convergence of meridians to the pole, it is possible to rewrite equation (1) in terms of quasi-Cartesian coordinates:

$$\frac{\partial \rho}{\partial t}+\frac{1}{R}\frac{\partial(\rho v_r)}{\partial\theta}+\frac{2}{R}\rho v_r+\frac{\rho v_\theta \operatorname{ctg}\theta}{R}=\alpha(\rho_0-\rho), \quad (1')$$

where v_r and v_θ are the radial (vertical) and meridional air-current velocities; θ is the colatitude; R is the distance to the center of the earth. The calculations were carried out for a model of a closed meridional circulation in the stratosphere, analogous to the known Murgatroyd model /19/ and described by the equations

$$\begin{aligned} v_r &= v_0 \exp\left[\left(\mu-\frac{2}{R}\right)(r-R_e)\right]\cos b(r-R_0)\frac{\sin aR(\theta-\theta_0)}{\sin\theta}, \\ v_\theta &= -v_0 b\exp\left[\left(\mu-\frac{2}{R}\right)(r-R_e)\right]\sin b(r-R_0)\frac{\cos aR(\theta-\theta_0)}{a\sin\theta}, \end{aligned} \quad (2)$$

where R_e is the radius of the earth; R_0 is the distance from the center of the earth to the center of the stream; v_r and $v_\theta=0$; $\mu=0.125\ \text{km}^{-1}$ is the reciprocal of the height of the homogeneous atmosphere. The circulation velocity was determined by the value of multiplier v_0, which was assumed in the two versions to be equal to 1 and $2\cdot10^{-7}$ km/sec, corresponding to vertical velocities at the region boundaries of up to 1 cm/sec and to horizontal velocities of up to 2.5 m/sec.

It was assumed for simplicity that

$$\alpha=\alpha_0 e^{0.5H}=6\cdot10^{-13}e^{0.5H}\ \text{sec}^{-1}, \quad (3)$$

$$\rho_0=10(H-10)^3 e^{-0.25H}\begin{cases}1+\sin 2\theta & \text{for } \theta<\frac{\pi}{4}\\ 1+(2-2\sin 2\theta) & \text{for } \theta>\frac{\pi}{4},\end{cases} \quad (3')$$

where $H=r-R_e$. Equation (3') thus made it possible to allow for the difference in the rates of photochemical reactions in the north and south belts. The solution of (1'), which is a first order partial differential equation, depends on the values of ρ, specified along some characteristics, in the case at hand, along the vertical and horizontal characteristics passing through the circulation center. Selecting the distribution of ρ in such a manner that the solutions in all four circulation sectors are identical at the corresponding boundaries, we found a distribution $\rho(z)$ which is continuous and independent of any arbitrary initial conditions. However, the solution was obtained only for the steady case. Figure 2 shows the corresponding curves of $\rho(z)$ for the 30 and 65° latitudes. The ozone accumulation in the polar latitudes in the layer below 20 km is extremely pronounced and very clearly expressed; it is compensated only to a small extent by a reduction at altitudes above 25 km.

The total amount of ozone at the 65° latitude in the 15—36 km layer reaches 0.410 cm for $v_0=2\cdot10^{-7}$ km/sec and amounts to almost 150% of the average ozone concentration in this layer (about 0.273 cm). At the same time the absorbed amount of ozone over this latitude exceeds by only 19% the average value of X for the entire hemisphere. Thus, a prolonged general meridional circulation, which concentrates ozone in the upper stratosphere in a poleward direction, would have produced in the relatively small circumpolar region even a greater ozone increase than that observed on the average in nature. Possibly the effect of the polar night (not considered by us), when the accumulation of atomic oxygen (O) required for ozone formation ceases, should substantially attenuate this circulation effect, if reference is had to extended processes. We recall that in individual cases the values of X measured in polar regions are much greater, for example, up to 0.656 cm (Dickson Island) in April 1958, so that most probably the circulation may sometimes result in an even greater ozone accumulation than under the average conditions considered above.

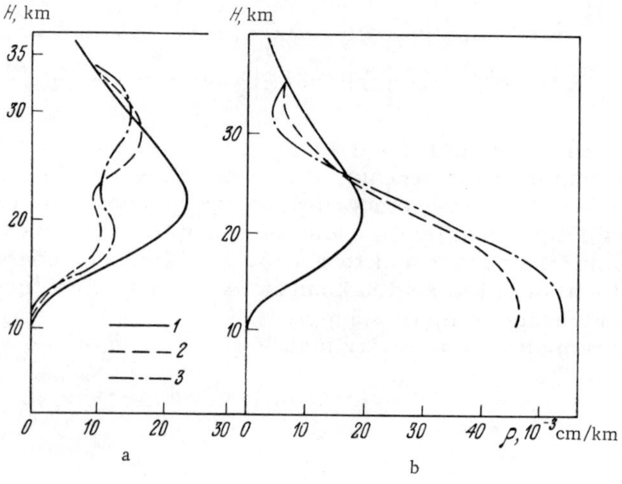

FIGURE 2. Modification of the vertical ozone profile for meridional circulation:

a — $\varphi = 30°$; b — $\varphi = 65°$; 1 — equilibrium curve; 2 — $v_0 = 1 \cdot 10^{-7}$ km/sec; 3 — $v_0 = 2 \cdot 10^{-7}$ km/sec.

The ozone accumulation rate decreases and the maximum concentration level moves upward as one passes from polar to moderate latitudes. A maximum corresponding to the principal photochemical maximum starts forming already from the 55—60° latitudes at the 24—25 km level. Here we also have a second, lower maximum, "introduced" by lower-lying air influx from higher latitudes. The principal maximum rises even higher as one approaches the tropics, and the total amount of ozone is then markedly smaller. It was calculated that in the 10—36 km layer at 20° the ozone concentration decreases to 0.207—0.208 cm, i. e., to 72—73% of the normal amount of ozone in this layer. In nature, however, X at 20° is appreciably greater, amounting to about 86% of the normal for the hemisphere. Thus, our circulation scheme emphasizes somewhat, also in this case, the effect of the latter on ozone changes. But then the agreement with calculations using the simpler one-dimensional model (see above) is much greater.

Traces of the lower, advective ozone maximum on the computed curves remain in the lower stratosphere also in tropical latitudes at altitudes of 16—18 km. The behavior of this lower maximum should have been subjected to a more detailed experimental study to determine, among others, to what extent its appearance is associated with the increasing meridional exchange in the tropical latitudes.

To obtain some idea about the ozone transport in the atmosphere by large-scale eddies, we used the known concept of macroturbulence /13/. The intensity of the latter precisely typifies the horizontal and vertical transport averaged over a relatively extended period of time. To make allowance for it, we supplemented the right-hand side of equation (1') by the terms

$$D_z \frac{\partial^2 \rho}{\partial z^2} + D_y \frac{\partial^2 \rho}{\partial y^2} - D_y \frac{\partial \rho}{\partial y} \left(\operatorname{tg} \frac{y}{R} \Big/ R \right), \tag{4}$$

where D_z and D_y respectively denote the coefficients of vertical and horizontal eddy transport, while y is the horizontal coordinate. We selected, according to the Palmer-Libby model /17/, $w = B - \lambda y$, where $B = 2$ cm/sec and $\lambda = 0.4 \cdot 10^{-18}$ sec^{-1} and where, consequently, the maximum ascending motion velocity at the equator is +2 cm/sec and the minimum descending velocity is −2 cm/sec at the pole. Using the continuity equation, we took the meridional circulation velocity in the form

$$v = f(z) \cos^{-1} \frac{y}{R} = (v_0 z - F) \cos^{-1} \frac{y}{R} \tag{5}$$

relative to the pole and yielding at 10 km a northern wind of ~ 3 m/sec and at the 40 km altitude a southern wind of ~3.45 m/sec, with the circulation center located at $\theta = 45°$ and $H = 24$ km. We also assumed that $D_z = 25$ m^2/sec (quite a high value) and $D_y = 10^6$ m^2/sec.

Figure 3 shows the computed steady vertical ozone distribution in the absence of ordered circulation, i.e., in the presence only of mixing by large-scale eddies. Here the boundary condition is the specified vertical ozone distribution at the equator, corresponding to its total content and equal to 0.160 cm in the 10—40 km layer. It is seen that the exchange by large-scale eddies perceptibly smooths out the vertical ozone concentration, without affecting the altitude of the maximum-concentration layer. The total amount of ozone in the 10—40 km layer there increases somewhat (up to 0.172 cm at 22.5°) and then decreases slowly poleward to 0.164 cm.

Thus, theoretically even very strong large-scale eddy exchange by itself results only in a moderate redistribution of ozone and in very small changes relative to its total concentration.

Figure 3b presents the ozone distribution in the presence also of meridional circulation, which most likely best corresponds to winter conditions. The maximum concentration layer in the tropics lies as high as the 32 km level; in the polar region it is located at about 21 km, i.e., approximately as in the preceding models, which assume simple settling. The total amount of ozone in the tropics dropped to 0.144 cm, and in the polar region increased to 0.191 cm, i. e., by approximately 20 % as compared with the initial distribution at the equator. The differences between the pole and equator in this model are thus smoothed out as compared with those actually observed (50 %).

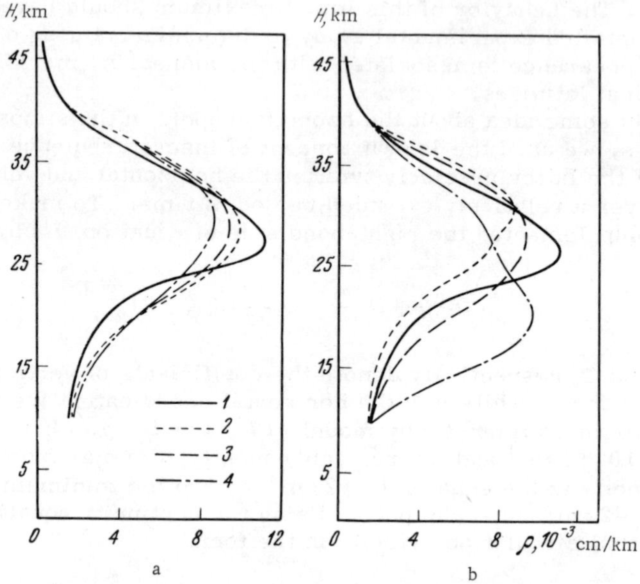

FIGURE 3. Modification of the vertical profile of ozone:

a— with large-scale eddy exchange; b— with allowance for meridional circulation; 1—boundary distribution; 2—φ = 22.5°; 3—φ = 45°; 4—φ = 90°.

This means that, as expected, large-scale eddies attenuate the contrasts and differences in the ozone concentration produced by the general circulation. It is seen by comparing the curves of Figures 2 and 3b that allowance for large-scale eddy exchange makes it possible to eliminate those features which are poorly verified by observation and which would have arisen if allowance were made for only ordered circulation, for example, the exaggeratedly sharp and very low ozone maximum at 65°. It also smooths out the very sharply expressed twin-peak vertical ozone distribution in the intermediate latitudes, mentioned previously.

CONCLUSIONS

1. The study of atmospheric ozone transport due to the general circulation (with ascending motions in the tropics and descending motions in the polar region) and of large-scale eddy phenomena is needed for understanding the causes of the observed geographic distribution of ozone and of its yearly variations. It may also be found useful for diagnozing atmospheric motions (particularly vertical), based on ozone observations.

2. Simple models of vertical motions, without allowance for horizontal circulation, yield for tropical latitudes (for ascending motions) a distribution which is close to that observed. The maximum-concentration level here lies quite high, and the total amount of ozone decreases rapidly in keeping with observations. The descent of this layer and particularly an increase in the ozone concentration with descending motions is not as evident and depends on a number of other circumstances.

3. The general meridional atmospheric circulation with vertical motion velocities of up to 1 cm/sec results in appreciable accumulation of ozone in polar latitudes (associated to a large extent with the convergence of upper air currents in the polar zone) and in a lowering of the ozone layer there. It also results in a marked reduction in the amount of ozone and in a rise of its layer in tropical latitudes.

It is significant that the existence of this meridional circulation accounts for the presence of twin-peak ozone concentrations, of which the lower maximum (below 20 km) is associated with air influx from the polar regions, while the upper (above 25 km) is due primarily to photochemical processes.

4. The existence of exchange by large-scale eddies smoothes out the effect of the general circulation, markedly attenuates the polar increase and descent of the ozone layer, etc. However, large-scale eddy transport alone cannot account for changes in the height of the ozone layer, nor for the observed changes in the concentration, shape of distribution curves, etc.

References

1. Alekseev, P.P., E.G. Shvidkovskii, et al. Raketnye issledovaniya atmosfery (Rocket Studies of the Atmosphere). — Meteorologiya i gidrologiya, No. 8. 1957.
2. Bezverkhnii, Sh. A. Ozonometricheskie dannye po Alma-Ate v sopostavlenii s nekotorymi meteorologicheskimi faktorami (Ozonometric Data for Alma-Ata as Compared with Some Meteorological Factors). — Trudy Kazakhskogo NIGMI, No. 5. 1955.
3. Bekoryukov, V.I. O raschete vliyaniya zamknutoi vozdushnoi tsirkulyatsii na ravnovesnoe raspredelenie ozona v zemnoi atmosfere (On Calculating the Effect of Closed Air Circulation on the Equilibrium Distribution of Ozone in the Terrestrial Atmosphere). — Geomagnetizm i aeronomiya, 5(3). 1965.
4. Berezin, V.M. and Yu.A. Shafrin. O raschete vertikal'nogo raspredeleniya atmosfernogo ozona (On the Calculation of the Vertical Distribution of Atmospheric Ozone). — Geomagnetizm i aeronomiya, 4(1). 1964.
5. Gushchin, G.P. Zakonomernosti gorizontal'nogo raspredeleniya i kolebanii vo vremeni soderzhaniya atmosfernogo ozona (Regularities of the Horizontal Distribution and Time Variations in the Content of Atmospheric Ozone). — In: Atmosfernyi ozon. Izdatel'stvo MGU. 1961.
6. Gushchin, G.P. Ozon i aerosinopticheskie usloviya v atmosfere (Ozone and Aerosynoptic Conditions in the Atmosphere). — Gidrometeoizdat. 1964.

7. Dobson, G. M. B., A. W. Brewer, and B. M. Cwilong. Meteorology of Lower Layers of the Atmosphere. Proc. Roy. Soc., A, Vol. 185. 1946.
8. Karimova, G. U. Nekotorye rezul'taty nablyudenii za ozonom v Arktike v 1958 g. (Some Results of Observations of Ozone in the Arctic in 1958). — In: Atmosfernyi ozon. Izdatel'stvo MGU. 1961.
9. Kuznetsov, G. I. Ozon i obshchaya tsirkulyatsiya atmosfery (Ozone and the General Circulation of the Atmosphere). — In: Atmosfernyi ozon. Izdatel'stvo MGU. 1961.
10. Lipatova, O. A. Nekotorye rezul'taty nablyudenii obshchego soderzhaniya atmosfernogo ozona v Irkutske (Some Results of Observation of the Total Atmospheric-Ozone Content in Irkutsk). — Trudy GGO, Vol. 141. 1963.
11. Mashkovich, S. A. and Ya. M. Kheifets. O nekotorykh kharakteristikakh zonal'noi tsirkulyatsii atmosfery nad severnom polushariem za periodom MGG (Some Features of the Zonal Circulation of the Atmosphere over the Northern Hemisphere during the IGY). — Trudy TsIP, Vol. 111. 1961.
12. Khrgian, A. Kh., G. I. Kuznetsov, and A. V. Kondrat'eva. Issledovanie atmosfernogo ozona (Atmospheric Ozone). — Izdatel'stvo Nauka. 1965. [IPST Cat. No. 1994, TT 67-51379. 1967.]
13. Shafrin, Yu. A. O makroturbulentnom perenose i meridional'noi tsirkulyatsii ozona (Large-Scale Eddy Transport and the Meridional Circulation of Ozone). — Izvestiya AN SSSR, Seriya Fizika Atmosfery i Okeana, 1(7). 1965.
14. Dobson, G. and A. Meetham. Atmospheric Ozone and Meteorology. — Quart. J. Roy. Meteorol. Soc., 60(256). 1934.
15. Götz, F. Der Stand des Ozonproblems. — Vorträge, Disskuss. der Tagung "Ozon". Tharandt. 1944.
16. Johansen, H. On the Relation Between Meteorological Conditions and Total Amount of Ozone over Tromsö. — In: Polar Atmosphere Symposium, Oslo, July 2—8. 1956; Proc. Method Section. New York, Pergamon Press. 1958
17. Libby, W. F. and C. E. Palmer. Stratospheric Mixing from Radioactive Fall-out. — J. Geophys. Res., 65(10). 1960.
18. Martin, G. W. and A. W. Brewer. Some Synoptic Aspects of European Results (on Ozone). — Sci. Proc. Int. Assoc. Met. Tenth Assembly, Rome. 1954.
19. Murgatroyd, R. J. and F. Singleton. Possible Meridional Circulations in the Atmosphere and Mesosphere. — Quart. J. Roy. Meteorol. Soc., 87(372). 1961.
20. Miyake, Y. and K. Kawamura. Studies of Atmospheric Ozone at Tokyo. — Sci. Proc. Internat. Assoc. Meteorol., Rome. 1954.
21. Normand, C. Atmospheric Ozone and Upper Air Condition. — Quart. J. Roy. Meteorol. Soc., 79(339). 1953.
22. Ozonesonde observations over North America, Vol. I. — U. S. Air Force. Cambridge Research Labs., Manscom Field, Mass. 1964.
23. Paetzold, H. K. On the New Investigation of the Ozone Layer and its Variations. — J. Geophys. Res., 59(3). 1954.
24. Ramanathan, K. R. and R. N. Kulkarni. Mean Meridional Distributions of Ozone in Different Seasons Calculated from *Umkehr* Observations and Probable Vertical Transport Mechanisms. — Quart. J. Roy. Meteorol. Soc., 86(368). 1960.

V. A. Dzhordzhio

JET STREAMS AS INDICATORS OF THE GENERAL CIRCULATION OF THE ATMOSPHERE

A number of problematic items pertaining to the general circulation of the atmosphere is examined, and the entire complexity of the present-day concept of the general circulation is pointed out. The author's goal is the selection of an element which is observable at all altitudes and which reflects the dynamics and energetics of the general circulation. Jet streams are suggested as such an element and arguments in their favor are presented.

During the past few years, when the extensive IGY and IQSY programs were carried out, our knowledge of the structure of the terrestrial atmosphere and of physical processes in it have expanded and deepened. They have also changed qualitatively. Instead of more or less bold guesses and hypotheses, we were supplied with direct measurements from satellites, rockets and space missiles to supplement aircraft and radiosonde observations. So much data are available that it cannot be even processed, and there is not even time to make all the necessary correlations and comparisons, in spite of the fact that the methods proper of processing and analyzing observations have changed: starting with the use of statistical devices which process punched cards and ending with complex specialized electronic computers.

The concept of the general circulation also changed. Not so long ago the general circulation was regarded as some averaged system of air currents which was called upon to explain the similarly averaged climatic pattern, and this also in the most general outline. What then are the changes?

First, studies of the general circulation make extensive use of the mathematics of thermohydrodynamics, which makes it possible (as follows from Mashkovich's article in the present collection) to construct various dynamic models, operative for several tens of days, for example, the Phillips and Smagorinsky models. Many studies of the Kibel'-Blinova school also confirm this thesis.

The modern theory of the general circulation has now become not so much a theory of some average states, as much as a theory of dynamic changes, transitions from one circulation state to another. These are not only seasonal changes, for example, from the winter average to the summer average pattern, but primarily of intraseason departures from dynamic stability, sharp changes in circulation regimes associated with Arctic and tropic invasions, and storm processes of a variety of kinds. The question of dynamic instability thus becomes one of the central problems of the modern theory of the general circulation.

Second, the theory of the general circulation utilizes the newest statistical tools: the theory of correlation and structural functions; spectral analysis; analysis of influence functions, etc., the use of which is extremely labor consuming.

Third, the theory of the general circulation is becoming more geophysical. In addition to the common meteorological elements: pressure, temperature and wind, the general circulation is associated with many other physical or geophysical parameters and processes: solar and terrestrial radiation; electronic density of the atmosphere; its electromagnetic field; cosmic rays; ozone; atomic oxygen and nitrogen; tritium and some products of thermonuclear fission, etc.

Instead of ordinary thermodynamics in the upper layers of the atmosphere one deals with photochemical thermodynamics, and instead of hydrodynamics one considers magnetodynamics.

Instead of high wind speeds in jet streams, which may be as high as 100—120 and even 140 m/sec in the dynamic region, the velocities encountered in the upper atmosphere are even higher. Analytical solutions for the lower and upper atmosphere should be combined into a single and ordered system of equations, the individual components of which take effect depending on the altitude.

During these years there were discovered new phenomena such as electron streams at 90 and more kilometers, earth radiation belts and many others. All of them are to a greater or lesser extent related to the dynamics of atmospheric processes within the widest sense of this word, i.e., to the general circulation of the atmosphere.

Alongside the utilization of purely geophysical processes and of phenomena related primarily to upper atmospheric levels in the study of the general circulation, we are continuously expanding our knowledge of the underlying surface, i.e., of the surface of continents and oceans and of their cloud cover.

Very extensive studies were carried out on the radiation and heat balances of the terrestrial and cloud surfaces, and methods were developed for their synoptic study by satellites and rockets.

Fourth, the space within which the general circulation is studied has become much larger. These studies have become global. At the same time we moved from the exosphere to the study of higher altitudes. It now becomes necessary to investigate the gaseous tail of the earth as well as the solar wind, i.e., corpuscular fluxes from the sun and their accompanying electromagnetic processes, and the capture of upper layers of the terrestrial atmosphere by the solar corona.

It should not be thought that the concept of the general circulation is extended upward artificially and prematurely. Even without reference to the commonly understood concept of a relationship between all the geophysical processes occurring in the atmosphere, it suffices to point out purely practical needs of the next several years. It is to be expected that in the next 5—7 years aircraft flying above 20—25 km will be developed and the need will arise to protect the craft from the corroding effect of ozone (rubber, for example, has poor resistance to prolonged exposure to an ozone atmosphere) and, much more important, to protect the crew and passengers from the possible effects of cosmic radiation. Cosmic rays may not act always and everywhere, but rather during individual periods and possibly over particular routes. One must know the corresponding regularities and be able to reliably forecast these dangerous regions. This problem is also contained within the concepts embraced by the common definition of the general circulation of the atmosphere.

The problem of long-range forecasting, although using the same equations as the general circulation, can be solved only on the basis of a multilevel and geophysical theory of the general circulation. The equations will change, although the coupling noted by Monin will definitely remain. In view of the above, the recent definition by Tucker /9/ to the effect that the general circulation is the integral influence of all the meteorological phenomena and their effect on large-scale atmospheric motion, is clearly unsatisfactory and reflects the past state of the science.

After the entire complexity of the modern concept of the general circulation of the atmosphere and its geophysical and dynamic nature have been described, one must select an element which could be observed at all altitudes, would reflect the dynamics and energetics of the general circulation, and be easily observable and lend itself to study. I suggest jet streams as such an element. Everybody knows what a jet stream is and hence I shall not discuss in detail their physical properties, for which the reader is referred to Chapter 6 of Part I of /3/. The principles of jet-stream classification are presented in /2/. The argument for selecting jet streams is now presented in brief.

1. Jet streams exist virtually in all spheres of the terrestrial atmosphere.

2. The modern radiosonde network, which grows from year to year, allows one to easily detect jet streams in the troposphere and lower stratosphere up to approximately 38—40 km. The existing synoptic rocket network will soon be expanded and then jet streams in the upper stratosphere, mesosphere and thermosphere will become accessible.

3. Jet streams constitute clusters of atmospheric kinetic energy, since they embrace quite large air masses which develop tremendous speeds.

4. The relationship between jet streams and high planetary frontal zones /1/ indicates that they are regions of the greatest concentration of not only kinetic, but also potential energy. Fronts associated with jet streams and frontal cloud systems carry copious supplies of a third kind of energy—the energy of condensation processes and of moisture instability. Finally, jet streams are regions of eddy and momentum concentrations. Certain features in the ozone distribution are also associated with them.

5. Jet streams may apparently be detected by meteorological satellites from characteristic cloud systems, and where there are no clouds they can be sensed from radiation measurements.

6. Jet streams are real winds with all the ageostrophic components and accelerations contained in them, produced by unsteadiness. It should be mentioned that the time has come to cease working with the baric field with its tempting accuracy in determining the pressure, apparent simplicity and sufficiency of the geostrophic approximation, and to start working with the real wind. The baric hypnosis must be overcome and new algorithms must be developed for numerical forecasting and for the theory of the general circulation.

7. Large-scale vertical motions, which change sign at the stream axes, develop precisely in the jet-stream zone.

8. Jet streams are closely associated with atmospheric baroclinic zones, by characteristic bunching of thermodynamic solenoids. These baroclinic

zones may start at the earth, but may also develop at some altitude above it, in the upper troposphere and also in the stratosphere.

9. Jet streams border on high-altitude troughs and ridges, and after diffluence also on high-altitude cyclones and anticyclones, usually designated as central. The dynamics of long waves can be easily followed from the configuration of the jet stream and the distribution in them of maximum-wind regions.

10. If the jet-stream type is determined precisely and different jet types are not confused, then it is possible in the first approximation to neglect changes in jet altitude along the stream and approximate it by means of a two-dimensional mathematical model.

11. Jet streams are easily modeled by physical as well as mathematical means.

12. When working with a multilevel model of the general circulation of the atmosphere, jet streams can be encountered at different stages, but they are still accessible to analysis. In the final count, these are several axes on a map along which one can mark out all the necessary data on the altitude, wind speed, kinetic energy, eddies, etc., and then supplement these by various geophysical characteristics of the general circulation. Hence one can plot on one map for a given day, using different colors, the entire multilevel set of jet streams: from the Arctic front streams at the bottom to mesospheric and thermospheric streams at the top.

We now consider in more detail the simultaneous analysis of the ensemble of jet streams.

At present a student of the general circulation has to deal with the surface-level synoptic map and satellite cloudiness map, which reflect the properties of the surface, as well as with high-altitude 850 to 10 mb constant-pressure maps and with results of synoptic rocket soundings. The meteorologist must take into account also geophysical data: the electron density; electromagnetic field intensity; chemical constituents; ozone; water vapor; cosmic radiation intensity; location in space of radiation belts, etc. How can one grasp at the same time this large volume of data, how should one clarify the dynamics of all this ?

It is here that some basis (a framework or skeleton which will be covered by other numerous data) comes in useful. This is the skeleton which will be provided by the three-dimensional (multilevel) ensemble of jet streams.

Work on ensembles of jet streams is now carried out in the USA and USSR, particularly in Kazakhstan in Soviet Central Asia. However these are only first steps, pertaining only to the troposphere and lower stratosphere, and it is still too early to talk about results.

In conclusion I would like to dwell on the uneven development of the science of the general circulation of the atmosphere.

The range of studies is here extraordinarily wide: from elementary statistical aeroclimatic work, to extensive studies, which can be carried out only by large institutes. Alongside fine methods of analysis of the dynamic instability of stratospheric jet streams and means of kinetic energy transport from one atmospheric level to another, an example of which is the work by Charney /5/, there exist bold, but sometimes unsubstantiated predictions. I shall now cite several examples of the latter.

Studies by Usmanov /4/ of the critical parallels and of the genesis of jet streams due to overheating of terrestrial deserts entirely lack proofs of the quite bold assumptions of the author. Serra /7, 8/ proves the symmetry of the baric field relative to the equator, although the author is forced to stretch some points beyond limit. The author forgets that in addition to the earth rotation, which may produce such a symmetry, there exists an asymmetrical thermal situation (the winter of the northern hemisphere occurs during the summer of the southern hemisphere, and conversely). Faust and Attmannspacher /6/ postulate the existence of two circulation systems, separated by the 20 km level and situated one above the other. Each of these systems has its own dynamics and its own heating level, whose energy is transferred to the maximum-wind layer at which large-scale vertical motions change sign. Hence they call this layer the null layer. The lower system has the terrestrial surface as the heating level with the maximum-wind layer located at the 10 km altitude. The upper system has the upper ozone-layer boundary as the heating level, with the maximum-wind layer located at approximately 55 km.

It thus follows that no jets may exist between the jet streams at the tropopause and stratopause levels. In fact, however, quite strong jet streams are actually encountered at altitudes between 20 and 35 km, which contradicts the above assertion. Jet streams are also encountered at the 20 km level, where a wind speed minimum should exist according to the theory of /6/.

Now I wish to present only some of the data of the Problematic Laboratory of Jet Streams of Tashkent University. The western wind over Gur'ev on 26 January 1962 at the 26 km level reached a speed of 93 m/sec, and over Aktyubinsk on 28 January 1962 at the 21 km level it was 88 m/sec. Over Mineral'nye Vody on 1 February 1962 at the 33.6 km level the northwestern wind reached a speed of 105 m/sec, while the next day over Gur'ev (27 km level) the wind speed in this direction was 127 m/sec. On 6 July of this year a western wind was recorded at the 26.6 km level over Alma-Ata with a speed of 91 m/sec. Hundreds of such cases can be cited.

These examples point to the need for a critical approach to many works on general circulation and for careful checking of any hypotheses against actual data.

References

1. Bugaev, V.A. and V.A. Dzhordzhio. Planetarnaya vysotnaya frontal'naya zona (The Planetary High-Altitude Frontal Zone). — Trudy TsIP, 25(50). 1961.
2. Dzhordzhio, V.A. K voprosu o klassifikatsii struinykh techenii (On the Classification of Jet Streams). — Trudy Tashkentskogo Gosudarstvennogo Universiteta, Vol. 270. 1964.
3. Rukovodstvo po kratkosrochnym prognozam Tsentral'nogo instituta prognozov (Instructions of the TsIP on Short-Range Forecasts) (Chapter 6), Part I. — Gidrometeoizdat. 1964.
4. Usmanov, R.F. O vliyanii vrashcheniya Zemli na obshchuyu tsirkulyatsiyu atmosfery (The Effect of the Earth Rotation on the General Structure of the General Circulation). — Trudy TsIP, Vol. 104. 1961.

5. Charney, J.G. and M.E.Stern. On the Stability of Internal Baroclinic Jets in a Rotating Atmosphere.—J. Atmos. and Terrest. Phys., 19(2). 1962.
6. Faust, H. and W.Attmannspacher. On a Connection Between Heatings Levels and Levels of Maximum Winds.—Space Research II. Proc. 2nd Internat. Space Sci. Sympos. Florence, Apr. 1961. Amsterdam, North-Holland Publ.Co. 1961.
7. Serra, A. Principio de Simetria.—Rev. Brasil. Geogr., No. 3, July—Sept. 1962.
8. Serra, A. Letter to the editor.—Monthly Weather Review, 92(9). 1964.
9. Tucker, G.B. The General Circulation of the Atmosphere.—Weather, 17(10). 1962.

P. D. Astapenko and S. S. Gaigerov

SOME FEATURES OF THE ATMOSPHERIC CIRCULATION OVER THE SOUTHERN HEMISPHERE

The regime of the troposphere and the structure of the stratosphere in the southern hemisphere are examined on the basis of rocket soundings from R/V "Ob" in the 65–69° S belt in 1957–1958 by comparison with the northern hemisphere. A quantitative estimate is presented of the intensity of atmospheric circulation in both hemispheres for an area contained between 90 and 180° E, and 70° S and 70° N from 500 mb maps obtained by calculating the circulation indexes for 1961–1963 by the Kats method.

The geographic features of the southern hemisphere affecting the atmospheric circulation and appearing at all latitudes, particularly in the moderate and southern belts, are well known. Unlike the northern hemisphere, the continents of the southern hemisphere occupy a relatively small area, particularly in middle and lower latitudes. Hence they have a less disturbing effect on the circulation.

At the same time, the existence in the southern polar region of a wide expanse of ice-covered and quite elevated continent serves during the entire year as a source of cold for the atmosphere and a cause of a horizontal temperature gradient, much more pronounced than in the northern hemisphere and directed from the equator to the South Pole. The cold source in the Antarctic extends in summer to the lower half of the troposphere, and in winter even further upward to the lower stratosphere. This is responsible for the high speed, stability and appreciable vertical extent of the latitudinal western current in moderate latitudes of the southern hemisphere, exceeding its counterpart in the northern hemisphere in all the above respects. To sum up, the southern hemisphere has a higher potential energy level of the planetary baric field of the moderate-latitude troposphere than the northern hemisphere, with large current velocities in the high-altitude planetary frontal zones.

As atmospheric sounding reached ever-increasing altitudes, it became known that many circulation features in the troposphere affect the regime of the winter stratospheric eddy. The circumpolar stratospheric cyclone in the southern hemisphere is more stable and symmetrical than its northern hemisphere counterpart. Below are presented some additional data on this. It is clarified that, on the one hand, the physicogeographic features of each hemisphere are felt up to very high altitudes, while on the other hand, reconstruction of stratospheric fields first occurs in the upper and middle stratosphere and only then in the lower stratosphere. This feature is clearly expressed also in the atmosphere of the southern hemisphere.

In considering the tropospheric processes of the southern hemisphere one must emphasize the role of the Antarctic and the Antarctic Ocean as cold sources for the atmosphere and the World's oceans. At the same time it should be remembered that the cooling effect of the Antarctic continent is attenuated by the fact that the continual intensive surface-level inversion isolates to a great extent the high-lying part of the troposphere from the effect of the intensively cooled part of the continent. Under the effect of gravity the cooled surface-air layers produce drainage winds, as a result of which the cold air is removed to the continent periphery, being heated during its rapid descent.

Of particular importance in the study of the general circulation is estimating the contribution of the southern hemisphere to the energy budget of the atmosphere as a whole and clarifying the problem of momentum transport through the equator. Definite results are obtained from a comparative estimate of the circulation in both hemispheres, presented in this article.

The regime of the troposphere in the southern hemisphere and the structure of the stratosphere. The location of continents and oceans in the southern hemisphere creates most favorable conditions for the development of zonal circulation. The isohypses on the 500 mb climatological maps pass almost parallel to the latitude circles. The altitude difference of the 500 mb surface in the 50—90°S belt is 42 decameters according to the January average and 59 decameters according to the July average maps. This is due to the relatively small variations in the radiation balance of the surface in moderate and polar latitudes of the southern hemisphere.

A comparison of the location of zero isotherms of the surface air layer illustrates the difference in the radiation and heat balance of the surface in the circumpolar regions of both hemispheres, which determines to a large extent the circulation features in a quite thick layer of the atmosphere. As is well known, the zero isotherms of the surface air layer of the winter and summer seasons in the southern hemisphere are almost symmetrical and their extreme positions differ by only 7° of latitude. In summer the temperature is below zero not only over the entire Antarctic continent, but also over a large part of the Antarctic Ocean. In the northern hemisphere the zero isotherm is located in summer over the Central Arctic due to strong heating of the air over the continents, while in winter, as a result of intensive cooling over the continents, it extends far to the south, particularly in Eastern Asia and in the east of North America.

It is known that the isohypse configurations of the average 500 mb constant-pressure maps and of the 50 and 10 mb winter maps are similar. The January 10 mb maps clearly show troughs oriented toward Eastern Asia and the east of North America. Constant-pressure 50 mb maps for the southern hemisphere show that the winter stratospheric depression has a more regular shape, which is also in agreement with the configuration of isohypses of the middle troposphere and the location of the zero isotherm in the surface layer of the southern hemisphere.

The effect of the physicogeographic features of the southern hemisphere on the atmospheric circulation up to appreciable altitudes is graphically demonstrated by the greater zonality of stratospheric processes.

For the Antarctic we only have data of several rocket soundings made in the 65—69°S belt from the research vessel "Ob" in 1957—1958.

Figure 1 presents data for summer (December—January) and autumn (April). The summer curve is similar to the corresponding curve for Hays Island. If one remembers that Franz Josef Land is located at higher latitudes (the difference relative to the location of the "Ob" in Antarctic waters is 5—11°), it can be said that the Antarctic stratosphere is heated in summer somewhat more than the Arctic stratosphere. A cooling of up to 27—29° is observed in autumn in the Antarctic at altitudes of 35—40 km. The cooling at these altitudes over Hays Island (i. e., at much higher latitudes) for the corresponding period is 30—35°. It can be assumed from this that the upper stratosphere is cooled more intensively at the Antarctic shores than the corresponding region of the northern hemisphere.

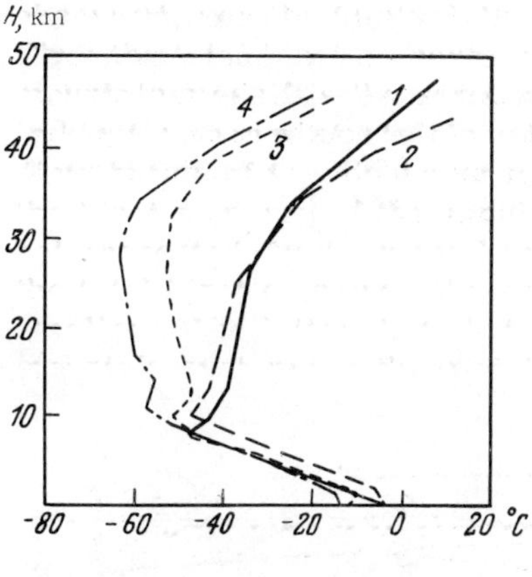

FIGURE 1. Average vertical temperature distribution from radiosonde and rocket data:

1 — summer, Antarctica ("Ob") 65—69° S (December 1957—January 1958); 2 — summer, Hays Island 80° N (June—July 1957—1959); 3 — autumn, Antarctic (April 1958); 4 — autumn, Hays Island (October 1958—1959).

The Antarctic stratosphere is intensively cooled in winter. For example, a temperature of −93° was measured at the rather low altitude of 21 km at the Amundsen-Scott station on 17 July 1958. During the polar night the cooling extends to increasingly lower stratosphere layers. The temperature minimum in the tropopause sets in at the end of winter in August—September. In spring the stratosphere is warmed very rapidly and, in addition, this appears to occur from the top downward.

The synoptic processes developing during spring reconstruction are clearly shown on the 50 mb map for 24 October 1958, presented in /15/. In this case the anticyclone ridge extended from New Zealand to Hallett Station, while the weakening stratospheric cyclone moved over Queen Maud Range and continued to the Weddell Sea. It can be concluded from this that synoptic processes upon spring reconstruction in the southern hemisphere are similar to stratospheric processes occurring in the northern hemisphere.

Usually spring reconstruction starts in October, and terminates in November in the formation of a stratospheric anticyclone over the Antarctic. Figure 2 shows the monthly average vertical section through Mirnyi—Vostok—South Pole—Byrd for October. The temperature field reconstruction in the lower latitudes is finished, while the lower stratosphere over the Central Antarctic still remains cold. The warm ridges propagate from the continent periphery, with the result that the greatest temperature contrasts and atmospheric jet streams move from the periphery to the internal regions of the Antarctic.

The spring warming dates are not fixed. In 1957—1962 warming started in the stratosphere quite early, in September. Wave activity developed at fronts on the periphery of the stratospheric cold region during this 1962 warming. Judging from observations at Mirnyi the development of ridges resulted in increasing the velocities of the western stratospheric jet stream to over 130 m/sec. This example illustrates the scheme for vanishing of the stratospheric cyclone; the cyclonic eddy is somewhat attenuated by the general warming of the stratosphere and the ridges which develop over moderate latitudes extend to the Antarctic. The stratospheric anticyclogenesis over the moderate latitudes of the southern hemisphere may be associated with appreciable solar elevations and also with the fact that, according to MacDowall, the ozone maximum in the southern hemisphere in spring (September—October) is located at 50—55°S. Over the Pole it appears much later, in November and December /16/.

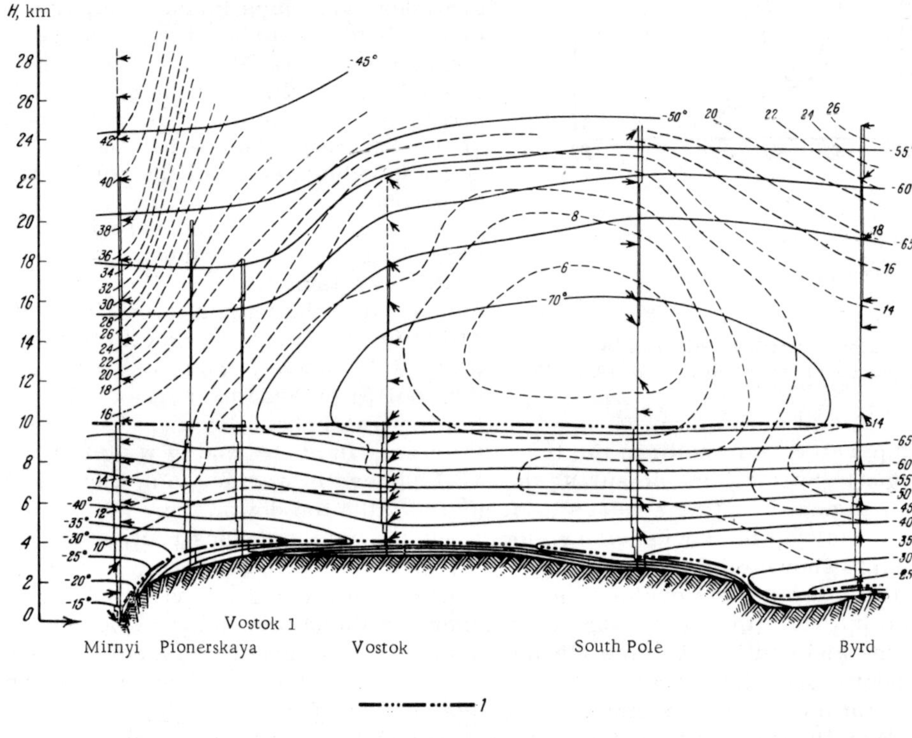

FIGURE 2. Vertical section through the stratosphere over the Antarctic along the Mirnyi—Vostok—South Pole—Byrd meridian:

1 — tropopause region.

Figure 3 compares various features of the stratospheric regime of both hemispheres. This is done by means of an aerological section constructed by the U.S. Weather Bureau along 75°W from pole to pole for January 1958 /17/. Drifting station observations and rocket data over Hays Island in the Arctic, as well as data of Antarctic stations and rocket soundings from the "Ob" /11/, were used in constructing the section.

As a result of winter warming, there is virtually no difference between the summer and winter poles in the Arctic stratosphere, and the stratospheric cold region in the northern hemisphere is shifted somewhat to the south, into 83—53°W. Unfortunately, we could not construct such a section for July due to the unavailability of rocketsonde data over the Antarctic. However, it appears doubtful that the temperature over the summer and winter poles in July will equalize, since no winter stratospheric warmings were observed in the Antarctic. Apparently, these do not exist here.

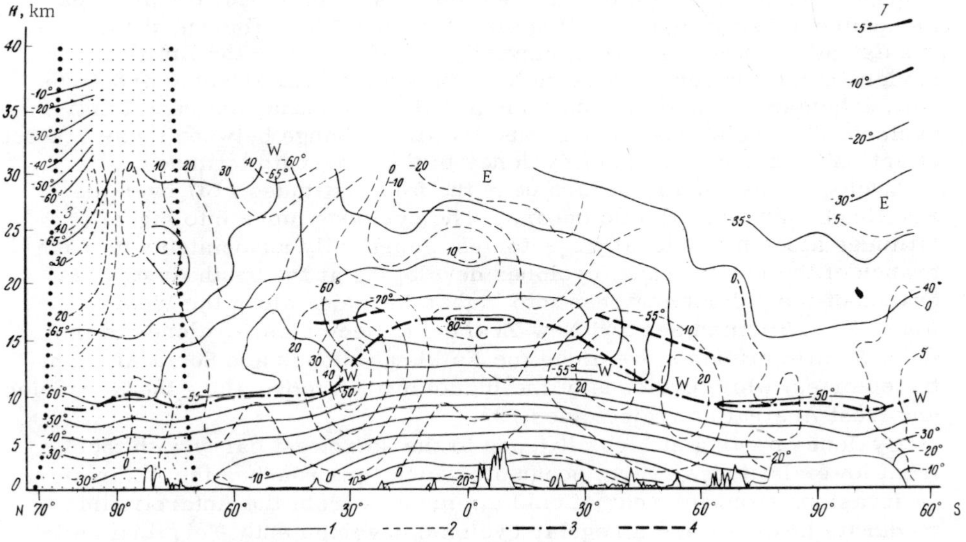

FIGURE 3. Vertical meridional section from the Arctic to the Antarctic:

1 — isotherms; 2 — isolines of the zonal wind component, in m/sec; 3 — tropopause; 4 — second tropopause.

Warmings in the lower layers of the stratosphere occur for a number of reasons before temperature rises in the lower layers. This is apparently due to the fact that descending motions in the clearly delineated stratospheric inversion produce more heating than in the lower stratosphere and mesosphere. The wind speeds in the upper stratosphere are high and, consequently, advective temperature changes are appreciable. In addition, the upper part of the ozone layer is heated very much by solar radiation. To this one should add the effect of the inclination of thermal and circulation systems in the stratosphere; these may be associated with different temperature changes in the layers above and below. However, this does not mean that stratospheric reconstruction occurs independently of the troposphere. It was emphasized before that the averaged schemes of the terrestrial thermobaric field of the stratosphere of the northern and southern hemispheres are in close agreement with the corresponding tropospheric fields and features of the underlying surface. This points to the existence of correlation between processes in these parts of the terrestrial atmosphere, although the nature of these correlations still remains unclear.

The atmospheric circulation over the southern hemisphere and its features. Due to the large size of the Antarctic continent and its great effect on the climate and atmospheric circulation conditions, the moderate-latitude region in the southern hemisphere is very limited from the south. The boundary of the Antarctic passes along the Antarctic convergence line, lying substantially to the north of the South Polar Circle. At points it reaches 50°S and only over individual segments of the Pacific does it reach 62°S. The moderate latitude belt in the southern hemisphere is thus narrower than in the northern hemisphere.

The circulation conditions in the troposphere of moderate latitudes of the southern hemisphere develop directly under the influence of the quasisteady centers of action, partially located outside the limits of the moderate belt proper. These include the subtropical elevated pressure belt, subantarctic depressions strip and the Antarctic continental anticyclone. The relationship and constant air exchange between these "centers of action" occur by means of cyclones in the moderate latitudes.

Cyclogenesis usually occurs near the forty latitudes /13/ over the Pacific, Indian and Atlantic oceans. The cyclones move into higher latitudes along meridional trajectories, genetically associated with each branch of the front. Thus, cyclones developing at the South Australian branch of the polar front pass into higher latitudes along the West Australian, Tasman Sea and New Zealand trajectories; cyclones of the West Pacific branch move along the Falkland Islands and South African trajectories, while cyclones of the Indian branch move along the Kerguelen and West Australian trajectories /2/.

Cyclones moving at high altitudes to the southwest carry with them, from lower latitudes, warm ocean air which thus reaches the Antarctic.

Invasions from the rear of cold air masses from the Antarctic into moderate latitudes are a regular cyclonic development. Terminal anticyclones supplying the subtropic oceanic maxima of the southern hemisphere, unlike their north hemisphere counterparts, do not become wider in winter than in the summer; conversely, their areas are somewhat smaller.

The polar-front jet stream in the southern hemisphere can be distinctly followed in the upper troposphere throughout the year. It is a planetary-scale formation, consisting of some long waves. The average wavelength usually varies from 60 to 120° longitude.

The number of long waves simultaneously typifies the intensity of subtropical maxima and the intensity of circulation, the speed at which surface baric systems and atmospheric fronts travel in moderate latitudes. When there are only a few waves, the surface cyclones and fronts move almost twice as fast as compared with the presence of the usual number of waves. The speed changes accordingly from 60—65 to 30—35 km/hr. The cyclonic velocities in the southern hemisphere are higher than in the northern, corresponding to the higher circulation energy of the southern hemisphere. The depth of the southern cyclones is also greater than in the northern hemisphere.

The development of cyclonic activity and the corresponding formation of the weather over land within the limits of South America and southern parts of the African and Australian continents has its special features.

Thus, cyclones over South America develop at moderate or polar fronts, and their trajectories for all seasons are of the same kind; to the north of the 50° parallel the majority of them move from southwest to northeast, following the steering current and the general direction of the polar front at the southeastern periphery of the subtropic anticyclone over the South Atlantic. Upon reaching the 35—30° latitude the cyclones are either filled or turn eastward and then southeast. The cyclones do not pass the Andes anywhere north of the 50° parallel. Cyclones from the Pacific arrive at the South American continent only by swinging past the southern tip of the continent; here they either rise into high latitudes along the South American or Falkland meridional trajectories into the Antarctic, or turn to the northeast, as stated above.

Cyclones also form over the continent, particularly during transition seasons, but the primary cyclogenesis center is located at the eastern shore of South America, at the La Plata estuary, where the western segment of the Atlantic branch of the polar front lies.

The depressions over Australia typically appear from the west, from the central regions of the Indian Ocean or from the shores of South Africa. They replace anticyclones and their appearance is first detected by a change in wind direction in Western Australia, where dry and hot northeastern winds are replaced by fresh cool sea-air currents from the southwest. In individual cases the cooling following a cold front at the rear of a meridionally aligned trough may occur quite sharply.

The passing of the so-called "meridional fronts" from west to east is typical of Australia, as well as of the Tasman Sea and New Zealand; these fronts usually occupy the col between the anticyclones, extending from northwest to southeast, forming an arc with the southern end at the center of some high-latitude depression and with the northern end extending to the tropics. The meridional front as a whole behaves as a cold front; it may develop steady and secondary cyclones, which subsequently flow into the main high-latitude cyclone, develop along it and move southward. It is precisely these kinds of synoptic situations which, according to some investigators (Kraus), cause 90% of autumn and spring rains in Southern Australia. There are no warm-type atmospheric fronts over Australia, although they are known to New Zealand synoptitians and are followed on weather maps of the Antarctic Ocean and the Antarctic, as well as of South Africa.

As a whole moderate latitudes of the southern hemisphere are colder than these latitudes of the northern hemisphere. This is particularly noticeable in summer. The cause for this is the combined effect of the Antarctic and of the Antarctic Ocean washing it. Cooling of air masses arriving from low and moderate latitudes and carrying heat and moisture supplies occurs over the Antarctic continent by means of precipitation, which amounts to approximately 190 mm/year over the entire Antarctic. A large amount of heat is carried away by the waters of the Antarctic Ocean, which annually melts a tremendous amount of Antarctic ice and icebergs, approximately estimated at 1500—2000 km^3/year. The volume of water in the Antarctic Ocean represents 22% of the water in the World's ocean, while its heat content represents only 10%, i. e., the Antarctic Ocean appreciably cools the oceanic waters and at the same time serves as a source for warming the atmosphere of the southern hemisphere. Exchange of water masses between high and low latitudes in the southern hemisphere occurs by means of a system of oceanic currents, which move cold surface Antarctic waters to the northwest, up to the region of the Antarctic convergence. Here they mix with warm water of subtropic origin, forming a strong western circulation current, or the so-called eastern drift. Warm deep waters, releasing their heat supply to Antarctic waters, are supplied to the south of the Antarctic shores from low latitudes.

To sum up, the Antarctic serves as a constant source of cold for the atmosphere and water of the World's ocean. It was calculated that the cooling effect of the Antarctic on the terrestrial atmosphere is 25% greater than the effect of the Arctic: the total heat supply at all levels, from the surface to the 100 mb level through the 70° parallel in the southern hemisphere, amounts to $13{,}233 \cdot 10^{21}$ cal/year, and in the northern hemisphere to $10{,}414 \cdot 10^{21}$ cal/year /20/.

Here we must again emphasize the significance of the circulation over the Antarctic in the planetary atmospheric circulation, the main feature of the former being the existence of a strong cold circumpolar cyclonic eddy in the troposphere and lower stratosphere in winter and the no less significant warm high anticyclone in the stratosphere in summer.

A quasisteady surface continental anticyclone, whose role in the mechanism of the general circulation is not too great, exists in the lowest layer of the Antarctic troposphere, above the ice dome of the continent, particularly above its eastern part. The main weather-forming factor in the Antarctic, as in moderate latitudes, is cyclonic activity, developing at atmospheric fronts and occurring in close association with the topography, distribution of land and water, and other surface features.

Active cyclonic activity in the Antarctic is observed during the entire year at all latitudes, to the South Pole.

A polar front is usually associated with the motion of depressions above the oceans surrounding the Antarctic continent. This occurs primarily along meridional trajectories. The "Arctic" front is associated with the motion of depressions along circular trajectories around the Antarctic continent. Cyclones which move over the continent are usually those which arrived from the north and are regenerated at the Arctic front; it is precisely these cyclones with which substantial weather deterioration at the inland parts of the continent is associated in both winter and summer.

Regions with the greatest recurrence of cyclones in the Antarctic are usually located at segments of relatively moderate elevations of the continent above sea level, as well as over wide shelf glaciers, which are at elevations of only several hundreds of meters above sea level. The latest, most complete maps of cyclonic recurrence in the Antarctic /12/ provide satisfactory verification of the above.

Regions at the Antarctic shores where the depressions are stationary and are slowly filled are located at the southern extremities of the principal segments of the Arctic front. The latter should not be represented by a single zone around the Antarctic, moving according to the location of the sea-ice limit, as was previously thought. There are at least four independent branches, moving onto the continent in the region where depressions are stationary: East Pacific, over the Weddell Sea, Atlantic-Indian and Indian-Pacific branches.

With respect to the tropical zone of the southern hemisphere one should agree with the conclusions of those /14, 19/ who believe that thermal factors play a decisive role in the formation of circulation, to a greater extent than for the rest of the hemisphere. Due to this, differences between this circulation and that of the northern tropical zone are less pronounced. The northward motion of the equatorial depression can also be ascribed to the lesser warming of oceanic regions of the southern hemisphere as compared with that of the northern.

The relationship between the development of processes over the southern and northern hemispheres. During the past few years the question of the energy budget of the atmosphere of the northern and southern hemispheres and of the existence of momentum exchange through the equator is extensively discussed in Soviet and Western literature. This question has a direct bearing on solving the problem of long-range forecasting of the development of macroprocesses and weather changes.

Independently of one another and on the basis of analysis of different initial data, Palmen, Riehl and Vuorela in 1958 /19/, Gruza in 1961 /9/ and a number of other investigators concluded that there exists a kinetic energy flow from the southern to the northern hemisphere in the system of large-scale eddies above certain latitudes.

Borisenkov in 1963 /3, 4/, estimating the components of the energy balance in the terrestrial atmosphere, noted that the difference in the kinetic energy of both hemispheres during the winter of the southern hemisphere is particularly great and amounts to $516 \cdot 10^{25}$ cgs units. Thus, according to the theory of conservation of absolute angular momentum, an intensive exchange of momentum between the hemispheres should exist at this time of the year.

In the same year Obasi /18/ arrived at similar conclusions whilst analyzing various features of the circulation in the southern hemisphere during the IGY. The energy of meridional motions in both hemispheres was found to be here, as in calculations by Gruza /9/, approximately the same.

In the meantime, quite well substantiated conclusions, confirming the interaction between atmospheric circulation of both hemispheres, can be obtained from objective analysis of processes in the Antarctic by comparison with processes occurring simultaneously in the Arctic. This was pointed out as early as in 1958 by Girs /7/, who presented a qualitative scheme for the possibility of such a relationship, after studying the circulation regularities in the earth polar regions.

We obtained quantitative characteristics of the atmospheric circulation which make it possible to judge not only about the relationships between circulations of the two hemispheres, but also about the contribution of zonal and meridional circulation components, as well as to some extent also about the role of the Antarctic in the energy balance of the terrestrial atmosphere. The main circulation indexes were the values of the Kats circulation indexes /10/, calculated for the 500 mb surface.

Due to the sparse network of aerological stations in the southern hemisphere we had to limit ourselves in calculating the indexes to one square, bounded by 90 and 180°E and 40 and 70°S, containing the bulk of stations with temperature and wind sounding of the atmosphere of New Zealand, Australia, Tasmania, of the subantarctic islands: Campbell, Auckland, Macquarie; Antarctic continent and shore: Vostok, Mirnyi, Wilkes, Dumond d'Urville, Hallett, McMurdo. This region is adjacent on the west to the Indian and Antarctic oceans, for which relatively many aerological observations are made (Antarctic shore stations such as Davis, Mawson and the island stations of Crozet, Kerguelen, New Amsterdam), ensuring reliability of high-altitude weather maps which serve as the starting information for index calculation.

TABLE 1. Circulation indexes in the southern and northern hemispheres, in decameters/deg-eq

Year	Month	Zonal		Meridional		General	
		southern hemisphere	northern hemisphere	southern hemisphere	northern hemisphere	southern hemisphere	northern hemisphere
1961	January	2.48	1.46	—	—	—	—
	February	2.30	1.07	—	—	—	—
	March	3.18	1.33	—	—	—	—
1962	January	2.31	1.17	0.84	1.32	2.76	0.89
	February	1.93	1.13	0.82	1.09	2.35	1.04
	March	2.22	1.27	0.87	1.19	2.55	1.08
	April	2.51	1.43	0.89	1.20	2.82	1.19
	May	2.30	1.23	0.84	1.06	2.73	1.16
	June	1.85	1.00	0.77	1.00	2.40	1.00
	July	2.33	0.83	0.86	0.80	2.70	1.04
	August	2.22	1.07	0.90	0.81	2.46	1.32
	September	1.97	1.30	0.80	0.94	2.48	1.38
	October	2.12	1.63	0.83	1.00	2.55	1.63
	November	1.91	1.66	0.92	0.93	2.08	1.78
	December	2.15	1.50	0.90	1.18	2.39	1.27
1963	January	2.41	1.10	—	—	—	—
	February	2.17	1.22	—	—	—	—

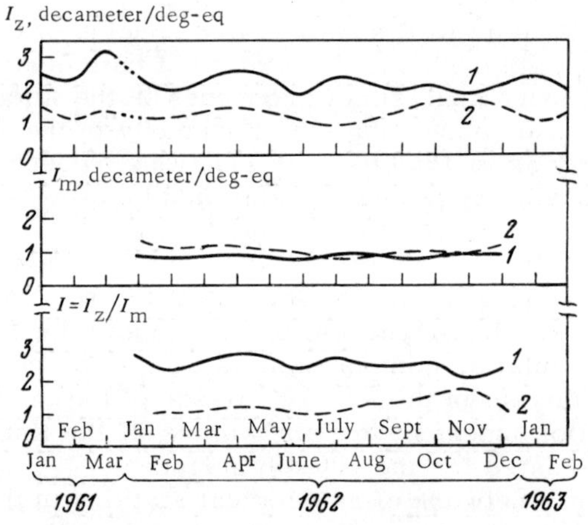

FIGURE 4. Graph of circulation indexes:

1 — southern hemisphere; 2 — northern hemisphere.

At the same time, the values of indexes for the northern hemisphere were calculated within the same limits, making it possible to compare corresponding quantities.

All the calculations were made for 17 months of 1961—1963 (Table 1 and Figure 4). The data clearly indicate that throughout the year the values of the general circulation index in the southern hemisphere exceed by 50—100% their counterparts in the northern hemisphere; the zonal circulation index, describing the intensity of the west-east transport of air masses, is in approximately the same proportion. The intensity of the latitudinal transport in the southern hemisphere, even in summer when the horizontal temperature gradients between high and low latitudes are not too pronounced, is greater in absolute magnitude than in the northern hemisphere in winter, during maximum temperature contrasts between the latitudes.

At the same time the values of meridional circulation indexes during almost all months of the year in both hemispheres are approximately the same, the value in the northern hemisphere being slightly higher.

It should be noted that the general and zonal circulation indexes in the southern hemisphere during transition seasons are a maximum and have a relatively low value in June, in the very midst of the southern polar night. If we recall the repeatedly noted "nucleus-free" nature of the Antarctic winter, the high recurrence there of advective temperature increases in winter and shifting of periods with extremely low temperatures toward the spring, then the great contribution of the circulation to the formation of the temperature regime of the south polar region and the impossibility of ascribing all the features of the yearly temperature march in this region to radiation factors alone will become obvious.

The great intensity of the zonal circulation of the southern hemisphere is a result of stable and high values of horizontal temperature gradients between the south-polar and equatorial regions, due to the continual existence of a cold troposphere over the Antarctic. At the same time one's attention is attracted by the fact that the absolute values of meridional circulation components in the northern and southern hemispheres are approximately equal, in spite of their different distributions of land and water masses. This implies that the nature of large-scale disturbances of the planetary zonal flux in the atmosphere is identical in both hemispheres and is less associated with the alternation of continents and oceans than one would believe.

Departures from zonality in both hemispheres, as shown by studies of the principal circulation modes for the past years, are definitely related, but this relationship does not by far always demonstrate itself as a simple and obvious correlation: periods of meridional invasions within the same time interval and at the same longitudes in the Arctic and Antarctic are not encountered frequently, and it is as yet impossible to establish regularities in their appearance. New studies on this question, which has an importance of its own, are needed.

The steps taken by us in this direction have as yet yielded little specific information; the work (which is quite time-consuming and for which insufficient raw data are available) is far from completion and reference may be had only to the most preliminary considerations of the problem. From the 16 months in 1959—1961 for which it was possible to determine the main circulation forms in both hemispheres, one can isolate cases when a gradual reduction in the intensity of the meridional circulation and an increase in the zonal circulation up to total predominance of the latter were observed in both hemispheres during several months. In such cases increases in meridionality were noted in both hemispheres; very good agreement between the occurrence times of predominant circulation forms was noted in 8 of the 16 months and satisfactory agreement existed in 2 of the months. However, no agreement between circulation forms over the different hemispheres was detected in more than a third of the months (6 out of 16). This can be illustrated by Figure 5, which shows the graph of recurrence of meridional circulation over the hemispheres.

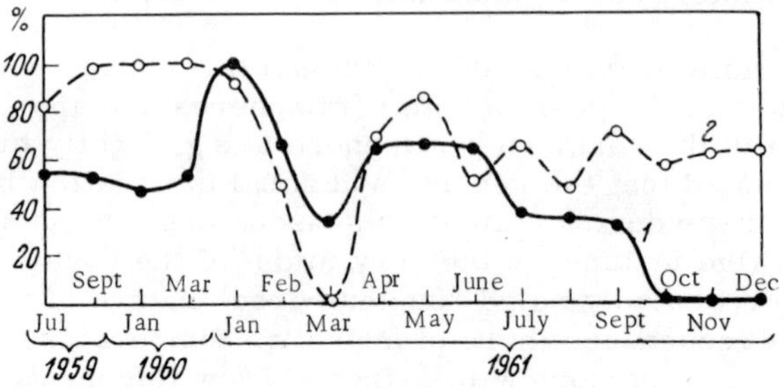

FIGURE 5. Graph of recurrence of meridional circulation forms:

1 — southern hemisphere; 2 — northern hemisphere.

It should be made clear that since the circulation forms were estimated for each hemisphere as a whole, while the circulation indexes were calculated only in one quadrant, the maximum values of the indexes do not always correspond to the greatest development of the corresponding circulation form. Moreover, in individual cases the meridional circulation is found to predominate over all the hemispheres when the horizontal circulation index in the Far Eastern (Australian) quadrant has a high value. This again emphasizes the amount of care that must be exercized in drawing conclusions from analysis of the presently available limited amount of information on the circulation over the southern hemisphere.

The above applies in full also to the conclusion drawn by Scherhag in 1958 /14/ from analysis of average zonal changes at the altitude of the 1000 mb surface over both hemispheres, the meaning of which reduces to the statement that "the seasonal air exchange between both hemispheres is limited to the zone between the equator and the 40° latitude." It may be regarded as established that some air masses heated in summer over low latitudes of the southern hemisphere flow into the circumpolar plateau; this is not denied even by Scherhag /14/.

CONCLUSION

During the IGY and subsequently, some success was achieved in clarifying the type of the general circulation in the troposphere of the southern hemisphere, but it must be admitted that atmospheric processes in this hemisphere have still not been sufficiently studied. Meteorological observations over oceans are carried out from an extremely small number of island stations, in the total absence of weather ships. The analysis of synoptic maps over a large part of the hemisphere is based on interpolation of meteorological fields to distances of the order of 1000 km, or extrapolation of a previously known position of baric formations on the basis of synoptic experience. Even less information is available on the aerological structure of the atmosphere of the southern hemisphere, and it was particularly important in connection with studies during the IQSY.

It is obvious that the comparative study of circulation up to high altitudes in the northern and southern hemispheres presents an excellent opportunity for clarifying the role of solar effects on atmospheric processes. It was noted above that the major physicogeographic features have their effect on the regime of the middle and upper stratosphere. Since the effect of the topography and the type of underlying surface is relatively constant, while the solar activity is variable, future research is quite promising.

Encouraging possibilities are provided by meteorological satellites, which carry television cameras for obtaining images of cloud systems and apparatus for radiation measurement.

References

1. Aver'yanov, V. G. Tsentral'naya Antarktika (Central Antarctic). — Trudy SAE, Vol. 30. 1963.
2. Astapenko, P. D. Atmosfernye protsessy v vysokikh shirotakh yuzhnogo polushariya (Atmospheric Processes in High Latitudes of the Southern Hemisphere). — Izdatel'stvo AN SSSR. 1960.
3. Borisenkov, E. P. Sezonnye preobrazovaniya energii v atmosfere severnogo i yuzhnogo polusharii (Seasonal Energy Transformations in the Atmosphere of the Northern and Southern Hemispheres). — Trudy AANII, Vol. 253. 1963.
4. Borisenkov, E. P. Fiziko-statisticheskie metody analiza i predvychisleniya meteorologicheskikh polei (Methods of Physical Statistics in the Analysis and Precalculation of Meteorological Fields). — Trudy AANII, Vol. 263. 1963.
5. Bugaev, V. A. Vliyanie Antarktiki na klimat yuzhnogo polushariya (Effect of the Antarctic on the Climate of the Southern Hemisphere).— Meteorologiya i gidrologiya, No. 1. 1961.
6. Gaigerov, S. S. Voprosy aerologicheskogo sostoyaniya tsirkulyatsii i klimata svobodnoi atmosfery v Tsentral'noi Arktike i Antarktike (Problems of the Aerological State of the Circulation and Climate of the Free Atmosphere in Central Arctic and Antarctic). — Izdatel'stvo AN SSSR. 1962.
7. Girs, A. A. O vzaimosvyazi mezhdu tsirkulyatsiei atmosfery v Arktike i Antarktike (On the Relationship Between the Atmospheric Circulation in Arctic and Antarctic). — Informatsionnyi byulleten' SAE, No. 2. 1958.
8. Girs, A. A. Osnovy dolgosrochnykh prognozov (Fundamentals of Long-Range Forecasts). — Gidrometeoizdat. 1960.
9. Gruza, G. V. Makroturbulentnost' v obshchei tsirkulyatsii atmosfery (Large-Scale Eddies in the General Circulation of the Atmosphere). — Gidrometeoizdat. 1961.
10. Kats, A. L. Sezonnye izmeneniya obshchei tsirkulyatsii atmosfery i dolgosrochnye prognozy (Seasonal Variations in the General Circulation of the Atmosphere and Long-Range Forecasts). — Gidrometeoizdat. 1960.

11. Khvostikov, I. A. editor. Rezul'taty raketnykh issledovanii atmosfery v period MGG i MGSS (Results of Rocket Sounding of the Atmosphere during the IGY and IQSY). — Trudy TsAO, Vol. 52. 1964.
12. Tauber, G. M. Okeanicheskie tsentry deistviya atmosfery v yuzhnom polusharii (Oceanic Centers of Action of the Atmosphere in the Southern Hemisphere). — Nauka. 1964.
13. Khromov, S. P. Osnovy sinopticheskoi meteorologii (Fundamentals of Synoptic Meteorology). — Gidrometeoizdat. 1948.
14. Scherhag, R. Probleme der allgemeinen Zirkulation. — Geophysica, Helsinki, 6(3/4): 539/557. 1958.
15. Farcas, E. Springtime Temperature Changes in the Antarctic Stratosphere, New Zealand. — J. Geol. and Geophys., Vol. 4. 1961.
16. MacDowall, J. Distribution of Atmospheric Ozone: A Preliminary Analysis of Some IGY Observations. — Nature, 187(4735). 1960.
17. Monthly Mean Aerological Cross-Sections Pole to Pole Along Meridian 75°W for the IGY Period. Washington. 1961.
18. Obasi, G. O. Poleward Flux of Atmospheric Angular Momentum in the Southern Hemisphere. — J. Atmos. Sci., 20(6). 1963.
19. Palmen, E., H. Riehl, and L. Vuorela. On the Meridional Circulation and Release of Kinetic Energy in the Tropics. — J. Meteorol., 15(3). 1958.
20. Rubin, M. J. Atmospheric Advection of the Antarctic Mass and Heat Budget. — Antarctic Res. Geophys. Monogr., No. 7. 1962.

E. M. Dobryshman

SOME FEATURES OF PRESSURE AND WIND FIELDS IN THE NEAR-EQUATORIAL BELT

A study is made of the pressure field in low latitudes. Values of the variance of the surface pressure and of the geopotential at 500 and 100 mb are presented. Normed correlation functions of these elements are calculated. The hydrodynamic equations for the equatorial belt are simplified and examples of exact solutions of the simplified system are given. The relationship between wind and pressure fields is analyzed.

Much attention has recently been paid to the circulation mechanism in tropical and equatorial latitudes. New methods of observation and special expeditions have considerably supplemented existing empirical data, analysis of which resulted in the detection of very many significant features of processes, such as the multilayer nature of the zonal flux, two-year periodicity, etc. These regularities are described in detail elsewhere /8, 10, 11, 21, 22, 24/.

The noticeable difference between processes in low and middle latitudes is ascribed, in particular, to the relationship between the pressure and wind fields. While, according to Dady's fitting remark, the quasigeostrophic hypothesis is the "alphabetical truth" of dynamic meteorology in middle latitudes /17/, the relationship between the wind and pressure in the tropical and, particularly, equatorial belts apparently cannot, in principle, be expressed by means of a linear operator. The question of the relationship between the wind and pressure is one of the key problems of meteorology. To study it one must analyze the structure of the pressure field. The present paper describes some results of analysis of the pressure field at low latitudes and examines simple hydrodynamic models enabling one to determine the wind field from the pressure field.

Statistical features of the pressure field at low latitudes. The statistical features of meteorological fields play an important role in the study of atmospheric processes. A knowledge of the statistical parameters frequently allows one to describe not only the "average" state of the quantity at hand, but in a number of cases to obtain a specific result pertaining to a given time instant and to a given point in space. The application of correlation and structural functions to objective analysis, i. e., calculation of values of the corresponding meteorological element at nodes of a regular grid from station observational data, is well known (see, for example, Gandin /3/ and Tatarskaya /13/).

The determination of reliable statistical parameters involves great difficulties, the main ones in the case at hand being as follows: 1) sparcity of observational data — the number of stations in low latitudes is very small; 2) variability of the pressure and temperature fields in low latitudes is small on the average and, as will be seen from the subsequent discussion, it is commensurable with the accuracy of the observations proper; 3) methods of calculating statistical characteristics were developed in detail for homogeneous and isotropic fields, while actual fields of the departures from average of meteorological elements (without mentioning the fields proper) near the equator can be regarded as homogeneous and particularly isotropic only in a very rough approximation. Hence care should be exercized when using the characteristics thus calculated. They can be used for determining the general features of the circulation mechanism, or when the statistical approach is the only one possible in describing the real process.

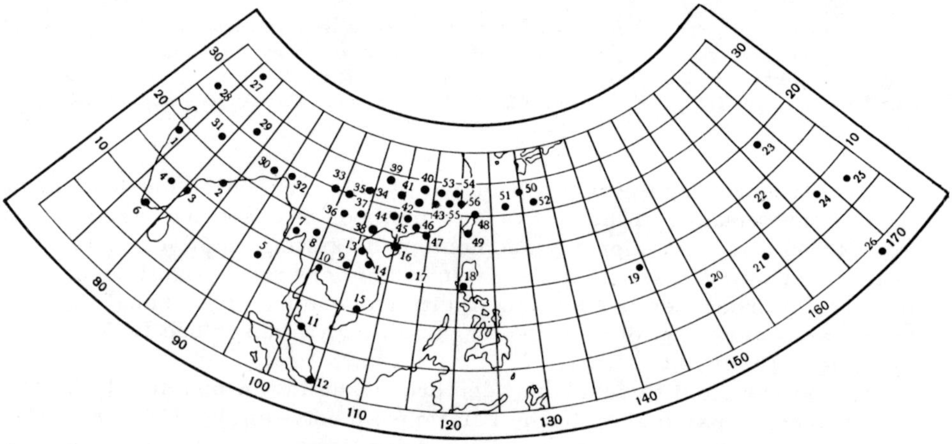

FIGURE 1. Distribution of stations, the data of which were used in calculating autocorrelation functions.

The above difficulties prevent one from making a study of statistical characteristics in low latitudes as comprehensive as that carried out for middle latitudes /1, 3, 5, 12, 14/. However, correlation functions were calculated for three levels as a first approach to the problem.

The starting information were data of 56 stations with continuous aerological sounding (Figure 1). Canton, the southernmost station, is located at 1.5°S. Twenty six stations are located in the belt up to 20°N, and 30 stations are situated between 20 and 30°N. Data on the pressure field were read from synoptic and constant-pressure maps at the 500 and 100 mb levels for June and July, 1962 —1963. A total of 60 situations were considered.

The distance between each station pair (i, j) was calculated with allowance for the earth sphericity. These distances were grouped at 300 km intervals and the average deviations from the mean were calculated for each graduation. The arithmetic mean of the element, determined from the same sampling, was taken as this mean.

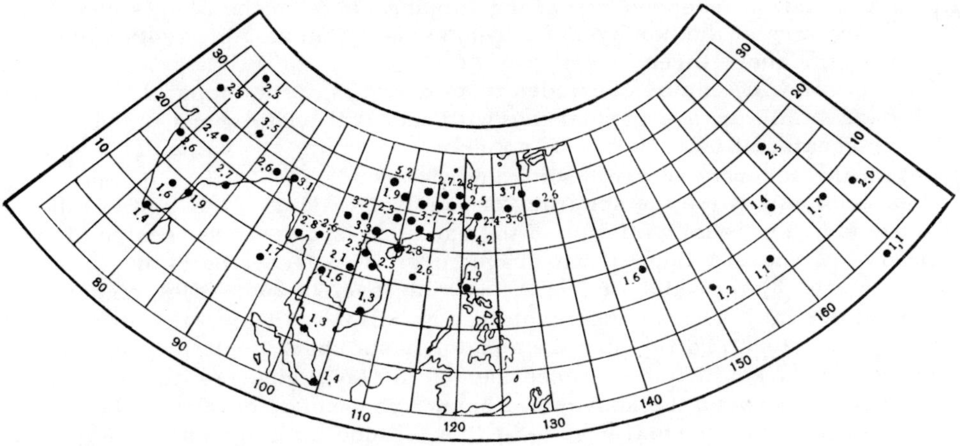

a

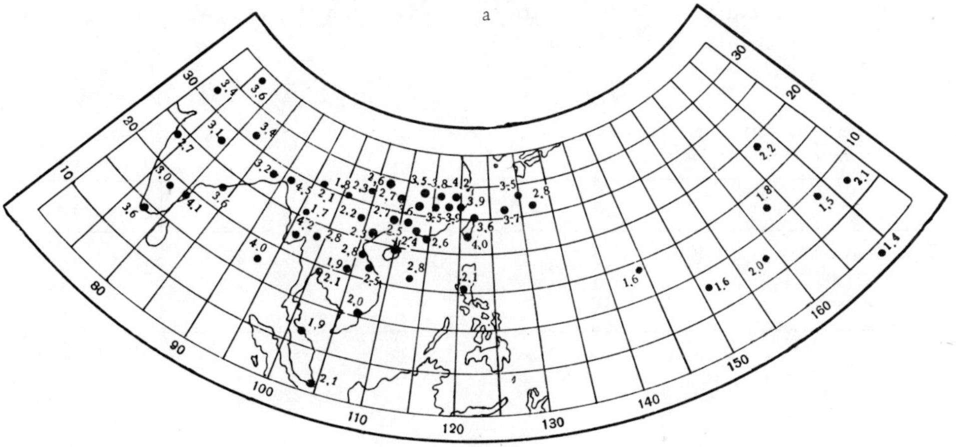

b

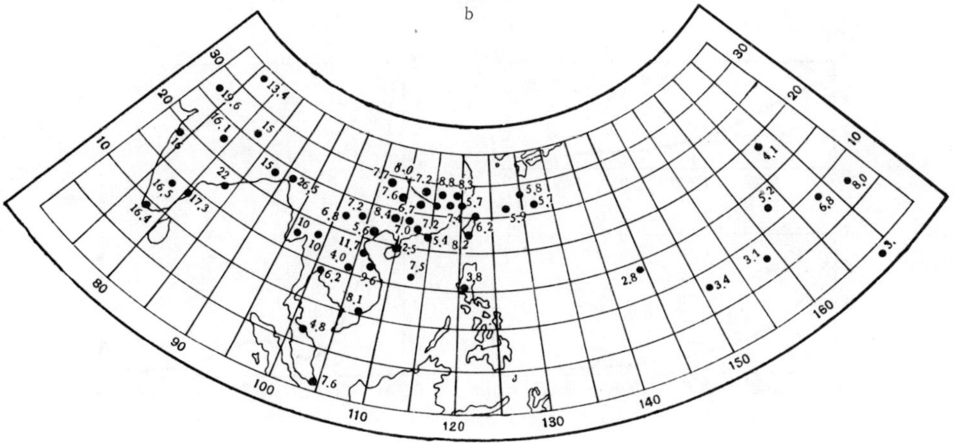

c

FIGURE 2. Value of σ at different levels:

a — surface; b — 500 mb level; c — 100 mb level.

As is known, independence of the variance $\tilde{\sigma}^2$ from the point's coordinates is a corollary of the homogeneity hypothesis. Figure 2 presents values of $\tilde{\sigma}$ for the above three levels.

The following can be concluded from analysis of the maps: 1) the values of $\tilde{\sigma}$ are small and are comparable with observational errors in the case of the equatorial belt. It can be noted by comparing the values of $\tilde{\sigma}$ with data on the accuracy of radio sounding presented in /9/ that on the average $\tilde{\sigma}$ exceeds only twice the accuracy of radiosonde data; 2) in spite of the above fact, the zonal variation of the variance is clearly delineated also in low latitudes. Values of $\tilde{\sigma}$ averaged over five-degree belts are presented in Table 1; 3) the value of $\tilde{\sigma}$ at the 100 mb level is 200—250% higher than at 500 mb. This may be explained as follows. It is well known that the temperature profile near the equator has a characteristic discontinuity at the 16—17 km altitude. Hence the 100 mb level changes its position, alternating between the above- and below-tropopause location. The difference in the thermal regime at this altitude for different stations affects the deviations from the mean, increasing the probability of the appearance of large deviations.

TABLE 1. Variability of the geopotential field (in decameters) at different levels near the equator averaged over five-degree belts (northern hemisphere)

Level	0—5°	5—10°	10—15°	15—20°	20—25°
Surface	1.2	1.6	1.6	2.5	3.1
500 mb	1.8	2.1	2.6	2.8	2.8
100 mb	5.4	7.1	9.3	12.5	11.3

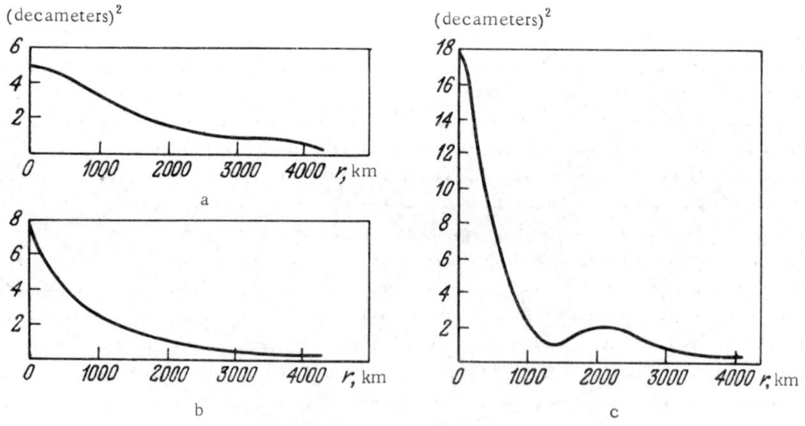

FIGURE 3. Smoothed autocorrelation functions at different levels, averaged over all the stations:

a — surface; b — 500 mb; c — 100 mb.

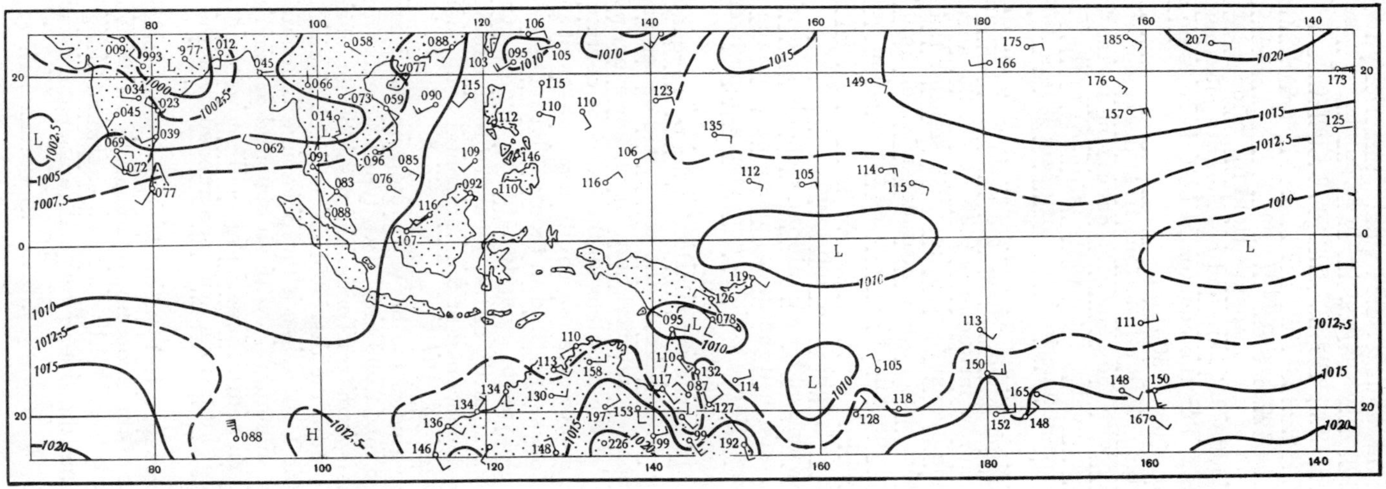

FIGURE 4. Pressure and wind fields at 0300 hrs on 9 June 1964.

Due to insufficient raw data and methodical errors in the calculation, the computed correlation functions had small fluctuations, which were eliminated by smoothing with respect to three points. These smoothed normed autocorrelation functions are shown in Figure 3. Comparing these functions with similar functions for middle latitudes quoted in works by Gandin /3/, Fortus /14/ and Yudin /16/, one notes certain differences. The correlation function of the pressure field for middle latitudes reverses sign at a distance of 2500—3000 km and reaches its maximum at approximately 4500 km. For the equatorial belt there is no sign reversal and zero is reached at 4000—4500 km.

At lower latitudes correlation functions first decrease at a higher rate than at middle latitude, with the reduction occurring faster at the higher rather than at the lower levels. At middle latitudes in the troposphere the normed autocorrelation functions for different levels differ little from one another.

The following should be noted. Since the ratio η of the rms observational error to the variance in low latitudes is 0.25—0.35, the accuracy of the objective analysis of the pressure field is much smaller than that in middle latitudes. We recall that η in middle latitudes is taken as 0.02—0.05 /3/. To improve analysis of the wind field outside the tropical belt use is usually made of wind data /2, 20/. This involves resorting to the hypothesis of quasigeostrophicity of large-scale motions.

Information on the wind field could also have been used in low latitudes for improving the analysis, had the relationship between the pressure and wind fields been known.* Hence we now consider simple hydrodynamic schemes of the circulation mechanism.

Hydrodynamic models of the circulation in the near-equatorial belt. In general it is difficult to determine the wind field on the basis of the pressure field in the near-equatorial belt. It is, however, possible to simplify the statement of the problem on the basis of certain features of the thermobaric field, and to obtain a solution for cases of practical interest.

It is known that the pressure field in tropical latitudes is quite spread out, with the isobars (and isohypses) close to latitude circles. Outside of the 5—7° equatorial belt the zonality of the pressure field can be attributed to the Rossby effect — dependence of the horizontal component of the Coriolis force on latitude /4, 19/.

The characteristic value of the baric gradient in the near-equatorial region amounts to 3—5 mb/1,000 km. Two typical baric situations are also observed: a quasisymmetric pressure distribution characteristic of the Pacific, and an asymmetric distribution, frequently encountered above the Indian Ocean. In the latter case a low pressure is located over the heated Asian continent in summer, with an elevated pressure existing over the ocean in the southern hemisphere. Figure 4 shows an example of a real map. Such pairs of situations are encountered over the Atlantic and Africa, respectively.

The fact that the pressure field is close to the zonal makes it possible to simplify the statement of the problem, by dropping all derivatives with respect to x, which is subsequently directed eastward along the equator. It is also well known that the linear vertical temperature variation is

* Of course, establishing a correlation between these fields is also of independent interest.

clearly linear in low latitudes (Figure 5). This means that it is possible to calculate the baric gradient using the polytropic model of the atmosphere:

$$\frac{1}{r\rho}\frac{\partial p}{\partial \varphi}=\frac{1}{r\rho_0}\frac{\partial p_0}{\partial \varphi}\left(1-\frac{\gamma z}{T_0}\right)+\frac{gz}{rT_0}\frac{\partial T_0}{\partial \varphi},$$

where p is the pressure; ρ is the density; γ is the lapse rate; T is the temperature; g is the acceleration of gravity; $r = 6.37 \cdot 10^6$ m is the radius of the earth; φ is the latitude; z is the vertical coordinate; subscript 0 denotes values at the surface.

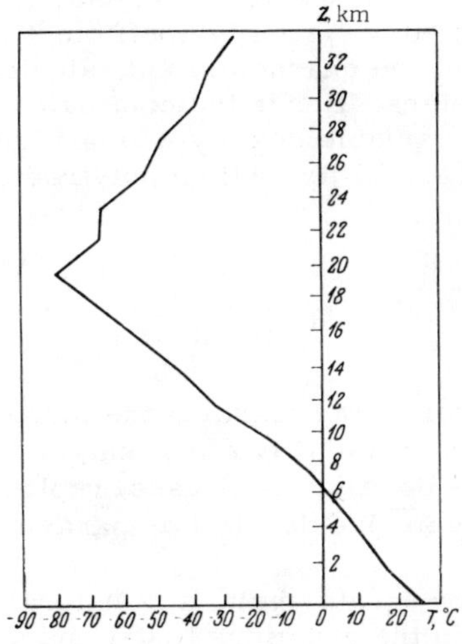

FIGURE 5. Vertical temperature distribution in Gonton at 0300 hrs on 5 May 1964.

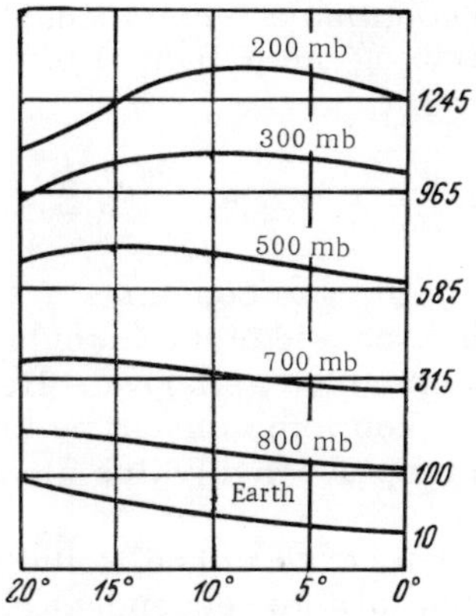

FIGURE 6. Zonal geopotential profiles at different altitudes /18/.

It is here of importance that the baric gradient is a linear function of the altitude. It can be represented in the form

$$\frac{1}{r\rho}\frac{\partial p}{\partial \varphi}=A(\varphi)+B(\varphi)z=A(\varphi)[1+zb(\varphi)]. \qquad (1)$$

Functions $A(\varphi)$ and $B(\varphi)$ are easily approximated, for example, by polynomials of appropriate degrees.

To illustrate the above considerations regarding the pressure distribution near the equator, we present in Figure 6 pressure curves constructed from data of /18/, where the pressure and wind fields from the equator to 40°N have been carefully analyzed. It is seen from the figure that the baric gradient increases with altitude; consequently $b(\varphi) > 0$.

With reference to the above, we write the starting system of hydrodynamic equations in the form

$$\left.\begin{aligned}
&\frac{\partial v_\lambda}{\partial t}+v_z\frac{\partial v_\lambda}{\partial z}+\frac{v_\varphi}{r}\frac{\partial v_\lambda}{\partial \varphi}+\frac{v_z v_\lambda}{r}-\frac{v_\varphi v_\lambda \operatorname{tg}\varphi}{r}=\\
&\qquad=2\omega v_\varphi\sin\varphi-2\omega v_z\cos\varphi+\nu\frac{\partial^2 v_\lambda}{\partial z^2};\\
&\frac{\partial v_\varphi}{\partial t}+v_z\frac{\partial v_\varphi}{\partial z}+\frac{v_\varphi}{r}\frac{\partial v_\varphi}{\partial \varphi}+\frac{v_z v_\varphi}{r}+\frac{v_\lambda^2 \operatorname{tg}\varphi}{r}=\\
&\qquad=-\frac{1}{\rho r}\frac{\partial p}{\partial \varphi}-2\omega v_\lambda\sin\varphi+\nu\frac{\partial^2 v_\varphi}{\partial z^2};\\
&\frac{\partial v_z}{\partial z}+\frac{v_z}{\rho}\frac{d\rho}{dz}+\frac{1}{r}\frac{\partial v_\varphi}{\partial \varphi}-v_\varphi\frac{\operatorname{tg}\varphi}{r}=0,
\end{aligned}\right\}\qquad(1')$$

where t is the time; v_λ, v_φ and v_z are velocity components along axes corresponding to the subscripts; $\omega = 7.29 \cdot 10^{-5}\ \sec^{-1}$ is the angular velocity of the earth; and v is the kinematic eddy viscosity coefficient.

As will be seen from subsequent analysis, the characteristic value of the altitude is of the order of tens of kilometers. Hence the continuity equation contains a term accounting for the variable density with altitude and describing the principal effect of compressibility. For a polytropic atmosphere

$$\frac{1}{\rho}\frac{d\rho}{dz}=-\frac{\gamma(g-R\gamma)}{g}\frac{1}{T_0-\gamma z},$$

where R is the gas constant.

We shall consider the circulation mechanism by regarding the thermodynamic variables as known. Hence the compressibility effect can be partially taken into account by (as is usually done in such cases) replacing $1/\rho\cdot d\rho/dz$ by its average value along the vertical. We denote this parameter by $-\sigma_1$.

Since reference is had to the narrow near-equator belt, it is natural to replace the latitude-dependent coefficients in the equations by the principal terms of their series expansion in powers of φ. Assuming that $x = r\lambda$, $y = r\varphi$, $\sin\varphi \approx \operatorname{tg}\varphi \approx y/r$, $\cos\varphi = 1$, and using the notation $u = v_\lambda$, $v = v_\varphi$, $w = v_z$, the starting system of equations takes the form

$$\left.\begin{aligned}
&\frac{\partial u}{\partial t}+v\frac{\partial u}{\partial y}+w\frac{\partial u}{\partial z}+\left(\frac{wu}{r}\right)-\left(\frac{uv}{r^2}\right)=2\omega\frac{y}{r}v-2\omega w+\nu\frac{\partial^2 u}{\partial z^2};\\
&\frac{\partial v}{\partial t}+v\frac{\partial v}{\partial y}+w\frac{\partial v}{\partial z}+\left(\frac{wv}{r}\right)+\frac{u^2 y}{r^2}=\\
&\qquad=-A(y)\left[1+b(y)z\right]-2\omega\frac{y}{r}u+\nu\frac{\partial^2 v}{\partial z^2};\\
&\frac{\partial v}{\partial y}+\frac{\partial w}{\partial z}+\left(\frac{y}{r^2}v\right)=\sigma_1 w.
\end{aligned}\right\}\qquad(2)$$

Parameters r, ω, ν and σ_1 are known; function $A(y)$ representing the characteristic value of the baric gradient should be specified, since reference is had to determining the motion from a specified pressure field.

We introduce the dimensionless quantities u, v, w, t, y and z. It will be remembered here that the horizontal component of the Coriolis force vanishes at the equator and hence the vertical component can be compensated only by inertia terms (or else by viscous terms). Then, the characteristic horizontal scale L should define a region in which the baric gradient force cannot be balanced by the horizontal component of the Coriolis force; in other words, L should define a region where the quasi-geostrophic hypothesis cannot be used.

Under these assumptions the dimensionless quantities are defined as

$$t = t_0 t' = \sqrt[3]{\frac{r}{2\omega A_0}}\, t'; \quad y = Ly' = \sqrt[3]{\frac{A_0 r^2}{4\omega^2}}\, y';$$
$$z = hz' = \frac{1}{2\omega}\sqrt[3]{\frac{A_0^2 r}{2\omega}}\, z', \quad \left.\begin{matrix} u \\ v \end{matrix}\right\} = u_0 \left\{\begin{matrix} u' \\ v' \end{matrix}\right\} = \sqrt[3]{\frac{A_0^2 r}{2\omega}} \left\{\begin{matrix} u' \\ v' \end{matrix}\right\};$$
$$w = w_0 w' = \frac{A_0}{2\omega}\, w'.$$

Since the vertical velocity should be of the order of 0.2 m/sec, then

$$A_0 = w_0 2\omega \approx (1.5 - 3.0)\cdot 10^{-5}\ \mathrm{m^2/sec}.$$

For high baric gradients the value of w_0 would have been unnaturally high. This may to a large extent explain the absence of high pressure gradients in the narrow near-equatorial belt.

Assuming $A_0 \sim 3 \cdot 10^{-5}\ \mathrm{m^2/sec}$, we find the characteristic values $t_0 \sim 10^5$ sec, $L \sim 450$ km, $h \sim 25$ km, $u_0 = v_0 = 4$ m/sec. It is easy to note that $L^2 = rh$, i.e., the width of the belt where it is impossible in principle to use the geostrostropic relationship is the geometric mean of the earth radius and the characteristic altitude.*

Upon conversion to dimensionless variables, the series expansion parameter

$$\varepsilon_1^2 = \frac{h^2}{L^2} = \frac{L^2}{r^2} < 0.01$$

appears in front of terms due to spherical geometry (these terms are enclosed in parentheses in system (2)). This means that the earth sphericity can be neglected when considering processes in the narrow equatorial belt, i.e., the latter can be treated as a cylinder (however, the coefficients in the equation vary markedly). The series expansion parameter

$$\varepsilon_2 = \frac{4\nu\omega^2}{A_0}\sqrt[3]{\frac{2\omega}{A_0^2 r}} \sim 10^{-4} - 10^{-5}\ \text{for}\ \nu = (50 - 5)\ \mathrm{m^2/sec}$$

will also appear in front of terms with eddy viscosity. This means that the boundary layer will be thin. "Adhesion" of air particles to the surface points to the fact that the profile varies markedly with altitude in this thin boundary layer.

Ignoring terms with small parameters and omitting bars over dimensionless quantities, we may rewrite the system in the form

$$\frac{\partial u}{\partial t} + v\frac{\partial u}{\partial y} + w\frac{\partial u}{\partial z} = yv - w;$$
$$\frac{\partial v}{\partial t} + v\frac{\partial v}{\partial y} + w\frac{\partial v}{\partial z} = -yu - \frac{\partial \Phi}{\partial y};$$
$$\frac{\partial v}{\partial y} + \frac{\partial w}{\partial z} = \sigma w,$$

* This conclusion applies to any planet with an atmosphere.

where

$$\sigma = \sigma_1 h \sim 0.5 - 0.8; \quad \frac{\partial \Phi}{\partial y} = \frac{1}{A_0}[A(y) + B(y) z].$$

Making the usual assumption that the vertical velocity vanishes at the surface, the continuity equation yields directly that

$$w = -e^{-\sigma z} \int_0^z e^{-\sigma z} \frac{\partial v}{\partial y} dz.$$

The second of the equations of the system yields

$$u = -\frac{1}{y}\left[\frac{\partial v}{\partial t} + v\frac{\partial v}{\partial y} - \frac{\partial v}{\partial z} e^{\sigma z} \int_0^z e^{-\sigma z} \frac{\partial v}{\partial y} dz\right].$$

Substituting this latter expression into the first equation, we derive a nonlinear partial integrodifferential equation in one unknown function v, the meridional velocity component. After simple transformations it is reduced to the form

$$\frac{\partial}{\partial t}\left\{-\frac{1}{y}\left[\frac{\partial v}{\partial t} + v\frac{\partial v}{\partial y} - \frac{\partial v}{\partial z} e^{\sigma z} \int_0^z \frac{\partial v}{\partial y} e^{-\sigma z} dz + \frac{\partial \Phi}{\partial y}\right] - \frac{y^2}{2} + z\right\} +$$

$$+ v\frac{\partial}{\partial y}\left\{-\frac{1}{y}\left[\frac{\partial v}{\partial t} + v\frac{\partial v}{\partial y} - \frac{\partial v}{\partial z} e^{\sigma z} \int_0^z \frac{\partial v}{\partial y} e^{-\sigma z} dz + \frac{\partial \Phi}{\partial y}\right] - \frac{y^2}{2} + z\right\} +$$

$$+ w\frac{\partial}{\partial z}\left\{-\frac{1}{y}\left[\frac{\partial v}{\partial t} + v\frac{\partial v}{\partial t} -\right.\right.$$

$$\left.\left. - \frac{\partial v}{\partial z} e^{\sigma z} \int_0^z \frac{\partial v}{\partial y} e^{-\sigma z} dz + \frac{\partial \Phi}{\partial y}\right] - \frac{y^2}{2} + z\right\} = 0,$$

from which it is seen that in the sense of motion along a trajectory lying in the y, z plane the quantity

$$-\frac{1}{y}\left[\frac{\partial v}{\partial t} + v\frac{\partial v}{\partial y} - \frac{\partial v}{\partial z} e^{\sigma z} \int_0^z \frac{\partial v}{\partial y} e^{-\sigma z} dz + \frac{\partial \Phi}{\partial y}\right] - \frac{y^2}{2} + z$$

should be retained.

We shall not consider here this unsteady case, since in the study of the circulation mechanism one should naturally refer to the steady case. Ignoring time derivatives one easily determines the solution of this equation:

$$-\left[v\frac{\partial v}{\partial y} - \frac{\partial v}{\partial z} e^{\sigma z} \int_0^z \frac{\partial v}{\partial z} e^{-\sigma z} dz + \frac{\partial \Phi}{\partial y}\right] - \frac{y^2}{2} + z = F\left(\int_0^z v e^{-\sigma z} dz\right), \quad (3)$$

where $F(\xi)$ is an arbitrary function of the argument

$$\xi = \left(\int_0^z v e^{-\sigma z} dz\right).$$

To determine this function, a corresponding condition must be specified. It is difficult to do this rigorously, since it is necessary to match the solution for the equatorial belt with that valid outside this narrow belt.

We shall consider simple examples of function $F(\xi)$ which lend themselves to physical interpretation, so that the solutions found have a physical meaning.

$$1.\quad F\left(\int_0^z ve^{-\sigma z}dz\right) = \text{const.}$$

It can be easily shown that in this case the zonal wind-velocity component is given by the expression $u = \text{const} + \frac{y^2}{2} - z$, where the constant represents the velocity of the zonal circulation at the equator at the earth surface. We denote it by u_0. Thus, $u = u_0 + (y^2/2) - z$.

The feasibility of using such a zonal velocity profile, at least for the lower troposphere, is verified by observations. Figure 7, taken from /23/, shows the zonal velocity profile at the equator under a near-symmetrical pressure distribution. For a symmetrical pressure distribution $\partial\varphi/\partial y$ should be an odd function of y.

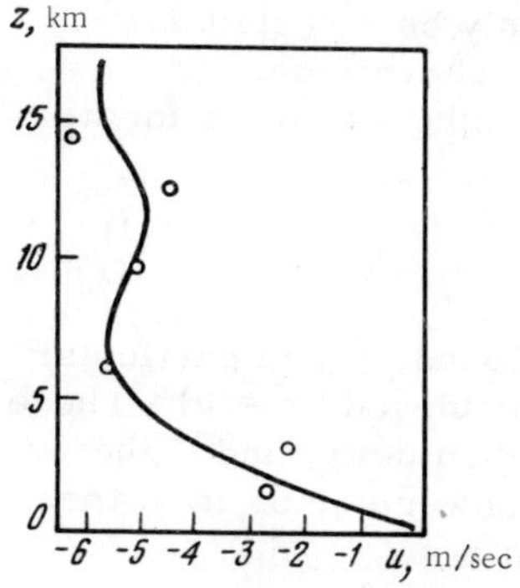

FIGURE 7. Profile of the equatorial zonal wind speed /23/.

The values of u are negative in the eastward direction.

We rewrite the equation for v in the form

$$v\frac{\partial v}{\partial y} - \frac{\partial v}{\partial z}e^{\sigma z}\int_0^z \frac{\partial v}{\partial y}e^{-\sigma z}dz + \left(\frac{y^3}{2} - zy + u_0 y\right) + \frac{\partial \Phi}{\partial y} = 0. \qquad (4)$$

Its solution should be analytical everywhere, including the equator. In addition, by virtue of symmetry, v should vanish at the equator. Thus, the left-hand side of the equation vanishes when $y = 0$. Accordingly we should have $\partial\varphi/\partial y = 0$. Due to the fact that this function is odd, $\partial\varphi/\partial y \sim y$ at the equator, or has a zero of an even higher odd order.

To solve equation (4) one must impose a condition on y. We shall require that at $y = 1$ the zonal wind be defined by the geostrophic relationship /6, 7, 15/, i.e., that

$$u|_{y=1} = -\frac{1}{y}\frac{\partial \Phi}{\partial y}\bigg|_{y=1} \quad \text{or} \quad u_0 + \frac{1}{2} - z = -\frac{\partial \Phi}{\partial y}\bigg|_{y=1}$$

(it is seen that linear dependence on the baric gradient is satisfied here).

All the conditions can be satisfied by setting, for example,

$$-\frac{\partial \Phi}{\partial y} = \left(u_0 + \frac{1}{2} - z\right)y^3\left(1 + \sum_{n=0}^{\infty} a_n y^{2n}\right)^*.$$

Here Φ will have a minimum at the equator if $u_0 < -\frac{1}{2}$.

* Specifying $\Phi_y = \frac{1}{2}y^3 - zy^3 + u_0 y$ we will have a minimum at the equator when $u_0 < 0$ and the solution for v will be independent of u_0.

There is reason in restricting oneself to the first, principal term of the expansion, since reference is had to the range $0 \leqslant y \leqslant 1$, where the components under the summation sign are small.* However, this limitation is not essential. Substituting the above value of $\partial\varphi/\partial y$ in equation (4) and seeking a solution for v in the form

$$v = y\sqrt{1 - \frac{y^2}{2}}\, v(z), \tag{5}$$

we obtain an ordinary nonlinear integrodifferential equation for function $v(z)$:

$$v^2 - \frac{dv}{dz} e^{\sigma z} \int\limits_0^z v(z)\, e^{-\sigma z} dz' = -u_0 + z. \tag{6}$$

Writing this equation formally at $z = 0$, we derive an assumption regarding the analytical nature of $v^2(0) = -u_0$, from which it is clear that $u_0 < 0$. In other words, for a symmetrical (and near-symmetrical) pressure distribution the zonal circulation component may only be directed eastward.** This fact has been verified by numerous observations.

We write $-u_0 = a^2$ and rewrite the equation governing $v(z)$ in the form

$$v^2 - \frac{dv}{dz} e^{\sigma z} \int\limits_0^z v(z')\, e^{-\sigma z'} dz' = a^2 + z. \tag{6'}$$

A solution of this equation in explicit form can be found only in particular cases, when the effect of compressibility is not considered ($\sigma = 0$). These cases are of physical interest, since σ is smaller than unity, and if the solution of equation (6') is sought in the form of a power series in σ, then the principal term of the expansion will not contain σ.

Thus, we present two examples of the exact solution of

$$v^2 - \frac{dv}{dz} \int\limits_0^z v(z')\, dz' = a^2 + z.$$

The first of the examples is for the case $a^2 = 0$, i.e., when the zonal circulation at the equator is zero. We note that the example quoted in the footnote on page 215 of a pressure distribution in the form of a combination of two terms of a series expansion in y also reduces to this case.

It is easy to check that the function

$$v = \pm\sqrt{\frac{n+2}{4}}\, z^{\frac{n}{2}}$$

is the solution of the equation

$$v^2 - \frac{dv}{dz} \int\limits_0^z v(z')\, dz' = z^n,$$

* Such an approximation of the pressure field at the equator obviously yields only the principal outlines of the phenomenon; it would have been more correct to specify a real pressure distribution taking one, for example, from /18/.

** It may also be proven rigorously that the first term in the left-hand side of the equation (in any case for small z) is larger than the second.

satisfying the condition $v(0)=0$. In our case $n=1$ and, consequently, $v=\pm\sqrt{\frac{3}{2}z}$.

Since the pressure attains a minimum at the equator, v is directed toward the latter. Consequently, the minus sign should be used in the northern hemisphere.

Figure 8 shows the distribution of the meridional velocity component at different altitudes for various values of a.

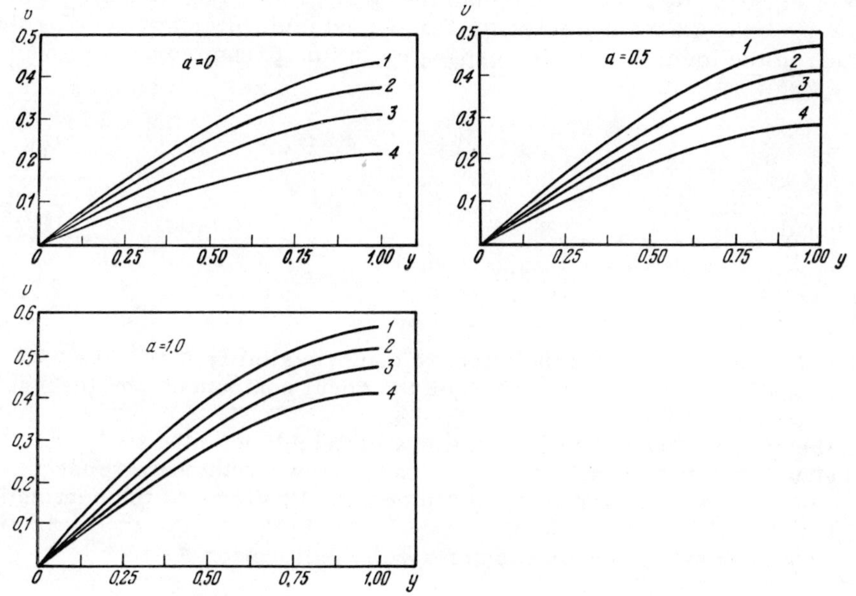

FIGURE 8. Zonal profiles of the meridional component of wind velocity v at different altitudes for various values of a:
1 — $z=1.0$; 2 — $z=0.75$; 3 — $z=0.5$; 4 — $z=0.25$.

The second example pertains to the opposite case, when the zonal flux t the equator is high, so that in the troposphere $a^2 \gg z$. Here v can be ound to a rough approximation from the simplified expression

$$\tilde{v}^2-\frac{dv}{dz}\int_0^z \tilde{v}(z')\,dz'=a^2.$$

One of its solutions is the function $\tilde{v}=a\cos Cz$, where C is an arbitrary constant determined by specifying a condition at some nonzero altitude or $m\pi/C$. It is convenient to specify the altitude (z_0) of the meridional current reversal (for $y\neq 0$). Then

$$C=\frac{\pi}{2z_0},$$

i. e.,

$$\tilde{v} = a\cos\frac{\pi}{2}\frac{z}{z_0}.$$

Schematically such a solution may point to the possibility of multiple changes in the direction of the meridional wind-velocity component with altitude. This is actually observed /18/.

To find a solution in the general case one can use the method of successive approximations, based on the fact that for small z the second term of equation (6') is smaller than the first. Assigning formally ε, the series expansion parameter, to the second multiplier and seeking the solution in the form of a series expansion in this parameter,* we find $v_0 = -\sqrt{a^2+z}$;

$$v_1 = \frac{e^{\sigma z}}{4(a^2+z)}\int\limits_0^z \sqrt{a^2+z'}\,e^{-\sigma z'}dz'. \qquad (7)$$

In general

$$v_n = -\frac{1}{2v_0}\left\{e^{\sigma z}\sum_{k=1}^{n-1}\frac{dv_k}{dz}\int\limits_0^z v_{n-k-1}e^{-\sigma z'}dz'\right\}.$$

It is obvious, first, that the effect of compressibility increases with altitude and, second, that the stronger the zonal component, the lesser the compressibility effect.

In the cases of symmetrical pressure distribution under study $v_{y=0} = 0$, and hence there is virtually no air exchange between the hemispheres. The value of w has an extremum at the equator by virtue of the continuity equation.

2. We consider a second case of specifying function $F(\xi)$:

$$F(\xi) = m\xi.$$

This means that the zonal flux satisfies the equation

$$u = \frac{y^2}{2} + z = m\int\limits_0^z ve^{-\sigma z'}dz', \qquad (8)$$

while the meridional velocity component is determined from

$$v\frac{\partial v}{\partial y} - \underline{\frac{\partial v}{\partial z}e^{\sigma z}\int\limits_0^z \frac{\partial v}{\partial y}e^{-\sigma z'}dz' + my\int\limits_0^z ve^{-\sigma z'}dz' + \frac{\partial \Phi}{dy}} + \frac{y^3}{2} - zy = 0.$$

For the symmetrical case at low z the contribution of the underlined term will be smaller than that of the first. Hence, one can again apply the method of successive approximations, taking as the zero approximation the expression

* In a number of cases, for example, when $\sigma = 0$, the convergence of this method is easy to prove /23/.

$$v_0 = \pm\sqrt{\Phi + \frac{y^4}{6} - \frac{zy^2}{2} + C(z)}.$$

The sign in front of the square root symbol is chosen depending on the direction of the baric gradient. For example, a low pressure zone is frequently observed in that part of the Indian Ocean located in the northern hemisphere, while an anticyclone prevails over this ocean in the southern hemisphere. The flow is northward and the plus sign should be used.

The arbitrary function of altitude $C(z)$ is determined by specifying the meridional circulation profile from observational data at some latitude. It is convenient to specify it at the equator.

Satisfying the condition $v = v_0(z)$ when $y = 0$, we find the zero approximation in the form

$$v_0(z, y) = \pm\sqrt{\Phi_{(y,z)} - \Phi_{(0,z)} + \frac{y^4}{6} - \frac{zy^2}{2} + [v^0(z)]^2}.$$

Subsequent approximations are calculated by quadratures:

$$v_{n+1}^2 = \int_0^y \left\{- my' \int_0^z v_{n-1} e^{-\sigma z'} + e^{\sigma z} \sum_{k=0}^{n-1} \frac{\partial v_k}{\partial z} \int_0^z \frac{\partial v_{n-k-1}}{dy} e^{-\sigma z'} dz'\right\} dy'.$$

The first few terms can be evaluated numerically quite easily using an electronic computer.

We note that in the given nonrigorous statement of the problem parameter m can be regarded as a free term, selecting it such that the zonal flow given by equation (8) is close to that observed at the latitude $y = 1$, where the wind may be regarded as geostrophic.

Without entering into a detailed analysis we note that in the case of an asymmetric pressure distribution the vertical velocities are almost an order of magnitude smaller than in the case of a symmetric distribution. The behavior of v with altitude is determined primarily by the behavior of $\partial\Phi/\partial y$; if, as a rule, the baric gradient increases with altitude, so does v.

References

1. Borisenkov, E. P. Fiziko-statisticheskie metody analiza i predvychisleniya meteorologicheskikh polei (Methods of Physical Statistics in the Analysis of Meteorological Fields). — Trudy AANII, Vol. 263. 1963.
2. Bykov, V. V. and G. P. Kurbatkin. Opyt ob''ektivnogo analiza aerologicheskikh dannykh (Experience in Objective Analysis of Aerological Data). — Izvestiya AN SSSR, seriya geofizicheskaya, No. 2. 1961.
3. Gandin, L. S. Ob''ektivnyi analiz meteorologicheskikh polei (Objective Analysis of Meteorological Fields). — Gidrometeoizdat. 1963. [IPST Cat. No. 1373, TT 65-50007. 1965.]
4. Dobryshman, E. M. Uchet peremennosti parametra Koriolisa pri issledovanii atmosfernykh protsessov (baroklinnaya model')

(Making Allowance for the Variability of the Coriolis Parameter in the Study of Atmospheric Processes (the Baroclinic Model)). — Trudy Vsesoyuznogo Meteorologicheskogo Soveshchaniya, Vol. 2. Gidrometeoizdat. 1963.

5. Dobryshman, E. M. Voprosy avtomatizatsii obrabotki tekushchei meteorologicheskoi informatsii i ob"ektivnogo analiza meteorologicheskikh polei (Problems of Automation in Processing Current Meteorological Information and in Objective Analysis of Meteorological Fields). — Trudy Simpoziuma po chislennym metodam prognoza. Gidrometeoizdat. 1964.

6. Dobryshman, E. M. O nekotorykh osobennostyakh tsirkulyatsii v troposfere v ekvatorial'noi oblasti (On Certain Features of the Circulation in the Troposphere at the Equator). — Meteorologiya i gidrologiya, No. 5. 1964.

7. Dobryshman, E. M. Issledovanie statisticheskikh kharakteristik polya v nizkikh shirotakh i opredelenie dvizheniya po polyu davleniya v ekvatorial'noi oblasti (Study of the Statistical Characteristics of the [Pressure] Field in Low Latitudes and Determination of the Circulation from the Pressure Field at the Equator). — Trudy MMTs, Vol. 7. 1964.

8. Dubentsov, V. R. O nekotorykh osobennostyakh tsirkulyatsii v troposfere i stratosfere vblizi ekvatora (Some Features of Tropospheric and Stratospheric Circulation Near the Equator). — Meteorologiya i gidrologiya, No. 12. 1963.

9. Zaichikov, P. F. Predvaritel'nye rezul'taty obrabotki nablyudenii II Mezhdunarodnogo sravneniya radiozondov (Preliminary Results of Processing the Observations of the 2nd International Comparison of Radiosonde [Data]). — Trudy TsAO, Vol. 22. 1957.

10. Kats, A. L. Vozdukhoobmen v tropicheskoi zone i ego svyaz' s obshchei tsirkulyatsiei atmosfery (Air Exchange in the Tropical Belt and its Relationship with the General Circulation of the Atmosphere). — Meteorologiya i gidrologiya, No. 2. 1963.

11. Kats, A. L. Dvukhletnaya tsiklichnost' v ekvatorial'noi stratosfere i obshchaya tsirkulyatsiya atmosfery (Two-Year Cyclic Variation in the Equatorial Stratosphere and the General Circulation of the Atmosphere). — Meteorologiya i gidrologiya, No. 4. 1964.

12. Smirnov, I. P. Prognoz indeksa tsirkulyatsii pri pomoshchi momentov svyazi dlya barotropnoi modeli atmosfery (Forecasting of the Circulation Index for a Barotropic Model of the Atmosphere Using Covariances). — Izvestiya AN SSSR, seriya geofizicheskaya, No. 9. 1963.

13. Tatarskaya, M. S. Ob"ektivnyi analiz nekotorykh elementov s ispol'zovaniem statisticheskikh kharakteristik (Objective Analysis of Certain Elements Using Statistical Characteristics). — Meteorologiya i gidrologiya, No. 4. 1963.

14. Fortus, M. I. K voprosu o prakticheskom osushchestvlenii statisticheskoi ekstrapolyatsii meteorologicheskikh polei (Concerning the Practical Implementation of Statistical Extrapolation of Meteorological Fields). — Izvestiya AN SSSR, seriya geofizicheskaya, No. 6. 1962.

15. Shvets, M. E. Ob osobennostyakh atmosfernykh dvizhenii v ekvatorial'noi oblasti (Features of Atmospheric Motions in the Equatorial Belt). — Trudy GGO, 33(95). 1952.
16. Yudin, M. I. Novye metody i problemy kratkosrochnogo prognoza pogody (New Methods and Problems of Short-Range Weather Forecasting). — Gidrometeoizdat. 1963.
17. Dady, G. L'état Actuel de la Météorologie Dynamique. — Météorologie, Oct., Dec. 1962.
18. Defant, F. and H. M. E. van de Boogard. The Global Circulation Features of the Troposphere Between the Equator and 40°N, Based on a Single Day's Data. I. — Tellus, 15(3). 1963.
19. Dobrischman, E. M. On Coriolis Force Variability in Prognostic Schemes. — Tellus, 16(1). 1964.
20. Gilchrist, B. and G. P. Cressman. An Experiment in Objective Analysis. — Tellus, 6(6). 1954.
21. Reed, R. J., W. J. Campbell, L. A. Rasmussen, and D. G. Rogers. Evidence of Downward Propagating, Annual Wind Reversal in the Equatorial Stratosphere. — J. Geophys. Res., 66(3). 1961.
22. Reed, R. J. and D. G. Rogers. The Circulation of the Tropical Stratosphere in the Years 1954—1960. — J. Atmos. Sci., 19(2). 1962.
23. Tucker, G. B. Zonal Winds over the Equator. — Quart. J. Roy. Meteorol. Soc., 90(386). 1964.
24. Veryard, R. G. and R. A. Ebdon. Fluctuations in Tropical Stratospheric Winds. — Meteorol. Mag., 90(1066). 1961.

V. R. Dubentsov

SOME FEATURES OF AIR CIRCULATION IN THE STRATOSPHERE AND MESOSPHERE

The article describes and analyzes the main circulation features in the stratosphere and mesosphere. The two-year cyclic variation in the distribution of zonal wind components in the tropical stratosphere is examined. Information obtained from rocket sounding of the wind distribution up to 100 km is presented; the monsoonal nature of circulation in both the stratosphere and mesosphere is noted. Zonal circulation models for different latitude belts from the surface up to 100 km are presented.

While a sufficiently complete picture of the predominant air-flow distribution in the troposphere and lower stratosphere has already been deduced, much still remains unknown about the currents in the upper stratosphere and mesosphere, the more so in the thermosphere, where only single observations were made.

The first wind-distribution model up to 120 km was suggested by Kellogg and Schilling in 1951 /9/ and obtained by correlating a quite limited quantity of actual data. The upper layers of the atmosphere were studied much later by Murgatroyd /13/. For the 30—60 km layer he used sound-propagation data, while for higher levels he used observations of meteor trails, luminous clouds and ionospheric drifts. Batten /7/ in 1961 studied critically Murgatroyd's data and showed that only observations of the motion of meteor trails yield satisfactory information regarding the wind. The meridional section of zonal wind components constructed by Batten differs appreciably from that of Murgatroyd.

In 1959 Borovikov, Golyshev and Kokin /1/ used rocket observations in the Atlantic and Pacific to compile two meridional sections up to 45 km. One of these sections (for July) gives some idea about the wind distribution over the Atlantic in the belt between 40°N and 30°S. They concluded that above 20 km the easterly winds of the northern hemisphere as if invade the southern hemisphere in summer, where they are detected at up to 15°S. Regretfully, observations of this kind were not repeated during subsequent years. Later investigations detected a two-year cyclic variation in the distribution of zonal circulation components in the stratosphere.

The development of rocket observations made it possible to accumulate a large amount of factual data on the temperature and wind in the stratosphere and mesosphere. This allowed a more complete study of the upper layers of the atmosphere to be made. The present paper correlates the data thus accumulated with a view to obtaining the general picture of the atmospheric structure and to detecting certain features of the circulation.

The two-year cyclic variation in zonal wind components in the stratosphere near the equator. A number of publications which appeared in 1959—1963 showed the presence of a two-year cyclic variation in zonal wind components in the equatorial stratosphere /14, 17/. Observations from R/V "A. I. Voeikov" and "Yu. M. Shokal'skii" made in 1960—1962 confirmed the existence of this two-year cyclic variation at the 180th meridian and made it possible to construct the air-current distribution for the lower 30 km /3, 4/. Wind data collected by stations in the Pacific from the equator to 70°N were analyzed by the author together with A. A. Unukova and showed that the two-year cyclic variation in the sequence of zonal wind components is detected clearly only up to 7—10°N. Both the two-year cycle and the ordinary yearly cycle with seasonal wind changes are observed in the 10—20°N latitude belt. To the north of 20°N the two-year cycle is damped out and to the north of 30°N it cannot be detected at all.

Processing of data collected at stations located directly at the equator showed that the duration of processes with westerly and easterly components of air current varies with altitude (Table 1).

TABLE 1. Duration (in months) of periods with easterly and westerly wind components and of its complete cycle at different isobaric surfaces

Wind component	Isobaric surface, mb							
	80	60	50	40	25	20	15	10
Easterly	9.8	9.9	11.2	11.9	12.2	13.2	15.2	15.3
Westerly	16.5	16.0	14.2	13.0	13.0	12.0	10.2	11.0
Complete cycle	25.3	25.9	25.4	24.9	25.2	25.2	25.4	26.3

As is seen, the duration of the period with an easterly component increases with altitude from 10 months at the 80 mb surface (18 km) to 15 months at 10 mb (30 km), while the duration of periods with westerly circulation accordingly decreases from 16 to 11 months. The total duration of a period is 25—26 months.

If it is assumed that the reduction in the period with the westerly component above 30 km takes place at the same rate as between 18 and 30 km, then this component should vanish in the 40—45 km layer, and easterly flow should prevail above it. In this case the current distribution in the equatorial troposphere and stratosphere can be represented as shown in Figure 1. This figure substantially refines the stratospheric circulation model at the equator described in /3, 4/.

It is seen from Figure 1 that westerly winds first appear in the 40—45 km layer, and then gradually extend downward. From the 40—42 km level they descend to the tropopause in 18 months (at an average rate of 1 km per 0.8 month), where they merge with the westerly current of the upper troposphere. The easterly fluxes descend much slower over a period of 32 months (at a rate of 1 km per 1.5 month).

Since the average velocity of the easterly component in the lower stratosphere is much higher than that of the westerly component, the easterly flux usually predominates on ordinary climatic averaging. This was why the alternation of fluxes in the equatorial atmosphere was detected only as late as the end of the fifties.

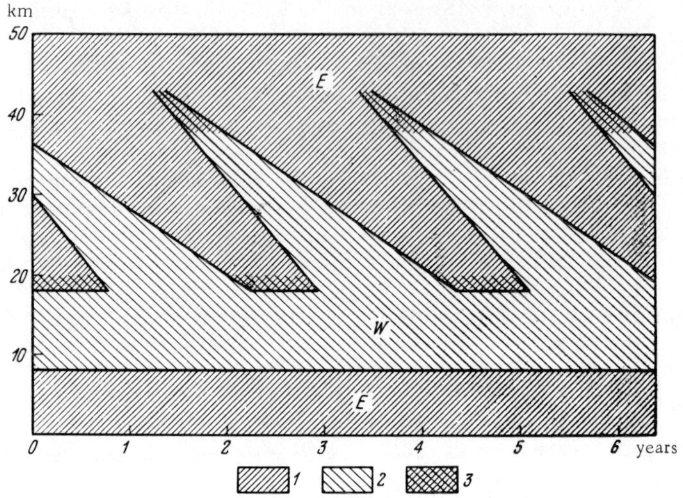

FIGURE 1. Two-year cyclic variation in the distribution of zonal wind components in the equatorial troposphere and stratosphere:

1 — easterly component; 2 — westerly component; 3 — region of unstable zonal components.

It should be noted that the westerly circulation is replaced by its easterly counterpart near the equator almost simultaneously at every meridian /17/. The westerly and easterly circulations in the stratosphere thus as if encircle the globe by a narrow band.

Knowing the date of appearance of the easterly or westerly component in the 30—40 km layer one can determine the date of appearance of the corresponding component also at lower levels (below 30 km and approximately up to 20 km) with sufficient accuracy and quite far in advance.

S e a s o n a l f e a t u r e s i n t h e d i s t r i b u t i o n o f z o n a l w i n d c o m p o n e n t s. Actual wind observations in the mesosphere and lower thermosphere are not numerous. Hence either the temperature must be calculated from the wind field or the wind from the temperature field. Here it is usually assumed that the observational wind data are more reliable. However, with a small amount of observations, the interpolation of the wind over data-sparse regions, as well as its extrapolation to higher levels, is less reliable than the extrapolation and the interpolation of the temperature. On the basis of these considerations the zonal wind components were calculated from the temperature field, data on which for winter and summer were obtained by the author in 1961 /2/.

Knowing the meridional component of the horizontal temperature gradient at different levels and the wind distribution at the lower level, it is possible, using the equation of the thermal wind, to determine the magnitude of the zonal wind component at higher levels.

The thermal wind equation for the zonal component is

$$\frac{\partial}{\partial z}\left(\frac{u}{T}\right) - \frac{-g}{T^2 2\omega \sin\varphi} \cdot \frac{\partial T}{\partial y}, \qquad (1)$$

where T is the temperature; u is the zonal wind component; z is the altitude; y is the horizontal coordinate, increasing northward; g is the acceleration of gravity; ω is the angular velocity of the earth; and φ is the latitude. Carrying out various transformations, equation (1) may be expressed in the form

$$u_1 = T_1\left(\frac{u_0}{T_0} - \frac{g\Delta z\, \Delta T}{2\omega \sin\varphi\, T^2 \Delta y}\right), \qquad (2)$$

where u_0 and T_0 are respectively the zonal component of the wind and the temperature at the lower level (z_0), while u_1 and T_1 are their counterparts at the upper level (z_1). Substituting the values y = 9.8 m/sec and ω = 7.3 · 10^{-5} sec^{-1}, we have

$$u_1 = T_1\left(\frac{u_0}{T_0} - \frac{1}{0.00332} \cdot \frac{\Delta T}{T^2 \sin\varphi}\right). \qquad (3)$$

This equation was used for calculating the zonal wind components. The calculations were made for the midpoints of 5-kilometer layers at 10° latitude intervals, starting with 30 km. The wind at the 25 km level and below was determined from IGY data (1957—1959).

The results thus obtained are plotted in Figure 2. The wind speed for the belt from the equator to 15°, where equation (3) does not yield satisfactory results, was extrapolated and the isolines are drawn as dashed lines. The section also shows the resultant wind obtained from rocket observations. It is seen by comparing the actual and calculated winds that agreement with respect to direction is quite satisfactory, with the exception of the 80 km level at 20°N in winter. According to observations at Johnston Island, quite a strong westerly wind is observed here, while calculations point to an easterly wind. Since these were single observations at these levels, it is quite possible that they were made during a time when the winter stratospheric-mesospheric westerly wind shifted somewhat to the south. It is also possible that the calculated boundary between the westerly and easterly winds actually lies somewhat more to the south than shown in the section.

The difference in the wind speed, particularly during winter, is much greater above the 50 km level. Comparison with the Murgatroyd and Batten sections showed that: in summer according to Murgatroyd weak westerly winds predominate at high latitudes in the 40—80 km layer, while the sections by this author and by Batten show weak easterly winds. As is known, a wide circumpolar anticyclone prevails at the 30 km level in summer in polar regions. This anticyclone is also observed at the 1 mb or 48 km level /16/. The direction of horizontal temperature gradients at 50 km shows that there is no basis for expecting a change in wind above 50 km. Hence easterly winds are more probable here in summer. The differences in winter in the 70—80 km layer are appreciable. Batten assumes that westerly winds invade the summer hemisphere up to 10°N from the winter hemisphere. However, he also assumes that this may be a random invasion of westerly winds into low latitudes. Batten's assertion of the existence of westerly winds is based on very little actual data on the wind above 60 km over Johnston Island (16°N). This author's section shows easterly winds above 40 km near the equator. The southernmost boundary of these winds in winter at the 50 km level is located at about 15°N.

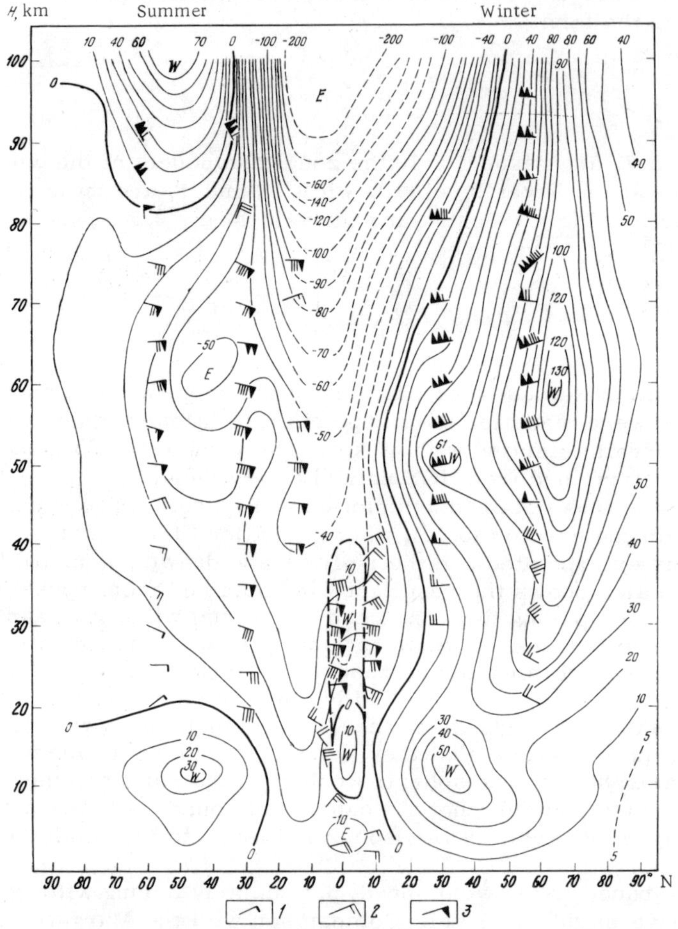

FIGURE 2. Vertical meridional section of zonal wind components for summer and winter:

E — easterly component; W — westerly component; isolines denote the wind speed in m/sec. The velocity of the resultant wind is determined from rocket observations: 1 — 2 to 3 m/sec; 2—5 m/sec; 3—25 m/sec.

The center of the westerly-wind maximum in winter in the Batten section is shifted a lot to the south. This was most probably due to the strong winds at Johnston Island. No westerly winds were noted during observations performed from R/V "Ob" near the equator, as well as in summer at the Island of Guam (13°N) up to the 80 km level. It may be assumed that strong easterly winds prevailing over Johnston Island in winter do not extend too far south and are little probable south of 10°.

A distinguishing feature of the section in Figure 2 are strong easterly winds in the equatorial zone; these gradually increase with altitude and, particularly in winter, move into high latitudes. Both westerly and easterly winds are shown in the equatorial zone for the 20 — 40 km layer with allowance for the two-year cyclic variation.

The second feature of this section is the location of the winter maximum of the westerly wind. Unlike the Batten and Murgatroyd sections, it shows two maxima: one (the higher) between 60 — 70°N and a second near 30°N. Actual wind data yield lower velocities than those given by calculation. The first maximum at the 60 — 70°N latitude corresponds to the so-called polar stratospheric current, the horizontal axis of which actually lies in the mesosphere and, possibly, even higher than shown in the section (70 — 80 km). The second maximum forms above the subtropical jet stream.

The upper boundary of the winter polar jet stream lies much higher than previously supposed. According to the section the maximum of the westerly wind is shown at the 60 km level, while according to actual wind data it should lie between 70 and 80 km. Strong westerly winds in the 50 — 90°N belt were detected up to at least 100 km. The horizontal extent of the westerly wind zone along the meridian decreases perceptibly with altitude starting with the 50 km level.

T h e m o n s o o n a l n a t u r e o f t h e c i r c u l a t i o n i n t h e s t r a t o s p h e r e a n d m e s o s p h e r e. The section in Figure 2 gives some idea of the distribution of the zonal wind components only in winter and summer. In order to be able to judge about the duration of the winter and summer circulation type and about the date when the winter circulation is replaced by the summer wind or vice versa, one must construct time sections for different latitude belts. Regretably, there exist very little observational data for constructing such sections. Pchelko /6/ showed that the transition of the wind from westerly to easterly in the lower stratosphere in spring usually occurs at the end of April or beginning of May. Upon processing rocketsonde data for White Sands (32°N) and Pt. Mugu (34°N), Miers /12/ established that the wind reversal in the atmosphere occurs at times close to equinox.

Time sections of zonal wind components were constructed from wind observations at a number of rocket stations /8/. For the lower part of the stratosphere (up to 30 km) use was also made of radiosonde data collected by stations located at different latitude belts of the Pacific basin. Relatively few observations were made above 30 km. The rockets rarely reached altitudes of 60 km. No regular month-to-month observations were made to the north of 45°N. The collected data made it possible to obtain a more complete idea about circulation variations in the 30 — 40° latitude belt up to 60 km. The circulation variations in the 60 — 110 km layer can be determined from Batten's time sections /7/, constructed for the 30 — 40° latitude zone in the northern hemisphere and from Rofe's section /15/, constructed from observations over Woomera (Australia, 31°S). It is seen by comparing these sections that the nature of the circulation in both the hemispheres in the 30 — 40° latitude zone during identical seasons is quite similar. All the above data were used for constructing the scheme of circulation variations for the 0 — 110 km layer (Figure 3).

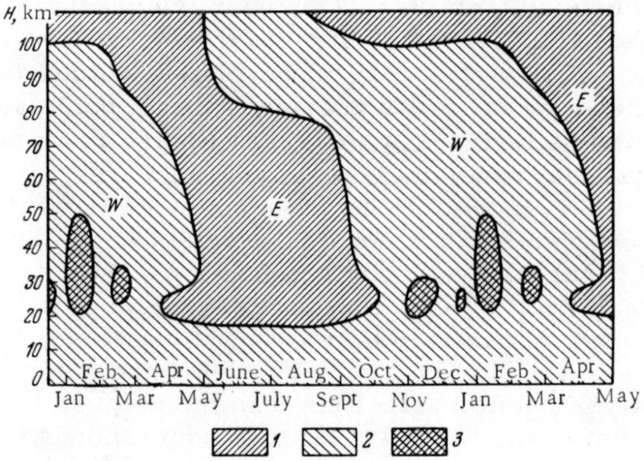

FIGURE 3. Distribution of zonal wind components by months in the 30—40° N belt:

1 — easterly component; 2 — westerly component; 3 — easterly component appearing episodically against the general background of the westerly flow.

It is seen from Figure 3 that the seasonality in the distribution of zonal circulation components or, as it is termed, the monsoonal nature of circulation in the stratosphere and mesosphere is quite pronounced. The figure shows that the circulation change in the stratosphere and lower mesosphere is preceded by a circulation change in the upper mesosphere and in the lowest part of the thermosphere. At 90—100 km the westerly circulation is replaced by the easterly wind on the average during the second half of February, while for the upper stratosphere this occurs only as late as in the first half of May, i. e., with a lag of 2.5—3 months. The replacement of the easterly by the westerly circulation in the upper mesosphere is even later, as compared with the lower stratosphere. Easterly transport at 90—100 km appears as early as at the end of May, while in the lower stratosphere this happens only in October, i. e., with a lag of 4—5 months. One's attention is attracted by the fact that easterly fluxes in the 17—75 km layer are observed during the warm season from May to September, while in the upper mesosphere these occur in spring (March—May); above 80 km westerly transport is observed above the summer easterly transport.

At the 23—26 km levels of the lower stratosphere the westerly fluxes are replaced by easterly winds somewhat earlier than in the upper stratosphere (at the beginning of April), and terminates somewhat later (mid-October). The circulation changes in the stratosphere and lower mesosphere are in satisfactory agreement with the ordinary seasonal temperature variation. The peculiar features observed in the distribution of zonal components in the 23—26 km layer can be apparently ascribed to the ozone distribution.

The ozone content in middle and high latitudes increases perceptibly in spring, with the maximum rise taking place precisely in the 20—25 km layer. This results in an earlier and more intensive heating of air at these latitudes and in an earlier reconstruction of the pressure fields at the 23—26 km levels. In autumn, despite the large reduction in ozone content, cooling at the maximum ozone-content level occurs with a certain lag, which results in the later reconstruction of the pressure fields.

Easterly winds move from the upper mesosphere into the lower stratosphere (from the 100 to the 30 km level) at an average rate of about 1 km/day, somewhat faster for the 100—65 km level at 1.7 km/day. Westerly currents move from the 100 to the 75 km level only at 0.2 km/day, in the 75—30 km layer this rate is 1.8 km/day, while on the average for the 100—30 km layer the rate is 0.5 km/day.

The duration of the easterly-circulation period at all levels below 100 km is much shorter than the duration of the westerly circulation period. At the 90 km level it amounts to only 3 months, at the 30—75 km level it is 4.5—5 months, and at the 25 km level it is about half a year.

The easterly circulation in the lower stratosphere (20—30 km) of high latitudes (60—70°N) is usually established somewhat earlier (beginning of April) and ends in mid-August. The duration of the easterly monsoonal circulation here does not exceed 4.5 months. Here the summer easterly fluxes are more stable than the westerly flows in the winter. Once the easterly flow is established in spring, it almost always remains undisturbed until autumn. Conversely, during the cold season the westerly circulation is frequently disturbed. From November to the end of March easterly circulation may prevail in the 20—35 km layer, and sometimes up to 45 km and possibly higher, for as long as 10—20 days and less frequently for an entire month. These depatures from westerly circulation are usually produced by temperature rises in the stratosphere of the middle and high latitudes and shifting of the stratospheric anticyclone, the latitudinal axis of which lies in midwinter near 25°N (with the exception of the Pacific, where it is located at 40°N), into higher latitudes.

As can be seen from Figure 3, the duration of the period with easterly circulation decreases only to the 100 km level. Above this level the period with easterly circulation increases perceptibly.

According to Manring et al. /11/, it follows also from observations of sodium clouds at 38°N that easterly fluxes prevail in summer above 110 km, while in winter (18 November 1959) easterly winds are detected only in the approximately 110—130 km layer.

The earlier circulation change in the upper atmosphere allows one to infer that the circulation in the lower mesosphere and in the stratosphere is apparently determined by processes occurring at higher levels. The short duration of observations, which were also quite irregular, do not allow a careful study of individual features of circulation changes. It is possible that circulation reversals at high altitudes do not always occur at the same time, which is also responsible for changes in the nature of the circulation at lower levels.

This average picture is definitely quite smoothed out. It is seen from analysis of observations for individual years for the 15—50 km layer that the time of circulation reversal within the limits of the belt under study (30—40°N) may vary highly from one year to another. Reversal of predominant circulation occurs in individual latitude belts also during a single year.

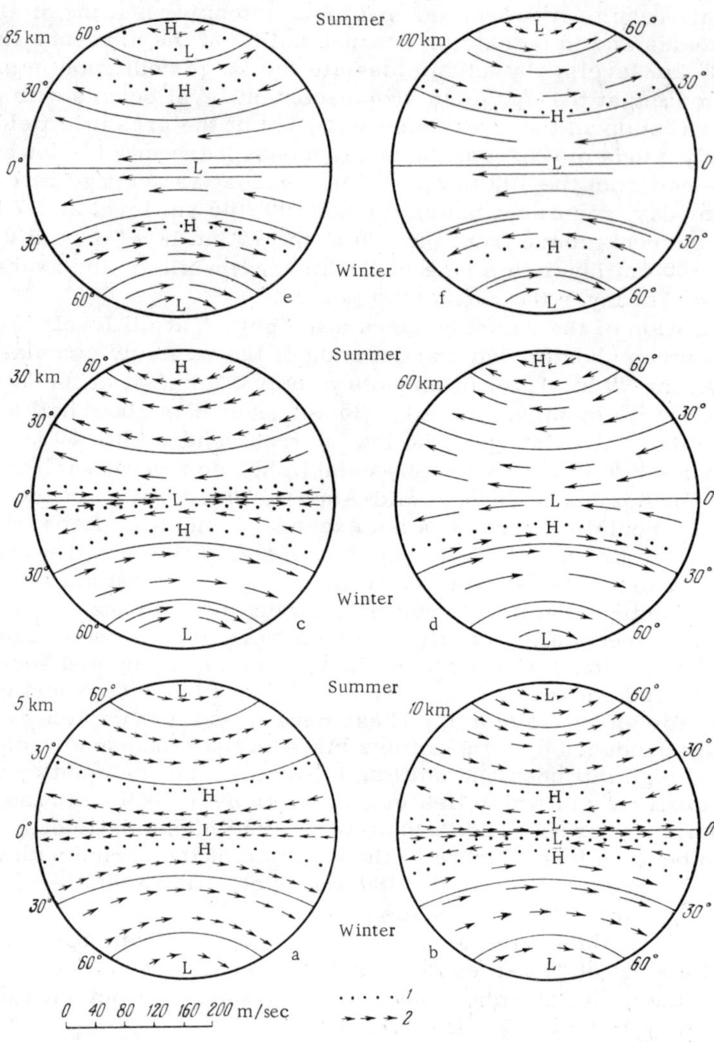

FIGURE 4. Schemes of summer and winter planetary zonal circulation at the 5, 10, 30, 60, 85 and 100 km levels:

1 — latitude circulation-reversal boundaries; 2 — flow direction (the length of arrows is proportional to the circulation speed).

Schemes of planetary zonal circulation. The available data make it possible to construct, in the first approximation, a scheme of the zonal circulation for different levels. Figure 4 gives a distribution of the zonal wind components for six levels, describing the most typical circulation layers. The upper half of the circles depicts the summer circulation of the northern hemisphere, with the lower half describing the winter circulation of this hemisphere. Since, as was confirmed by observations over Woomera (31°S) and in the 30 — 40° belt, the nature of circulation in the northern and southern hemispheres is quite similar during identical seasons, these schemes may be extended also to the entire globe. These schemes show the location of high- and low-pressure regions.

Easterly circulation prevails in the lower and upper troposphere of high and middle latitudes (Figure 4a and b). At low latitudes, on the periphery of subtropical anticyclones at the 5 km level, the easterly circulation predominates, extending in summer from the equator to 25 — 30° latitude, and only to 15° in winter. A similar pattern is also observed at the 10 km level, but a narrow belt of westerly winds appears at the equator and up to 5° to both sides of it. Accordingly, easterly winds in the winter hemisphere are observed only in a relatively narrow belt, approximately from 5 to 15° latitude.

The nature of circulation at the 30 km level (Figure 4c) in the winter hemisphere changes relatively little as compared with the lower levels, while in the summer hemisphere the pressure fields are radically reconstructed, resulting in replacement of zonal by easterly fluxes (as was shown above, this reconstruction starts above the 17 km level). During years when, in conjunction with the two-year cycle, an easterly circulation is observed at the equator, it moves into the southern winter hemisphere up to 25°S. During years with westerly circulation, the easterly winds of the northern and southern hemispheres at the equator separate into two independent flows. In accordance with this, Figure 4c shows for the equatorial belt alternately westerly and easterly winds.

The two-year cyclic variation of zonal wind components at the equator should no longer be detected above 40 km, and the nature of circulation in the upper stratosphere and lower mesosphere simplifies markedly.

Easterly circulation prevails at the 60 km level (Figure 4d) in the summer hemisphere, and it penetrates also into the southern winter hemisphere up to 20 — 25° latitude. Maximum velocities of the easterly component are observed at the equator. A secondary, but a weaker easterly-wind maximum is observed in summer at 40° latitude. Maximum westerly-component speeds are observed in winter at 60 — 65° latitude.

The nature of circulation in the upper mesosphere of the summer hemisphere changes somewhat at the 85 km level. As is seen from Figure 4e, easterly circulation appears in the 40 — 70° latitude belt of the northern hemisphere; this circulation extends with higher altitudes to the north and south, and at the 100 km level it covers the entire latitude belt from 35° to the pole (Figure 4f). Above 60 km in the winter hemisphere easterly winds from the equator gradually move into middle latitudes. The boundary between them at 85 km lies at about 40° latitude, and at 100 km it lies at about 50°. The highest easterly component velocities at the 85 and 100 km levels are at the equator, while the highest westerly circulation velocities in the summer hemisphere are between 40 and 60° latitude and in the winter hemisphere between 60 and 70°.

The above circulation schemes for the 85 and 100 km levels differ somewhat from those of Kohansky /10/, who used, in addition to rocket wind data, observations of the motion of meteor trails at Jodrell Bank (53°N, 2°W) and Adelaide (35°S, 138°E).

The differences between these circulation schemes pertain primarily to the location of the latitude boundaries of regions with westerly and easterly circulation and are not of major significance. More significant and basic differences exist at the 80—85 km level in winter. In the scheme shown in Figure 4e the westerly component is shown only in high and middle latitudes (from the pole to 35—40°), while in the Kohansky scheme for the 80 km level it extends over the entire winter hemisphere and the westerly winds are replaced at the equator by easterly winds of the summer hemisphere. According to the winter circulation scheme the pressure at the equator should be high, while according to the summer circulation it should be low. Since the schemes are applicable to both the northern and southern hemispheres, it is difficult to imagine the simultaneous existence at the equator of a low pressure from the side of the northern summer hemisphere and a high pressure from the side of the southern winter hemisphere. Kohansky corrects this contradiction to some extent by introducing secondary circulations at the equator. Unfortunately, no factual data on the wind in the mesosphere at the equator are available and hence it is impossible to check the validity of this hypothesis.

For the 30 km level the circulation schemes are in agreement, in the general outline, with those of Kats /5/. Some differences are encountered only at low latitudes. In the Kats schemes the anticyclone in the low latitudes of the winter hemisphere is shown as an extension of the polar anticyclone of the summer hemisphere. Actually a low-pressure band with its axis at the equator proper, separating the anticyclones of the northern and southern hemispheres, usually exists in the equatorial stratosphere attendant to easterly winds, which should have been emphasized in the schemes. The scheme presented by Kats shows that unlike the winter of the northern hemisphere, no high-pressure region similar to the North Pacific anticyclone exists in low latitudes of the southern hemisphere. Actually a high-pressure region is observed in winter in the northern hemisphere not only over the Pacific, but also over the entire hemisphere, although it is true that in other regions it lies much more to the south (about 25°N). It is seen by comparing observational data in the stratospheres of the northern and southern hemispheres, that the circulations in both hemispheres during similar seasons are quite similar. Thus, there is sufficient basis for assuming that a stratospheric anticyclone exists in winter at low latitudes also in the southern hemisphere.

In conclusion we note that the sparse network of rocketsonde stations and their low ceiling make it impossible at present to obtain a complete and sufficiently reliable idea about the major features of stratospheric and mesospheric circulation, its variability, and the more so of relationships between different atmospheric layers. The schemes of zonal circulation in different latitude belts, particularly for levels above 50—60 km, presented in this paper, give only some idea regarding the main circulation features. These schemes will most likely require extensive revision when the number of wind and temperature observations in the mesosphere and lower thermosphere increases.

It may be assumed with reference to the earlier zonal-circulation sign reversal in the lower thermosphere, that the nature of the wind distribution in the mesosphere is determined to a large extent by processes occurring above the mesosphere. The nature of circulation in the stratosphere, particularly in its lower part, depends also on the ozone concentration and distribution. The relationships noted by Kohansky /10/ between the variability of the wind in the upper and lower layers of the atmosphere (92, 16 and 5 km) again show the importance of studying the upper layers of the atmosphere. Hence the task of setting up a network of rocket stations which would carry out observations up to at least 100 km should be regarded as one of the utmost urgency. Radiation observations from satellites are just as important.

Bibliography

1. Borovikov, A. M., G. I. Golyshev, and G. A. Kokin. Nekotorye kharakteristiki stroeniya atmosfery yuzhnogo polushariya (Some Structural Features of the Atmosphere of the Southern Hemisphere). — Meteorologiya i gidrologiya, No. 3. 1962.
2. Dubentsov, V. R. Osnovnye osobennosti raspredeleniya temperatury v atmosfere v razlichnye sezony (v sloe 0—100 km) (Major Features of the Temperature Distribution in the Atmosphere during Different Seasons (in the 0—100 km Layer)). — Meteorologiya i gidrologiya, No. 9. 1961.
3. Dubentsov, V. R. O nekotorykh osobennostyakh tsirkulyatsii v troposfere i stratosfere vblizi ekvatora (Some Features of the Circulation in the Troposphere and Stratosphere Near the Equator). — Meteorologiya i gidrologiya, No. 12. 1963.
4. Dubentsov, V. R. Nekotorye osobennosti tsirkulyatsii v severnoi i tsentral'noi chastyakh Tikhogo okeana po materialam nablyudenii na ekspeditsionnykh sudakh "A. I. Voeikov" i "Yu. M. Shokal'skii" (Some Features of the Circulation in the North and Central Pacific from Observations Made from R/V "A. I. Voeikov" and "Yu. M. Shokal'skii"). — Trudy TsIP, Vol. 137. 1964.
5. Kats, A. L. Dvukhletnyaya tsiklichnost' v ekvatorial'noi stratosfere i obshchaya tsirkulyatsiya atmosfery (Two-Year Cyclic Variation in the Equatorial Stratosphere and the General Circulation of the Atmosphere). — Meteorologiya i gidrologiya, No. 6. 1964.
6. Pchelko, I. G. Ob obrazovanii i evolyutsii letnego stratosfernogo antitsiklona v severnom polusharii (The Genesis and Evolution of the Summer Stratospheric Anticyclone in the Northern Hemisphere).— Meteorologiya i gidrologiya, No. 12. 1959.
7. Batten, E. S. Wind Systems in the Mesosphere and Lower Ionosphere. — J. Meteorol., 18(3). 1961.
8. Data Report of the Meteorological Rocket Network Firings, Nos. 1—24. 1960—1963.
9. Kellogg, W. W. and G. F. Schilling. A Proposed Model of the General Circulation in the Upper Atmosphere. — J. Meteorol., 8(8). 1951.
10. Kohansky, A. Circulation and Temperature at 70 to 100 km Height. — J. Geophys. Res., 68(1). 1963.
11. Manring, E., J. Bedinger, H. Knaflich, and R. Lynch. Upper Atmospheric Wind Profiles Determined from Three Rocket Experiments. — Geophys. Corp. Amer. Rept. Contract NAS, 215, Bedford. 1961.

12. Miers, B. I. Zonal Wind Reversal Between 30 and 80 km over the Southwestern United States. — J. Atmos. Sci., 20(2). 1963.
13. Murgatroyd, H. I. Winds and Temperatures Between 20 km and 100 km. — Quart. J. Meteorol. Soc., 83(358). 1957.
14. Reed, R. I. and D. G. Rogers. The Circulation of the Tropical Stratosphere in the Years 1954—1960. — J. Atmos. Sci., 19(2). 1962.
15. Rofe, B. Australian Sounding Rocket Experiments. — Techn. Note SAD, 127. July 1963.
16. The Meteorological Rocket Network. An Analysis of the First Year in Operation. — J. Geophys. Res., 66(9). 1961.
17. Veryard, R. G. and R. A. Ebdon. Fluctuations in Tropical Stratospheric Winds.— Meteorol. Mag., 90(1066). 1961.

M. A. Petrosyants

EFFECT OF OROGRAPHY ON THE GENERAL CIRCULATION OF THE ATMOSPHERE

This paper is concerned with the effect of orography on the general circulation of the atmosphere. In particular, it treats the effect on the average wind fields, average geopotential of the 700, 500, 300 and 200 mb isobaric surfaces and energy of large-scale eddies during different months and seasons of the year. Questions of the thermal effect of mountain systems on the general circulation and the role of these systems in the formation of synoptic elements (such as cyclones and jet streams) are also examined.

The extensive mountain systems of our planet, such as the Central Asian mountain massif,* North American Cordillera and South American Andes, as well as glacial elevations such as Greenland and the Antarctic, comprise 1/3 — 1/4 of the thickness of the troposphere. In the horizontal direction these mountain systems are commensurable with the characteristic dimensions of synoptic elements, i.e., dimensions of planetary high-altitude frontal zones, cyclones and anticyclones, ridges and troughs. It is therefore natural that such elevations have a marked effect on atmospheric motion. Two principal mechanisms of this effect can be isolated. Mountain systems serving as gigantic obstacles exert a mechanical effect on atmospheric fluxes, producing various dynamic effects. Mountains, being a highly rugged surface with different radiative properties, projecting high into the atmosphere, have a thermodynamic effect on the atmospheric streams.

The direction and speed of air currents in each geographic region are subjected to changes and can be differently oriented relative to a mountain system. It is natural that the dynamic effect of mountain systems will be different depending on the nature and direction of the incoming air flow. Synoptic-scale air currents are very varied.

Quantitative allowance for the effect of orography on atmospheric motions at any instant results in an unsteady problem. This orographic effect is considered in different hydrodynamic-forecasting models in some form or another, which makes allowance for the terrain /6, 8, 11, 12, 25, 26/. However, allowance for terrain in a numerical forecasting model still does not give one an idea of the effect of orography on the general circulation. In fact, such an understanding of the problem would mean that the term "general circulation of the atmosphere" applies to an infinite sequence of synoptic-scale motions with all the large and small characteristic and random formations.

* This term as used here includes the Tien Shan, Altai, Pamir-Altai, Hindu Kush, Kunlun, Tibet and the Himalayas.

The description and clarification of the effect of orography on the general circulation thus depends on the meaning assigned to this concept. Usually the term "general circulation of the atmosphere" is used to denote the ensemble of planetary-scale motions (commensurable with the dimensions of continents and oceans) forming the principal circulation features of the atmosphere, namely, basic circulation cells, atmospheric centers of activity, climatic ridges and troughs, climatic fronts, regions with maximum cyclone and anticyclone recurrence. Here, as a rule, one considers the average fields of the pressure, geopotential, wind, temperature and humidity for months or seasons of the year /4, 14, 15, 27, 34, 53, 56/. The limited number of means for clarifying atmospheric motions is responsible for the fact that the average climatic fields are now regarded as being solely characteristics of the main flow, while the entire variety of motions which are observed are represented as large-scale eddies described by some system of variables: energy of pulsatory motion, transport of pulsatory angular momentum, density of meridional heat flux, etc. /16, 40, 75/. Many studies of the theory of the general circulation use an even simpler approach, regarding the zonal circulation, expressed by the angular velocity of atmosphere, treated as a solid, as the main atmospheric motion (circulation index). The climatic pattern of air currents and of the atmospheric pressure field is obtained by calculating small perturbances of the zonal circulation, produced by nonuniform heat influx from continents and oceans, effect of the topography, etc. /7, 8, 34, 39, 44, 45, 62/.

The effect of the orography on the general circulation is the average statistical result of dynamic and thermal effects of the orography on the air currents, representing the climatic distribution of meteorological fields and synoptic elements. It is here assumed that the nonuniform heat influx from continents and oceans during different seasons produces a certain seasonal air-current pattern, upon which the effect of the orography is imposed.

If one compares the sequence of the daily distribution of air-current systems at the globe with the climatic-average sequence, one will find similarity between most general circulation features: low pressure at the poles, high pressure in subtropical latitudes and predominance of westerly circulation in moderate latitudes. As a whole the air-current pattern at a given instant will differ from the monthly-average and the more so from the seasonal-average. This circumstance prompted the inference /20, 31, 52/ of schemes of the general circulation, regarding the latter as a sequence of changes of a small number of characteristic types of atmospheric circulation.

If the general circulation is interpreted in the above manner, then the orographic effect on it should be regarded as a result of dynamic and thermal effects of the terrain on the macrostructure of air currents and other meteorological fields of each circulation mechanism and on the sequence of transitions from one circulation type to another.

We now describe the effect of orography on the average fields of meteorological elements. Since a large quantity of aerological data on the western hemisphere was not available to this author, the greatest amount of attention will be paid to the Central Asia mountain massif and much less will be said about the Cordilleras and Greenland.

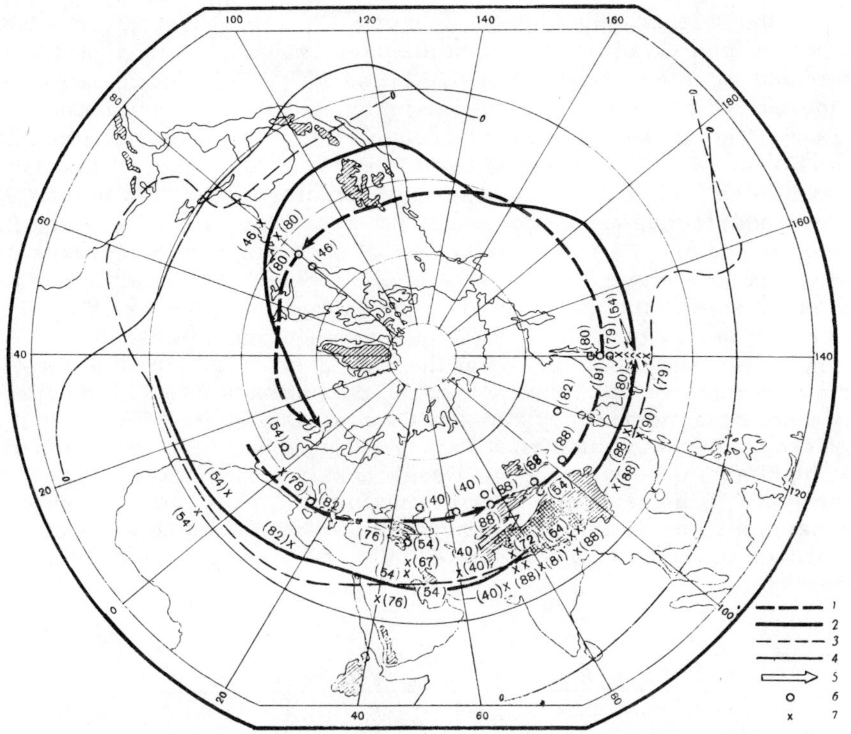

FIGURE 1. Axes of the planetary wind maximum and the location of isotach-maximum centers along some meridional sections.

Axis of westerly-wind maximum: 1 — in July; 2 — in January; zero isotach: 3 — in July; 4 — in January; 5 — axis of orographic wind maximum; isotach maximum centers: 6 — for summer; 7 — for winter.

We shall examine the dynamic effect of the orography, i. e., its effect on the climatic distribution of the wind and geopotential fields. The orographic effect may differ during different seasons, since the air-stream structure changes over the year with changes in the solar radiation flux and changes in the heat sources and sinks on the terrestrial surface. It also depends on the geometry of the mountain system. Thus, it is natural that the mountain ranges of the North American Cordillera and South American Andes will have a different dynamic effect on the seasonal air-current pattern than the mountain ranges of Eurasia and mountains of East Siberia and Indochina which are aligned latitudinally and develop into a gigantic triangle.

It is known that the most characteristic feature of the average circulation of the atmosphere is the existence of a general planetary easterly-wind maximum in moderate latitudes. In winter it is closer to the equator and in summer it is closer to the poles. This seasonal shift is due to causes common to the entire globe /36/. The effect of continents and oceans reduces to the fact that the seasonal shift of maximum zonal geostrophic wind from winter to summer in the northern hemisphere is 15° latitude, while in the southern hemisphere it is only 3° /83/. Figure 1 depicts the location of the axis of the principal planetary zonal-wind maximum for winter and summer from data of NIIAK and /74/. This figure also presents the location of centers of isotach maximum, taken from certain meridional sections. The greatest shift in the general planetary wind maximum in the eastern hemisphere is observed over Eurasia in the region of the Caucasus, the Armenian Highland and the Zagros mountains, as well as in the Central Asian mountain massif region, and in the western hemisphere in the Rocky Mountains region. To clarify the effect of orography on the average wind field, we compare two meridional sections located sufficiently close to each other so that the radiation and circulation conditions are the same, but one chosen in a region with and the other without mountains. It is seen by comparing the wind distribution at the 60 and 75° meridians that the effect of the orography on the average wind field expresses itself in the appearance, above the mountain systems of Central Asia at the 13–14 km level, of an orographic wind maximum during all seasons of the year /42, 45, 49/. This maximum does not undergo a seasonal shift. In summer it is located to the south of the general planetary maximum; in autumn and spring the two maxima coincide when the general planetary maximum is located at the latitude of the mountain systems; here the orographic maximum is the strongest.

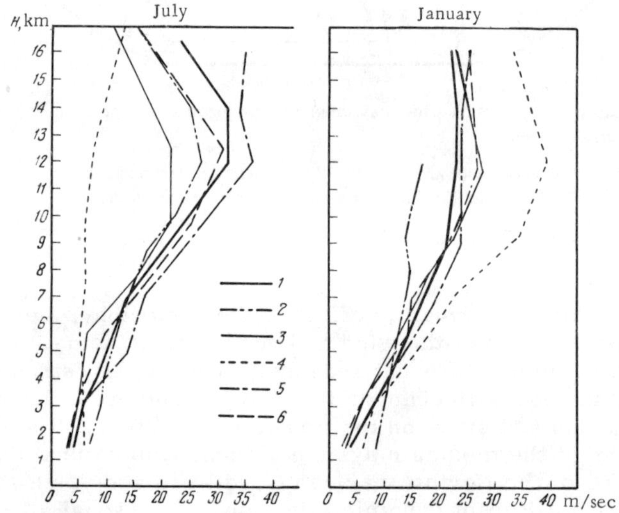

FIGURE 2. Average wind-speed profiles for plain (Tashauz, Aral Sea), submontane (Tashkent, Delhi) and intermontane (Dzhalal-Abad, Khotan (Hotien) [China], Kashgar (Shufu) [China]) stations:

1 — Tashkent; 2 — Aral Sea; 3 — Khotan; 4 — Delhi; 5 — Dzhalal-Abad; 6 — Kashgar.

In winter the orographic wind maximum lies to the north of the general planetary maximum. The mean velocity at the maximum wind axis increases from plains to mountains during all seasons; at the upper boundary of mountain system (mountain friction layer /51/) the wind speed decreases from plains to mountain systems. Figure 2 shows average long-term wind profiles for January and July at several stations located to the south, north and inside the Central Asian mountain massif. The orographic wind maximum is quite pronounced also here.*

We now examine the horizontal circulation pattern. In winter the main mountain systems of the northern hemisphere lie in a region with westerly transport. The air currents in the east flow below the mountain tops, and roll over the mountains when encountering them. Here the lower the isobaric surface in the middle troposphere, the more the effect of rolling over mountain tops is expressed, i.e., the higher the horizontal velocity components directed northward and southward determining the flow over the tops.

If the effect of this flow is neglected and it is assumed that an air particle must rise from the foothills to the middle altitude of the mountain ranges, then for Central Asia, for example, at a 10 m/sec transport and characteristic slopes of 0.03 — 0.1°, the boundary condition for the vanishing of the component normal to the mountain surface yields a vertical velocity of the order of 30 — 100 cm/sec, i.e., values which are quite unreal for the orderly rise of air over the entire mountain country. The rolling of air over mountaintops is most pronounced on maps of the prevailing and resultant wind at the 700 and 500 mb levels. However, it is also well expressed at the 850 mb surface, outside the limits of local circulation. For illustration we present maps of prevailing winds over Central and Hither Asia,** obtained at SANIGMI (Figures 3 through 6). The northern branch of the flow past the Central Asian mountain massif forms an anti-cyclonic flow system at the northern boundary of the range, the southern part of the flow forms a cyclonic circulation system along the southern mountain boundary (Figure 3b, c). The role of rolling over the mountain tops decreases above the mountain surface and the overrunning effect increases. This is shown by the prevalence of westerly winds over mountain country, starting from the 300 mb surface (Figure 3d) and above, as well as by the increase in wind speed from plains to mountains at the wind-maximum level.

In the western hemisphere the mountain ranges of the North American Cordillera extend more than 5,000 km along the meridian. In the bottom layer they interfere with westerly winds of moderate latitudes over large distances. Therefore, predominance should be held here by flow from lower levels, expressed by the appearance of a ridge flow system over a large part of the mountain range.

The mountain ranges of Alaska are aligned along the latitude, in Canada and Mexico they are oriented from northwest to southeast and are almost meridional only in the USA; hence the northern part of the westerly flow encounters the mountain systems earlier than its central part, and the southern part of the flow reaches the mountain system when the air currents located more to the north already start flowing over the ranges. A peculiar

* Unfortunately, these wind-field features were not reflected in maps and latitude sections of the NIIAK Atlas, since it did not use data from temperature and wind sounding stations located inside the Central Asian mountain massif.

** [Hither Asia refers to the southwestern part of Asia, including the Bosporus, Mediterranean and Red Sea to the Plateau of Iran.]

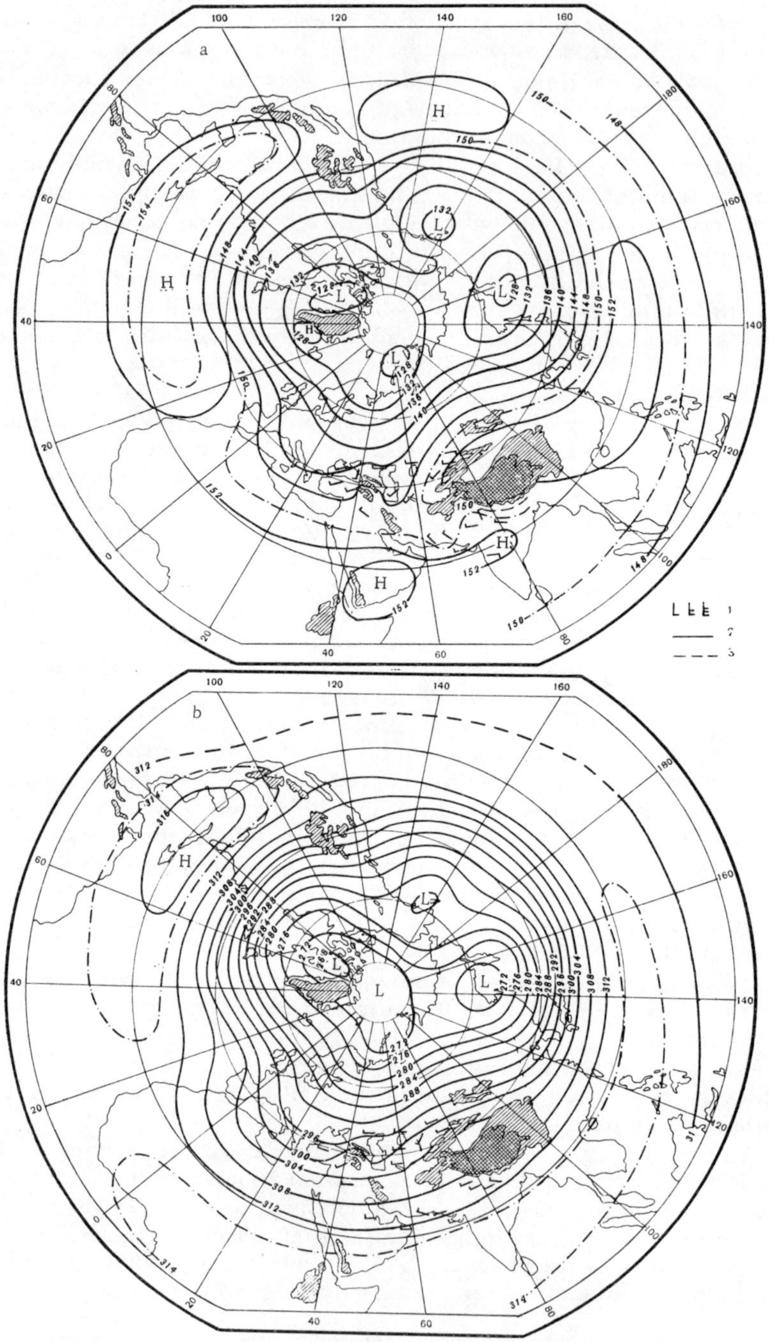

FIGURE 3. Prevailing wind over Central and Hither Asia and mean isohypse maps (January):

a — 850 mb; b — 700 mb; c — 500 mb; d — 300 mb; L and H denote regions of low and high pressures; 1 — prevailing wind; 2 — isohypses in decameters; 3 — intermediate isohypses in decameters. The legend given here is also applicable to Figures 4 through 6.

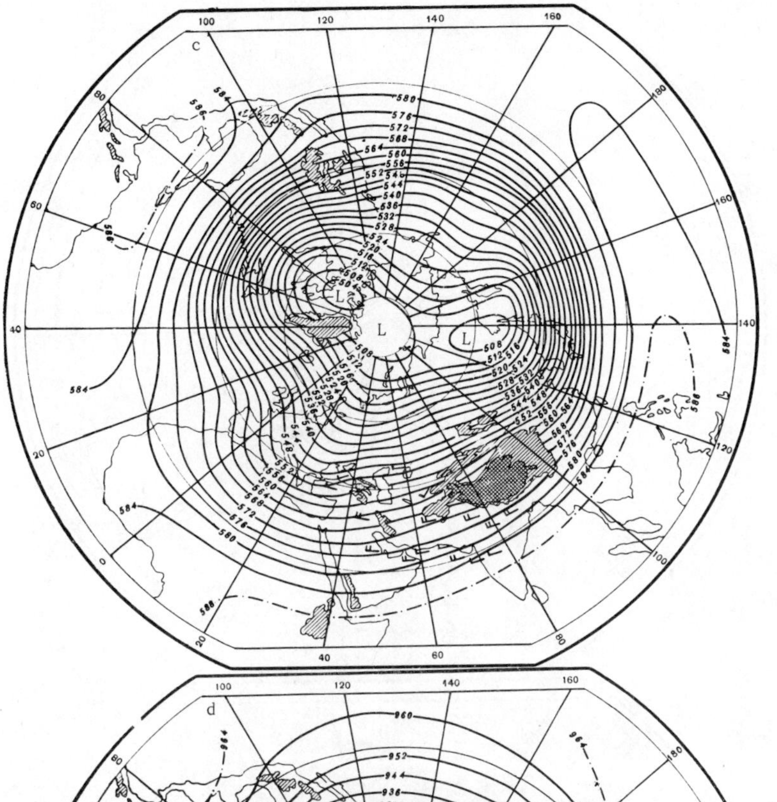
c
100
120
140
160
180
160
140
120
100
80
60
40
20
0
20
40
60
80
L
L
L
580
576
572
568
564
560
556
552
548
544
540
536
532
528
524
520
516
512
508
504
584
588
d
960
952
944
936
928
920
912
904
896
888
880
872
864
856
848
840
844
964
968
L
L

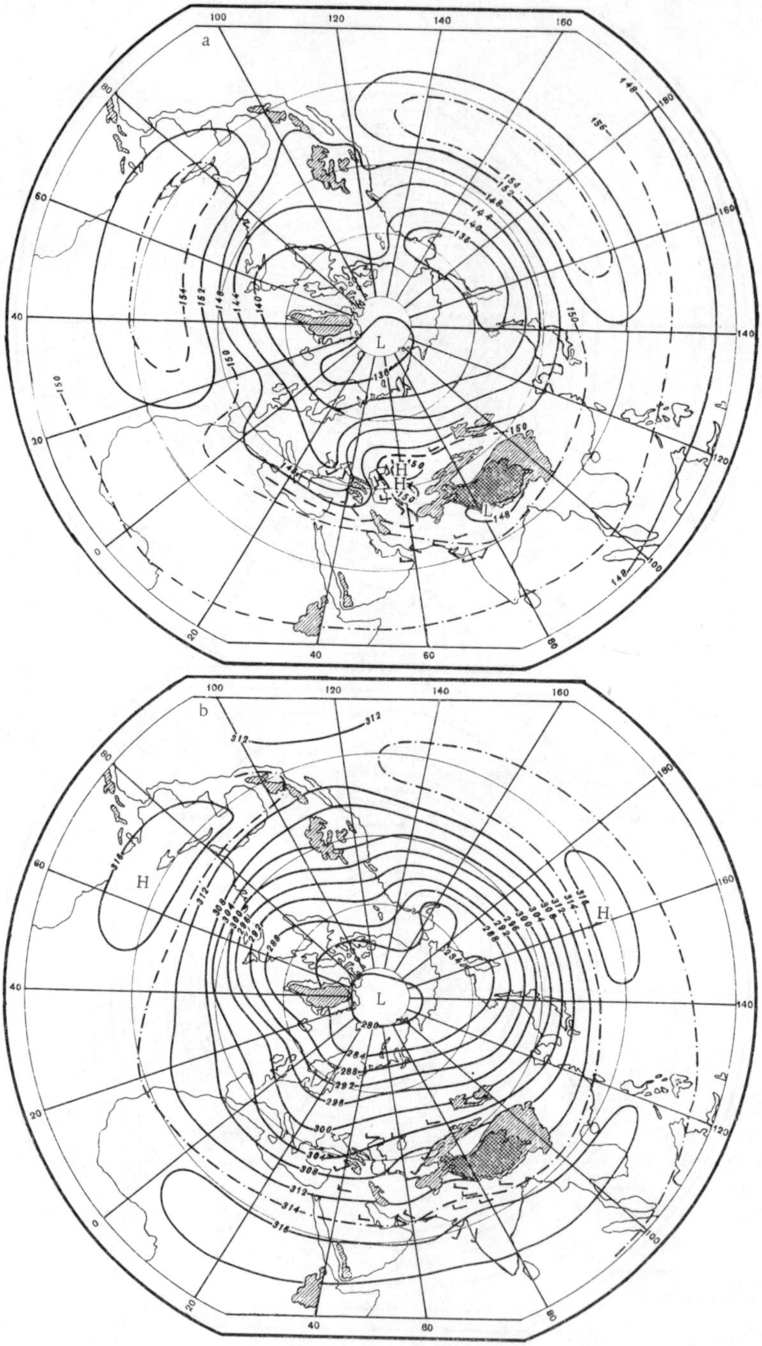

FIGURE 4. Prevailing wind over Central and Hither Asia and mean isohypse maps (April): a — 850 mb; b — 700 mb; c — 500 mb; d — 300 mb.

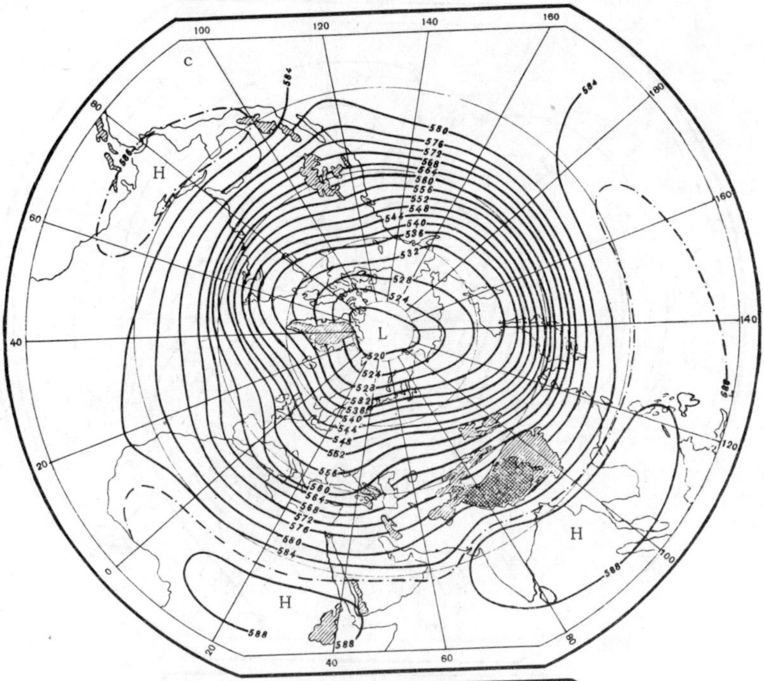
c
H
L
H
H
100
120
140
160
180
160
140
120
100
80
60
40
20
0
20
40
60
80
584
588
580
576
572
568
564
560
556
552
548
544
540
536
532
528
524
520

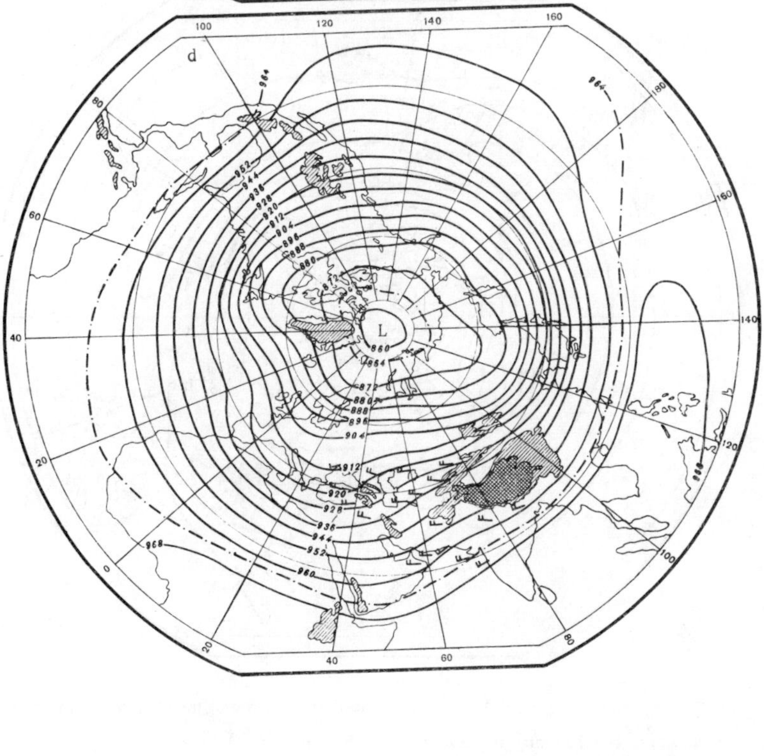
d
L
964
952
944
936
928
920
912
904
896
888
880
872
864
860
968
960

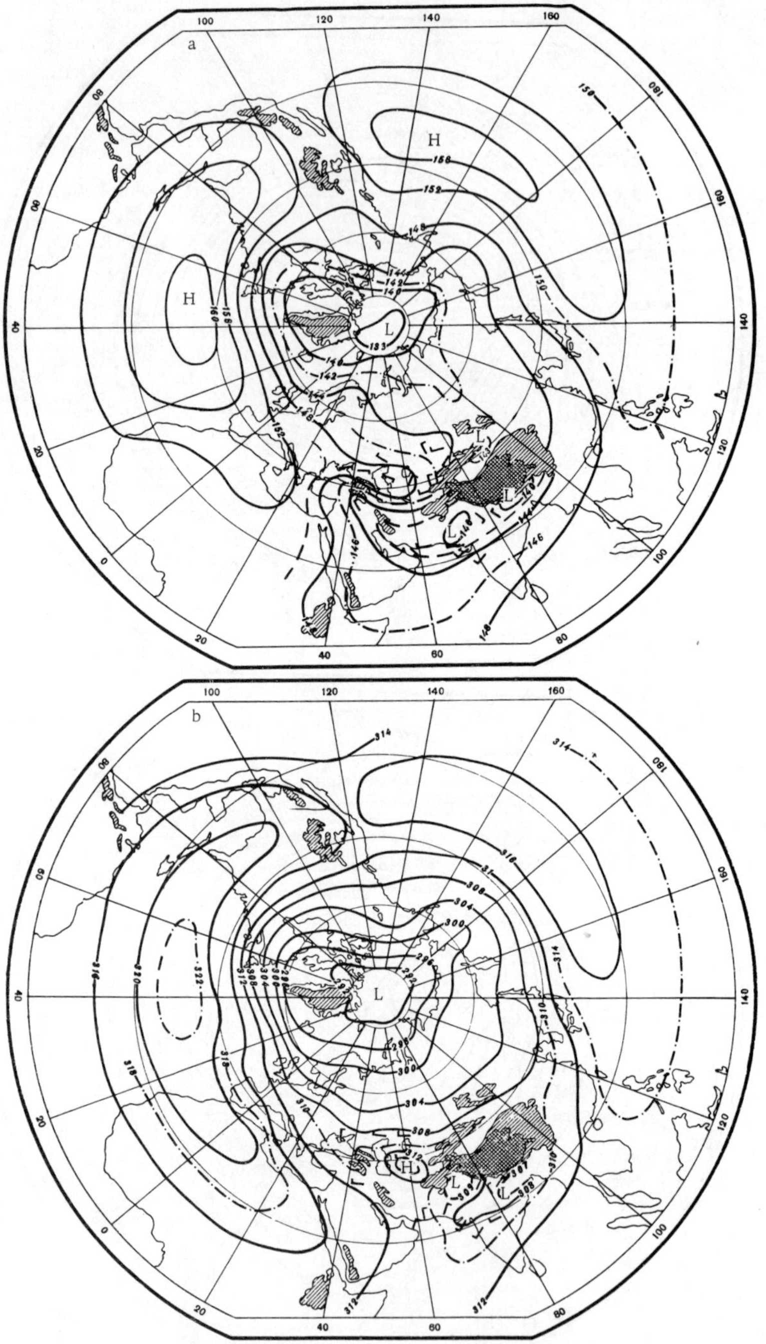

FIGURE 5. Prevailing wind over Central and Hither Asia and mean isohypse maps (July): a — 850 mb; b — 700 mb; c — 500 mb; d — 300 mb.

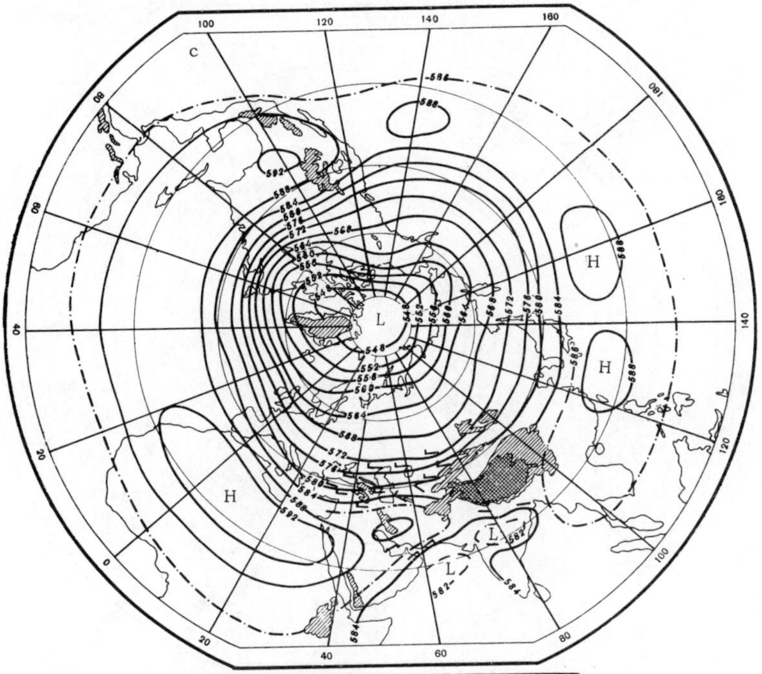

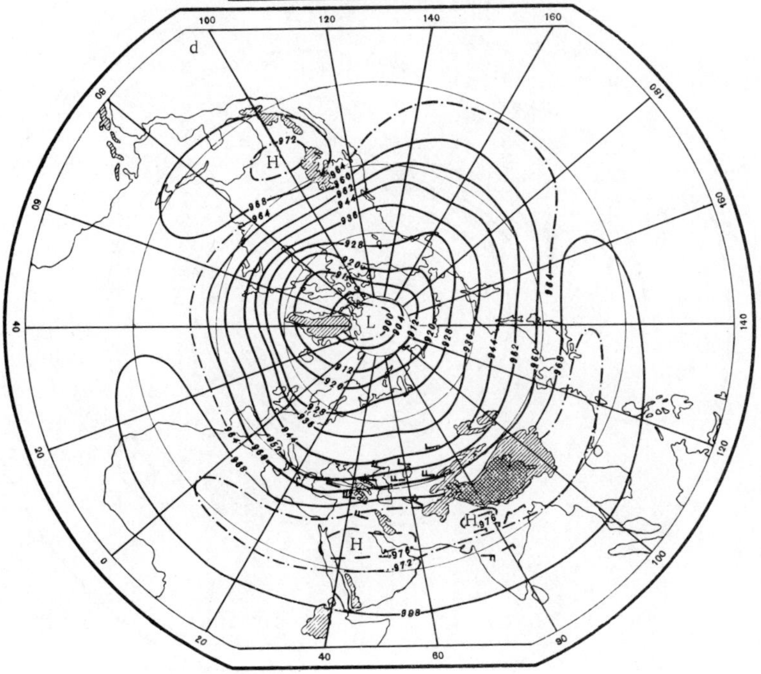

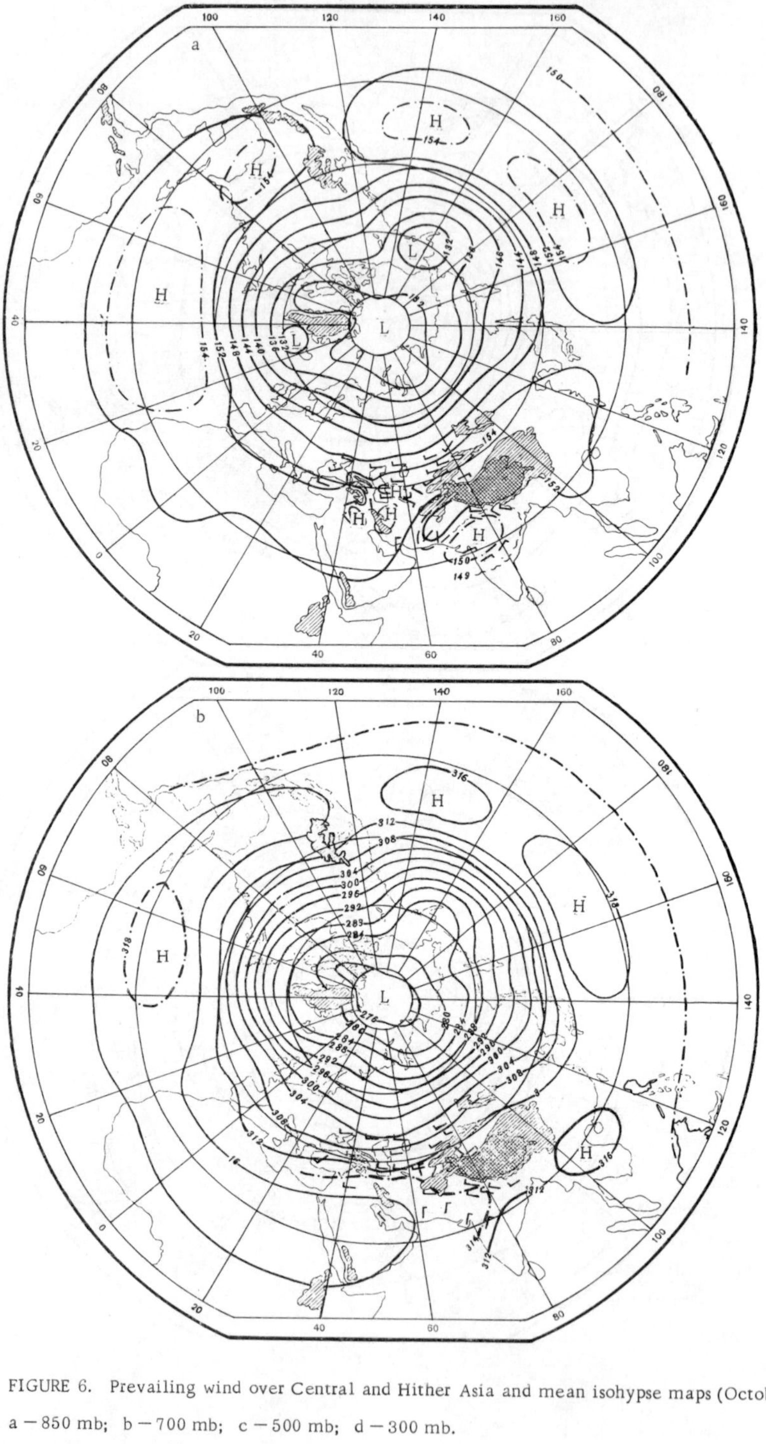

FIGURE 6. Prevailing wind over Central and Hither Asia and mean isohypse maps (October): a — 850 mb; b — 700 mb; c — 500 mb; d — 300 mb.

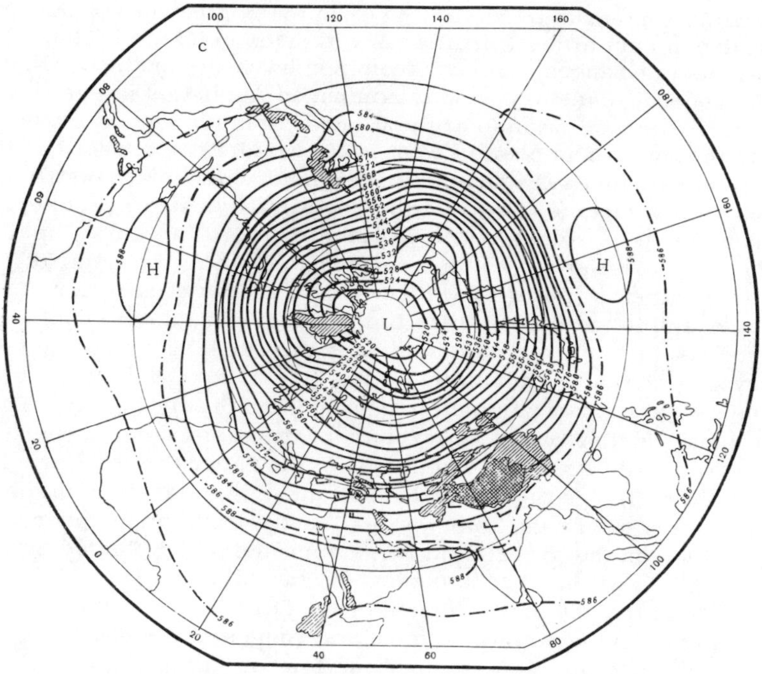
c
100
120
140
160
180
160
140
120
100
80
60
40
20
0
20
40
60
80
H
H
L
584
580
576
572
568
564
560
556
552
548
544
540
536
532
528
524
520
586
588

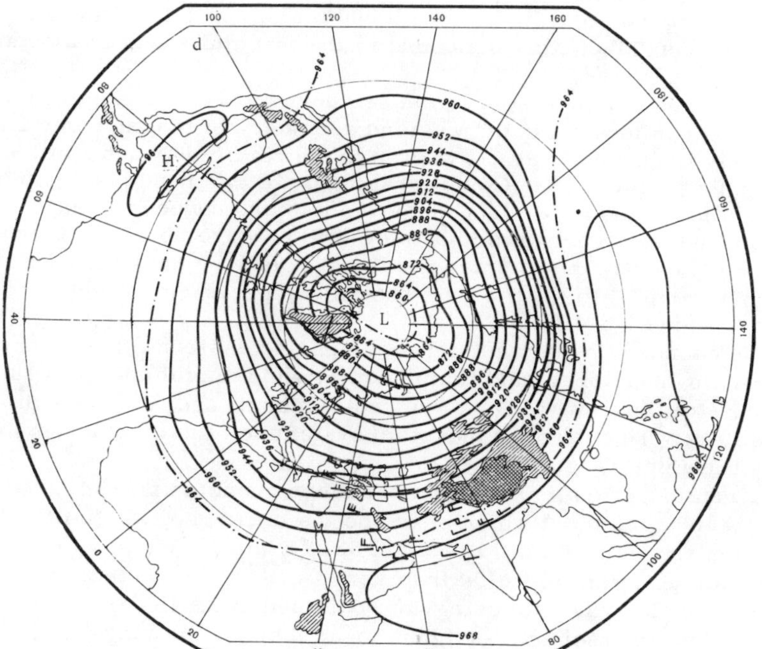
d
100
120
140
160
180
160
140
120
100
80
60
40
20
0
20
40
60
80
H
L
964
960
952
944
936
928
920
912
904
896
888
880
872
864
860
968

streamline system is formed: a ridge in the northern part and a trough in the southern part of the Cordilleras with axes oriented in the same manner as the mountain ranges, namely, from northwest to southeast (Figure 3b, c).

The general planetary wind maximum in April lies not only at the latitude of the Cordilleras, but also at the Central Asian mountain massif. Flow around this massif is clearly seen in the middle troposphere at the 700 and 500 mb surfaces (Figure 4b, c); signs of this flow are still apparent at the 300 mb surface (Figure 4d). The air flow pattern in the planetary boundary layer is somewhat more complicated. Two anticyclonic eddies are found to exist at the 850 mb surface in front of the Central Asian mountain systems (Figure 4a). These are eddies of orographic origin. It is impossible to ascribe them to the effect of radiative cooling, since this cooling in April is far less intense than in January and no eddies are observed in the latter month.

The westerly transport pattern changed markedly in July. The high subtropical North African anticyclone of the eastern hemisphere shifts northward, while its ridge extends from the Sahara to the Kara-Kum Desert /55/. Hence the Central Asian mountain massif is in the way of northwesterly air flows, which prevail at the eastern periphery of the anticyclone. In the lower 5 km these currents are deflected by the mountain ranges to the east and become zonal (Figure 5b, c). This effect is also noticeable at the 300 mb surface (Figure 5d). Some lows are formed at the leeward (southern) side of mountain ranges at the 850 mb surface (Figure 5a); one (the largest) of these lows is also retained at the 700 mb surface (Figure 5b). Starting with 500 mb and above anticyclonic circulation appears over Tibet, this being due to the thermal effect of mountains /21, 66/.

In the western hemisphere the southern part of the North American Cordillera is located at the western periphery of the subtropical anticyclone. Westerly transport prevails also to the north of 40°N and flows only onto the northern part of the Cordilleras. This current intersects mountains in the middle and upper troposphere and thus produces a ridge in the streamline field only to the north of 40°N (Figure 5b,c). In the first kilometer apparently the air masses do not roll over the Rocky Mountains, but flow parallel to them along the eastern periphery of the Pacific anticyclone (Figure 5a).

The boundary separating the western and eastern circulations shifts southward in October, and the planetary wind maximum appears at the latitudes of the Central Asian mountain massif and of the North American Cordillera. The effect of mountain systems on the air currents is similar to that in April (Figure 6a—d).

Mountain systems also affect the mean geopotential field. Many studies have been carried out which describe the mean geopotential fields for different time segments /1, 29, 52, 53, 74, etc./. We shall consider here the mean geopotential fields from maps in /1/, corrected over India, Hither and Central Asia using data obtained at SANIGMI. It is quite evident that the isobaric surfaces located below the ridge level of mountain ranges break off on the mountain slopes. Hence the isohypses of these surfaces envelope large mountain massifs. This pattern is observed at the 700 mb surface (Figures 3b, 4b, 5b, 6b), as well as over a large part of the Central Asian mountain massif and at the 500 mb surface (Figures 3c, 4c, 5c, 6c).

The effect of a mountain massif above mountain ranges is expressed by the appearance of a zone of large isohypse gradients along the northern and southern boundaries of the massif (winter, Figure 3d) or along the northern boundary (spring, summer and autumn, Figures 4d, 5d, 6d). The seasonal reconstruction of the geopotential field changes the isohypse deformation pattern. Thus, the Central Asian mountain massif in winter is located in the forward part of a large-scale high-altitude climatic trough (Figure 3c, d), while in summer it lies in the forward part of a large-scale high-altitude ridge (Figure 5c, d). Here an orographic ridge is formed to the north of the mountain massif and over its northern part, with an orographic trough forming to the south. These two orography-induced formations have their highest amplitude over the North and South American Cordilleras (Figure 4b); however, this pattern is quite pronounced also over the Central Asian mountain massif. The geopotential field up to the 2 km level is very complicated. The isohypses still bend around the mountains, but, in addition, there appear a number of anticyclonic (Figures 4a and 6a) and cyclonic (Figure 5a) centers, which are the result of the dynamic effect of the mountain massif and of the thermal effect of the surface on the tropospheric air flow.

The effect of mountain systems on the geopotential field described above is seen not only on average climatic maps, but also on maps of individual synoptic processes. All maps of individual synoptic processes show the bending of isohypses of lower isobaric surfaces around mountains and of the zone of large isohypse gradients along northern and southern boundaries of the range on high isobaric surfaces.

The predominance of certain forms of the baric field in the middle troposphere above mountain massifs also affects the daily departure of the geopotential from monthly-average values. Table 1 lists values of $\bar{\sigma}_n$, the standard deviation of the geopotential over the mountain systems of Central Asia and North America, in front and past these systems. The value of $\bar{\sigma}_n$ was obtained by averaging σ_n at the node points of the 10—60°W, 60—110°E, 110—160°E by 25—45°N grid for the Central Asian mountain massif and of the 150—125°W, 125—100°W, 100—75°W by 20—45°N grid for the North American Cordillera from data of /1/ for the eastern hemisphere and of /74/ for the western hemisphere. The standard deviations of the scalar and vectorial wind velocities were calculated in the same manner.

It is seen from Table 1 that the variability of the 700 mb geopotential over the North American Cordillera and of the 500 mb over the Central Asian mountain massif is smaller than in front of and behind the mountains. This type of variability is due to the fact that all kinds of baric formations may exist in front of and behind mountain systems in the middle troposphere, while forms of the baric field, related to the orographic deformation of the isohypse field (an orographic ridge in the northern part and a trough in the southern part), predominate over mountain country at isobaric surfaces located near mountain surfaces. Hence the geopotential field in the middle troposphere over mountain systems is less variable. In the upper troposphere (200 mb), where there is no flow around the mountains and orographic deformation of isohypses is not as definite, the variability of the geopotential field over mountain country is no longer the smallest, while over the Central Asian mountain massif in summer it is even the greatest.

It is interesting to note that the type of 500 mb variability over the Cordilleras in winter is the same as that of isobaric surfaces located close to mountains (700 mb), while in winter, on the other hand, it is the same as that of isobaric surfaces located high above the mountains (200 mb). It can be concluded from here that the effect of mountains on the deformation of the geopotential field is not identical in winter and summer.

TABLE 1. Standard deviation of the geopotential $\bar{\sigma}_n$ (decameters), scalar $\bar{\sigma}_v$ and vectorial $\bar{\sigma}_{\vec{v}}$ wind velocity (m/sec) in front of (A), above (B) and behind (C) the mountains

Isobaric surface, mb	Meteorological element	Month	Central Asian mountain massif, 25–45°N			North American Cordillera, 20–45°N		
			A	B	C	A	B	C
			10–60°E	60–110°E	110–160°E	125–150°W	100–125°W	75–100°W
700	Geopotential	January	—	—	—	8	6	7
		July	—	—	—	4	3	4
	Scalar velocity	January	—	—	—	7	5	6
		July	—	—	—	4	3	4
500	Geopotential	January	8	6	8	11	10	11
		July	5	4	5	5	6	6
	Scalar velocity	January	9	7	10	9	9	9
		July	6	5	6	5	5	5
200	Geopotential	January	12	11	10	14	12	12
		July	10	11	10	10	10	8
	Scalar velocity	January	14	12	14	12	12	12
		July	10	9	11	9	8	9
	Vectorial velocity	January	20	19	20	21	20	19
		July	18	16	20	16	15	15

The type of wind-field variability over the Central Asian and North American mountains is different. The smallest wind variability in winter and summer in Central Asia is observed over the mountain massif proper as well as over isobaric surfaces lying close to the mountains (500 mb) and those located high above them (200 mb). In North America this kind of variability is noted only at the 700 mb surface, i.e., the surface located in the mountain-friction layer. Starting with the 500 mb surface and above, the variability of the wind over mountain country is no less than behind it.

The dynamic effect of mountain systems on the wind and geopotential fields extends over large distances along the flow. This question was considered in a large number of theoretical studies. It was shown in /46/ that the pressure disturbance in the atmosphere, produced by large-scale orography (the dimensions of which require making allowance for the Coriolis force), decreases exponentially with altitude. According to Musaelyan /44/ a system of planetary waves, whose intensity and spatial location depends on the magnitude of the circulation index, arises behind large mountain massifs due to westerly transport. Reference /37/ considers the effect of the Himalayas and Tibet on the general circulation with the aid of a more exact statement of the boundary condition for the vertical velocity at the mountain surface (flow around and over mountains). The calculated streamline pattern for a high value of the circulation index reflects to some extent the actual streamline pattern at the 700 mb surface in January only along the southern boundary of the mountain massif. According to calculations a trough should prevail over the northern part of the massif and to the north of it, while actually a climatic orographic ridge prevails there. The calculated summer streamline pattern, on the other hand, is in better agreement with the real situation over the northern part of the massif and is incorrect over the southern part. Allowance for nongeostrophicity of the wind /38/ improves the agreement between the calculated and real streamline patterns at a level equal to the mountain altitude. However, the streamline pattern calculated for levels above the mountain top is different from that actually observed. Apparently, better agreement with fact is obtained from the streamline pattern obtained in /65/.* It makes allowance for geopotential perturbances due to the presence of heat sources and sinks, as well as due to orography. However, it is not clear from /65/ how far the dynamic effect of mountains extends along the flow. The contribution of the terrestrial orography to the formation of the average climatic pressure field at sea level was calculated in /45/. Here corrections for terrain effect were introduced into the climatic pressure field, calculated on the assumption that the temperature distribution in the atmosphere is known /8/; this also made allowance for climatic heat sources and sinks. Corrections for terrain improved the calculated surface pressure field /62/ in the Icelandic and Aleutian lows and in the Rocky Mountains region. However, the pressure field over Eurasia was less like that actually observed. As is seen, the question of theoretical study of the orographic effect on the pressure and wind fields has not yet been resolved.

The distribution of differential characteristics of the real wind in the region of mountain systems is of great interest. We present here only the distribution of the mean divergence of the real wind (in 10^{-7} sec^{-1}) in winter and summer in the Central Asian mountain massif (Figures 7a—c and 8a—c).

It should be immediately noted that the average divergence in mountain regions may be much higher than over level surfaces to the north and south of them. The effect of mountain systems is different at isobaric surfaces pertaining to atmospheric layers below the mountain tops (700 mb, Figures 7a and 8a), at mountaintop levels (500 mb, Figures 7b and 8b) and above the mountains (300 mb, Figures 7c and 8c). Air mass accumulation occurs in winter in front of the mountains of the Central Asian mountain massif. The convergence region extends over the mountain massif from northwest and southeast. Divergence extends to internal mountain regions, as well as to the Ganges Plain in India. Divergence past the mountain massif is observed to the northeast and convergence to the east and southeast of it (Figure 7a). This distribution of the divergence and convergence agrees with the pattern of air flows past the Central Asian mountain massif in the front part of the large-scale trough. Accumulation of air masses at mountaintop level (500 mb) occurs over the Caucasus, Kopet Dagh, Zagros, Plateau of Iran and Hindu Kush. Divergence is

* Cited from /24/.

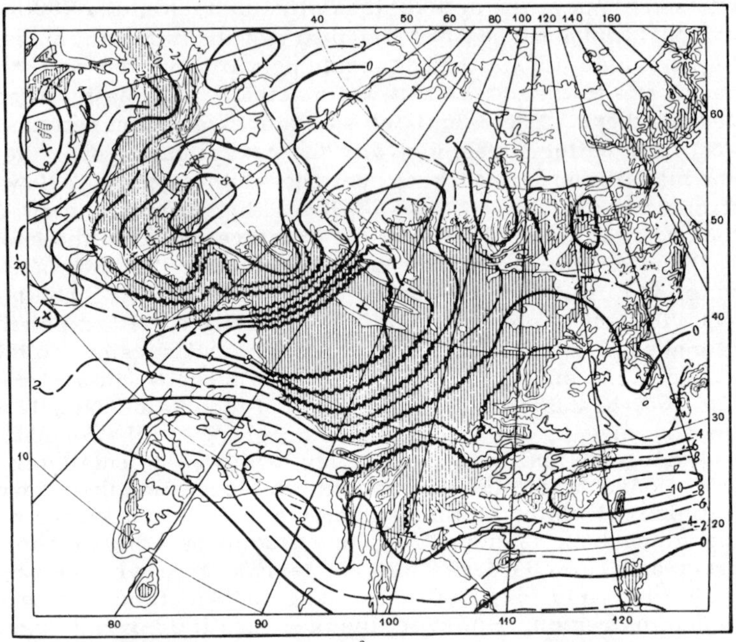
40
50
60
80
100
120
140
160
60
50
40
30
20
10
80
90
100
110
120

a

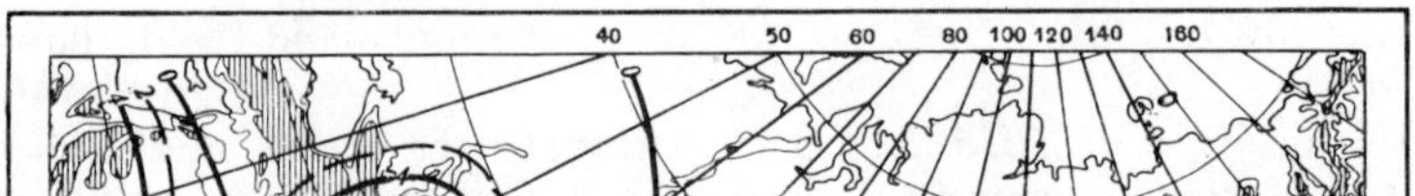
40
50
60
80
100
120
140
160
60
50
40
30
20
10
80
90
100
110
120

b

observed in air flows around the Central Asian mountain massif from the north and south, while past the mountain massif, where the northern and southern branches of the flow converge, a convergence region is located (Figure 7b). In the upper troposphere (300 mb) the convergence regions embrace from the north and south the entire belt of mountain systems of Central Asia, from the Caucasus and the Armenian Highland in the west to the East Siberian and Indochina mountains in the east. Divergence regions are located on the leeward side of high mountain chains — to the east of the Armenian Highland, Elburz and Zagros, to the east of Hindu Kush and Solomon mountains, as well as over the Tarim Basin and Northern Tibet, i. e., in the internal parts of the mountains, to where air influx is limited due to its accumulation at the outer periphery of the mountains (Figure 7c).

As was pointed out above, the seasonal air flow pattern in summer differs from that in winter and northern and northwestern slopes are those oriented windward. Hence the convergence region below mountain systems (700 mb, Figure 8a) and at mountaintop level (500 mb, Figure 8b) is located to the north of the entire belt of mountain ranges of Central Asia, while the divergence region is observed on the leeward side, i. e., to the south of the mountain system.

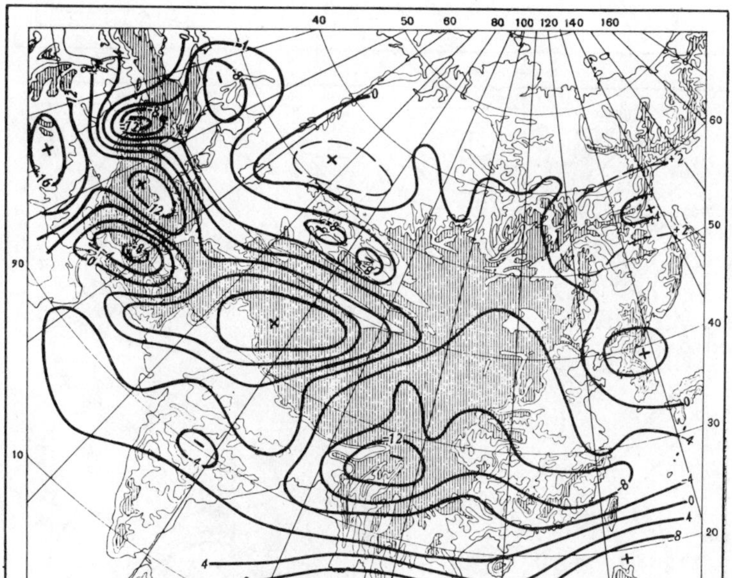

c

FIGURE 7. Map of the average divergence over the Central Asian mountain massif in winter:

a — 700 mb; b — 500 mb; c — 300 mb. + denotes positive divergence and – denotes negative divergence, in 10^{-8} sec^{-1}.

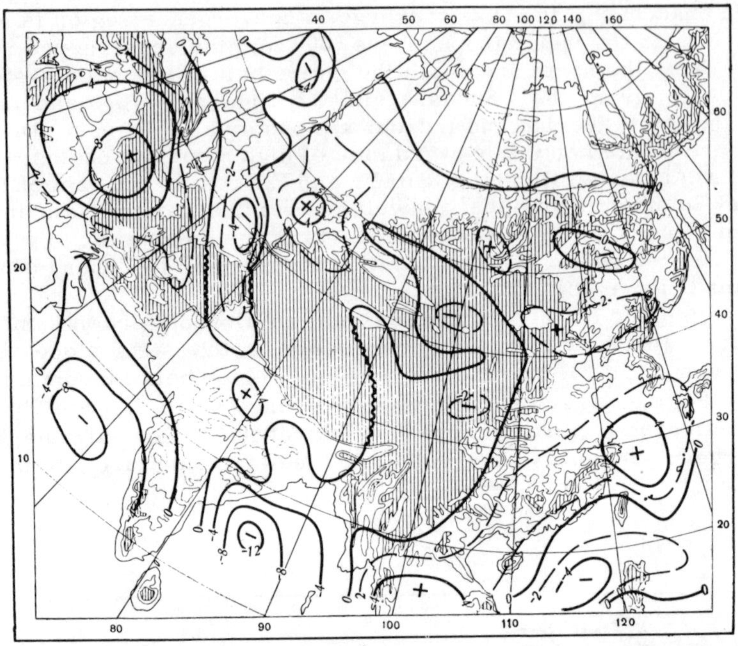
40
50
60
80
100
120
140
160
60
50
40
30
20
20
10
80
90
100
110
120

a

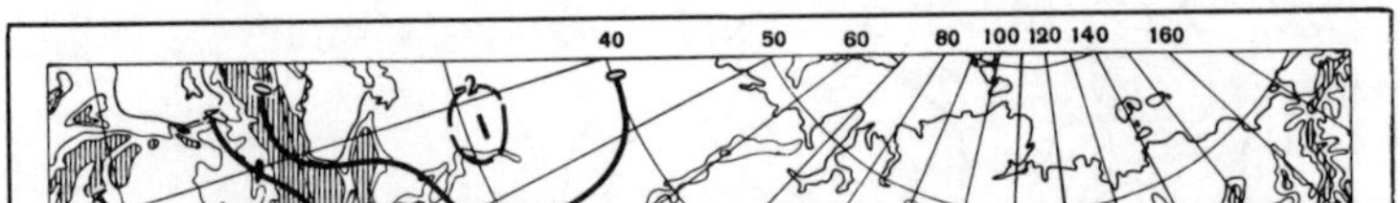
40
50
60
80
100
120
140
160
60
50
40
30
20
20
10
80
90
100
110
120

b

The convergence region in the upper troposphere (300 mb, Figure 8c) is also located to the north of the mountain ranges of Central Asia, with the divergence region lying more to the south. However, the air currents in the upper troposphere are more zonal; flow onto the Central Asian mountain massif results in the fact that the entire western part of the massif is occupied by a region of air-current convergence, while the leeward side serves as the divergence region.

Our understanding of the mechanical effect of the orography will be incomplete if we do not mention its contribution to the energetic state of the atmosphere. Orography affects the balance of atmospheric kinetic energy. First, the kinetic energy of atmospheric motions is expended as work done against the gravitational force attendant to the ordered rise of air masses onto mountain elevations. Second, the dissipation of kinetic energy by friction in air flows above mountainous countries is appreciably greater than over other surfaces, because mountains, from the viewpoint of planetary-scale motions, represent a macrorough surface. The existence of a mountain friction layer thicker than the boundary layer over the plains is clearly shown on wind profiles over plains and mountains (see Figure 2). Finally, mountains may affect changes in the kinetic energy of the atmosphere, creating a predominance of positive or negative pressure force, as a result of correlation between the values of the pressure and air-current divergence on the windward and leeward sides of mountain systems.

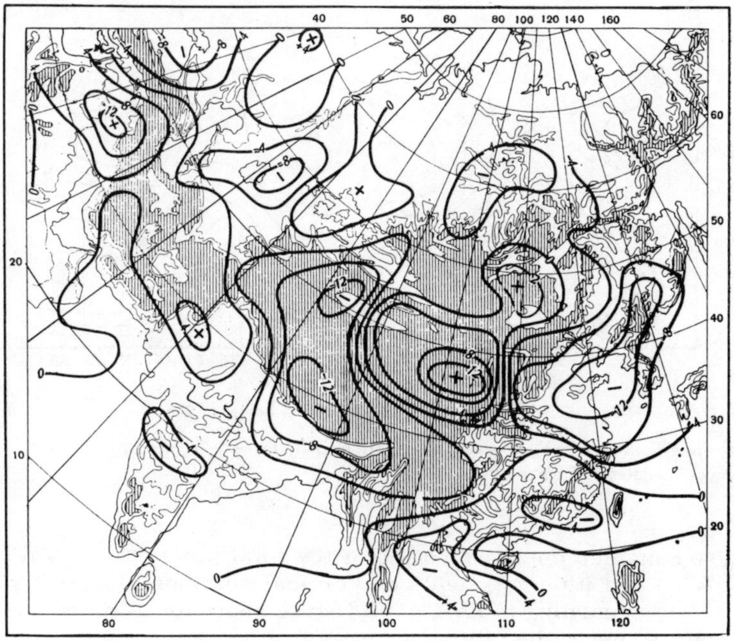

c

FIGURE 8. Map of average divergence over the Central Asian mountain massif in summer:

a — 800 mb; b — 700 mb; c — 300 mb.

The energy state of the atmosphere can be described by the kinetic energy of the average motion E and the kinetic energy of pulsations or of large-scale eddies b, total kinetic energy E, or kinetic energy of the zonal E_λ and meridional E_v motions. Clearly

$$E = \mathrm{E} + b = E_\lambda + E_v.$$

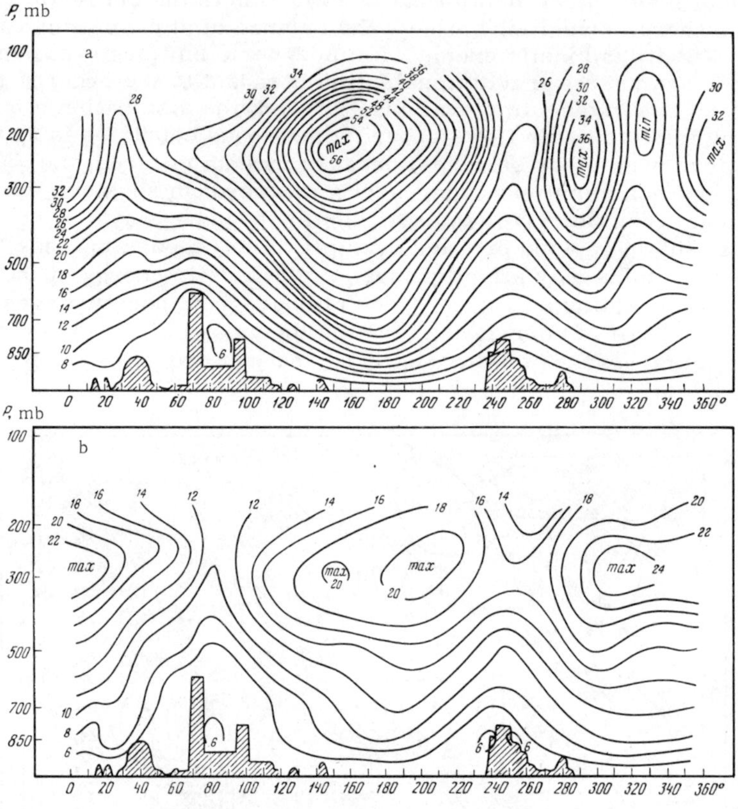

FIGURE 9. Kinetic energy distribution along 40°N in January 1958:

a — total kinetic energy of geostrophic motion ($\sqrt{E}$, m/sec); b — energy of geostrophic eddies ($\sqrt{b}$, m/sec).

We now consider the distribution of the total kinetic energy of the geostrophic wind and its components on vertical sections along 40°N in January and July 1958 according to data of /18/ (the sections do not show values of kinetic energy, but rather the square root of twice the corresponding kinetic energy). Figure 9a shows the distribution of total kinetic energy in the vertical direction along 40°N in January 1958. Three kinetic energy maxima can be noted in the upper troposphere: over the Pacific, and over the eastern and western Atlantic seaboards. There are three kinetic energy minima: in Hither Asia, the Rocky Mountains and the Atlantic Ocean. These maxima and minima reflect the location of climatic troughs and ridges and of the planetary wind maximum (Figure 1) in a section along 40°N. However, in the middle troposphere the lowest kinetic energy levels are observed over the Central Asian mountain massif and the Rocky Mountains. Thus the total kinetic energy at the 500 mb level over the Central Asian mountain massif is the same as over the Pacific at 900 mb. The reduction in kinetic energy starts already over the Alps and the Caucasus. The distribution of the kinetic energy of the mean geostrophic wind (E) and of the kinetic energy of the zonal component of the geostrophic wind (E_λ) is not presented due to space limitations; it is similar to the distribution of the total kinetic energy, the only difference being that the components are smaller. The latter distributions also show a reduction in the energy over mountain systems in the middle troposphere.

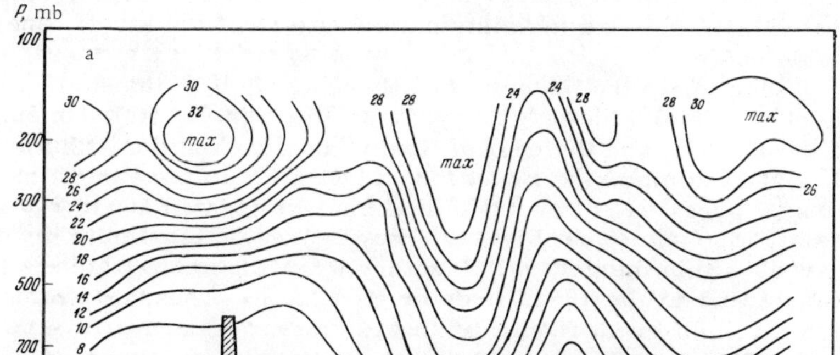

FIGURE 10. Kinetic energy distribution along 40°N in July 1958:

a — total kinetic energy of geostrophic motion ($\sqrt{E}$, m/sec); b — energy of large-scale geostrophic eddies ($\sqrt{b}$, m/sec).

The orographic effect is clarified most distinctly by examining the distributions of large-scale eddy energy (Figure 9b). The energy of large-scale eddies represents the kinetic energy of geostrophic motions associated with the departure of daily flow systems from their monthly average pattern, i.e., associated with cyclonic activity /59, 60/. As can be seen from Figure 9b, maximum energy of large-scale eddies is observed in January over the Pacific and Atlantic at the 300 mb level, while deep minima of large-scale eddy energy, extending over the entire depth of the troposphere, exist over the mountain systems of Asia and North America. The causes of the above were discussed previously. In fact, no matter what the baric formation which develops in front of mountain systems, it is affected by the orography once it reaches the mountains and acquires an orography-induced shape of isohypses — a ridge to the north and a trough to the south of the mountains. In conjunction with this the entire variety of forms and intensities of baric formations is not encountered over the mountains. Hence the kinetic energy of large-scale eddies here is at a minimum. The distribution of the kinetic energy of meridional motions (E_v), although more complicated, still repeats the distribution pattern of the kinetic energy of large-scale eddies.

During summer in the upper troposphere along 40°N there also exist three maxima of the total kinetic energy: over the Central Asian mountain massif, the Pacific, and the east of North America and the Atlantic Ocean, where a partial minimum is observed in the latter to the east of the Rocky Mountains (Figure 10a). Now these maxima are located at the 200 mb level and possibly also somewhat higher. They represent maximum kinetic energy areas within the general planetary wind maximum which is located in summer either at the 40° latitude or close to it. Thus, in summer the picture is much different from that in winter over Asia, North America and the Atlantic. Conversely, the kinetic energy minima in the middle troposphere are located in both summer and winter over the mountain systems of Asia and North America. The distribution of E and E_λ in general is similar to that of E. Hence mountain systems in winter and summer in the middle troposphere act as kinetic energy sinks. However, as will be shown below, in summer the mountain massifs serve as a heat source for the atmosphere. As a result a baroclinical region is produced, for example, at the northern boundaries of the Central Asian mountain massif /10, 49/, and this is one of the reasons for the appearance of a wind maximum over mountain country and consequently for the kinetic energy maximum (Figure 10a). Hence in summer the mountain systems are on the one hand obstacles with a macrorough surface projecting high into the atmosphere and thus absorbing kinetic energy, and on the other hand they serve as heat suppliers to the atmosphere, thus acting as regions of conversion of potential into kinetic energy. Here the absorption of kinetic energy decreases as one moves away from the mountain surfaces, while the effect of the baroclinic zone, demonstrating itself in the thermal wind, will accumulate with altitude. As a result the highest vertical kinetic energy gradient in the upper half of the troposphere is observed over the Central Asian mountain massif.

The distribution of large-scale eddy energy in July (Figure 10b) is very similar to that in January, except that an additional maximum appears at the 200 mb level on the leeward side of the Central Asian mountain massif.

On the other hand, minima of large-scale eddy energy are observed in summer and winter over the mountain systems proper of Asia and America. The cause of this is the same as in winter.

The above features of the distributions of E, E, b, E_λ and E_v over mountain systems are very clearly seen on average energy distribution maps at the corresponding isobaric surfaces /17/, which are not presented here due to space limitations.

It appears that the kinetic energy distribution of the real wind is in general of the same nature as the previously described distribution for the geostrophic wind. In particular, the reduction in kinetic energy of mountain massifs is verified by profiles of the actual wind shown in Figure 2. However, some important details are not reflected in the latitude sections of the geostrophic wind. Thus, in particular, the wind at the January maximum level (13 — 13.5 km) from plains to mountain systems of Central Asia increases and an orographic maximum exists over mountain systems /42, 43, 49/. This maximum is not reflected on sections of the geostrophic kinetic energy for January (Figure 9a). It is possible that this is due to the departure from geostrophicity over mountain systems.

We shall now consider the thermal effect of mountain systems on the general circulation. It should be emphasized that until recently no clear line was drawn between the large-scale effect of mountain systems (as a whole) on the temperature field and local temperature features in different types of terrain. The item of interest from the viewpoint of the general circulation is the large-scale effect of mountain systems.

A number of investigators studied either the average temperature field at isobaric surfaces, or the average temperature field of different atmospheric layers /52, 53, 54, 57, 58/. However, it is difficult to determine the effect of orography using average temperature distribution maps, since a drop or rise in temperature at some latitude may result from several factors. In order to clarify the kind of effects exerted by mountain systems, we shall compare the average temperatures at vertical sections along meridians, one of which passes over plains and the other sufficiently close to mountains.

Mountain systems may affect the temperature field in two ways: by acting on the dynamics and vertical motion of air and by changing the radiation factor. The effect of each factor on the temperature field above mountains depends on the direction of the incoming air flow, season of the year, time of day, type of underlying surface (snow, rocks, Alpine meadows, etc.) and other conditions.

Let us consider the meridional sections of temperature field differences over plains and mountains, taken from /50/. It is more convenient to consider the differences since this eliminates the natural temperature variation with changes in latitude and altitude, observed over even the most uniform surface. In addition, it follows from comparing the temperature gradient between the section and zonal temperature gradient inside the continent, that the temperature field is subjected to the orographic effect if the former gradient markedly exceeds the latter.

The temperature differences for July (Figure 11c) show that in summer the air over mountain systems of moderate and subtropical latitudes is warmer than at the same altitude over plains. The temperature in the mountains may be found to be lower than at the same altitudes over level land only at individual locations. This is due to the local effect of particular types of terrain or the presence of a snow (glacier) surface. The large-scale effect of mountain systems in summer thus demonstrates itself in the fact that they serve as heat sources with attendant formation of a warm-air layer above them. Above this warm layer the air over the mountains may be colder than over plains where ascending motions, resulting in adiabatic cooling, predominate.

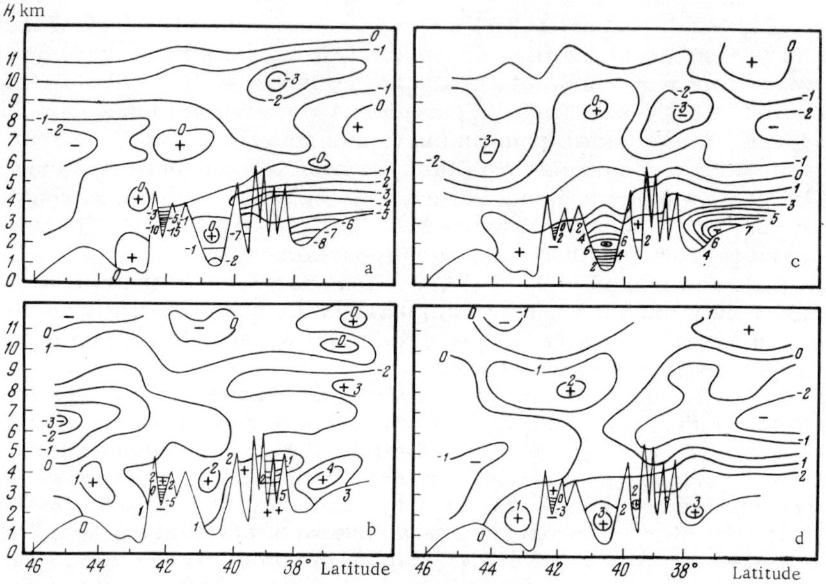

FIGURE 11. Meridional vertical section of the temperature difference (degrees) for mountains (75°E) and level land (60°E):

a — January; b — April; c — July; d — October.

During months when the mountain surfaces are free of snow, they serve as heat sources and a layer of warm air prevails over them (Figure 11b, c). In winter (Figure 11a), when the mountain systems are covered with snow, the air above them cools. However, the effect of cooling on the temperature field decreases rapidly with altitude, and the air at the mountain-range level is somewhat cooler than over plains. The effect of mountains as large-scale heat sinks thus extends only to the mountaintop level. Apparently, if the mountaintops are entirely free of snow, they will serve as heat sources even in winter.

We also recall the baroclinic zone, which arises in summer along the northern boundary of the Central Asian mountain massif and in winter at the Hindu Kush and the Himalayas. This zone is of great importance in processes of orographic frontogenesis /10/.

In conclusion, we consider briefly the role of mountain systems in the formation of certain features in the distribution of synoptic elements, for example, cyclones and jet streams. Maps of the frequency of ground-level cyclones and anticyclones presented in /52, 53/ show that mountain systems are regions with minimum recurrence of cyclones and maximum recurrence of anticyclones. It should be noted that these formations are frequently fictitious, since [according to these calculations] the isobaric surfaces lie underground. In general, analysis of the surface-level synoptic situation over large mountain systems of Asia and America is conditional. Apparently, this is responsible for the fictitious maxima of cyclones at the ground in Greenland presented in Table 2, constructed from data of /4/.

TABLE 2. Frequency of cyclones and anticyclones at the surface in front of (A), above (B) and behind (C) mountains, in %

Month	Baric formation	Central Asian mountain massif			Rocky Mountains			Greenland		
		A	B	C	A	B	C	A	B	C
January	Cyclones	25	12	17	32	28	26	31	42	35
	Anticyclones	19	19	14	6	26	29	15	18	13
July	Cyclones	34	29	31	24	25	28	33	35	32
	Anticyclones	10	7	8	2	10	10	16	12	20

TABLE 3. Frequency of cyclonic centers in front of (A), above (B) and behind (C) mountains, in %

Season	Latitude belts, deg	Greenland			Rocky Mountains			Central Asian mountain massif		
		A	B	C	A	B	C	A	B	C
		100°W–60°W	60°W–20°W	20°W–20°E	160°W–130°W	130°W–100°W	100°W–70°W	10°E–70°E	70°E–130°E	130°E–170°W
Winter	80–70	48.2	18.1	28.7	—	—	—	—	—	—
	70–60	47.7	27.2	18.7	11.2	7.4	36.6	—	—	—
	60–50	—	—	—	10.3	6.3	15.0	29.2	24.5	53.4
	50–40	—	—	—	1.5	2.8	7.3	33.3	6.8	13.7
	40–30	—	—	—	—	—	—	16.9	—	0.5
Spring	80–70	57.6	26.0	37.5	—	—	—	—	—	—
	70–60	28.5	22.5	21.2	16.6	8.0	22.2	—	—	—
	60–50	—	—	—	13.1	9.6	18.1	29.0	23.2	56.1
	50–40	—	—	—	2.4	8.0	13.0	44.7	9.5	18.1
	40–30	—	—	—	—	—	—	19.9	0.6	2.5
Summer	80–70	49.1	30.6	35.3	—	—	—	—	—	—
	70–60	46.9	40.8	37.0	14.9	10.9	35.2	—	—	—
	60–50	—	—	—	20.8	13.0	13.9	35.2	42.7	47.9
	50–40	—	—	—	2.9	7.0	4.7	42.1	16.5	21.6
	40–30	—	—	—	—	—	—	13.3	2.5	5.5
Autumn	80–70	70.7	22.2	28.0	—	—	—	—	—	—
	70–60	38.1	32.8	21.1	17.9	7.4	26.7	—	—	—
	60–50	—	—	—	13.8	7.8	14.1	26.1	24.5	38.2
	50–40	—	—	—	0.5	5.1	8.3	34.5	5.8	9.0
	40–30	—	—	—	—	—	—	13.2	0.6	1.1

It makes sense to compare the frequency of cyclones and anticyclones over mountains and in front of them at isobaric surfaces located above the mountaintops. Table 3 lists the frequency of cyclonic centers at the 500 mb surface over an area of 100,000 km^2 in front of (A), above (B) and behind (C) the mountain systems of the northern hemisphere for different seasons /42/. It is seen from examining the table that the smallest number of cyclones occurs above mountain systems. Cyclones at 500 mb in front of mountain massifs are encountered as frequently as behind them.

Finally, we note the contribution of mountain systems to the formation of jet streams /10/. It is shown by calculation of the frequency of jet-stream axes in summer along the 0, 20, 40, 60 and 75°E meridians /63/ when a baroclinic zone forms at the northern boundaries of mountain systems, that their number in the 40—45°N latitude belt increases sharply over the Caucasus, Pamir-Altai and Tien Shan, while decreasing over the deserts of Central Asia.

It follows from the above study that the earth orography affects the general circulation of the atmosphere. Hence any theory of the general circulation which makes no allowance for the actual location of large mountain systems cannot explain significant processes of atmospheric circulation.

References

1. Aeroklimaticheskii atlas severnogo polushariya (An Aeroclimatic Atlas of the Northern Hemisphere), Vol. I. — Leningrad. 1961.
2. Aeroklimaticheskii atlas severnogo polushariya. Temperatura, plotnost', vlazhnost' (An Aeroclimatic Atlas of the Northern Hemisphere. Temperature, Density, Humidity). — Leningrad. 1963.
3. Aeroklimaticheskii atlas kharakteristik vetra severnogo polushariya (An Aeroclimatic Atlas of the Wind Characteristics of the Northern Hemisphere). — Moskva. 1963.
4. Abramova, K. N. Povtoryaemost' tsiklonov i antitsiklonov razlichnoi intensivnosti nad severnym polushariem (Frequency of Cyclones and Anticyclones of Different Strengths over the Northern Hemisphere).— Trudy NIIAK, Vol. 19. 1962.
5. Belinskii, N. A. Dinamicheskaya meteorologiya (Dynamic Meteorology). — Leningrad, Gostekhizdat. 1948.
6. Belousov, S. L. and V. V. Bykov. Ob uchete vliyaniya gor pri prognoze baricheskogo polya (Making Allowance for the Effect of Mountains in Forecasting the Baric Field). — Izvestiya AN SSSR, seriya geofizicheskaya, No. 9. 1957.
7. Blinova, E. N. Gidrodinamicheskaya teoriya voln davleniya i tsentrov deistviya atmosfery (A Hydrodynamic Theory of Pressure Waves and Atmospheric Centers of Action).— Doklady AN SSSR, No. 39. 1943.
8. Blinova, E. N. K voprosu ob opredelenii davleniya na urovne morya (Concerning the Determination of Pressure at Sea Level). — Doklady AN SSSR, 92(3). 1953.
9. Boiko, A. P. Gidrodinamicheskii dolgosrochnyi prognoz polei davleniya dlya vsei sfery Zemli s uchetom orografii severnogo i yuzhnogo polusharii (Hydrodynamic Long-Range Forecasting of

Pressure Fields for the Entire Globe with Allowance for the Orography of the Northern and Southern Hemispheres). — Trudy MMTs, Vol. 2. 1964.

10. Bugaev, V. A. Obrazovanie struinogo techeniya v atmosfere pod vliyaniem gornykh massivov Srednei i Tsentral'noi Azii (Formation of an Atmospheric Jet Stream Due to the Effect of Mountain Ranges of Central and Soviet Central Asia). — Meteorologiya i gidrologiya, No. 5. 1958.

11. Buleev, N. I. and G. I. Marchuk. Vliyanie prizemnogo treniya na evolyutsiyu meteorologicheskikh elementov v svobodnoi atmosfere (Effect of Surface-Level Friction on the Evolution of Meteorological Elements in the Free Atmosphere). — In: Voprosy dinamicheskoi meteorologii. Gidrometeoizdat. 1960.

12. Bykov, V. V. K voprosu o vliyanii gor na izmenenie davleniya v srednei troposfere (Concerning the Effect of Mountains on Pressure Changes in the Middle Troposphere). — Meteorologiya i gidrologiya, No. 4. 1955.

13. Vitvitskii, G. N. Tsirkulyatsiya vozdukha nad Kitaem (Air Circulation Over China). — Izvestiya AN SSSR, seriya geograficheskaya, No. 2. 1956.

14. Gandin, L. S., D. L. Laikhtman, L. T. Matveev, and M. I. Yudin. Osnovy dinamicheskoi meteorologii (Fundamentals of Dynamic Meteorology). — Gidrometeoizdat. 1955.

15. Girs, A. A. Osnovy dolgosrochnykh prognozov pogody (Fundamentals of Long-Range Weather Forecasting). — Gidrometeoizdat. 1960.

16. Gruza, G. V. Makroturbulentnost' v obshchei tsirkulyatsii atmosfery (Large-Scale Eddies in the General Circulation of the Atmosphere). — Gidrometeoizdat. 1961.

17. Gruza, G. V. Integral'nye kharakteristiki obshchei tsirkulyatsii atmosfery (Integral Characteristics of the General Circulation of the Atmosphere). — Gidrometeoizdat. 1965.

18. Gruza, G. V. and M. A. Petrosyants. O roli orografii v energeticheskom sostoyanii atmosfery (The Role of Orography in the Energy State of the Atmosphere). — Trudy SANIGMI, 23 (38). 1965.

19. Gutman, L. N. Primenenie metoda dlinnykh voln v zadache obtekaniya gor (Use of the Method of Long Waves in the Problem of Flow Past Mountains). — Doklady AN SSSR, 115 (3). 1957.

20. Dzerdzeevskii, B. L., V. M. Kurganskaya, and Z. M. Vitvitskaya. Tipizatsiya tsirkulyatsionnykh mekhanizmov v severnom polusharii (Classification of Circulation Mechanisms in the Northern Hemisphere). — Trudy NIU, seriya II, No. 21. 1946.

21. Dzhordzhio, V. A. and M. A. Petrosyants. Letnii antitsiklon nad Tibetom (The Summer Anticyclone Over Tibet). — Doklady AN Uzbekskoi SSR, No. 8. 1950.

22. Dorodnitsyn, A. A. Nekotorye zadachi obtekaniya nerovnostei Zemli vozdushnym potokom (Some Problems of Air Flow Past Terrestrial Obstacles). — Trudy GGO, Vol. 31. 1940.

23. Dorodnitsyn, A. A. Vozmushcheniya vozdushnogo potoka, vyzvannye nerovnostyami na poverkhnosti Zemli (Air Flow Disturbances Produced by Obstacles on the Terrestrial Surface). — Trudy GGO, Vol. 23. 1938.

24. Dorodnitsyn, A. A. Vliyanie rel'efa zemnoi poverkhnosti na vozdushnye techeniya (Effect of Terrestrial Topography on Air Flows). — Trudy TsIP, 21(48). 1950.
25. Tu Hsin-Yüan. Utochnennaya skhema ucheta nerovnostei zemnoi poverkhnosti v chislennykh metodakh prognoza (A Refined Scheme for Taking into Account Terrestrial Surface Roughness in Numerical Forecasting Methods). — Trudy GGO, No. 99. 1955.
26. Tu Hsin-Yüan. Prostranstvennaya zadacha chislennogo prognoza baricheskogo polya s uchetom vliyaniya vysokogornykh plato (The Three-Dimensional Problem of Numerical Forecasting of the Pressure Field with Allowance for the Effect of Mountainous Plateaus). — Trudy GGO, No. 99. 1959.
27. Yeh Tu-Cheng and Chu Pa-Chen. Nekotorye vazhneishie voprosy obshchei tsirkulyatsii atmosfery (Some Major Problems of the General Circulation of the Atmosphere). — Leningrad. 1961.
28. Yeh Tu-Cheng and Ku Chen-Ch'ao. Vliyanie Tibetskogo nagor'ya na atmosfernuyu tsirkulyatsiyu i na pogody Kitaya (Effect of the Tibetan Highland on the Atmospheric Circulation and on the Weather in China). — Izvestiya AN SSSR, seriya geograficheskaya. 1956.
29. Zastavenko, L. G. Nekotorye osobennosti baricheskogo polya v yanvare i iyule nad severnym polushariem (Some Features of the Baric Field in January and July over the Northern Hemisphere). — Trudy NIIAK, Vol. 14. 1961.
30. Kats, A. L. Kolichestvennaya kharakteristika gorizontal'nykh sostavlyayushchikh obshchei tsirkulyatsii atmosfery v severnom polusharii (A Quantitative Description of Horizontal Components of the General Circulation of the Atmosphere in the Northern Hemisphere). — Meteorologiya i gidrologiya, No. 2. 1955.
31. Kats, A. L. Sezonnye izmeneniya obshchei tsirkulyatsii atmosfery i dolgosrochnye prognozy (Seasonal Variations in the General Circulation of the Atmosphere and Long-Range Forecasts). — Gidrometeoizdat. 1960.
32. Kibel', I. A. K voprosu o perevalivanii tsiklona cherez gornyi khrebet (Concerning the Motion of Cyclones over Mountain Ranges). — Trudy NIU, seriya 1, No. 30. 1947.
33. Kibel', I. A. Prostranstvennaya zadacha obtekaniya nerovnostei poverkhnosti zemli vozdushnym potokom (The Three-Dimensional Problem of Air Flow Past Terrestrial Surface Obstacles). — Doklady AN SSSR, 100(2). 1955.
34. Kochin, N. E. and B. I. Izvekov. Dinamicheskaya meteorologiya (Dynamic Meteorology). — Part 2. Moskva. 1937.
35. Kochin, N. E. O vliyanii rel'efa zemli na volny na poverkhnosti razdela dvukh mass zhidkosti raznoi plotnosti (Effect of the Terrestrial Topography on Waves at the Interface Between Two Fluid Masses of Different Density). — Trudy GGO, Vol. 14. 1937.
36. Kochin, N. E. Prostranstvennaya zadacha o volnakh na poverkhnosti razdela dvukh mass zhidkosti raznoi plotnosti, vyzyvaemykh nerovnostyami dna (The Three-Dimensional Problem of Waves at the Interface Between Two Fluid Masses of Different Density Produced by Unevenness of the Bottom). — Trudy GGO, Vol. 28. 1938.

37. Kurbatkin, G. P. Vliyanie krupnykh orograficheskikh prepryatstvii na zapadno-vostochnyi potok (Effect of Large Orographic Obstacles on West-East Transport). — Izvestiya AN SSSR, seriya geofizicheskaya, No. 2. 1958.
38. Kurbatkin, G. P. Uchet negeostrofichnosti v prostranstvennoi zadache obtekaniya gor (Allowance for the Departure from Geostrophicity in the Three-Dimensional Problem of Flow Past Mountains). — Izvestiya AN SSSR, seriya geofizicheskaya, No. 4. 1959.
39. Mashkovich, S. A. O sezonnykh izmeneniyakh struinykh techenii (Seasonal Variations in Jet Streams). — Meteorologiya i gidrologiya, No. 6. 1956.
40. Monin, A. S. O makroturbulentnom obmene v zemnoi atmosfere (Exchange by Means of Large-Scale Eddies in the Terrestrial Atmosphere). — Izvestiya AN SSSR, seriya geofizicheskaya, No. 4. 1956.
41. Morozova, M. I and M. A. Petrosyants. Povtoryaemost' tsiklonov i antitsiklonov na AT_{500} nad severnym polushariem (Frequency of Cyclones and Anticyclones at 500 mb over the Northern Hemisphere). — Trudy TsIP, Vol. 106. 1961.
42. Morozova, M. I., M. A. Petrosyants, and O. N. Chernysheva. Osobennosti vozdushnykh techenii nad Pamirom i Zapadnym Tian'-Shanem (Features of Air Flows over the Pamir and Western Tien Shan). — Meteorologiya i gidrologiya, No. 9. 1959.
43. Morozova, M. I., M. A. Petrosyants, and O. N. Chernysheva. Srednee pole vetra nad gornymi sistemami Srednei i Tsentral'noi Azii (The Average Wind Field over Mountain Systems of Central and Soviet Central Asia). — Trudy Instituta matematiki im V. I. Romanovskogo, No. 25. Tashkent. 1962.
44. Musaelyan, Sh. A. Planetarnye orograficheskie volny v zapadnom potoke (Planetary Orographic Waves in Western Flow). — In: Voprosy dinamicheskoi meteorologii. Izdatel'stvo AN SSSR. 1960.
45. Musaelyan, Sh. A. Ob uchete vliyaniya nerovnostei rel'efa poverkhnosti zemli pri opredelenii davleniya na urovne morya (On Making Allowance for the Effect of Obstacles of the Terrestrial Topography in Determining the Sea-Level Pressure). — AN SSSR, seriya geofizicheskaya, No. 5. 1958.
46. Musaelyan, Sh. A. Prostranstvennaya zadacha obtekaniya nerovnostei zemnoi poverkhnosti s uchetom sferichnosit zemli (The Three-Dimensional Problem of Flow Past Terrestrial-Surface Obstacles Allowing for Earth Sphericity). — Doklady AN SSSR, 103(3). 1955.
47. Petrosyants, M. A. O masshtabe orograficheskikh vliyanii na sinopticheskie protsessy Srednei Azii (The Scale of Orographic Effects on Synoptic Processes in Central Asia). — Izdatel'stvo UGMS Uzbekskoi SSR, No. 27. Tashkent. 1951.
48. Petrosyants, M. A. K voprosu ob obrazovanii Sibirskogo antitsiklona (Concerning the Formation of the Siberian Anticyclone). — Doklady AN Uzbekskoi SSR, No. 2. 1951.

49. Petrosyants, M. A., M. I. Morozova, and O. N. Chernysheva. O vliyanii gornykh sistem Pamiro-Altaya i Tyan'-Shanya na strukturu vozdushnykh techenii (Effect of Mountain Systems of Pamir-Altai and Tien Shan on the Structure of Air Flows). — Materialy konferentsii po itogam MGG (1960) i meteorologicheskogo izucheniya Antarktidy (1959). Moskva. 1961.
50. Petrosyants, M. A., O. I. Subbotina, and S. G. Chanysheva. O vliyanii orografii Srednei Azii na srednee pole temperatury (Effect of the Orography of Central Asia on the Average Temperature Field). — Trudy SANIGMI, 23(38). 1965.
51. Petrosyants, M. A. Izmenchivost' vetra nad gornymi raionami Srednei Azii (Variability of the Wind over Mountain Regions of Central Asia). — Trudy SANIGMI, 15(30). 1963.
52. Pogosyan, Kh. P. Sezonnye kolebaniya obshchei tsirkulyatsii atmosfery (Seasonal Variations in the General Circulation of the Atmosphere). — Trudy TsIP, Vol. 1. 1947.
53. Pogosyan, Kh. P. Obshchaya tsirkulyatsiya atmosfery (General Circulation of the Atmosphere). — Gidrometeoizdat. 1959.
54. Pogosyan, Kh. P. Struinye techeniya v atmosfere (Jet Streams in the Atmosphere). — Gidrometeoizdat. 1960.
55. Sarymsakov, T. A., V. A. Bugaev, and N. A. Dzhordzhio. O vliyanii kvazipermanentnogo vysotnogo antitsiklona nad Severnoi Afrikoi na tsirkulyatsiyu v Srednei i Perednei Azii (The Effect of the Semi-Permanent High-Altitude Anticyclone over North Africa on the Circulation in Central and Hither Asia). — Byulleten' AN Uzbekskoi SSR, No. 12. 1947.
56. Haltiner, G. J. and F. L. Martin. Dynamical and Physical Meteorology. — McGraw-Hill Book Company, Inc. 1957.
57. Khanevskaya, I. V. Raspredelenie temperatury nad severnym polushariem letom (Summer Temperature Distribution over the Northern Hemisphere). — Trudy NIIAK, Vol. 9. 1963.
58. Khanevskaya, I. V. Osnovnye cherty zimnego polya temperatury v svobodnoi atmosfere nad severnym polushariem (Main Features of the Winter Temperature Field in the Free Atmosphere over the Northern Hemisphere). — Trudy NIIAK, Vol. 14. 1961.
59. Khromov, S. P. Mussony v obshchei tsirkulyatsii atmosfery (Monsoons in the General Circulation of the Atmosphere). — In: A. I. Voeikov i sovremennye problemy klimatologii. Gidrometeoizdat. 1956.
60. Khromov, S. P. Osnovy sinopticheskoi meteorologii (Fundamentals of Synoptic Meteorology). — Gidrometeoizdat. 1948.
61. Ch'ao Chi-P'ing. Dinamika vozmushchenii s ogranichennoi amplitudoi, obuslovlennykh orografiei v baroklinnom poyase zapadnykh vetrov (Dynamics of Disturbances with Limited Amplitude, Due to Orography, in the Baroclinic Belt of Westerly Winds). — Tsisyan Syuebao, No. 28. 1957.
62. Chekirda, A. Z. K gidrodinamicheskoi teorii tsentrov deistviya atmosfery (On the Hydrodynamic Theory of Atmospheric Centers of Action). — Izvestiya AN SSSR, seriya geofizicheskaya, No. 7. 1957.
63. Chernysheva, O. N., M. A. Petrosyants, M. I. Morozova, and V. A. Dzhordzhio. Geograficheskoe raspredelenie struinykh techenii po kartam maksimal'nogo vetra (Geographic Distribution

of Jet Streams from Maximum Wind Maps). — Trudy Vsesoyuznogo meteorologicheskogo soveshchaniya, Vol. 3. 1963.

64. Chu Yün-T'img. Ob uchete dinamicheskogo vliyaniya gornykh massivov v nelineinoi zadache dolgosrochnogo prognoza meteorologicheskikh elementov (On Making Allowance for the Dynamic Effect of Mountain Massifs in the Nonlinear Problem of Long-Range Forecasting of Meteorological Elements). — Izvestiya AN SSSR, seriya geofizicheskaya, No. 12. 1959.

65. Chu Pao-Chen. Statsionarnye vozmushcheniya poyasa zapadnykh vetrov istochnikami i stokami tepla i orografiei (Stationary Disturbances of the Westerly-Wind Belt Due to Heat Sources and Sinks and to the Orography). — Tsisyan Syuebao, No. 28. 1956.

66. Chuchkalov, B. S. Sinopticheskie usloviya perenosa vozdukha Indiiskogo mussona cherez Tibetskoe nagor'e (Synoptic Conditions for the Transport of Air of an Indian Monsoon over the Tibetan Highland). — Trudy VNMS, Vol. 3. Leningrad. 1963.

67. Bannon, J. K. Note on the Structure of the High Altitude Strong Wind Belt in the Middle East in Winter. — Quart. J. Roy. Meteorol. Soc., 80(344). 1954.

68. Boffi, G. A. Los Effectos de los Andes on la Circulation Aerea Sobre la Parte Austral de sur America. — Rev. Meteorol., 9(34). 1950.

69. Bolin, B. On the Influence of the Earth's Orography on the General Character of the Westerlies. — Tellus, Vol. 2. 1950.

70. Brooks, C. E. P., C. S. Durst, N. Carruthers, D. Dewar, and J. S. Savier. — Geophys. Mem. Meteorol. Office London, No. 85. 1950.

71. Charney, J. G. and A. A. Eliassen. Numerical Method for Predicting the Perturbations of the Middle Lätitude Westerlies. — Tellus, Vol. 1. 1949.

72. Chaudhury, A. M. On the Vertical Distribution of Wind and Temperature over Indo-Pakistan Along the Meridian 76°E in Winter. — Tellus, 2(1). 1950.

73. Che Chao Koo. Der dynamische Einfluss des Hochlandes von Tibet auf die Ost-Asiatische Zirculation. — Acta phys. Acad. sci. hung., Vol. 4. 1955.

74. Crutcher, H. L. Upper Wind Statistics Charts of the Northern Hemisphere. — U. S. Office of Naval Operations, NAVAER, Vol. 1. 1959.

75. Defant, A. Die Zirculation der Atmospäre in den gemässigten Breiten der Erde. — Geogr. Ann., No. 3. 1921.

76. Frost, R. Upper Air Circulation in Low Latitudes in Relation to Certain Climatological Discontinuities. — Profess. Notes Meteorol. Office, 7(107). 1953.

77. Gambo, K. The Topographical Effect upon the Jet Stream in the Westerlies. — J. Meteorol. Soc. Japan, Vol. 34. 1956.

78. Games, R. W. February Cross-Section along the Greenwich Meridian. — Meteorol. Mag., 80(954). 1951.

79. Gillchrist, A. Upper Wind in the Tropics and Subtropics. — MRP, No. 795, L. Met. Res. Com. 1953.

80. Kochanski, A. Cross-Sections of the Mean-Zonal Flow and Temperature along 80°W. — J. Meteorol., 12(2). 1955.
81. Koteswaram, P., G. R. V. Raman, and S. Parthasarthy. The Mean Jet Stream over India and Burma in Winter. — Indian J. Meteorol. and Geophys., 4(2). 1953.
82. Landsberg, H. Klimatische Meridional-Abschnitte der freien Atmospäre. — Meteorol. Taschenbuch, Linke F. 1953.
83. Mintz, Y. The Observed Zonal Circulation of the Atmosphere. — Bull. Amer. Meteorol. Soc., Vol. 35. 1954.
84. Murakami, T. The Geographical Effect upon the Stationary Upper Flow Patterns. — Papers Meteorol. and Geophys., Vol. 7. 1956.

O. A. Drozdov

THE GENERAL CIRCULATION OF THE ATMOSPHERE AND THE HYDROLOGIC CYCLE

The article examines correlations between the fluctuation in the amount of precipitation and variation in temperature levels by seasons using data of 37 USSR stations for the period 1894—1915, as well as the relationship between precipitation anomalies and gradients of temperature anomalies along the meridians and parallels from maps of anomalies for the northern hemisphere during the period 1891—1940. Analysis of the above relationships brings to light a number of features regarding the role of the general circulation in the formation of the precipitation regime of individual regions.

The role of phase transitions of water vapor in the formation of the general circulation of the atmosphere is determined primarily by the energetic significance of this process. As is known /2/, the influx of heat of condensation when heating the atmosphere is, on the average, a factor of 4.5 times the heat supplied as a result of heat exchange by means of large-scale eddies, while in tropical latitudes and during the warm season in moderate latitudes this ratio is even more impressive (below the equator this factor is about 10—15). It is known that turbulent heat transfer is the real heat source in the atmosphere, while condensation is compensated for by evaporation which, although pertaining to the active surface, in the final count also affects the cooling of the atmosphere. However, in examining the energy balance in the earth—atmosphere system, and comparing the heat released or bound in accordance with the condensation—evaporation difference with the heat actually transported by atmospheric circulation, one notes the pronounced predominance of the former process over the latter in the mean annual cycle for latitudes below 50° and in summer, obviously, also for higher latitudes. Below the equator this ratio is close to 2, and below 30°N it is close to 8. Clearly, this does not provide a complete description of the role of each process in forming the general circulation.

Heat can be converted into kinetic energy only when the intensity of the thermal process varies over a given area. Actually the pole-equator and the ocean-land temperature differences are produced by long-time, energy-maintained pressure differences, associated with the velocities of horizontal air currents. Thermal condensation also participates in this process. In implicit form it amplifies the pressure differences between the cold (dry) and warm (humid) sections due to differences in the specific heats of water vapor and dry air. In explicit form it decreases the differences due to the radiation distribution, since the release of heat of condensation exceeds the heat expended on evaporation at high latitudes and only below the equator does the release of heat of condensation aid in additionally heating the already relatively warm air. Conversely, condensation processes should amplify the circulation in contrasts between the oceans and land (where participation falls primarily over warm regions).

The release of heat of condensation is most significant for amplifying the circulation between regions of elevated and reduced pressures of any scale. In this case additional heating should occur by virtue of the release of latent heat in low-pressure regions. However, it is compensated for, and sometimes even overcompensated, by screening of the radiation by thick continuous clouds. At the same time the cooling of the rising saturated air is attenuated as compared with that of nonsaturated air by the magnitude of the difference between the dry and wet adiabatic lapse rates.

As was shown above, the role of release of the heat of condensation in atmospheric circulation processes is quite complicated and these processes were not rigorously subjected to theoretical or empirical studies. Some exception is the estimate of the role of release of the heat of condensation in maintaining the ascending fluxes proper and in the formation of precipitation. Thus, estimating the minimum relative humidity at the ground for which the fall of precipitation still depends upon the temperature, we find that this limit drops with a rise in temperature; for southern European USSR it varies from 53% at temperatures of about 10° to 27% humidity at temperatures above 32° /8/. Obviously, this effect is also the combined result of several processes. To some extent this situation may have been affected by the fact that low relative humidity is observed at the surface during high temperature more frequently than during low temperatures.

Further, a high temperature at the ground serves as an indirect proof of the fact that it drops rapidly with altitude (although by far not always), but apparently the most important factor here is the altitude of the condensation level, which is directly related to the relative humidity. The higher the condensation level, the more difficult is it for vertical motions to reach it if the free-atmosphere lapse rate is less than the dry-adiabatic, which is usually the case. Above the condensation level it suffices that the lapse rate be higher than the wet-adiabatic; the latter depends on the temperature, which is responsible for the above dependence. The higher the condensation level, the greater is the probability that precipitation, once formed, will evaporate without reaching the ground. But a high temperature and a rise in the condensation level in this case will aid in rapid evaporation of the precipitation and thus may reduce the quantity of the latter. This is precisely the situation which must be taken into account in the study of the hydrologic cycle, and one should not expect an equal quantity of precipitation for the same relative humidity but at higher temperatures. Consequently, the latter factor is subordinate.

It must also be remembered that the predominance of ascending or descending motions in a given region results in a marked change in the atmospheric moisture content. It is particularly perceptible, even on long-term maps, that the quantity of potentially precipitated moisture decreases in anticyclonic regions, where descending motions predominate. The same factor reduces throughout the year the moisture content in tropical deserts by approximately a factor of 2 as compared with the surrounding areas.

The latter is observed even in summer when a thermal low is observed over deserts below the 500 mb level. This circumstance casts doubt on the accuracy of radiosonde readings /7, 8/.

Much less attention was accorded to processes proceeding in the opposite direction, i. e., increasing atmospheric moisture content in regions with predominance of ascending motions. Thus the moisture content in summer over the European continent is higher than over the Atlantic. However, such a rise in moisture content cannot be ascribed to simply increased evaporation over Europe, since the amount of precipitation in this region exceeds the evaporation. The quite substantial rise in moisture content during summer in the monsoon region at the southeastern shores of Asia is apparently of the same origin, although it is at present difficult to present an exhaustive analysis of this phenomenon. Vertical air motions can thus result, within the general circulation of the atmosphere, in a peculiar transformation of the circulation and indexes are required to allow for vertical motions of this kind. Sometimes the average baric field is used for this purpose, but many of its features are lost in averaging, although by far not all of its aftereffects are actually compensated for. Hence the quantity of precipitation, particularly, averaged over an area, can in itself serve as one of the indexes of the general circulation, signifying the availability of ascending motions with large vertical extent, independent of their horizontal scales and of compensation by motions in the opposite direction. At the same time the amount of precipitation proper interacts in a quite complicated manner not only with the vertical currents of the general circulation, but also with a number of other factors, affecting both the precipitation and somewhat also the vertical atmospheric motions proper. In the study of the hydrologic cycle these factors are estimated by semiempirical methods, and the impression is created that the circulation factors proper are neglected. Actually this is far from being true.

The formation of precipitation, i. e., release of heat of condensation, requires the simultaneous existence of the following conditions: 1) presence of a definite quantity of moisture in the atmosphere; 2) existence of ascending motions (of dynamic, eddy and thermal natures); 3) presence of thermal instability of the atmosphere and a certain altitude of the condensation level, which can be characterized by the relative moisture and the dew point deficit. The condensation level also depends on the total moisture content of the atmosphere, since the wet-adiabatic lapse rate decreases with temperature.

The simplest form of correlation, in which any given quantity is simultaneously a function of several factors and is proportional to their intensity, is expressed by the product of the above factors:

$$r = \alpha w \psi (H), \tag{1}$$

where r is the amount of precipitation (usually monthly values are used); α is a proportionality factor, depending on the intensity of ascending motions in the atmosphere, including also the process of vertical exchange; $\psi(H)$ is some function of the relative humidity of the atmosphere (most frequently averaged over all the altitudes with allowance for the moisture content of different atmospheric layers); w describes the presence of moisture in the air mass in which the water vapor condenses and which participates in the upward vertical motion. This quantity frequently is not identical with the moisture content of the vertical column of air in the region where precipitation occurs. The validity of replacing data on the moisture content of the specific air layer participating in the formation of precipitation by the total moisture is determined solely by the assumption of uniform vertical moisture distribution at the center of air-mass formation. This assumption has empirical validity in the case of averaging over extended time or area. The value of α can also be regarded as a factor, or determined on the basis of objective features of the baric field /3, 8/. The value of $\psi(H)$, determined primarily by the probability of precipitation, is obtained empirically for a number of areas, although attempts were made to obtain theoretical estimates of this quantity.

Due to the fact that the surface-level humidity is a function of random local evaporation sources and of the falling precipitation, attempts are made in practice to use either vertically-averaged humidity data, or the humidity at characteristic atmospheric levels. Cyclonic processes are best related to the humidity at the 700 mb level (although a relationship is retained also with the 850 mb surface). Local precipitation apparently is best related to the humidity at lower levels. The form of the relationship $\psi(H)$ for constant w, obtained by averaging over a month or more, is quite unusual and is not always properly treated. For a high relative moisture the probability of precipitation (and its amount) is virtually independent of H. For average humidities (50%), corresponding approximately to the humidity of the forest-steppe zone, the probability of precipitation changes rapidly as a function of humidity. At low humidities this relationship may again become weak. However, here it is not accurately determined due to the low precipitation probability.

The question arises naturally whether the above form of $\psi(H)$ is determined by the real dependence of the probability of precipitation on the altitude of the condensation level, or whether this is the result of the type of humidity distribution curve for different average values of the latter. To resolve this question one must consider data for individual days and the direct dependence on H of the precipitation probability. It is found that in this case the form of the relationship changed markedly. The dependence of the amount of precipitation on H in this case is close to linear. The precipitation probability changes most rapidly as early as at $H = 67\%$, when the air rises by 400—650 m (instead of 800—1250 m), corresponding to a humidity change from 50 to 100%. The change in the precipitation probability with altitude for lower humidities is approximately linear. The straight line should have intersected the abscissa near the value 38%, but this relationship becomes somewhat nonlinear at such low humidities. Thus, for the range of $H = 40-80\%$ and above the precipitation probability is not proportional to the altitude, though it retains an overall approximate linearity, while the inflection points of this curve for $H = 50\%$, using monthly average data, depend on features of the humidity distribution, which aids or interferes with precipitation. Since precipitation involves the release of large amounts of energy in the atmosphere, so affecting the horizontal and vertical components of the general circulation, particularly in cyclonic regions, the hydrologic cycle and atmospheric circulation should interact with one another in a complex manner.

The interaction between the hydrologic cycle and the general circulation can be followed, in particular, using relationships involving precipitation totals over very large areas. It is known that individual rains, particularly during summer in semidesert regions, have a quite local nature, but even the monthly anomalies of precipitation distribution show also relationships pertaining to rather large areas, while the secular variation shows the existence of these regularities over extremely large areas. It is natural that these regularities are related to the corresponding fluctuations in the general circulation.

These fluctuations can be partially detected by changes in the frequency of corresponding circulation forms and indexes; due to our inability to properly describe the intensity of a number of circulation phenomena, they are in part reflected only indirectly by fluctuations of the temperature, amount of precipitation, cloudiness, etc. This is especially the case with respect to variations in systems with long-term cycles. One of them attracted at its time the attention of E. Brückner. These variations in the amount of precipitation in a cycle occurring over several decades are related, but opposite in sign, to variations in the temperature level. Many authors doubt the very existence of this cycle, since temperature variations do not always correspond to fluctuations in precipitation. However, these secular precipitation variations were actually verified by observation and, in addition, they tie in quite well with the secular march of the meridional temperature gradients. At the same time it is of practical interest to determine to what extent the relationships which are found in connection with the secular variation are observed during specific years. For this, correlation factors between temperature and precipitation anomalies were calculated for 37 stations in the USSR (the qualitative correlation factor between the signs of the anomalies was calculated, and then it was recalculated for the ordinary correlation on the assumption of a normal distribution curve).

The reduction in accuracy brought about by this method was compensated by grouping the months according to seasons (April—May, June—August, September—October and November—March). The results of these calculations were found to be in satisfactory agreement with published data. For example, for the spring and summer months in Kirov the correlation with temperature anomalies has a factor of 0.64, while the corresponding correlation factor for 1894—1915 in Vyatski for June was found to be 0.6305 /1/, which is virtually identical with this author's calculations.

The relationship between temperature and precipitation anomalies is determined, first, by the cloudiness: in winter, for a positive precipitation anomaly in moderate belts it is usually warmer than for a negative anomaly; in autumn and spring the same may be observed in polar latitudes; in summer and during transitional seasons the signs of the correlation between the temperature and precipitation anomalies should be opposite. Second, the sign and degree of relationship may depend on the need for heat advection in humid but cold regions, or for cold advection for producing temperature contrasts in dry and warm regions.

It was found from calculations that in winter the sign of the correlation between the temperature and precipitation anomalies is usually opposite to that suggested by Brückner (with the exception of Transcaucasia, Central Asia, East Siberia and the Far East). The highest positive correlation factors, which are on the average above 0.5 (up to 0.77), are found in the Volga-River and Cis-Ural regions (Table 1), the lowest are found in Transcaucasia (Tbilisi — 0.57) and Chita Region (Nerchinskii Zavod — 0.66). Regions of high correlations of different sign lie at the same latitude and are produced primarily by general-circulation factors.

TABLE 1. Correlation factors between the temperature and amount of precipitation

Station	Correlation factor			
	April—May	July—August	September—October	November—March
Kola	0.29	-0.24	-0.49	—
Syktyvkar	-0.32	-0.57	-0.20	—
Troitsko-Pechersk	0.35	-0.55	0.50	—
Tallin	0.55	-0.59	-0.39	—
Leningrad	0.49	-0.41	-0.39	0.30
Vologda	-0.49	-0.71	0.39	—
Kirov	-0.63	-0.63	-0.43	—
Kaliningrad	-0.15	-0.59	-0.25	—
Riga	0.50	-0.49	-0.41	—
Moscow	-0.60	-0.64	-0.30	0.49
Kazan'	-0.57	-0.80	0.25	0.55
Sverdlovsk	-0.43	-0.47	-0.35	0.34
Kiev	-0.60	-0.60	-0.60	0.25
Khar'kov	-0.63	-0.55	-0.63	—
Voronezh	-0.62	-0.69	-0.66	—
Oktyabr'skii Gorodok	-0.02	-0.41	0.08	—
Ufa	-0.45	-0.68	-0.64	—
Orenburg	-0.61	-0.75	-0.34	0.77
Malyi Uzen'	-0.70	-0.71	-0.41	0.63
Odessa	-0.62	-0.63	-0.47	0.32
Temir	-0.71	-0.57	-0.55	—
Uil.	-0.71	-0.65	-0.65	—
Sevastopol'	-0.58	-0.44	-0.50	—
Sukhumi	-0.46	-0.63	-0.42	-0.32
Tbilisi	-0.63	-0.61	-0.44	-0.57
Kazalinsk	-0.44	-0.58	-0.42	—
Ashkhabad	-0.62	-0.57	-0.27	—
Tashkent	-0.70	-0.32	-0.36	-0.20
Tselinograd	0.49	-0.75	-0.44	—
Alma Ata	-0.68	-0.68	-0.50	—
Barnaul	-0.57	-0.71	-0.65	0.39
Zaisan	-0.41	-0.80	-0.43	—
Turukhansk	-0.20	-0.43	0.39	—
Tara	-0.25	0.17	0.35	—
Irkutsk	-0.39	-0.32	0.27	-0.43
Nerchinskii Zavod	-0.49	-0.47	-0.52	-0.66
Vladivostok	-0.34	-0.34	-0.53	-0.39

As expected, positive correlation in spring is found primarily in the north, but maximum correlation factors due to circulation causes were obtained for the Baltic Sea region and the Tselina Territory (about 0.5), with the lowest factors occurring for the interfluvial regions of Volga and Ural and Central Asia (about -0.7). The boundary of regions with substantially negative correlation ($|r| > 0.5$) rises toward Vologda and Kirov, Barnaul and Nerchinskii Zavod. Thus, copious precipitation in the Baltic-Sea region, Omsk Region and Tselina Territory requires warm advection, while in steppe, forest-steppe and some forest regions of the European USSR, East Siberia and Far East, and even subtropic regions, cold advection is needed. The correlation between temperature and precipitation anomalies in summer are opposite everywhere, but one's attention is attracted by a region with a very high opposite correlation in the Volga-River region and in Kazakhstan (up to −0.80), pointing to the large role of Arctic invasions in the formation of precipitation in these regions. The distribution of regions with positive and negative correlation in autumn is somewhat similar to that in spring: positive correlation in the north is observed only to the east of the White Sea and extends to the east of Enisei. The region with very negative correlation extends over Ukraine, Central Chernozem Area, Northern Caucasus, Lower Volga River region. At the same time the region of positive correlation between temperature and precipitation extends from north to the central Volga-River region. Negative correlation is pronounced in the belt from Alma Ata through Semipalatinsk to Barnaul (to the west of the mountain system in the vicinity of this belt). The negative correlation region in the Far East is not as pronounced as in spring. Analysis of the above anomalies thus makes it possible to establish a number of features in the role of the general circulation in forming the precipitation regime of individual areas (see Table 1).

Interesting results are also obtained by analyzing the relationship between anomalies of precipitations with gradients of temperature anomalies. These relationships were initially obtained from data on the secular variation of precipitation at individual stations. Then this was carried out for anomaly regions over the entire USSR (Table 2). The USSR was subdivided into three longitude regions (European part, western and eastern Asian USSR), but since the relationships were found, in the first approximation, to be of the same type, the results were combined to reduce random errors. As in the preceding case, the qualitative correlation between the gradients of temperature anomalies and precipitation anomalies were calculated first, and then the results were recalculated to obtain a numerical correlation assuming normal distribution of the variables. This eliminated a number of difficulties not only of a technical nature, but also in selecting scales for calculating the gradients, and it also became possible to obtain a better estimate of the effect of the direction of gradients on the magnitude of the correlation.

For simplicity cases were considered of anomalies with predominance of gradients along the meridian and along parallels. In cases when the anomaly boundaries occupied an intermediate position the results were conventionally referred to both the above directions. These calculations were carried out by A. S. Grigor'eva from anomaly maps of the northern hemisphere compiled at the GGO for the 1891—1940 period. Averaging of these quantities over such a long period was justified by the fact that no substantial differences in the type of correlation were detected for the periods 1891—1915 (up to warming of the Arctic) and 1926—1940 (after warming).

TABLE 2. Correlation factors between gradients of temperature anomalies and the amount of precipitation over the USSR (1891–1940)

Direction of gradients	Jan.	Feb.	Mar.	Apr.	May	June	July	Aug.	Sept.	Oct.	Nov.	Dec.
Along the meridian	0.82	0.85	0.82	0.66	0.55	0.43	−0.20	−0.20	0.44	0.87	0.88	0.91
Along the parallel	0.64	0.70	0.84	0.89	0.77	0.82	0.88	0.78	0.84	0.85	0.74	0.72

When the region with negative temperature anomaly lies along the meridian the total temperature gradient to the north of the point where precipitation is determined increases. In this case the correlation with the amount of precipitation should be positive. When the negative temperature anomaly region lies to the south of the point of precipitation measurement, the opposite should happen. The correlation of precipitation anomalies was better with smaller than with large gradients. This indirectly points to the nonlinear relationship between precipitation and temperature gradients. However, this was neglected in the first approximation and the results were combined assuming a positive increase in the gradient. It can be noted in the annual variation of these correlations that from October to March their values are above 0.82 (up to 0.91), while in July and August the correlation drops to zero (in any case within the limits of errors, and even lower). The latter appears strange at first sight, while actually it does not imply that the temperature gradients have no effect, but rather that it is local. Large meridional air transports in the absence of conditions for rapid replenishing of the cold in the north result in equalization of the temperature in the north and south in cyclonic regions and in their accentuation under purely zonal circulation forms. The above is illustrated by correlation factors between the temperature gradients along the latitude and anomalies in the amount of precipitation. Here these correlations are also of different sign: positive when the cold region lies to the west of the precipitation-measurement location and negative when it lies to the east of the point where precipitation is determined. The former case is most frequently associated with a cyclonic situation and the latter with an anticyclonic situation.

It is natural that in the former case the temperature gradients aid in producing precipitation, while in the latter they do not. Since in the latter case the correlation with the precipitation field is found to be even closer than in the former, the impression is created that temperature gradients under anticyclonic conditions interfere with the formation of precipitation. This is very unlikely: most probably the high temperature contrast simply demonstrates the intensity of the anticyclonic circulation. At the same time, the forward part of the cyclone where one would expect direct correlation between the temperature gradients and precipitation should exist under conditions of this location of temperature gradients. As in the preceding case, the nonlinearity in the correlation between the precipitation anomaly and zonal temperature gradients was neglected. The final correlations are relatively weak in winter, when the negative temperature anomalies in the west, upon addition to the average temperature field, may sometimes signify an overall reduction in the temperature gradient. From December to February the correlation of temperature gradients with precipitation is expressed by factors below 0.80 (0.64 in January), while during other months they fluctuate from 0.74 to 0.89; here the correlation remains quite significant during the central summer months.

Comparing the above results with correlations between precipitation and temperature anomalies we can conclude that a significant correlation between temperature and precipitation fields is obtained in central as well as in peripheral parts of temperature anomalies. In the first case the amount of precipitation is related to the temperature proper, in the second it is related to its gradients, while in intermediate regions it is related to both factors, each of which separately makes a smaller contribution. The general circulation thus produces a complicated system of relationships between the temperature anomaly fields. It may be assumed that these relationships have not by far been sufficiently studied and their investigation may be of practical significance for a number of problems, in particular, for describing long-term climate fluctuations, etc.

It is quite important to estimate the extent to which the previously enumerated, quite high statistical correlations between the precipitation and temperature (including its gradients) fields are retained in the secular variation, since the latter is related to the secular variation of atmospheric circulation. The secular variation of precipitation would then serve as a quite substantial index of changes in the atmospheric circulation proper. This is important because the present characteristics of the secular circulation variations either cannot be calculated for an extended period, or one must be restricted to the recurrence of forms of zonal and of various versions of meridional circulation, without estimating their intensity; this does not present the complete picture of circulation changes and makes necessary the introduction of additional indexes /4, 5/.

It was shown in /6/ that the correlation between the precipitation and temperature anomalies is retained when using long-term average data. The correlation between the secular precipitation variation and the secular variation of meridional gradients in a given region is particularly close (clearly, with the exception of warm months, when this correlation was found during former times, while the cold reserves in the Arctic Sea region in summer were much higher than at present). It is interesting that only two principal variations in meridional gradients were observed in moderate latitudes of the northern hemisphere: the Atlantic (Europe and Asia to the west of the Baikal) and Pacific (Eastern Asia and America). The Pacific sector still leads the Atlantic in temperature and precipitation variations by approximately a quarter wave, as verified also by some data on temperature variations in the Atlantic /9/. This precipitation variation remains in some places also in summer, when it is not related to the march of meridional temperature gradients.

In regions further to the south (mainly in subtropical regions) the precipitation and temperature gradient variations have a shape opposite of those in the north; the latitude at which the sign of precipitation variation changes is not the same every month. The impression is created that this feature is related to the distribution of ice in the Arctic Sea region. In particular, when the Arctic warms up, the center of its ice concentration shifts to the east, which affects the value of meridional temperature gradients in moderate latitudes, as well as the characteristic of atmospheric circulation.

An attempt was made to carry out a similar examination of the secular variation of zonal gradients and its relation with the precipitation (for the time being for the limited region contained within 30 — 90°E). It was found that the zonal gradients also have a secular variation. This variation was found to be most perceptible in regions with significant secular variation of the amount of precipitation (in the eastern part of European USSR), but here the variations of zonal gradients are much smaller than those of the meridional gradients. Thus, if we refer temperature differences to 5° of the meridian (555 km), then variations in zonal gradients, averaged over decades in the secular variation, yield for the cold season values not exceeding 0.7 — 0.8° as compared with 1.2 — 2.4° obtained for variations of the meridional temperature gradients. The nature of the variation in zonal gradients is somewhat different from that of the meridional, and the effect of changes in the former are hardly noticeable against the background of the effect of the latter. The variation of zonal gradients in South Siberia and southern European USSR are opposite in type.

During the warm season the meridional gradients are low, the correlation between them and the secular variation of precipitation is also small or nonexistent; the variation in zonal temperature gradients is somewhat greater (up to 0.8 — 0.9° per 5° of meridian) and its correlation with the amount of precipitation in the southeast is becoming perceptible. The secular precipitation variation as well as the variation of zonal gradients have during this time compensation regions, associated with the scale of circulation regions. The northwest and southeast of the above region have opposite variations in precipitation, with regions to the west and east of the Kazan' meridian being opposite to the type of gradient variation.

Since formation of interlatitude gradients involves participation of temperature conditions of high latitudes, while very large temperature changes occur in the latter due to variations in the ice conditions, a correlation should exist between temperature conditions in the Arctic and precipitation at lower latitudes. Such a correlation is actually observed, but it is quite complicated and extends along the latitude over tremendous distances, sometimes to the equator and further down, although they form a system of zones with different correlation signs. At the same time the atmospheric circulation at high latitudes was found to be such that there may be no relationship between temperatures at different latitudes, while several virtually independent regions form in the Arctic-Sea region. One may attempt to isolate such regions in accordance with their natural physicogeographical features, but this approach is disappointing. The number of independent regions in the Arctic is found to be different at various times of the year.

Since the intention is to follow the effect of temperatures in the Arctic over quite different areas, it is convenient to average over a large time interval, during which the temperature field can produce sufficiently stable forms of atmospheric circulation. Here exchange by large-scale eddies succeeds in transmitting the effect of temperature conditions in the Arctic over quite large distances. However, since the winter and summer conditions in the Arctic, particularly at moderate latitudes, are markedly different, averaging was carried out over warm (April—October) and cold (November—March) periods in accordance with previous studies of precipitation and the type of relationships detected above. The correlation factors obtained in the temperature field between 11 stations at the southern periphery of the Arctic Sea region, at which observations are available over sufficiently long periods of time, are presented in Table 3. It can be concluded from this table that in winter there exist approximately three independent regions in the temperature field at different longitudes of the Arctic Sea region (the number of degrees of freedom is sometimes greater than three). By their physicogeographic conditions these three regions may be: 1) Atlantic—Barents—Kara; 2) Chukotsk—Alaska; 3) Canada—Greenland. Stations of the East Siberia Sea and of the Laptev Sea still exhibit quite close relationships with the first and second regions. Stations of the Canadian Archipelago are just as closely related to the second and third regions. In summer these relationships do not extend over such large distances; here there is no tendency for their restoration at large distances, as would have been the case had the correlation function been of the wave type (the total number of degrees of freedom is somewhat less than six). With respect to natural conditions these regions may be: 1) Atlantic—Barents; 2) Kara; 3) East-Siberian; 4) Chukotsk—Alaska; 5) Canada—West Greenland; 6) East Greenland—West Atlantic.

TABLE 3. Long-term ranges of temperature variations during observational years

Station	Cold season		Warm season	
	number of years	range, deg	number of years	range, deg
Spitsbergen	37	17.6	36	5.7
Malye Karakuly. . . .	51	15.0	40	5.6
Dikson	33	11.4	32	4.7
Kyusyur (Bulun)	32	8.6	30	3.8
Wrangel Island	24	6.0	22	3.6
Anadyr'	34	8.5	29	2.7
Nome	42	7.7	41	4.9
Barrow	39	6.9	39	5.8
Upernavik	70	13.8	68	4.1
Ivigtut	71	9.6	70	3.3
Angmagssalik	61	8.8	58	4.9

One thus obtains for the summer twice the number of regions arising due to formation of intermediate zones between the principal winter regions. Whether the number of these regions is again different during transitional seasons is still unclear. The above regions should at the same time form zones of specific atmospheric circulation. The lack of temperature relationships outside the first zone of positive correlation shows that virtually independent circulation sectors exist in these regions.

The situation with respect to correlation between temperatures inside some Arctic sector and precipitation at stations of lower latitudes is somewhat different. The correlation factors are here relatively low and exceed ±0.5 only infrequently. This is natural if one considers that purely local weather factors are of great significance in the formation of the precipitation regime. Formally such factors cannot be regarded as significant, particularly if it is remembered that the series are linked. But the general nature of the correlation extends over large areas which are outside the range of influence of given local factors, with regular alternation of regions with positive and negative correlation along the latitude. This author attempted to check whether this correlation results from temperature and precipitation fluctuations during individual years, or from the universality of the secular variation. It was found that correlations for individual years can be found only to the north of moderate latitudes, while larger distances are affected only by the universality of the secular variation. It is quite difficult to provide a statistical verification of this type of correlation, since long-term cycles may have tendencies to possess one or another sign over a given time period /10/. Consequently, one may investigate the reality of such correlations reaching the equator or passing through it by assuming a certain physical model of the phenomenon.

In any case one cannot ascribe these relationships to the prevalence of a given circulation type: they can be produced only by long-time cycles extending over a wide latitude range.

The general temperature rise near the warm region in the Arctic should increase the precipitation by increasing the moisture content; more to the south this is compensated by the reduction in the velocity of western transport (from the oceans). In winter this is affected by anticyclogenesis of monsoonal nature and the shift in cyclonic trajectories to the northern seas.

In subtropical latitudes, as a rule, these correlations become again direct corresponding to shifts in the circulation zones. To some extent this factor appears also in regions more to the south, although an indirect effect of temperature contrasts under favorable circulation conditions is possible. It is important to note that here one also detects correlation with certain sectors of the polar region independent of temperature (in tropical latitudes virtually with all sectors, and in moderate latitudes with sectors lying to the west of the region where precipitation is estimated); here the correlations with different regions may be of different sign. Here correlations with the nearest Arctic regions are not always the closest. Thus, the relationship between precipitation in the Mediterranean region (in the cold season) and the temperature of the Atlantic—Barents region is direct, while the relationship between the precipitation of this region with the temperature field of Greenland is an inverse one; here it is the stronger, which is natural if one considers its more westerly location relative to the Mediterranean. Since the temperature and the amount of precipitation are different types of quantities with different units, correlations between them cannot be strictly linear. Most frequently departures from linearity lie outside the working field of the correlation graphs, but sometimes the curvilinear nature of these correlations is quite pronounced and reflects the interaction of factors affecting precipitation under given temperature conditions in the Arctic (in particular, the amount of shift in the circulation zones, depending on which the region under study may be located in a cyclonic or anticyclonic zone).

It should be noted that the year-to-year temperature fluctuations in different parts of the Arctic region are not the same, and in winter they are much greater than in summer. Naturally, the higher the temperature variations over a period of many years, the closer the relationships (at least within a season) to be expected between the temperatures at corresponding parts of the Arctic and the amount of precipitation of lower latitudes. Information on temperature variations of Arctic stations is presented in Table 3.

The table shows that differences in temperature variations by Arctic regions are quite large; however, considering climatic changes which are particularly pronounced in winter and the short period of observations at certain stations, it may be said that the range for stations at Dikson, Wrangel Island, Kyusyur and Anadyr' are on the low side, but they agree in general with the ranges of their surrounding stations with longer observation records (Malye Karmakuly, Nome), so that the distortion apparently is not too large. According to the tabulated data, the temperature variations are strongest in winter in the Atlantic—Barents sector of the Arctic. This is followed by the Greenland region and then by the Chukotsk—Alaska region.

The greatest temperature variations in summer arise in the Atlantic—Barents region, followed by Chukotsk—Alaska, and East Greenland regions. The Kara, East Siberian Sea and West Greenland regions seem to have a much smaller variability. Apparently, functions of the effect of the Arctic-temperature field on the precipitation field are related similarly.

Depending on the temperature in polar regions, the precipitation totals can vary within the limits of the overall range of their variation: in the Philippines, compared with Upernavik and Spitsbergen during the warm season and with Nome during the cold season (more than 500 mm); in the Zonda archipelago, compared with Ivigtut in the warm season (more than 1000 mm) and with Ivigtut and Angmagssalik in the cold season (more than 400 mm); in India and Burma, compared with Spitsbergen in the warm season (more than 1000 mm); in Ethiopia, compared with Spitsbergen in the warm season (more than 400 mm); in Guiana, compared with Spitsbergen, Ivigtut and Nome in the warm season and with Nome, Upernavik, Angmagssalik and Dikson in the cold season (up to 500 mm and more).

In such dry regions as Syria and Palestine [Israel] changes in the precipitation total of up to 250 mm and more in the cold season were noted as compared with Angmagssalik, Ivigtut and Spitsbergen. The signs of these correlations vary for different groups of stations. Thus, in the regions enumerated above the correlation between precipitation in the Philippines with temperature anomalies of the warm season at Spitsbergen and Upernavik is direct, while that with anomalies of the cold season takes the form of an inverse relationship. Stations of the Zonda archipelago in the warm season yield a direct correlation with Ivigtut and an inverse relationship with Ivigtut and Angmagssalik in the warm season. The correlation between precipitation in India with the temperature anomaly at Dikson in the warm season is direct, and is inversely related with that in Burma. In Ethiopia in the warm season the correlation with Spitsbergen is inverse, and in Guiana it is direct; that between Guiana, and Nome, Ivigtut and Upernavik is inverse. The correlation with Spitsbergen (in the west) in the cold season is direct, and with Dikson (in the east), Nome, Ivigtut and Angmagssalik is inverse. In Syria and Israel in the cold season an inverse correlation exists with Angmagssalik, Ivigtut and Spitsbergen.

Temperatures at different parts of the Arctic Sea region thus also yield during different seasons correlations of different sign with the precipitation field of certain regions, pointing to the complicated nature of the effects of features of the general circulation of the atmosphere on precipitation fields.

As a rule the highest correlations for the amount of precipitation in northern parts of moderate latitudes are obtained with the temperature of the nearest situated parts of the Arctic. More to the south such correlations are obtained with the temperature of Arctic sectors located to the west of the region under study (although situated farther away).

Finally correlation between precipitation of different regions with the temperature of virtually all the sectors of the Arctic is noted in the intertropical zone; this points to the duration of transmission of these effects by the general circulation and partial shifting of these effects. As a whole these correlations show that, in the process of the general circulation of the atmosphere, and probably also of the hydrosphere, there exists a complicated series of factors determining in part the aftereffects of the process and in part the circulation proper.

References

1. Boyarskii, A. Ya., V. N. Starovskii, V. I. Khotimskii, and B. S. Yastremskii. Teoriya matematicheskoi statistiki (Theory of Mathematical Statistics). — 2nd edition. Gosudarstvennoe sotsial'no-ekonomicheskoe izdatel'stvo. 1931.
2. Budyko, M. I. Teplovoi balans zemnoi poverkhnosti (Thermal Balance of the Terrestrial Surface). — Gidrometeoizdat. 1956.
3. Gal'tsov, A. P. Issledovanie nekotorykh zakonomernostei mirovogo raspredeleniya osadkov kak osnova genetiko-klimaticheskogo raionirovaniya zemnogo shara (A Study of Some Regularities in the World Distribution of Precipitation as a Basis of Geneticoclimatic Zoning of the Globe). — Trudy VNMS, Vol. 4. Gidrometeoizdat. 1962.
4. Girs, A. A. Osnovy dolgosrochnykh prognozov pogody (Fundamentals of Long-Range Weather Forecasting). — Gidrometeoizdat. 1960.
5. Dzerdzeevskii, B. L. Mnogoletnyaya izmenchivost' obshchei tsirkulyatsii atmosfery i klimata kak osnova klimaticheskogo prognoza (Long-Term Variability of the General Circulation of the Atmosphere and the Climate as a Basis for Climatic Forecasting). — Trudy I nauchnoi konferentsii po obshchei tsirkulyatsii atmosfery. Gidrometeoizdat. 1962.
6. Drozdov, O. A. O svyazi razlichnykh storon kolebanii klimata (On the Relationship Between Different Aspects of Climatic Variations). — Trudy GGO, Vol. 84. 1958.
7. Drozdov, O. A. Karty vlagosoderzhaniya atmosfery severnogo polushariya (Maps of the Moisture Content of the Atmosphere of the Northern Hemisphere). — Trudy I nauchnoi konferentsii po obshchei tsirkulyatsii atmosfery. Gidrometeoizdat. 1962.
8. Drozdov, O. A. and A. S. Grigor'eva. Vlagooborot v atmosfere (The Hydrologic Cycle in the Atmosphere). — Gidrometeoizdat. 1963.
9. Petrov, L. S. O strukture kolebanii klimata Arktiki v poslednie desyatiletiya (On the Structure of Variations in the Climate of the Arctic during Past Decades). — Vestnik LGU, No. 6, sektsiya geologii i geografii, Vol. 1. 1959.
10. Slutskii, E. E. Statisticheskii eksperiment kak metod issledovaniya (Statistical Experiments as a Method of Investigation). — Geofizika, 5(7). 1935.
11. Deacon, E. L. Water Vapor over the Sahara and Tiros III Observations. — J. Atmos. Sci., 20(6). 1963.

B. L. Dzerdzeevskii

LONG-TERM FLUCTUATIONS OF THE GENERAL CIRCULATION OF THE ATMOSPHERE AND OF THE CLIMATE AND THE MAJOR REGULARITIES OF PLANETARY CIRCULATION

Definitions are suggested of the following concepts: variations and fluctuations of the weather and climate and changes of climate, as well as atmospheric circulation, planetary circulation and general circulation. Classification of large-scale processes is considered as a method of studying atmospheric circulation. Basic principles are presented in the classification of processes over the northern hemisphere, developed by the author and his co-workers. Trajectories of baric formations on surface maps and the location of isohypses at high altitudes (up to the 100 mb surface) on maps of the northern hemisphere are analyzed. Problems are examined of statistical processing of long-term data on circulation characteristics and some results are presented of processing a 60-year series of these data and of establishing climatic epochs of the 20th century. Energy characteristics are presented of elementary circulation mechanisms in the northern hemisphere, as well as the energy distribution for extreme months. Criteria for establishing boundaries of circulation and climatic epochs are discussed. The coupling of circulation in the northern and southern hemispheres is considered. The question is posed of the possible use of disturbances in the terrestrial magnetic field as an indicator of the relationship between fluctuations of planetary circulation and variations in all kinds of solar activity.

Interest in the study of climate variations has increased sharply during recent years. One obvious reason for this is the accumulation of climatological data and the increased capability for analyzing them, in particular, by the extensive use of computers. This was instrumental in a perceptible development of dynamic climatology and studies of the theory of climate. An even more important reason are the sharply increased needs, which are due to the increased extent of planning in the national economy. Allowance for climatic variations in developing methods of long-range (month, season) weather forecasts and in compiling the forecasts proper is becoming increasingly obvious. Finally, one should mention work on weather modification. If one actually has to deal with pronounced alterations, extending over large areas, the approach to them is by means of their modification of the general circulation of the atmosphere and of the climate. This implies that neglect of the study of the natural variations of the climate may have quite undesirable consequences.

In the majority of existing definitions (of which there are more than fifty) climate is regarded as the ensemble of weathers, expressed by a continuous long-term variation. Frequently reference is had here to the weather of a 24-hour period, although other intervals (from the instantaneous weather to that of a year, decade or greater number of years) are also quite frequently used; in the latter case we no longer deal with weather but with climate.

In every case, with the exception of the instantaneous weather, one encounters not only periodic (daily, yearly), but also aperiodic changes. Frequently these are called oscillations, although this term should be applied only to phenomena of periodic nature; aperiodic weather and climate variability should be regarded as their fluctuation. To avoid frequently encountered misunderstandings, it is useful to introduce the following terminology*: a) weather and climate oscillations should be regarded as referring to periodic variability of the corresponding weather of climatic indicators; b) the term weather and climate fluctuations should be applied to aperiodic variability in their indicators, observed within periods of different length (including many-year, secular periods); c) the term climate change should be applied only in the case of radical changes in the characteristics of the climatic regime, produced by changes in climate-formation processes. The onset of noncorrespondence of climatic indicators to previously established boundaries of climatic regions, requiring a new climate zoning scheme, may be regarded as external signs.

In discussing the relationship between weather, climate and atmospheric processes it is useful to recall the similarity of correlations between the weather and individual synoptic processes to those between the climate and general (planetary) circulation of the atmosphere. Although the direct effect of the surface increases markedly in the latter case, the climate-forming importance of the circulation is not lost. This becomes very clear in the study of climate fluctuations, since it is precisely the circulation which is the least stable of the primary and secondary climate-forming factors. It is hence natural to first compare the variability of the climatic regime with this factor, which varies within wide limits.

The relationship between the climatic regime and circulation was best expressed in their time by Voeikov /12—14, 75/ and Shaw /69/. The climatic classification of Alisov /1/ and the entire system of long-range weather forecasting developed by Multanovskii /37/ and his school were also well expressed. Similar views were published by Bergeron /54/, as well as in /16, 20, 55, 57, 71/.

The study of large-scale circulation processes is most frequently commenced with a classification, which reduces the entire variety of processes to a small number of groups, characterizing the generalized features of individual processes. This situation remains valid irrespective of whether reference is had to qualitative (from synoptic charts) or quantitative (indexes, values of gradients) characteristics. Classifications of large-scale processes for the purpose of long-range weather forecasting are quite well known /37, 52, 62/. The classification of Lir /32—35/ is based entirely on a dynamic approach. Others commence by classifying individual processes and gradually cover the entire northern hemisphere /3, 7, 8, 15—18, 22/.

In addition, systems were developed for classifying large-scale processes on the basis of their estimated intensity with the aid of various indexes

* The author wishes to recall similar and even more differentiated terminology in the English language: variations, oscillations, fluctuations, change of climate; and particularly in the German language: Kimaabweichung, Oszillation, Fluktuation, Klimaverwerfung, Klimaschwankungen, Klimawechsel, Klimaänderung.

/4, 9, 10, 23, 28/. One should also mention examples of regional classification of synoptic processes /5, 26/, work in this direction having been carried out recently in Poland /66, 74/ and Bulgaria /46/.

Such a large variety of classifications needs to be combined and a single classification of large-scale processes over the hemisphere created /17/. Although the known classifications are based on a variety of criteria, in the overwhelming majority of cases they reflect the relationship and interaction of zonal and meridional processes. Similar principles are also used for classifying atmospheric processes, undertaken by this author and his co-workers /22/. It is based on the zonality and meridionality of the cyclonic and anticyclonic trajectories.

We found it possible to reduce the entire variety of atmospheric processes observed daily over the northern hemisphere to 13 typical schemes, which are combined into four basic groups: zonal processes (without clearly expressed invasion and blocking actions); departure from zonality (invasions and blocking action along any one direction); meridional (several invasions and blocking actions); developing cyclonic action in polar regions, coinciding with predominance of zonal transport over a large part of the hemisphere. In addition, the detailed statistical study carried out by Chaplygina /47/ shows that the typical circulation mechanisms isolated by us are sufficiently objective.

A list of studies containing details of classifications of circulation mechanisms and results of investigations is given in /19, 20/.

The following are reasons for the use of trajectories of baric formations in defining classification criteria: 1) In the overwhelming majority of cases the direction of motion of well developed extratropical cyclones and anticyclones on the 1000 mb map reflects the configuration of isohypses in the middle troposphere and, consequently, describes the main transports and the general outlines of the circulation mechanisms over the entire hemisphere. 2) In their general outline these large-scale circulation processes prevail for several (usually 3—5) days. The external feature of the completion of operation of one mechanism and the onset of operation of another is the change in the location and directions of trajectories, which in the majority of cases is accompanied by sign reversal of baric fields over large areas. 3) The grouping of baric-formation trajectories on the 1000 mb map is in satisfactory agreement with the development and location of long waves in the zonal flow and with the blocking processes associated with them. 4) The relatively short series of available high-altitude maps of the northern hemisphere does not provide for sufficient statistical reliability for conclusions regarding the stability of isolated large-scale processes. A much longer series of surface (1000 mb) maps is available, and their use, in conjunction with the previously noted relationship between the trajectories and the principal transports, makes statistical analysis possible. At present a partial comparison was made of all the typical mechanisms using maps for the 850, 700, 500, 300 and 200 mb levels (Figures 1 and 2) (daily TsIP bulletins).

The onset of a new circulation mechanism, established from trajectories on the surface-level map, in the majority of cases is seen even clearer on 500 mb maps and the circulation type at this level usually prevails during the entire activity of this large-scale process.

The constant pressure maps of lower levels for the first day of a period are in satisfactory agreement with the composite ground-level kinematic map, but the agreement with higher-altitude maps is poorer. As time passes the agreement with higher-level maps gradually increases. This is frequently accompanied by deterioration in the agreement with the surface-level map. Schematically this is depicted as follows:

	850	700	500	300	200 mb
First day of period:	very good	good	satisfactory	poor	very poor
Last day of period:	very poor	poor	good	good	very good

i. e., the process develops from the bottom upward. Sometimes the situation is expressed as follows:

	850	700	500	300	200 mb
First day:	very poor	poor	good	very good	very good
Last day:	very good	good	satisfactory	poor	very poor

i. e., the process develops from the top down.

During winter in the polar region the sign of the field is most frequently opposite to that at surface level (low level anticyclone), but the general outline of the typical process over the hemisphere is usually retained.

5) At present a calendar of changes in elementary circulation mechanisms for 61 years (1899—1959) has been compiled.

The data used for statistical analysis were the frequency of some given typical process and the duration (in days) of its action. All the calculations and conclusions were made for the hemisphere as a whole and separately for each of six sectors. Data by months, seasons and years were used in each case. Temperature, precipitation and pressure data from the station network of the entire globe (according to the World Weather Record) were processed simultaneously.

We recall that the isolation of elementary circulation mechanisms was carried out in order to classify the circulation characteristics for the entire hemisphere; zonality and meridionality were determined from the predominant air fluxes (mainly in the middle troposphere) over the entire hemisphere, and not from surface-level winds over a limited area. These features of the present classification which differ from those of other systems should be remembered when using the data and results of this author's studies.

To facilitate regional studies the large-scale processes were classified with reference to geographic location of blocking actions and zonal transports, i. e., by sectors. However, the main classification principles were retained also in this case.

Fluctuation processes are studied by methods of mathematical statistics. These methods are also successfully applied to the study of fluctuations in the atmospheric circulation and in the climate. Smoothed and integral curves are extensively used for this purpose, although some scientists sharply disapprove of their use, because incorrect use of smoothed and particularly integral curves may result in the erroneous "discovery" of

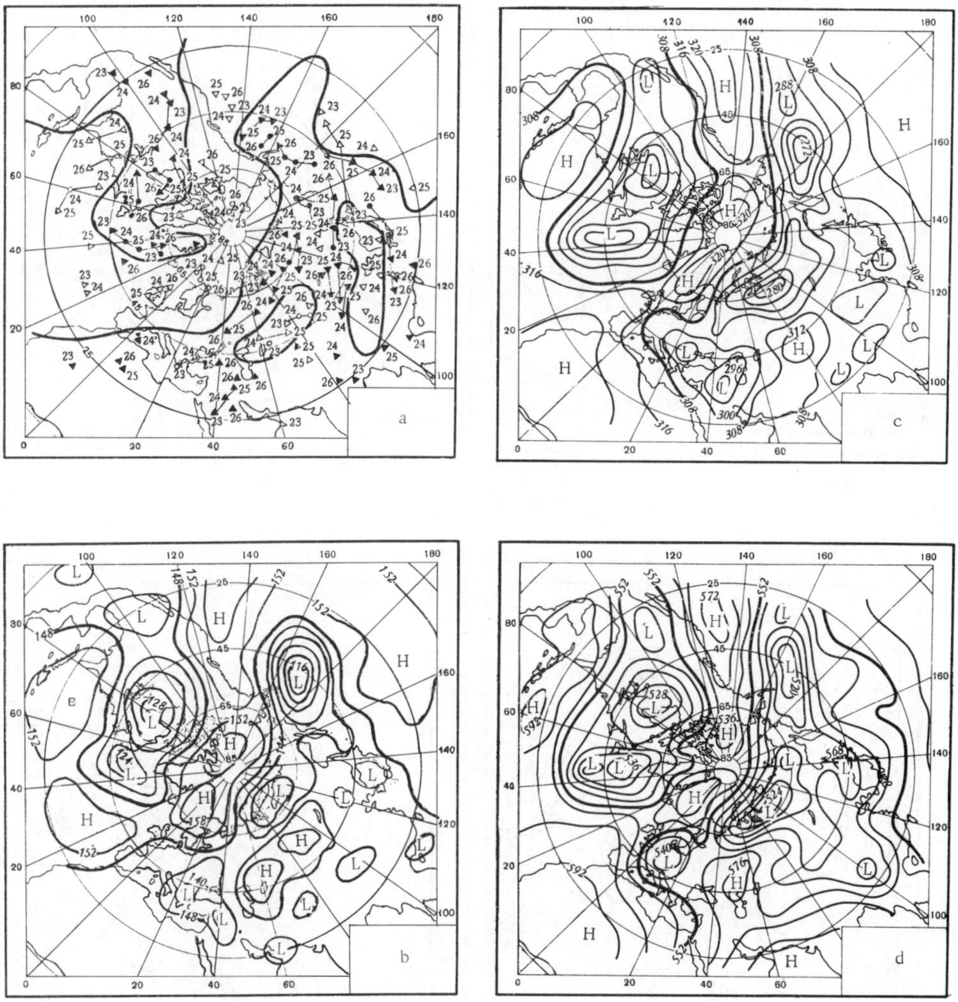

FIGURE 1. Type 9a elementary circulation mechanism on 23–26 May 1956:

a — composite kinematic map (trajectories of baric centers); b, c, d — constant-pressure maps of a central day of the period (24 May 1956); b — 850 mb; c — 700 mb; d — 500 mb. Dark triangles denote troughs; light triangles denote ridges.

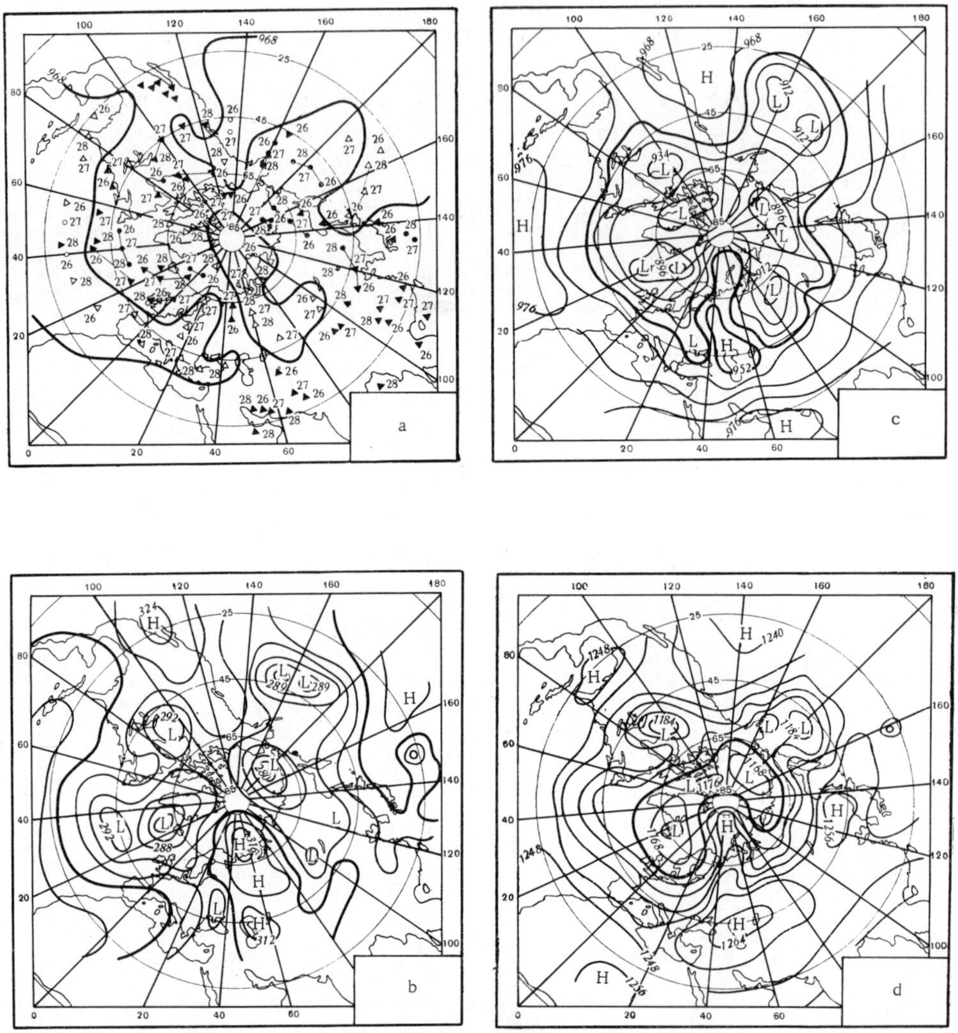

FIGURE 2. Type 4c elementary circulation mechanism on 26–28 July 1958:

a — composite kinematic map (trajectories of baric centers); b, c, d — constant-pressure maps of a central day of the period (27 July 1958); b — 700 mb; c — 300 mb; d — 200 mb.

nonexistent periods or cycles (the Sludskii effect) and in the loss of actual cycles. To avoid this, it is useful to construct smoothed curves with respect to periods of different duration, the latter being determined by the problem at hand. It is even more useful not to be limited by the formal graphical method, but to accompany it at all times by a physical verification and to check the conclusions by several methods. Conformance to these rules is mandatory in this case.

A great deal of data on statistical processing of climatic information for the problem of interest here can be found in the Proceedings of the Rome Symposium of the UNESCO and the World Meteorological Organization on climatic changes /56/. Using these data one may suggest the following sequence of operations, needed for reaching valid conclusions on climate fluctuations: a) construction of sliding averages; b) comparison of averages over different periods with one another and each of them in turn with the long-term average; c) study of orthogonal polynomials; d) determination and comparison of time intervals between pronounced features by different criteria; e) derivation of periodograms and correlograms.

Upon analysis of the pertinent data this author found that two circulation and climatic periods ("epochs"), perceptibly differing from one another, existed during the 20th century. Meridional circulation predominated during the first of the periods, while zonal currents prevailed during the second (Figure 3) (which corresponds to directions of motion of baric centers and primary transports over the hemisphere). This was accompanied by a similar variation in the climatic indexes. This can also be illustrated by the durations of operation of circulation mechanisms during the first and second epochs (Table 1).

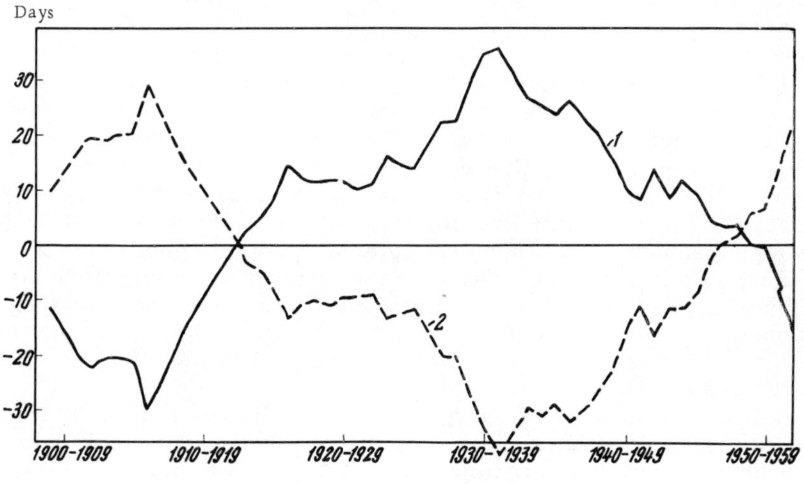

FIGURE 3. Long-term meridional and zonal components of atmospheric circulation over the northern hemisphere (10-year sliding averages):

1 — zonal circulation; 2 — meridional circulation.

TABLE 1. Duration of zonal and meridional circulation, in days

Epoch	Cold season		Warm season	
	zonal	meridional	zonal	meridional
I	133	1158	430	669
II	435	916	638	554
Difference	302	79	208	115

In spite of the marked fluctuations of the monthly average and even of yearly average values of temperature and precipitation in different circulation and climatic epochs, the average epochal values are different; they differ from the long-term average. The average values of climatic indexes during each epoch serve as its common "background" characteristic and are located at different levels relative to the long-term average.

The above can be illustrated by a number of curves of the average temperature and precipitation totals at different stations from data for a month, season, year. It should be noted first that their shape in all cases is close to that of zonal and meridional circulation components for these periods. Two average, but different levels of fluctuations of circulation components and climatic data for the first and second epochs are clearly seen /20, 21/.

Although fluctuations for individual years are appreciable, their values, which are the deviations from the average for the epoch, are found to be much smaller than deviations from the long-term average and this reduces the anomalous nature of these years (Figures 4 through 6). However, in every case the differences of the temperature as well as precipitation between the epochs were quite large, perceptibly changing the climatic properties of each epoch from the viewpoint of the needs of the national economy and its planning.

Changes in the circulation and climatic regimes caused by changes in the epoch have still another feature of great significance to forecasting. Here reference is had not only to extremely long-range forecasts, but to ordinary long-range predictions, i. e., for a month or season.

Many studies have shown that changes of epochs, as a rule, are accompanied by changes in pre-established geophysical and heliogeophysical relationships. Here not only the closeness of the correlation, but frequently also its sign changes. Consequently, determination of boundaries of circulation and climatic epochs acquires an even greater significance.

An important aspect of studies of climate fluctuations, determining to a large extent the prospects of developing a theory of the climate, are calculations of energy of large-scale processes. We make reference here to the use of climatological methods for such computations, i. e., for calculating the quantity of energy and of the alteration of its individual forms not for individual processes (typhoon, limited segment of a frontal surface, etc.) and not over average climatic quantities, but continuous series of energy calculations for all processes varying in their development, for the entire period during which the climate fluctuations are studied. In other words, series of computations of atmospheric-process energy are required which could be compared with the long-term variation of components of the atmospheric circulation and values of climatological elements.

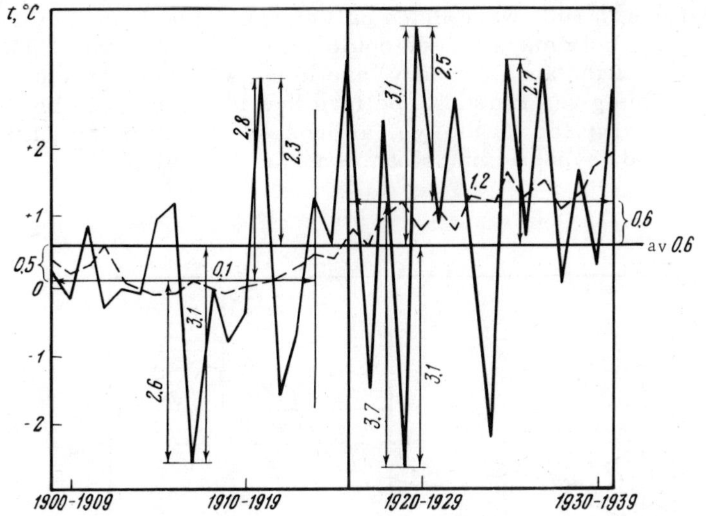

FIGURE 4. Comparison of average air temperatures in January at the Copenhagen station for the 1st and 2nd circulation epochs with the long-term average, and of the yearly fluctuation with epochal and long-term averages.

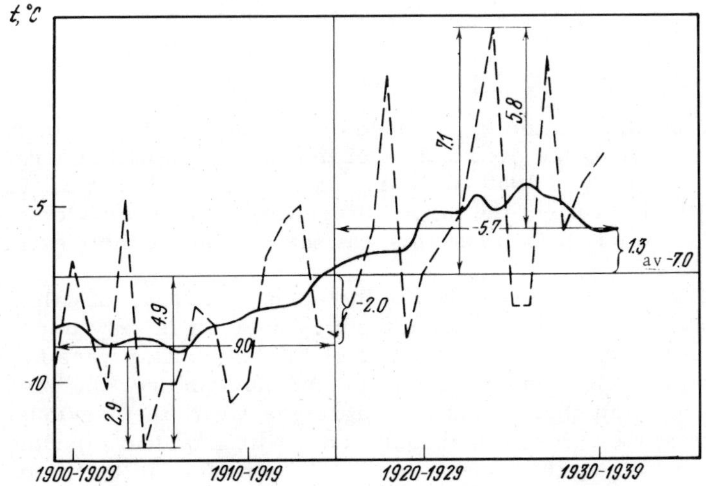

FIGURE 5. Comparison of average air temperatures in March at the Angmagssalik station for the 1st and 2nd circulation epochs with the long-term average, and of the yearly fluctuations with epochal and long-term averages.

We present two illustrations of the use of energy computations directly related to the circulation over the hemisphere. One of these are results of careful calculation of the absolute eddy, carried out by Newton and Palmen /68/ for 12 December 1957 (Figure 7). According to our system this day is classified as having type 12bw circulation.

The calculations in /68/ were made only for the American sector of the hemisphere, and both maps (of the composite-kinematic type 12bw circulation and of the absolute eddy) are in satisfactory agreement in this region. This gives reason to believe that it will be possible to calculate the energy for all the typical processes and for the entire period covered by synoptic maps of the northern hemisphere.

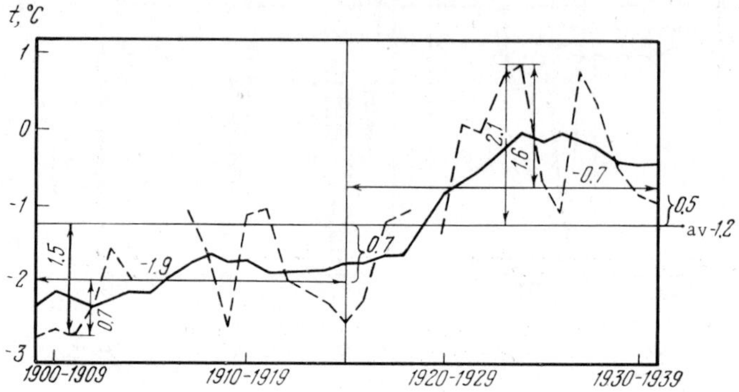

FIGURE 6. Comparison of yearly average air temperatures at the Angmagssalik station for the 1st and 2nd circulation epochs with the long-term average, and of the yearly fluctuations with the epochal and long-term averages.

The second illustration is provided by two maps of the northern hemisphere showing the distribution of internal and kinetic energies for the cold January of 1950 and warm January of 1961 (Figure 8a and b).* The maps clearly reflect the large differences in the energy distribution for the cold and warm months; this was the reason for their extreme nature.

The energy distribution shown in the maps is a result of calculations for each day, and not for averaged monthly data. It is very important that substantial changes in energy and of its distribution were found here when one typical elementary circulation mechanism over the hemisphere was replaced by another. This is in agreement with the previously cited comparison of the calculated absolute eddy with a typical circulation mechanism. This again verifies in principle the feasibility of comparative analysis of distribution and transformation of energy in different climatic epochs.

However, climatic calculations of various forms of energy of atmospheric large-scale processes have just started and no energy criteria for epoch changes were established. For this, other indicators must be used. The question therefore is: what are the criteria which should be used for this? The problem is complicated by the fact that the duration of epochs is not the same and varies widely. Hence boundaries cannot be established

* The calculations were carried out by L. V. Khmelevskaya, a graduate student of the Institute of Geography of the USSR Academy of Sciences.

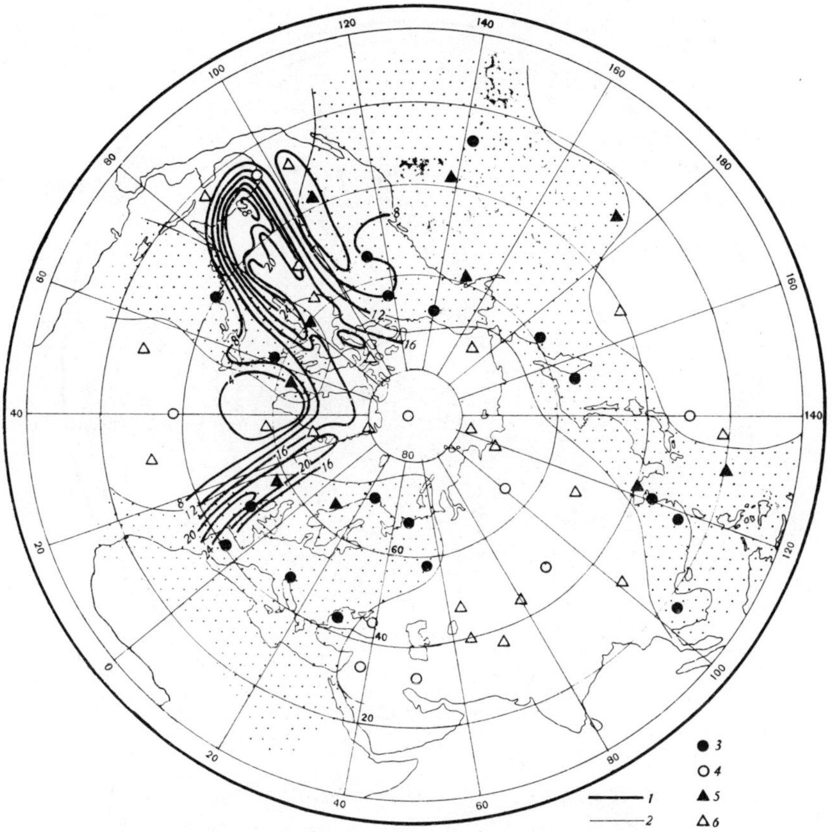

FIGURE 7. Superimposed maps of type 12 bw circulation mechanism for 11–14 December 1957 (after Dzerdzeevskii) and the distribution of the absolute eddy shear for 12 December 1957 (after Newton and Palmen):

1 — isolines with reference to eddy shear; 2 — demarcation lines between fields of cyclonic and anticyclonic activity; 3 — cyclonic centers; 4 — anticyclonic centers; 5 — troughs; 6 — ridges.

on the basis of any one given criterion, and the largest possible number of these should be used. The following may be regarded as fundamental: a) changes in the principal components of the general circulation, changes in the frequency and duration of meridional and zonal circulation components; b) changes in the closeness or sign of correlation between circulation processes and the weather as well as with other geophysical or heliogeophysical phenomena; departure from the nature of climatic "stability" as defined by Rosini /56/; c) use of "peaks" on the curve of different circulation and climatic indexes. Attention should be paid to their magnitude relative to the long-term data on the amplitude at the given location. As a rule, the relationship between them should be as follows: the magnitude of individual fluctuations relative to the epoch average should be smaller than the difference between the averages of both epochs; d) as a rule, transition from one epoch to another does not

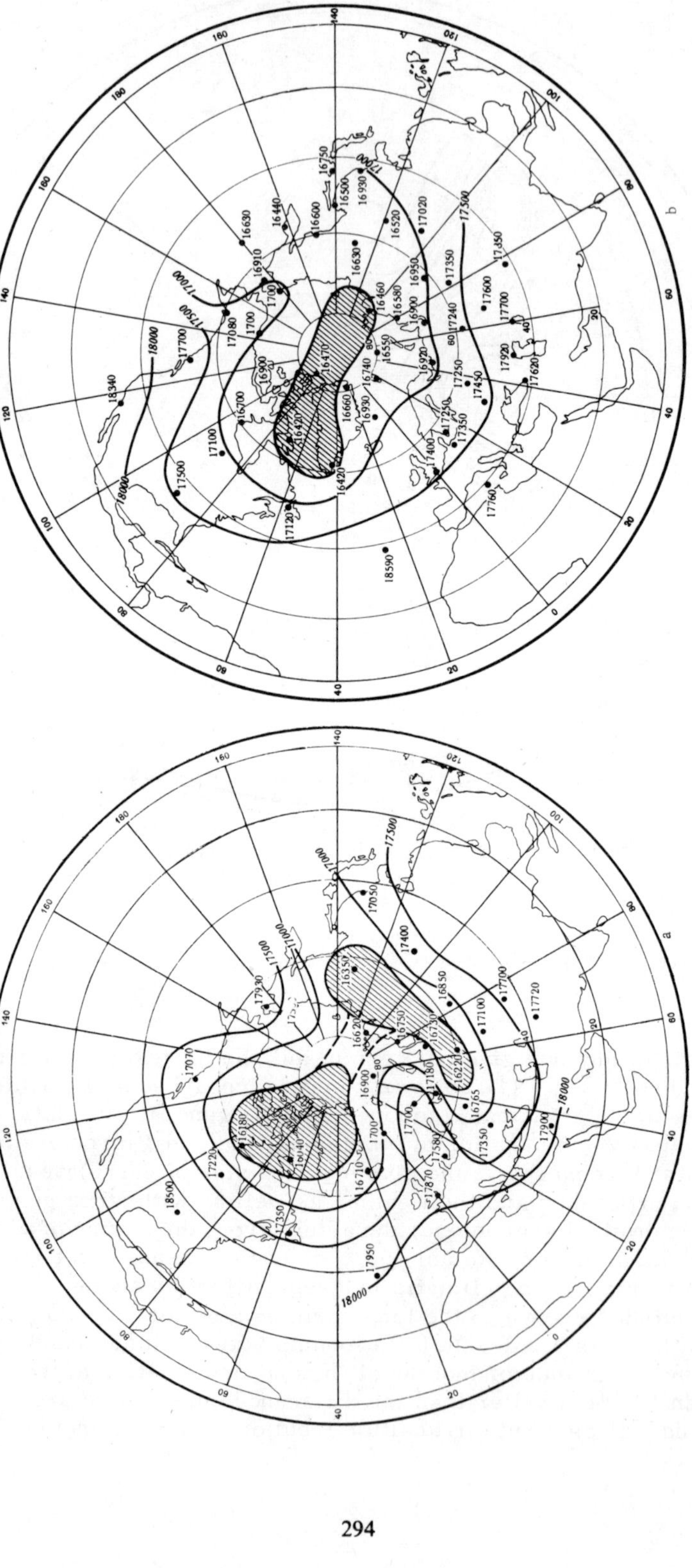

b

a

occur as a sharp jump (which is usually inherent to individual fluctuations), but it results in new qualitative characteristics of the following epoch as a result of a regular accumulation of quantitative changes; e) according to studies of Rubinshtein /41/, the day-to-day variability of air temperature changes markedly from one epoch to another; consequently, this quantity can also be used for our purposes; f) the method for determining the warmness and moderateness of climate suggested by Bailey deserves attention. He made use of effective temperatures. Also in this case allowance should be made for the geographic location of the station (marine climates with their smaller oscillation amplitudes are more sensitive to changes in heat influxes). Here the reaction of vegetation to changes in the effective temperature is also more pronounced; g) the rapid response of vegetation (particularly forest growths) to climatic fluctuations is observed not only in warm marine climates. Substantiated examples of this can be found in /43, 44, 59, 63, 64, 67/. All these references point to a sharp, climatic-fluctuation induced latitudinal shift in the extent of various tree species, and consequently to a perceptible change in the species composition of the forest. Nevertheless, this slowly acting feature can be used primarily for subsequent analysis and description of past epochs; h) much use can be made of data on many other natural phenomena, particularly those with seasonal variations. Among these are: dates of freezing and thawing of lakes and rivers; the rates at which these phenomena occur, defined as the difference between the time of the first appearance of slush pools and the formation of a stationary ice cover, and as the difference in time between similar freezing (or thawing) phases at different latitudes; dates of formation and thawing of the snow cover; first and last frosts; duration of the frost-free period of each season as a whole; depth of soil freezing, etc. In a number of cases such indirect information may be very useful, since they yield integral and quite stable characteristics.

The feasibility should be considered and examined of using for our purposes yet another index: the time change in the level of the World's ocean in its different parts. Periodic and aperiodic changes in level were observed, depending on changes in atmospheric pressure /6, 24, 60/. The study of periodic level variations commensurable with respect to time with circulation and climatic epochs is quite expedient, since this may yield a new representative index.

Results obtained by processing long-term data, particularly of quantitative observations, are frequently improved by using relative values, expressed as deviations from the long-term "average" or from the corresponding yearly values, instead of the directly measured absolute values. For example, precipitation data may be found much more representative when their seasonal distribution over different epochs is not compared by sums of absolute amounts during months or ten-day periods, but by the percentage of these quantities to the yearly sum.

We now return to the question of the extent to which the above comments aid in solving the main problem, namely, forecasting the regime of the following epoch. Here it must be remembered that at present not only a physically substantiated method, but even systematized and generalized methods of such forecasts are not available. In addition, methods of ordinary long-range (for month and season) forecasts cannot be regarded as fully tested and generally accepted. Nevertheless, they are usually issued, and this is sensible, since the method being used is simultaneously developed and perfected.

There is some basis for expecting that a substantiation in principle of extra long-range forecasts (for the following epoch) will come earlier than that for ordinary long-range forecasts. This contradictory assertion is justified (in a given case) by the greater weight of the statistical-stochastic methods, incomparably greater duration of the forecasting period and the feasibility of determining the nature of the epoch by its start, as well as by the fact that extra long-range forecasts can predict only background information. The success in developing and substantiating methods of extra long-range forecasts, as well as the substantiation of their practical feasibility, depends on the development of a rigorous theory of the general circulation of the atmosphere and of a theory of the climate. Studies of climate fluctuations for the purpose of determining their genesis will definitely aid in developing the theory of the climate and by virtue of this also of methods of extra long-range forecasts.

The main difficulties here consist in the fact that increasing the forecasting time to several decades requires particular attention to the role of solar activity (as a direct agent) and of its variations. This means that forecasting of solar activity becomes the primary objective. Here two kinds of difficulties are encountered: a) forecasting processes at the sun proper, prediction of fluctuations and changes in its activity over short and long periods; b) determining the mechanism of transmission of these solar activities to the earth, or, more precisely, the mechanism by which these are received by the terrestrial atmosphere.

Both these difficulties are formidable, but nevertheless some success can be claimed in both directions. As is known attempts are made to forecast solar activity and its effect. Mechanisms of heliogeophysical couplings are sought just as persistently. Marked progress in this work was made during the IGY and IQSY, and in conjunction with the use of artificial satellites. Determination of the mechanism of heliogeophysical couplings can be perceptibly advanced by comparative analysis of atmospheric circulation over both hemispheres.

The author made no use of the generally utilized term "general" (circulation) for the following reasons: the term "general circulation of the atmosphere" should describe the circulation in the entire terrestrial atmosphere, i.e., over the entire globe and in the entire thickness of the atmosphere. The term "general" does not fully satisfy this condition; however, the term "general circulation of the atmosphere" came into excessively widespread use even when reference is had to synoptic processes over relatively small areas.

It would have been useful to assume the following terms: a) macroscopic processes (over some area); b) atmospheric circulation (over the northern or southern hemisphere); c) planetary circulation (of the terrestrial atmosphere).

The importance of comparative study of atmospheric circulation over both hemispheres is due to the tremendous differences in the nature of their surface and of the latter's effect on the atmosphere. Hence circulation processes common to both hemispheres and reflecting regularities of the planetary circulation are produced by factors which do not act on the atmosphere by means of the surface or not only by means of it, but directly on the terrestrial atmosphere, primarily on its upper layers.

On the basis of these considerations circulation processes which are similar in both hemispheres should be sought near the terrestrial surface, i.e., in those places where its direct effect is clearly demonstrated and where the observed demonstrations of the mechanism of planetary circulation evidently reflect powerful effects, which are easier to detect and study.

Comparative analysis of circulation above the hemispheres is made difficult by the insufficient amount of data on the southern hemisphere and by the fact that data which were collected prior to the IGY and IQSY were not systematic. At the same time the need for discovering frequently occurring (typical) processes, reflecting sufficiently stable features of planetary circulation, is evident. For this purpose it was necessary to analyze and correlate all the scattered results for all the previous years in the southern hemisphere.

The most complete and systematized data are those of Astapenko /2/, who also classified circulation processes in high latitudes of the southern hemisphere on the basis of the same principle, namely, the relationship between the meridional and zonal circulation components.

Comparison of typical circulation models /21/ made it possible to establish similarity between a number of circulation mechanisms in both hemispheres and thus to verify the existence of certain planetary-circulation processes.

It is quite evident that one cannot expect an unequivocal answer to the question whether circulation is formed solely due to surface effects or only due to other effects. The fact that the surface has a direct effect on processes up to high altitudes was shown, for example, by Pogosyan /39/, who analyzed the formation of high ridges. Kessenikh and Bulatov /31/ studied the well defined continental effect in the location of the F_2 layer and in the electron concentration at its level. Nadubovich /38/ established the influence of shores on the location of Aurora Borealis. One can also cite other studies verifying these conclusions.

However, a study of the principal regularities of the planetary circulation which do not entirely depend on the direct effects of the surface should aid in obtaining a more reliable estimate of the contribution of various factors to the development of the circulation mechanism and, consequently, should facilitate the construction of the theory of planetary circulation of the terrestrial atmosphere and of the theory of the climate.

Here one should again emphasize the tremendous importance which is acquired in this context by the study of fluctuations of the planetary circulation and of the climate; it is precisely these which will make possible a more graphical determination of the variable effects on the atmosphere, whereas the surface is conservative and its effects are stable.

Many studies were made recently of heliogeophysical relationships. The reader is referred to the two books by Eigenson /48, 49/ and monographs by Rubashev /40/ and Sazonov /42/. One should mention also several other investigations in this field, involving attempts to use relationships between atmospheric processes and solar activity for weather forecasting /53/, as well as to clarify the mechanism by which solar effects are received by the terrestrial atmosphere. It should be noted that in these investigations increasing attention is being paid to the earth magnetic field, some authors regarding this relationship as functional while others consider it purely correlational. Flohn /61/, upon establishing that Aurora Borealis coincides with axes of high ridges, concludes that the pressure distribution in the troposphere is controlled by the external or internal magnetic field and allows for the possibility of a relationship between the daily periodicity of the stray electric ring current in the ionosphere with the asymmetry of the tropospheric polar eddy. Willett /72/ established a relationship between the maximum frequency of the polar anticyclogenesis and blocking action with oscillations in solar activity, but he assumes that differences in the direction of invasions depend on the location of magnetic poles and lines of forces and the state of the magnetic field, which in turn depends on ultraviolet radiation. Barta /51/ established the chain: corpuscular radiation—magnetic field—atmospheric circulation. Changes in atmospheric pressure are related to the corpuscular and ultraviolet radiation in /58/. Williams /73/ postulates the following sequence of solar-effect transmission: corpuscular radiation—air current obeying the Faraday law—increasing condensation—increasing wind strength—increasing evaporation from oceans—amplifying the electrical charge of water, which again increases the evaporation. Sugiura /70/ studied sinusoidal oscillations of magnetic waves in the region of Aurora Borealis which result in the formation of hydromagnetic waves and heating up of the upper atmosphere. The possibility of heating the upper atmosphere by hydromagnetic waves is also postulated by Ershkovich /25/, although Kulkarni /65/ attributes the reason to heating of the ozone layer.

The relationship between atmospheric circulation and the magnetic field is described by Yakovlev /50/. One should also mention the two studies by Kessenikh /29, 30/ of the possible coupling mechanism between the synoptics of the ionosphere and troposphere, and of the relationship between ionospheric dynamics and the dynamics of the atmosphere as a whole. The paper by Kazimirovskii /27/ on the general circulation of the ionosphere is worthy of mention. In conclusion there exist several studies by Rubashev (1958—1960), who demonstrated statistically the satisfactory correlation between the state of the magnetic field and our typical circulation mechanisms.

There is thus sufficient basis for the claim that the search for the mechanism by which the terrestrial atmosphere incorporates variable solar effects is rapidly developing. Obviously, the relationships now being established cannot by far be regarded as reliable. Nevertheless, the results are promising and encourage studies in this direction, the more so since a certain regularity is observed which does not fit in with some of these relationships. Reference is had here to the comments at the start of this article regarding heliogeophysical and tropospheric relationships, which differ when passing from one circulation and climatic epoch to another. This was pointed out by Vitel's /11/, Sleptsov-Shevlevich /45/ and others. To some extent the earlier "law of accentuation" of baric centers, due to Fedorov and Vize, has a similar basis.

Consideration of fluctuations of planetary circulation can be aided by the recent work of Maksimov and Sleptsov /36/ on 11-year variations in average atmospheric pressure. They present two maps of the spatial distribution of these variations in the northern and southern hemispheres.

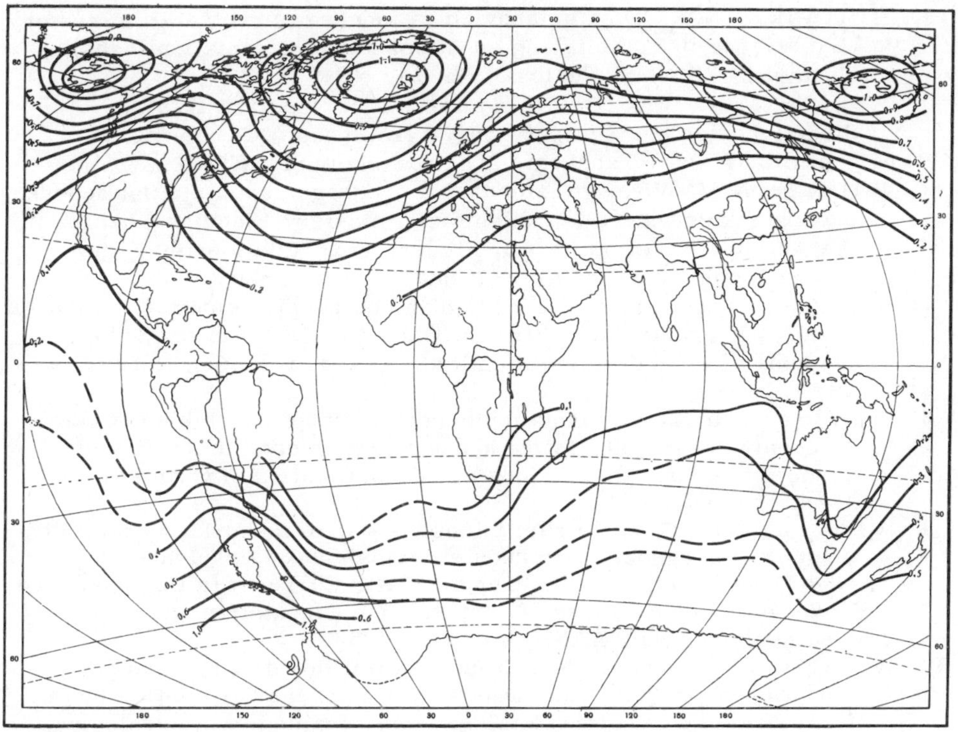

FIGURE 9. Distribution of amplitudes of 11-year variations in the average atmospheric pressure (after Maksimov and Sleptsov).

When these data are plotted on the World map, one easily sees the common planetary nature of the features under study (Figure 9). It should also be mentioned that the location of axes of wave distortions of isolines coincides satisfactorily with locations of polar invasions and blocking actions, used when classifying the circulation mechanisms of both hemispheres. This again emphasizes the relationship of both these systems with the planetary circulation and shows that it is permissible to extend conclusions regarding heliogeophysical relationships obtained earlier for the northern hemisphere to the entire terrestrial atmosphere.

References

1. Alisov, B. P., O. A. Drozdov, E. S. Rubinshtein, I. A. Berlin, and V. M. Mikhel'. Kurs klimatologii (A Course in Climatology). — Gidrometeoizdat. Part 1. 1952; Part 2. 1954.
2. Astapenko, P. D. Atmosfernye protsessy v vysokikh shirotakh yuzhnogo polushariya (Atmospheric Processes in High Latitudes of the Southern Hemisphere). — Izdatel'stvo AN SSSR. 1960.
3. Baidal, M. Kh. Kompleksnyi makrotsirkulyatsionnyi metod dolgosrochnykh prognozov pogody (Complex Large-Scale Circulation Methods of Long-Range Weather Forecasting). — Gidrometeoizdat. 1961.

4. Belinskii, N. A. Opyt ustanovleniya indeksa tsirkulyatsii atmosfery (Experience in Establishing an Atmospheric Circulation Index). — Trudy NIU GUGMS, Series 5, No. 14. 1946.

5. Bugaev, V. A., V. A. Dzhordzhio, E. M. Kozik, et. al. Sinopticheskie protsessy Srednei Azii (Synoptic Processes in Central Asia). — Tashkent, AN Uzbekskoi SSR. 1957.

6. Butorin, N. V. Vekovye izmeneniya srednego urovnya Atlanticheskogo okeana i ikh svyaz' s tsirkulyatsiei atmosfery (Secular Changes in the Average Level of the Atlantic and their Relationship with Atmospheric Circulation). — Izdatel'stvo AN SSSR. 1960.

7. Vangengeim, G. Ya. Materialy dlya postroeniya sinopticheskikh metodov kharakteristiki klimata (Data for Constructing Synoptic Methods of Climate Classification). — Izvestiya GGO, Nos. 2—3. 1933.

8. Vangengeim, G. Ya. Opyt primeneniya sinopticheskikh metodov k izucheniyu i kharakteristike klimata (Experience in the Use of Synoptic Methods in the Study and Classification of Climate). — Moskva. 1935.

9. Vangengeim, G. Ya. Uchet intensivnosti vozdushnykh techenii dlya diagnoza i prognoza osobennostei atmosfernoi tsirkulyatsii i pogody (Allowing for the Strength of Air Currents in Diagnosis and Forecasting of Features of Atmospheric Circulation and of the Weather). — Trudy AANII, Vol. 253. 1963.

10. Vitel's, L. A. Ob opredelenii indeksa tsirkulyatsii po dannym sinopticheskogo kataloga (Determination of the Circulation Index from Data of a Synoptic Catalog). — Meteorologiya i gidrologiya, No. 5. 1947.

11. Vitel's, L. A. O vozmozhnoi prichine izmenenii solnechno-atmosfernykh svyazei (A Possible Cause of Changes in Relationships Between the Sun and the Atmosphere). — Meteorologiya i gidrologiya, No. 7. 1960.

12. Voeikov, A. I. Sootnoshenie (korrelyatsiya) mezhdu osadkami v Gavanne i na yuge Velikobritanii (Relationship (Correlation) Between Precipitation in Havana and in the South of Great Britain). — Meteorologicheskii vestnik, Nos. 8—9. 1914.

13. Voeikov, A. I. Klimaty zemnogo shara, v osobennosti Rossii (Climates of the World, in Particular of Russia). — St. Peterburg. 1884; Izbrannye sochineniya, Vol. 1. Izdatel'stvo AN SSSR. 1948.

14. Voeikov, A. I. Atmosfernaya tsirkulyatsiya, raspredelenie atmosfernogo davleniya, vetrov i osadkov po poverkhnosti zemnogo shara (Atmospheric Circulation and Distribution of Atmospheric Pressure, Winds and Precipitation over the Terrestrial Surface). — Izbrannye sochineniya, Vol. 2, Izdatel'stvo AN SSSR. 1949.

15. Girs, A. A. Tipovye kharakteristiki osnovnykh raznovidnostei form atmosfernoi tsirkulyatsii v kholodnoe vremya goda (Features of the Principal Varieties of Forms of Atmospheric Circulation during the Cold Season). — Problemy Arktiki, No. 7. 1959.

16. Girs, A. A. Osnovy dolgosrochnykh prognozov pogody (Principles of Long-Range Weather Forecasts). — Gidrometeoizdat. 1960.

17. Girs, A. A. O sozdanii edinoi klassifikatsii makrosinopticheskikh protsessov severnogo polushariya (Creation of a Unified Classification of Large-Scale Synoptic Processes of the Northern Hemisphere). — Meteorologiya i gidrologiya, No. 4. 1964.
18. Dzerdzeevskii, B. L. Tipizatsiya atmosfernykh protsessov v severnom polusharii kak metod kharakteristiki sezonov (Classification of Atmospheric Processes in the Northern Hemisphere as a Way of Characterizing Seasons). — Doklady GOIN, No. 42. 1945.
19. Dzerdzeevskii, B. L. Mnogoletnyaya izmenchivost' obshchei tsirkulyatsii atmosfery i klimata kak osnova klimaticheskogo prognoza (Long-Term Variability of the General Circulation of the Atmosphere and of the Climate as a Basis of Climatic Forecasting). — Trudy I nauchnoi konferentsii po obshchei tsirkulyatsii atmosfery. Gidrometeoizdat. 1962.
20. Dzerdzeevskii, B. L. Sinoptiko-klimatologicheskaya kharakteristika severnogo polushariya (A Synoptic-Climatic Description of the Northern Hemisphere). — Trudy Vsesoyuznogo nauchnogo meteorologicheskogo soveshchaniya, Vol. IV. Gidrometeoizdat. 1963.
21. Dzerdzeevskii, B. L. Analiz mnogoletnei izmenchivosti kharaktera obshchei tsirkulyatsii atmosfery i pokazatelei klimaticheskikh elementov na poverkhnosti zemnogo shara (Analysis of the Long-Term Variability of the Nature of the General Circulation of the Atmosphere and of Indexes of Climatic Elements on the Terrestrial Surface). — Geofizicheskii sbornik, No. 14. Nauka. 1964.
22. Dzerdzeevskii, B. L., V. M. Kurganskaya, and Z. M. Vitvitskaya. Tipizatsiya tsirkulyatsionnykh mekhanizmov v severnom polusharii i kharakteristika sinopticheskikh sezonov (Classification of Circulation Mechanisms of the Northern Hemisphere and Characterization of Synoptic Seasons). — Gidrometeoizdat. 1946.
23. Dzerdzeevskii, B. L. and A. S. Monin. Tipovye skhemy obshchei tsirkulyatsii atmosfery i indeks tsirkulyatsii (Typical Schemes of the General Circulation of the Atmosphere and the Circulation Index). — Izvestiya AN SSSR, seriya geofizicheskaya, No. 6. 1954.
24. Duvanin, A. I. Kolebaniya urovnya morya (Variations in Sea Level). — Leningrad. 1951.
25. Ershakovich, A. I. O vozmozhnosti nagreva verkhnei atmosfery gidromagnitnymi volnami (On the Possibility of Heating the Upper Atmosphere by Hydromagnetic Waves). — Izvestiya AN SSSR, seriya geofizicheskaya, No. 3. 1963.
26. Ismailov, I. A. Tipizatsiya sinopticheskikh protsessov, obuslovlivayushchikh vtorzhenie kholoda v Azerbaidzhan (Classification of Synoptic Processes Responsible for Invasions of Cold into Azerbaidzhan). — Doklady AN Azerbaidzhanskoi SSR, 20(5). 1964.
27. Kazimirovskii, E. S. K voprosu ob obshchei tsirkulyatsii v ionosfere (Concerning the General Circulation in the Ionosphere). — Geomagnetizm i aeronomiya, 2(6). 1962.
28. Kats, A. L. Sezonnye izmeneniya obshchei tsirkulyatsii atmosfery i dolgosrochnye prognozy (Seasonal Changes in the General Circulation of the Atmosphere and Long-Range Forecasts). — Gidrometeoizdat. 1960.

29. Kessenikh, V. N. Vozmozhnye mekhanizmy svyazi mezhdu sinoptikoi ionosfery i troposfery (Possible Mechanisms of the Relationship Between the Synoptics of the Ionosphere and Troposphere). — Issledovanie ionosfery i meteorov. Sbornik statei, No. 2. Izdatel'stvo AN SSSR. 1960.
30. Kesennikh, V. N. Nekotorye voprosy svyazi dinamiki ionosfery s dinamikoi atmosfery v tselom (Some Problems of the Relationship Between the Dynamics of the Ionosphere and the Dynamics of the Atmosphere as a Whole). — Trudy Fizichesko-tekhnicheskogo instituta, Vol. 41. 1962.
31. Kesennikh, V. N. and N. D. Bulatov. Kontinental'nyi effekt v geograficheskom raspredelenii elektronnoi kontsentratsii sloya F_2 (The Continental Effect in the Geographical Distribution of the Electron Concentration of the F_2 Layer). — Doklady AN SSSR, 45(250). 1944.
32. Lir, E. S. Printsipy klassifikatsii sinopticheskikh protsessov (Principles of Classifying Synoptic Processes). — Zhurnal geofiziki, 1(1—2). 1931.
33. Lir, E. S. Osnovnye cherty zimnego sinopticheskogo protsessa v severnom polusharii (Basic Outlines of the Winter Synoptic Process in the Northern Hemisphere). — Zhurnal geofiziki, 2(3—4). 1932.
34. Lir, E. S. Osnovnye cherty atmosfernoi tsirkulyatsii v teplom polugodii (Principal Features of the Atmospheric Circulation in the Warm Season). — Zhurnal geofiziki, 3(4). 1933.
35. Lir, E. S. Tipy sezonnykh tsirkulyatsii atmosfery nad Evraziei i Atlantikoi (Types of Seasonal Circulations of the Atmosphere over Eurasia and the Atlantic). — Meteorologiya i gidrologiya, Nos. 1, 3—7. 1936.
36. Maksimov, I. V. and B. A. Sleptsov. K izucheniyu odinnadtsatiletnikh variatsii atmosfernogo davleniya v Antarktike (On the Study of Eleven-Year Variations of Atmospheric Pressure in the Antarctic). — Byulleten' SAE, No. 43. 1963.
37. Mul'tanovskii, B. P. Osnovy sinopticheskogo metoda dolgosrochnykh prognozov pogody (Fundamentals of the Synoptic Method of Long-Range Weather Forecasts). — Moskva. 1933.
38. Nadubovich, Yu. A. Vliyanie beregovogo effekta na geograficheskoe raspredelenie dug i polos siyanii po nablyudeniyam russkoi polyarnoi ekspeditsii 1900—1902 gg (Effect of the Shoreline on the Geographic Distribution of Arcs and Streamers of Auroras from Observations of the Russian Polar Expedition in 1900—1902). — Geomagnetizm i aeronomiya, 3(3). 1963.
39. Pogosyan, Kh. P. Obshchaya tsirkulyatsiya atmosfery (The General Circulation of the Atmosphere). — Gidrometeoizdat. 1958.
40. Rubashev, B. M. Problemy solnechnoi aktivnosti (Problems of Solar Activity). — Nauka. 1964.
41. Rubinshtein, E. S. Mezhdusutochnaya izmenchivost' temperatury vozdukha v SSSR (Day-to-Day Variability of Air Temperature in the USSR). — Trudy Vsesoyuznogo nauchnogo meteorologicheskogo soveshchaniya, Vol. 4. Gidrometeoizdat. 1963.
42. Sazonov, B. I. Vysotnye baricheskie obrazovaniya i solnechnaya aktivnost' (High-Altitude Baric Formations and Solar Activity). — Gidrometeoizdat. 1964.

43. S k r y a b i n, M. P. Vekovye tsikly prirodnykh uslovii i borovaya lesnaya rastitel'nost' lesostepi (Secular Cycles of Natural Conditions and the Pine Forests of the Forest-Steppe Region). — Trudy Voronezhskogo zapovednika, Vol. 3. 1949.
44. S k r y a b i n, M. P. Smeny porod v Usmanskom boru i zadachi zapovednika po sokhraneniyu i uluchsheniyu lesnogo massiva (Changes in Species in the Usman' Pine Forest and Aims of the National Forest [Administration] in Maintaining and Improving Forest Areas). — Trudy Voronezhskogo gosudarstvennogo zapovednika, Vol. 7. 1957.
45. S l e p t s o v - S h e v e l e v i c h, B. A. Korotkie vozmushcheniya solnechnoi aktivnosti i vozmozhnye prichiny neodnoznachnosti ikh svyazi s kolebaniyami gidrometeorologicheskikh elementov (Short Perturbances in Solar Activity and Possible Causes for the Lack of Unequivocal Correspondence Between them and Fluctuations of Hydrometeorological Elements). — Problemy Arktiki i Antarktiki, No. 14. 1963.
46. Tipove vreme nad B''lgariya (Standard Time in Bulgaria) (joint authorship).— Trudy na Instituty po khidrologiya i meteorologiya, Vol. 7. 1960.
47. C h a p l y g i n a, A. S. Statisticheskii analiz cheredovaniya tipov tsirkulyatsii atmosfery (Statistical Analysis of the Alternation of Atmospheric Circulation Types). — Izvestiya AN SSSR, seriya geofizicheskaya, No. 12. 1961.
48. E i g e n s o n, M. S. Ocherki fiziko-geograficheskikh proyavlenii solnechnoi aktivnosti (Outlines of Physicogeographic Manifestations of Solar Activity). — L'vov, Izdatel'stvo L'vovskogo gosudarstvennogo universiteta. 1957.
49. E i g e n s o n, M. S. Solntse, pogoda i klimat (Sun, Weather and Climate). — Gidrometeoizdat. 1963.
50. Y a k o v l e v, B. A. Svyaz' magnitnykh bur' s atmosfernoi tsirkulyatsiei (Relationship Between Magnetic Storms and the General Circulation).— Trudy I nauchnoi konferentsii po obshchei tsirkulyatsii atmosfery. Gidrometeoizdat. 1962.
51. B a r t a, G. Über die Sekularbewegung des magnetischen Zentrums und der magnetischen Pole der Erde. — Z. Geophys., 24(4/5). 1961.
52. B a u r, F. Einführung in die Grosswetterforschung. 2 Aufl. 1942.
53. B a u r, F. Grosswettervorhersagen durch extraterrestrische Sonnenbeobachtungen. — Naturwiss. Rundschau, 16(7). 1963.
54. B e r g e r o n, T. Richtlinien einer dynamischen Klimatologie. — Meteorol. Z., Vol. 47. 1930.
55. B r e z o w s k y, H. H., H. F l o h n, and P. H e s s. Some Remarks on the Climatology of Blocking Action. — Tellus, 3(3). 1951.
56. Changes of Climate. — Proc. Rome Sympos. UNESCO and World Meteorol. Organ. 1963.
57. D a m m a n n, W. Klimatologie der atmosphärischen Störungen über Europa. — Erdkunde, 14(3). 1960.
58. D u e l l, B. and G. D u e l l. The Behavior of Barometric Pressure during and after Solar Particle Invasions and Solar Ultraviolet Invasions. — Smithsonian Misc. Collect., 110(8). 1948.
59. E r k a m o, V. On Plant-Biological Phenomena Accompanying the Present Climatic Change. — Fennia, Vol. 75. 1952.

60. Fairbridge, R.W. Mean Sea Level Related to Solar Radiation during the Last 20,000 Years. — In: Changes of Climate. UNESCO. 1963.
61. Flohn, H. Die Zirkulation der Atmosphäre in den Polargebieten. — Polarforschung, 3(1). 1951.
62. Hess, P. and H.I. Brezowsky. Katalog der Grosswetterlagen Europas. — Ber. Dtsch. Wetterdienstes US-Zone, No. 33. 1952.
63. Keränen, J. On Temperature Changes in Finland during the Last Hundred Years. — Fennia, Vol. 75. 1952.
64. Keränen, J. Die Temperaturschwankungen in Finnland seit dem Jahre 1830. — Meteorol. Abhandl., 2(4). 1955.
65. Kulkarni, R.H. Relation Between Atmospheric Ozone and Geomagnetic Disturbances. — Nature, 198(4886). 1963.
66. Lityńska, Z., J. Lityński, I. Morozowska, B. Osuchowska, and M. Wodzińska. Typy cyrkulacji atmosferycznej w Europie i ich zastosowanie przy opracowywaniu prognoz 5-dniowych dla Polski. Cz. I. — Wiadom. sluzby hydr. i meteorol., 4(3). 1958.
67. Mikola, P. The Effect of Recent Climatic Variations on Forest Growth in Finland. — Fennia, Vol. 75. 1952.
68. Newton, C. and E. Palmen. Kinematic and Thermal Properties of a Large Amplitude Wave in the Westerlies. — Tellus, 15(2). 1963.
69. Shaw, N. Manual of Meteorology, Vol. II. — Cambridge. 1936.
70. Sugiura, M. Evidence of Low-Frequency Hydromagnetic Waves in the Exosphere. — J. Geophys. Res., 66(12). 1961.
71. Tsuchiya, I. An Analysis on the Relationship Between General Circulation and Climatic Fluctuation. Pt. I. — J. Meteorol. Soc. Japan, 41(5). 1963.
72. Willett, H.C. The Patterns of Solar-Climatic Relationships. — Ann. N.Y. Acad. Sci., Vol. 95, Art. 1. 1961.
73. Williams, D. Sunspot Cycle Correlations. — Ann. N.Y. Acad. Sci., Vol. 95, Art. I. 1961.
74. Wodzińska, M.B., B. Osuchowska, I. Morozowska, J. Lityński, J. and Z. Lityńska. Typy cirkulacji atmosferycnej w Europie z zastosowaniem do pizciodniowych prognoz pogody dla Polski, Cz. II. — Wiadom. sluzby hydrolog. i meteorol., 8(5). 1961.
75. Wojeikow, A. Die Winde Nordasiens. — Z. Österreich. Ges. Meteorol., 7(6). 1872.

A. A. Girs

EPOCHAL VARIATIONS IN ATMOSPHERIC CIRCULATION IN THE NORTHERN HEMISPHERE AND CORRESPONDING CHANGES IN CHARACTERISTICS OF THE ATMOSPHERE AND HYDROSPHERE

The article examines the principal results of studies of epochs observed in the development of the atmospheric circulation of the northern hemisphere during the past 75 years. In addition, data are presented on certain features of the state of the atmosphere and hydrosphere during each epoch and predictions are made on the nature of the forthcoming epoch. The role of solar activity in causing the appearance of epochs is discussed in great detail. The author's point of view on this question is also formulated.

To explain the causes of climate fluctuations in various regions of the earth, and also to develop methods of long-range and extra long-range hydrometeorological forecasts, one must investigate the regularities in the development of the general circulation of the atmosphere for both relatively short and extended periods of time.

The leading Soviet scientist, G. Ya. Vangengeim, was the pioneer in the study of the general circulation with this aim in mind. This classification of large-scale synoptic processes includes 26 types of elementary synoptic processes /4/, combined somewhat later /5, 6/ into three forms of atmospheric circulation: western *(W)*, eastern *(E)* and meridional *(C)*.

The present author showed in 1947 /25/ that these forms, established from surface-level data of the Atlantic-Eurasiatic sector, have corresponding radically different high large-scale deformation fields also in the entire thickness of the troposphere. Here forms *E* and *C* represent the meridional stage of the circulation. Consequently, if the atmospheric circulation is subdivided only into two forms, zonal and meridional, then the processes can be combined. The concept of a "high" and "low" index is introduced in /49/.*

However, it is seen by comparing the geographic localization of high-level ridges and troughs peculiar to forms *E* and *C* (Figure 1) that they are in almost complete opposition to each other. The geographic distribution of positive and negative temperature and pressure anomalies is also found to be almost opposed. This difference in *E* and *C* forms cannot be neglected in developing a method of long-range weather forecasting.

It is of importance that the Vangengeim classification can be applied to the study of changes in atmospheric circulation and weather for both short and extended periods of time /24/.

* [This reference is not quoted at the end of the paper.]

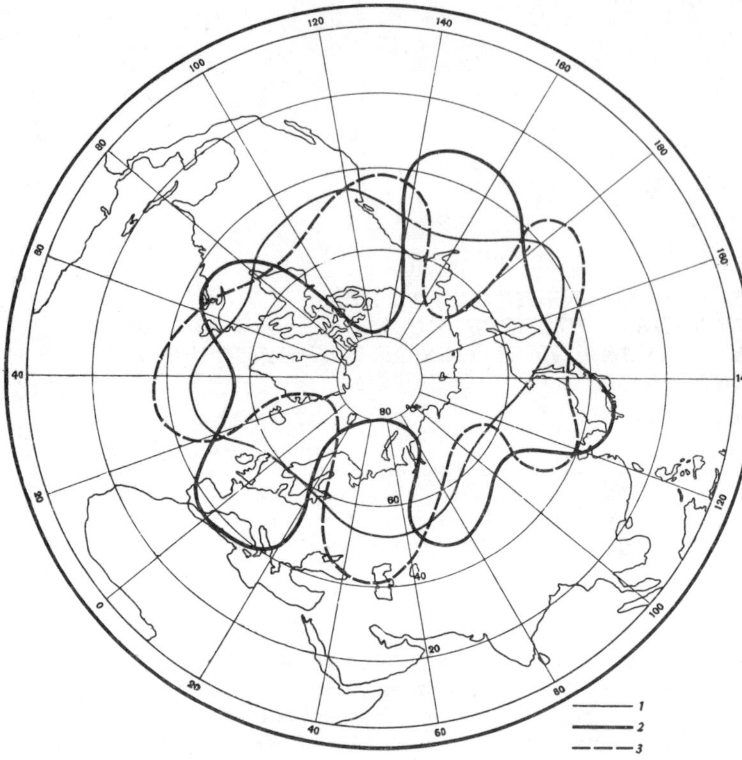

FIGURE 1. Location of high-altitude (500 mb) ridges and troughs during large-scale processes:

1 — W_Z; 2 — E_{M_2}; 3 — C_{M_1}.

It was established by Vangengeim from a study of long-term circulation changes that epochs exist in the development of circulation /7/. The present author clarified, in a number of publications, the main causes for the appearance of epochs and the most general regularities of epochal and intraepochal circulation variations in the atmosphere and hydrosphere of the northern hemisphere /8—30/.

A circulation epoch is understood to span a period of time during which the frequency of processes of one (or two) of the three above circulation forms exceeds the average, while the frequence of processes of the other forms was below average. On the basis of the above, the large-scale processes observed from 1891 to 1960 were subdivided into the following five circulation epochs:

1891—1899		1900—1928		1929—1939		1940—1948		1949—1960
$W+C$	→	W	→	E	→	C	→	$E+C$

As mentioned above, forms W, C, E were established from the Atlantic-Eurasiatic (first) sector of the northern hemisphere. In order to extend the concept of circulation forms and epochs to the entire northern hemisphere, it became necessary to study the nature of these processes in the Pacific-American (second) sector, observed there during the prevalence of forms W, C, E in the first sector.

Applying the Vangengeim classification principles, the author classified processes observed in the second sector from 1900 to 1946 /22, 23/. As in the first sector, they are subdivided into three types: one zonal (Z) and two meridional (M_1 and M_2). The latter, in a manner similar to that of forms C and E, are differentiated by the geographic location of high-altitude ridges and troughs and their corresponding surface-level fields of pressure and temperature anomalies. In connection with this the type of synoptic processes in the northern hemisphere can be represented by nine types of large-scale processes, which reflect the different combinations of processes in the first and second sectors and represent varieties of forms W, C, E in the northern hemisphere. These large-scale processes are: W_Z, W_{M_1}, W_{M_2}, E_Z, E_{M_1}, E_{M_2}, C_Z, C_{M_1} and C_{M_2}.

As a result of studying circulation epochs in the northern hemisphere the author /13, 20/ was able to present the above scheme of epochs in the form

$$\begin{array}{ccc} 1900\text{—}1928 & & 1929\text{—}1939 \\ (W_{Z+M_1+M_2} + C_{M_1} + E_Z) & \longrightarrow & (E_{M_2+M_1} + C_{M_2}) \\ 1940\text{—}1948 & & 1949\text{—}1960 \\ (C_{M_2+Z+M_1} + W_{M_2}) & \longrightarrow & (E_{Z+M_1+M_2} + C_Z) \end{array}$$

It follows from these schemes that form W epochs, which underwent anomalous development in 1900—1928, were most frequently combined with processes C in the second term, although the frequency of combinations C_{M_1} (W_{M_1}) and C_{M_2} (W_{M_2}) was also above average.

At the same time the frequency of forms E and C against the background of anomalous development (within the epoch) of form W processes was also above average during some years. However, their frequency for the entire epoch was on the whole below average. Hence only form W processes are used for designating the epoch. Nevertheless, it was found in examining the frequency of individual varieties of large-scale process ($E_Z, E_{M_1}, E_{M_2}, C_{M_2}$, $C_Z, C_{M_{1,2}}$) that forms C_{M_1} and E_Z were still quite developed. Hence C_{M_1} and E_Z are also added to the main designation of the epoch, which is $W_{Z+M_1+M_2}$. Here the E processes can be regarded as "nuclei" of the circulation of a new epoch, which episodically appear against the background of the main processes of the given epoch, which is of form W.

Circulation forms E, which underwent anomalous development in the first sector, in 1929—1939 were combined primarily with M_2 processes (E_{M_2}) in the second sector, although the frequency of E_{M_1} was also somewhat above average. Hence the epoch was described as $E_{M_2+M_1}$. At the same time the frequency of variety C_{M_2} also was above average against the background of these principal processes. These varieties serve as "nuclei" of the features of the following epoch. The designation of characteristic circulations during the other two epochs, 1940—1948 and 1949—1960, should be understood in the same way.

The above limits of circulation epochs were obtained from analysis of yearly data on the frequency of the different circulation forms. It was shown by examining the long-term frequency of these types during individual months /12/, that the same limits of epochs are peculiar to the majority of months of a given year. However, during some months (October, March, etc.) they differ markedly from the limits established using yearly data /17/. It is important to note here that the circulation form which is anomalously developed during a given epoch does not become evident in the same manner during different months of a year and has its own specific properties during each epoch /21/.

Thus, during the epoch with western circulation /21/ form W processes developed most extensively during spring-summer (May-September) and somewhat in winter. Conversely, the frequency of these processes in September and March was below average. Consequently, form W did not appear at all in these months, during which form E processes developed and predominated during the 1929—1939 epoch.

It follows from this that nuclei" of the new epoch (E) did not appear during all the months, particularly in October and March. It is typical that during the following epoch, i. e., eastern circulation epoch, E processes did not appear at all during these months. The principal processes of epoch E developed primarily in July and August /21/. Their frequency also was above average in November, December and January. During the remaining months (February, March, June, October), they were weaker, i. e., the E-circulation epoch did not appear at all.

The meridional circulation epoch (1940—1948) was evident primarily in autumn and spring. During summer, on the other hand, forms E and C alternated. During the subsequent epoch of combined $E+C$ circulation development of eastern circulation was observed in spring and summer and of meridional circulation in autumn.

Above we considered the characteristics of epochs from data on the frequency of circulation forms (W, C, E) and their varieties in the northern hemisphere (W_Z, W_{M_1}, W_{M_2}, E_Z, etc.).

The anomalous development of processes of a specific atmospheric circulation form during this epoch should have produced certain anomalies in the characteristics of the atmosphere and hydrosphere. Since these two media are in constant and close relationship, the nature of these anomalies should also be similar to a large extent.

It was shown in a number of studies that the pressure and temperature distributions, as well as the distributions of their anomalies, is peculiar to forms W, C, E which were observed during short time intervals (5—10 days, a month) and reflects fields of anomalies forming in the northern hemisphere. In other words, if we were to construct a map of average pressures for the W epoch (1900—1928) and compare it with the map of average values for the entire series of observations (i. e., with "average" maps), then the map of pressure anomalies for the epoch will be similar to the typical map of form W, constructed from a number of cases of processes of this form, observed over short intervals. In conjunction with this the epochal background of pressure and temperature anomalies in the northern hemisphere can be typified by typical maps of these anomalies constructed for each of the nine types of large-scale processes /22, 23/. Typical maps can also be utilized for determining the sign of anomalies of various features of the hydrosphere, arising during the given epoch due to the atmospheric processes /11/.

As an illustration of such use of typical maps we now consider maps for type E processes and for the eastern circulation epoch. It can be seen from maps of /22/ that under processes of this form (E_Z, E_{M_1}, E_{M_2}) a region with positive pressure anomalies forms in the Volga basin, i. e., anticyclogenesis processes and precipitation shortage are brought about here. This results in reducing the amount of water discharged by the Volga into the Caspian Sea, and consequently reduces the level of the latter. This is aided by accelerated evaporation, associated with a positive air-temperature anomaly. In this case one would expect a sharp reduction in the level of the Caspian Sea; this is actually the case /10/.

It also follows from these typical maps that formation of a positive pressure anomaly in the Volga-River region during form E processes is accompanied by a negative pressure anomaly in the region of the Baikal drainage basin (particularly during E_{M_1} and E_{M_2}). Activation of cyclogenesis in this region produces a precipitation surplus, which results in a rise in the level of the Baikal and an increase in the discharge into the Angara /11, 14/.

It can also be concluded by examining typical E form maps that these processes are accompanied by a deepening of the Icelandic low and by propagation of the ridge of the Siberian anticyclone from east to west. As a result, high baric gradients are produced at the eastern Atlantic seaboard and winds of the southern quarter increase in strength. The latter aid in moving warm Atlantic water into the northern latitudes, and this in turn additionally heats the already warm air supplied to the same region from the south. This produces a positive anomaly of the air temperature and of the temperature of the surface water layer in the Greenland and Barents seas. However, these features of form E processes should also form the background of the eastern circulation epoch. This is precisely the process which brought about the Arctic warming epoch, which was most evident in 1929—1937 /11, 14/.

A similar analysis makes it possible to explain features in the atmospheric and hydrospheric processes of other regions in the northern hemisphere during the eastern circulation epoch. Such an analysis /11/ can also be applied to typical maps of the two other atmospheric circulation forms (W and C) and their varieties in the northern hemisphere both for cold and warm seasons /13/. Only brief conclusions are presented here.

The following is typical of the western (W) circulation epoch.

1. Increasing frequency of zonal component of air motions in the troposphere, increased zonal-circulation intensity in the entire hemisphere (in the case of W_Z) or in its Atlantic-Eurasia sector (in the case of W_{M_1} and W_{M_2}).

2. Weakening of interlatitude exchange is responsible for the decisive role of the radiation factor in formation of air temperature anomalies: negative temperature anomalies predominate in northern latitudes and positive anomalies in moderate latitudes. The latitude of the zero isabnormal depends on the intensity of the zonal circulation and on the given combination of form W processes of the first sector with those of the second ($W_Z W_{M_1}$ or W_{M_2}).

3. Latitude trajectories of baric formations prevail. This is responsible for influx of water to the eastern shores of the Atlantic and Pacific, with a corresponding rise in level and appropriate reduction in the western parts. The Baltic-Sea level increases /44/, the salinity of Baltic water is reduced /45/, the Gulf of Finland and Neva River water levels rise more frequently, and the thawing and freezing dates of the latter approach average /42/. The water volume in the Irtysh and Ishim increases and that of the Tobol decreases /33/; the flow in the Dnieper at Kiev is low /36/. The amount of precipitation in the northern Volga-River valley increases and that in the south of the Volga decreases; the Caspian-Sea level is unstable with insignificant oscillations and a small tendency toward increasing. The level of the Baikal and the discharge into the Angara decrease /11/. The Arabian Sea level at Bombay and that of the western Mediterranean rise /11/. The water temperature at the Kola meridian decreases /11/, and the amount of ice in the Barents Sea increases.

4. Accumulation of cold in the Arctic and a reduction in the removal of ice from the Arctic Sea (by way of the Greenland Sea) aid in increasing the total amount of ice in the Arctic and cause additional interference with sailing over various segments of the Northern Sea Route. The geographic location of these segments depends on the variety (W_Z , W_{M_1} or W_{M_2}) of the most frequently appearing W-form processes (during the epoch of 1900 — 1928 these were W_Z and W_{M_1}).

5. The isolines of tropopause altitudes lie basically in a zonal direction, with minimum tropopause altitude, as a rule, over the Arctic /29/.

6. A planetary high-altitude frontal zone and its associated jet stream are most frequently aligned zonally in the moderate belt of the hemisphere. Since the meridional circulation components are rather low, the jet stream only infrequently penetrates into high latitudes (on ridge tops) and into low latitudes (in the southern part of troughs).

The interaction between the different belts of the hemisphere is weak and the ridges and troughs of climatological centers of action are developed primarily in the zonal direction (to the west and east).

Tops of high-altitude subtropic ridges penetrate less frequently into northern latitudes, while troughs of moderate latitudes penetrate into subtropic and equatorial latitudes. This reduces to a minimum sharp fluctuations in the altitude of the tropopause and temperature changes in the Arctic and subtropic stratosphere. Western winds of the moderate zone appear less frequently over near-equatorial trade winds, while the Arctic anticyclone is less frequently converted into a high-altitude baric formation.

7. Positive temperature anomalies are most probable in Kazakhstan, particularly in its southeastern part /1, 2/.

8. Mild winters with air temperatures 3 — 4° above average (in January) prevail in the Ukraine /3/. The amount of precipitation is close to average (80 — 90%) and is much below average in the south of the Ukraine (up to 47%)

In summer the air temperature is near or somewhat above average (approximately +0.8°). Near-average precipitation is observed only in the western part of the Ukraine (approximately 98%), and in its remaining regions it amounts to 50 — 75% of the average.

9. Negative air temperature anomalies prevail in eastern and southern parts of China /47/ and the thunderstorm activity weakens, while an

elevated temperature background is observed in the northern parts of China, with a precipitation shortage in the Peking and Shanghai regions and a reduction in the level of the Hwang Ho and Yangtze rivers.

10. According to observations in Seoul (South Korea) summer seasons with precipitation shortages occur here during the *W*-circulation epoch, negative air-temperature anomalies are more frequent in the Pyongyang region (particularly in summer); the rice yield on the northeastern shore of Korea is below average /46/.

11. The area of Arctic glaciers decreases /35/, although relatively slowly (2—3 m/year). The same tendency is observed also in changes of Altai glaciers /1/.

12. The water content of rivers in the northern, central and southeastern parts of European USSR (Northern Dvina, Pechora, Sula, Volga, Kama, Bol'shoi and Malyi Uzen', Don, Kuban') and somewhat of rivers of Asian SSSR (Northern Sos'va, Irtysh, Amu-Darya and Syr-Darya) increases. The correlation between the water content of the above rivers and the frequency of form *W* processes lies in the range 0.74—0.87 /84/.

The water content of the Vuoksa, Usa, Western Dvina, Dnieper, Southern Bug, Terek, Kura, Ob', Enisei, Angara, Lena and Kolyma decreases. The correlation factor between the water content of rivers and circulation characteristics ranges from 0.46 (Vuoksa River) to 0.98 (Usa River).

13. The amount of precipitation deposited in the Cis-Ural region /38/ is much greater than in Trans-Ural. It has a maximum (250—300 mm) in the Vishera and Kos'va river basins. The precipitation gradually decreases to the south and southeast. Under form *W* processes about 40—48% of yearly precipitation is deposited in the Upper Kama river basin (up to Perm'), while the precipitation in the Tobol river basin amounts to only 23—29% of the yearly total.

14. The January air temperature increases /33/ at the Upernavik, Malye Karmakuly, Vienna, Lugansk, Tbilisi, Kazalinsk, etc. stations. At the same time a reduction in the temperature background is observed in Vladivostok, Shanghai, Bermuda Islands, Freetown, Alice Springs, etc.

The following is typical of the eastern (*E*) circulation epoch.

1. Reduction in the intensity of zonal circulation and amplification of interlatitude exchange over the entire hemisphere (in the case of varieties E_{M_2} and E_{M_1}) or only in its first sector (in the case of E_Z).

2. Air temperature and pressure anomalies are produced primarily by advective and dynamic factors.

Since large-amplitude stationary waves are observed in the troposphere under processes of this form (see Figure 1), regions of negative temperature anomalies and positive pressure anomalies are observed beneath eastern parts of high-altitude ridges at the surface, while regions with positive temperature anomalies and negative pressure anomalies form beneath their western parts.

The geographic location of these ridges and troughs over the hemisphere, as well as of their associated anomaly regions, is determined by the kind of *E*-form processes (E_Z, E_{M_1} or E_{M_2}).

3. The trajectories of baric formations usually have a large meridional component. Here one notes a steadying and amplification of the principal atmospheric centers of action in their long-term average locations /31/. Only the Siberian winter maximum shifts to the west of its average location.

All this results in prolonged maintenance of specific air flow directions in different regions of the hemisphere and formation of regions with large anomalies of atmospheric and hydrospheric characteristics /22, 23, 11/.

4. A tendency is observed to a systematic increase or reduction in air temperature in certain regions of the hemisphere, presented on typical temperature anomaly maps and on graphs of its long-term variations /43/ by individual stations of the northern hemisphere.

5. Isolines of tropopause altitudes are usually located meridionally, forming ridges and troughs, the location of which is shown in Figure 1. Due to this, unlike the *W*-form epoch, frequent and substantial increases in tropopause altitude may be observed in the polar region and reductions may occur in moderate and subtropic latitudes.

6. Jet streams "meander," swinging around ridges (see Figure 1) from the north and around troughs from the south. This is responsible for activation of cyclonic activity in the polar region, up to its central part.

7. The interaction between different zones of the hemisphere is much more active and stronger than during the *W*-circulation epoch. The marked meridional development of high ridges and troughs aids in a more frequent rise of temperature in the Arctic troposphere and stratosphere and in development of the Arctic anticyclone. At the same time penetration of polar troughs into tropical latitudes and genesis of westerly winds above surface-level trades is frequently observed, which aids in weakening the latter.

8. Cold winters prevail in the Ukraine /3/ (the January temperatures are 2—3° below average and in a number of cases even as much as 6—8° below). The air temperatures in July are, as a rule, above average by 2—3° and in individual cases by as much as 6°. The precipitation is above average, particularly in western and southern regions of the Ukraine.

9. A reduced temperature background prevails stably in Kazakhstan /35/; negative anomalies exceed 2°. The amount of precipitation markedly exceeds average, particularly in the northern, central, eastern and southeastern parts of the area, being below average in the western parts.

10. A higher temperature background and increasing thunderstorm activity are observed in eastern China /47/. The annual amount of precipitation in the Peking region is below average and in the Shanghai region above average. The level of the Yangtze and Hwang Ho rivers rises and, as a rule, exceeds the average.

11. A higher temperature background and increasing amount of precipitation is observed in the Pyongyang (North Korea) region in summer /46/. The rice yield on the northeastern shore is above average.

12. The Arctic glaciers recede faster than during the form *W* epoch: up to several tens of meters per year /35/. The same tendency is observed also in Alpine glaciers /1/.

13. The water content of rivers /37, 25/ of the northern, central and southeastern parts of European USSR and somewhat of Asian USSR (Northern Dvina, Pechora, Sula, Volga, Kama, Bol'shoi and Malyi Uzen', Don, Kuban', Northern Sosva, Irtysh, Amu-Darya and Syr-Darya) decreases. The water content of the Vuoksa, Usa, Western Dvina, Dnieper, Southern Bug, Terek, Kura, Ob', Enisei, Angara, Lena and Kolyma rivers increases /34, 36/.

14. Unlike the *W*-form epoch, a precipitation shortage exists at the Middle and Southern Urals /48/. Here the entire Cis-Ural area and the northern Trans-Ural (headwaters of Tura and Tavda rivers) receive a

minimal amount of precipitation (22—29% of the yearly total), while the headwater of the Tobol gets the maximum amount of precipitation (40—45%).

15. The Neva thawing dates during this epoch /42/, as a rule, are earlier, while the freezing dates are later. The salinity of the Baltic Sea increases /45/.

16. The level of the Caspian Sea as well as of the eastern Atlantic and western Pacific decreases. The level of the Mediterranean and Arabian seas, western Atlantic, Baikal, eastern Pacific (at San Francisco), and the Atlantic at the eastern seaboard of South America increases /11, 27/.

17. The water temperature at the Kola meridian increases and the amount of ice in the Barents Sea decreases /11, 27/.

The following is typical of the meridional (C) circulation epoch.

1. Reduction in the intensity of zonal circulation and increasing inter-latitude exchange over the entire hemisphere (in the case of varieties C_{M_1} and C_{M_2}) or only in its first sector (in the case of C_Z). The geographic location of high-altitude ridges and troughs (see Figure 1) during the C epoch is almost opposite to that of the E epoch. In connection with this the geographic location of regions with pressure and temperature anomalies are also in opposition. This is seen from typical anomaly maps /22, 23/, as well as from the shape of curves of long-term air-temperature variations at individual stations of the northern hemisphere /43/. However, the location of surface-level temperature and pressure anomalies relative to high-altitude ridges and troughs is the same as during the E epoch.

2. Isolines of tropopause altitudes, jet streams and interaction between different zones of the hemisphere during the C epoch are basically similar to those of the E epoch.

3. The Icelandic and Aleutian lows, as a rule, are weak, and high-pressure regions with closed isobars are frequently observed in these regions even in winter.

4. Cold, low-snow weather prevails in the Ukraine in winter /3/ (the January temperature is 3—6° below average). In summer cool dry weather is encountered most frequently (the July temperature is 1° below average).

5. The January air temperatures in Kazakhstan /2/ are markedly above average. The −14° isotherm shifts far to the north, to the boundaries of Kazakhstan.

6. Precipitation excess is observed in China /47/ in the Peking region and a precipitation shortage exists in the Shanghai region. The level of the Yangtze and Hwang Ho rivers rises.

7. The Upernavik glacier in western Greenland recedes markedly /30/. The Nordenskjöld Glacier (Spitsbergen) recedes more slowly (2—3 m/year, similar to the W-form epoch). The Alpine glaciers advance quite rapidly /1/.

8. The water content of the Kuban', Kura, Amu-Darya, Syr-Darya, Ob' and Lena rivers increases /34/. The correlation factor in this relationship ranged from 0.56 to 0.88.

9. A gradual reduction in precipitation from north to south is observed in the Ural region /48/ (similar to that of the W epoch); however, during the C epoch these differences between the north and south are smaller. The greatest part of the yearly precipitation is deposited in the Belaya river basin, upper parts of the Tavda, Tura and Iset rivers (36—39%), with the minimum (28—34%) devolving upon the left-bank tributaries of the Kama (Vishera, Kos'va, Chusovaya) and headwaters of the Tobol (the Uil, Ayat, Ubagan, etc. rivers).

10. The freezing dates of the Neva River come close to average, while the thawing dates (as during the *E* epoch) are earlier than average. However, the value of this anomaly during the *C* epoch is smaller than during the *E* /42/.

11. The level of the Caspian Sea rises and of the Baikal drops /11/. The level of the Mediterranean rises and that of the Arabian Sea drops.

12. The water temperature at the Kola meridian rises and the ice content of the Barents Sea increases /11, 27/.

These are the main features of the state of the atmosphere and hydrosphere of the northern hemisphere observed during different atmospheric circulation epochs. They point to the fact that forms *W*, *C*, *E* of atmospheric circulation and their varieties and circulation epochs are real, large stages in the development of the circulation of the terrestrial atmosphere. It is precisely for this reason that the changes in frequency of these forms are closely related to large changes not only in the nature of atmospheric processes and weather regime, but also large changes in important characteristics of the hydrosphere.

This also proves that the above regularities in the epochal variations of forms *W*, *C*, *E* can serve as a reliable basis in the study of causes of variations in the climate and in the state of the hydrosphere of various regions of the northern hemisphere, and also for constructing integrated methods of long-range and extra long-range hydrometeorological forecasts. The above characteristics of epochs have, in addition, direct forecasting significance, since the predicted circulation forms of a forthcoming epoch serve as a forecast of the state of the atmosphere and hydrosphere.

However, regularities of epochal variations in circulation forms are of interest not only for these extra long-range forecasts for the forthcoming epoch. Allowance for the epochal background is of great significance also in developing methods of long-range forecasting for shorter periods: for a year, season and month /16, 19, 24, 28/.

Before considering the nature of the atmospheric circulation of the forthcoming epoch, we shall briefly examine the causes of the appearance of circulation epochs and epochal variations in the atmospheric circulation.

The atmospheric activity can be represented in general outline as a process of continuous variations in its zonal and meridional stages. Among factors known at present one may mention some /29/ which are due to zonal circulation. These are primarily nonuniformity of the latitude distribution of the thermal energy and the earth's rotation. These constantly acting factors thus produce the principal form of atmospheric circulation, the western form (*W* according to the Vangengeim classification), which apparently would have predominated entirely if not for the simultaneous existence of other factors producing departures from zonality and transition to meridional types of large-scale processes. These factors include nonuniformity of distribution and heating of the land and seas, cyclonic activity at tropospheric fronts, solar activity and other factors of terrestrial and cosmic origin.

Simultaneous transformation of zonal to meridional, and vice versa, states of the atmosphere over the entire hemisphere is rather infrequent /30/. Most frequently this transformation is observed at one of its sectors and then, by virtue of the interrelationship between these processes, starts in another sector after a time lag.

The long-term average ("normal") frequency of various versions of variations in the nine types of initial large-scale processes (W_Z, W_{M_1}, W_{M_2}, E_Z, E_{M_1}, E_{M_2}, C_Z, C_{M_1}, C_{M_2}), the duration of which ranges from 3 to 12 days, is described in references /20, 30/ by the present author. These data on the frequency of various variations show the manner in which the long-term annual frequency of processes ("average") of various types forms.

Suppose that during some specific year (or over a number of years) the frequency of processes, for example, of form *W* exceeded the average (was more than 127 days). This means that the epoch of development of form *W* processes (1900—1928) has arisen due to reduction in the frequency and stability of type *E* and *C* meridional processes, increasing their transformation into *W* at an above-average rate and increasing the stability of the latter forms. In this case the genesis of form *E* and *C* epochs can be attributed to increasing stability and frequency of these processes (*E* and *C*) at the expense of above-average frequent variations of zonal processes (*W*) into meridional ones and increasing the stability of the latter /20/.

In the opinion of this author /20, 21/, extended anomalies in the development of atmospheric circulation can be ascribed primarily to the increasing or decreasing effect of external factors and in the first place of long-term changes in solar activity (11 and 80—90 year cycles), the angular velocity of the earth and the motion of its poles /37—41/.

As was shown by numerous studies, all these factors result in departure from zonality in the atmospheric circulation, which is produced primarily by surface factors, and consequently aid in the appearance of meridional processes. In other words, they are instrumental in reducing the stability of form *W* processes, increasing the frequency of their transformation into form *E* and *C* processes, appearance of meridional processes in the atmosphere at locations where zonality existed up to the instant of amplification of the effect of the external factor, and increasing the stability of meridional processes produced by terrestrial surface factors up to this instant of time. It is therefore natural to assume that during years of activation of external factors in the terrestrial atmosphere the frequency of meridional processes (*E*, *C*, M_1, M_2) increases sharply while that of zonal processes (*W*, *Z*) decreases, i. e., epochs originate of forms *E* and *C* which are most frequently combined with M_1 and M_2 meridional processes in the Pacific-American sector of the hemisphere.

During years with lesser effect of external factors the zonal circulation peculiar to the terrestrial atmosphere is less frequently and less strongly disturbed by external effects. Hence the stability of zonal processes increases, while the frequency of their transformation into meridional processes decreases sharply. All this results in an above-normal frequency of form *W* processes for a number of years which then comprise the western circulation epoch. Here this form will most frequently occur together with zonal processes (*Z*) in the second sector.

Thus western circulation epochs arise primarily due to the effect of terrestrial factors, which are most pronounced during these years in connection with the marked reduction in the externally produced changes.

On the other hand, epochs of eastern and meridional circulations arise primarily due to external factors, activation of which markedly suppresses the contribution of terrestrial surface factors in forming tendencies of long-term changes in the frequency of atmospheric circulation forms.

It should be noted that the prediction made by the author ten years ago /8/ that we are about to enter a western circulation epoch which, if it is similar to the epoch already observed (1900—1928), will be typified by development of zonal and partially meridional (M_1) processes in the second sector, has been verified by new data /20/.

It can also be concluded from these data that zonal circulation actually developed during the past ten years. However, this as yet occurred only in the Pacific-American sector, while form *E* and *C* processes continued developing in the Atlantic-Eurasia sector.

However, as a result of interaction between processes in both sectors, the zonal circulation developing in the second sector should be instrumental in a more frequent transformation of form *E* and *C* processes into form *W* processes, and consequently in the appearance of outlines of the new epoch also in the first sector. This should also be promoted by the reduction in solar activity, which started after 1957. The more so since, as is assumed in heliogeophysics, this reduction is due not only to 11-year cycles, but also serves as the start of a quite steep branch of reduction in the secular (80—90 year) solar activity cycle.

Usually outlines of a new epoch do not grow monotonically, but in more or less pronounced spurts. The first such spurt occurred in 1955—1956, when the frequency of *W* processes was above average for four months. After these decreased in 1957—1958 they again became amplified in 1959. Even a stronger spurt occurred in 1961, when the average was exceeded during eight months. The number of days per year with processes of this form (*W*) also exceeded the average for the first time in 13 years. A new reduction in the frequency of form *W* processes was observed in 1962—1963.

The previously mentioned forecasting considerations /20/ can be presented in somewhat more detail on the basis of the above prediction by workers in heliogeophysics and by assuming that the duration of the solar activity cycle is equal to its average value (11 years).

The following conclusions may be drawn from these quite approximate starting data, as well as from consideration of some general regularities in the interaction of processes of the first and second sectors of the northern hemisphere. Zonal processes apparently will continue developing to the end of the descending branch of the current 11-year cycle (up to 1965) in the second sector. This will be instrumental in the appearance of increasingly new spurts of form *W* processes in the first sector upon the activation of *E* and *C* processes. Zonal processes will most probably undergo a reduction in the second sector approximately from 1965 to 1978 (time of growth of the 11-year solar activity cycle).

Since this 11-year cycle will lie on the descending branch of the secular cycle, its growing branch should not be steep and possibly will produce only insignificant weakening of zonality in the second sector. As to the first sector, here the processes of the second sector apparently will not aid in activating the *W* form.

It should, however, be remembered that zonal processes developed actively during the past ten years in the second sector; by virtue of interrelationship with the second sector these processes should change the nature of *E* and *C* processes in the first sector, creating a tendency toward increasingly pronounced production in them of form *W* features. This is verified by analysis of observed data. At the same time, the development of type E and C meridional processes in the first sector should have changed the form of zonal processes in the second sector, and impart to them a tendency toward increasing formation of meridional-process features, which is also verified by observations.

Consequently, irrespective of the effect of external factors and solely as a result of interaction between processes, conditions were created from 1955 to 1964 for activation of form W processes during the subsequent 6—7 years in the first sector and of M_1 and M_2 processes in the second sector.

Thus, if one considers only the effect of the solar activity (the growing branch of the 11-year cycle), then a reduction in the zonal circulation should be expected at the hemisphere. However, the preceding circulation development at the hemisphere during the past ten years prepared conditions for activation of form W processes in the first sector. It is possible that as a result of combination of these two factors, the development of western circulation may continue in the first sector in 1965—1973, in spite of the retarding effect of the external factor. As to the second sector, here both the above factors aid in reducing zonality and increasing meridionality.

The period following 1973 occurs during a drop in the new 11-year cycle, which should be even more intensive than the drop in the preceding cycle, since a descending branch of the secular cycle can exist only if the preceding comments are true. Then the effect of the external factor will be reduced. Consequently, zonal processes will again be activated in the second sector, and this will aid in activating form W processes in the first sector, producing systematic "bursts" of the latter. However, also in this case one must consider a second factor, namely, the type of processes over the hemisphere during the preceding years, 1965—1973. If form W then develops in the first sector and forms M_1 and M_2 in the second, then by virtue of interaction this will be instrumental in creating conditions for the development of zonality in the second sector and of meridional E and C processes in the first.

Combining the two factors it may be assumed that the development of primarily zonal processes in the second and meridional (E and C) in the first is most probable for 1973—1978. However, from time to time these "regulating" processes in each sector should be disturbed by processes of the other sector, thus producing "bursts" of circulation of the following period (1979—1984).

It is at present impossible to present more specific and substantiated details of such forecasts. This is primarily due to the unavailability of any quantitative indicators of the relationship between the role of internal and external factors, and the more so of data on changes in their role with time.

The above results of studies, which typify to a certain extent the present state of the problem at hand, are considered by the author as only the first stage of investigations on this quite complicated and very important question.

Nevertheless, work carried out until now allows one to maintain that: 1) the process of epochs in the development of atmospheric circulation definitely exists and is responsible for large-scale changes in major characteristics of the state of the terrestrial atmosphere and hydrosphere; 2) a major role in the formation and transformation of circulation epochs is played by external factors, which not only directly affect the nature of atmospheric processes, but also change in a certain manner the role of the surface-level factors. Hence further studies of epochs should be based on consideration of an as wide as possible ensemble of factors of terrestrial and cosmic origin and should have as its purpose clarification of the change of their role with time; 3) the presence of circulation epochs in the development of atmospheric circulation requires a new approach to problems of climate classification, of constructing "averages" of various characteristics of the atmosphere and hydrosphere, producing climatic descriptions and handbooks, investigating the causes of climatic fluctuations in different regions of the globe, constructing forecasting relationships, setting up studies to do with the regime of the hydrosphere and many other meteorology and oceanology problems; 4) the study of regularities of epochal transformations of atmospheric circulation forms brings us directly to the construction of scientifically substantiated methods of integrated, extra long-range hydrometeorological and climatic forecasts. Hence particular attention in subsequent studies should be paid to the investigation of the continuous process of epoch transformation, establishing stages in these transformations and clarifying the physical factors responsible for them; 5) the background of an epoch has an important, frequently decisive effect on the type of circulation background of individual years and their seasons, which are a part of the given epoch and serve as stages in its formation. Hence subsequent refinement of existing methods, and the development of new methods, of long-range hydrometeorological forecasts is basically impossible without studying the large-scale processes of this stage as stages of circulation epochs and epochal transformations of circulation; 6) the epoch property which was found in the development of atmospheric circulation of the northern hemisphere undoubtedly exists also in the southern hemisphere. Hence a more complete study of the causes of genesis of circulation epochs and epochal transformations requires the investigation of epochs of the southern hemisphere and their relationship with those of the northern hemisphere.

References

1. Baidal, M. Kh. Epokhal'nye osobennosti atmosfernoi tsirkulyatsii i svyazannye s nimi yavleniya (Epochal Features of Atmospheric Circulation and their Related Phenomena). — Trudy Kazakhskogo NIGMI, Vol. 10. 1959.
2. Baidal, M. Kh. Kompleksnyi makrotsirkulyatsionnyi metod dolgosrochnyk prognozov pogody (Integrated Large-Scale Circulation Method of Long-Range Weather Forecasting). — Gidrometeoizdat. 1961.
3. Buchinskii, I. E. Atmosfernaya tsirkulyatsiya i asinkhronnost' kolebanii klimata na Ukraine (Atmospheric Circulation and the Asynchronism of Climatic Fluctuations in the Ukraine). — Trudy Vsesoyuznogo nauchnogo meteorologicheskogo soveshchaniya, Vol. 4, Gidrometeoizdat. 1963.

4. Vangengeim, G. Ya. Opyt primeneniya sinopticheskikh metodov k izucheniyu i kharakteristike klimata (Experience in the Use of Synoptic Methods in the Study and Classification of the Climate). — Gidrometeoizdat. 1935.
5. Vangengeim, G. Ya. Dolgosrochnyi prognoz temperatury vozdukha i vskrytiya rek (Long-Range Forecasting of Air Temperature and of River Thawing). — Trudy GGI, Vol. 10. 1940.
6. Vangengeim, G. Ya. O kolebaniyakh atmosfernoi tsirkulyatsii nad Severnym polushariem (On Fluctuations of Atmospheric Circulation over the Northern Hemisphere). — Izvestiya AN SSSR, seriya geograficheskaya i geofizicheskaya, No. 5. 1946.
7. Vangengeim, G. Ya. Osobennosti atmosfernoi tsirkulyatsii v razlichnykh epokhakh i kolebaniya klimata (Features of the Atmospheric Circulation in Different Epochs and Fluctuations of Climate). — Trudy II geograficheskogo s"ezda, Vol. 2. Geografgiz. 1948.
8. Girs, A. A. K voprosu ob izuchenii obshchei tsirkulyatsii atmosfery (Concerning the Study of the General Circulation of the Atmosphere).— Izvestiya AN SSSR, seriya geograficheskaya, No. 4. 1955.
9. Girs, A. A. Mnogoletnie preobrazovaniya form atmosfernoi tsirkulyatsii i izmeneniya solnechnoi aktivnosti (Long-Term Variations in Atmospheric Circulation Forms and Changes in Solar Activity). — Meteorologiya i gidrologiya, No. 10. 1956.
10. Girs, A. A. Epokhal'nye preobrazovaniya form atmosfernoi tsirkulyatsii i svyazannye s nimi kolebaniya urovnya Kaspiiskogo morya (Epochal Transformation of Atmospheric Circulation Forms and Related Fluctuations in the Caspian Sea Level). — Izvestiya AN SSSR, seriya geograficheskaya, No. 1. 1957.
11. Girs, A. A. Mnogoletnie preobrazovaniya form atmosfernoi tsirkulyatsii i svyazannye s nimi kolebaniya urovnya okeanov i morei (Long-Term Variation in Atmospheric Circulation Forms and Associated Fluctuations in the Level of Oceans and Seas). — Materialy konferentsii po probleme "Vzaimodeistvie atmosfery i gidrosfery v severnoi chasti Atlanticheskogo okeana," No. 1. Gidrometeoizdat. 1958.
12. Girs, A. A. Osobennosti mnogoletnikh kolebanii tsirkulyatsii atmosfery v otdel'nykh mesyatsakh goda (Features in the Long-Term Fluctuations of Atmospheric Circulation During Individual Months of the Year). — Meteorologiya i gidrologiya, No. 12. 1958.
13. Girs, A. A. Nekotorye kharakteristiki epokh v razvitii form atmosfernoi tsirkulyatsii, svyazannye s uchetom raznovidnostei (Certain Characteristics of Epochs in the Development of Atmospheric Circulation Forms, Associated with Consideration of their Varieties). — Meteorologiya i gidrologiya, No. 8. 1960.
14. Girs, A. A. Osnovy dolgosrochnykh prognozov pogody (Fundamentals of Long-Range Weather Forecasts). — Gidrometeoizdat. 1960.
15. Girs, A. A. O prichinakh i proyavleniyakh mnogoletnikh kolebanii tsirkulyatsii atmosfery (The Causes and Effects of Long-Term Fluctuations in the Circulation of the Atmosphere). — Izvestiya AN SSSR, seriya geograficheskaya, No. 6. 1960.

16. G i r s , A. A. Osnovnye rezul'taty issledovanii mnogoletnikh kolebanii obshchei tsirkulyatsii atmosfery primenitel'no k probleme sverkhdolgosrochnykh gidrometeorologicheskikh prognozov (Principal Results of Studies of Long-Term Oscillation of the General Circulation of the Atmosphere as Applied to the Question of Extra Long-Range Hydrometeorological Forecasts). — Trudy I nauchnoi konferentsii po obshchei tsirkulyatsii atmosfery. Gidrometeoizdat. 1962.
17. G i r s , A. A. and N. F. P o l y s a l o v a . Osobennosti raspredeleniya anomalii davleniya i temperatury v Arktike v razlichnykh tsirkulyatsionnykh epokhakh (Features of the Distribution of Pressure and Temperature Anomalies in the Arctic during Different Circulation Epochs). — Trudy AANII, Vol. 240. 1961.
18. G i r s , A. A. Vnutrigodovye preobrazovaniya form atmosfernoi tsirkulyatsii i ikh prognoz (Variations in Atmospheric Circulation Forms within a Year and their Prediction). — Trudy Vsesoyuznogo nauchnogo meteorologicheskogo soveshchaniya, Vol. 3. Gidrometeoizdat. 1963.
19. G i r s , A. A. Nekotorye osobennosti vnutrigodovykh preobrazovanii form atmosfernoi tsirkulyatsii i ikh prognosticheskoe znachenie (Some Features of Variations in Atmospheric Circulation Forms within a Year and their Forecasting Significance). — Trudy AANII, Vol. 253. 1963.
20. G i r s , A. A. Vnutriepokhal'nye preobrazovaniya form atmosfernoi tsirkulyatsii i ikh prichiny (Variations in Atmospheric Circulation Forms within Epochs and their Causes). — Trudy AANII, Vol. 255. 1963.
21. G i r s , A. A. Osobennosti vnutrigodovykh preobrazovanii makrosinopticheskikh protsessov v razlichnykh tsirkulyatsionnykh epokhakh (Features of Variations in Long-Term Synoptic Processes within a Year in Different Circulation Epochs. — Trudy AANII. 1963.
22. G i r s , A. A. Tipovye kharakteristiki osnovnykh raznovidnostei form atmosfernoi tsirkulyatsii v kholodnoe vremya goda (Typical Characteristics of Principal Varieties of Atmospheric Circulation during the Cold Season). — Problemy Arktiki i Antarktiki, No. 7. 1959.
23. G i r s , A. A. Tipovye kharakteristiki osnovnykh raznovidnostei form atmosfernoi tsirkulyatsii v teploe vremya goda (Typical Characteristics of Principal Varieties of Atmospheric Circulation during the Warm Season). — Problemy Arktiki i Antarktiki, No. 2. 1960.
24. G i r s , A. A. O sozdanii edinoi klassifikatsii makrosinopticheskikh protsessov severnogo polushariya (On Creating a Unified Classification of Large-Scale Synoptic Processes of the Northern Hemisphere). — Meteorologiya i gidrologiya, No. 4. 1964.
25. G i r s , A. A. K voprosu izucheniya osnovnykh form atmosfernoi tsirkulyatsii (Concerning the Study of Principal Forms of Atmospheric Circulation). — Meteorologiya i gidrologiya, No. 3. 1948.
26. G i r s , A. A. Vertikal'naya struktura, formirovanie i preobrazovanie osnovnykh tipov atmosfernoi tsirkulyatsii (Vertical Structure, Formation and Transformation of the Principal Types of Atmospheric Circulation). — Trudy AANII, Vol. 33. 1951.

27. Girs, A. A. Ledovidost' Barentsova morya pri razlichnykh vidakh vnutrigodovykh preobrazovanii form atmosfernoi tsirkulyatsii (The Ice Content of the Barents Sea Corresponding to Different Forms of Variation in Atmospheric Circulation Forms within a Year). — Trudy LGMI, Vol. 20. 1964.
28. Girs, A. A. K voprosu o napravlennosti i koordinatsii issledovanii po probleme dolgosrochnykh meteorologicheskikh prognozov (Concerning the Directions and Coordination of Studies Pertaining to Long-Range Meteorological Forecasts). — Problemy Arktiki i Antarktiki, No. 14. 1963.
29. Girs, A. A. Osnovy dolgosrochnykh prognozov pogody (Fundamentals of Long-Range Weather Forecasts). — Gidrometeoizdat. 1960.
30. Girs, A. A. Obshchie zakonomernosti vzaimodeistviya razlichnykh form tsirkulyatsii atmosfery v severnom polusharii (General Regularities in the Interaction of Various Forms of Atmospheric Circulation in the Northern Hemisphere). — Izvestiya AN SSSR, seriya geograficheskaya, No. 6. 1963.
31. Girs, A. A. Sostoyanie nekotorykh tsentrov deistviya atmosfery severnogo polushariya v razlichnykh tsirkulyatsionnykh epokhakh (The State of Certain Atmospheric Centers of Action in the Northern Hemisphere during Different Circulation Epochs). — Meteorologiya i gidrologiya, No. 10. 1961.
32. Katalog makrosinopticheskikh protsessov (Catalog of Large-Scale Synoptic Processes). AANII. 1964.
33. Kovzel', A. G. Svyaz' mnogoletnikh kolebanii stoka rek s tipami atmosfernoi tsirkulyatsii (Correlation Between Long-Term Variations in the Riverine Discharge and Types of Atmospheric Circulation). — Meteorologiya i gidrologiya, No. 11. 1960.
34. Kochukova, T. N. Kolebaniya godovogo stoka rek SSSR (Variations in the Yearly Discharge of USSR Rivers). — Trudy GGI, 50 (104). 1955.
35. Krenke, A. N. Raionirovanie lednikov Arktiki v svyazi s tsirkulyatsiei atmosfery (Division of Arctic Glaciers into Regions in Connection with Atmospheric Circulation). — Glyatsiologicheskie issledovaniya. Sbornik statei, No. 9. Izdatel'stvo AN SSSR. 1963.
36. Kryzhanovskaya, A. B. Dolgosrochnye prognozy kharakteristik stoka Dnepra (Long-Range Forecasts of the Characteristics of the Dnieper River Discharge). — Izdatel'stvo AN Ukrainskoi SSR. 1957.
37. Maksimov, I. V. Dvizhenie polyusa Zemli i mnogoletnie izmeneniya kontinental'nosti klimata Evropy (Motion of the Terrestrial Poles and Long-Term Changes in the Continentality of the European Climate). — Doklady AN SSSR, 93 (5). 1953.
38. Maksimov, I. V. Nutatsionnaya tsirkumpolyarnaya baricheskaya volna v atmosfere Zemli (The Nutational Circumpolar Baric Wave in the Terrestrial Atmosphere). — Doklady AN SSSR, 100 (1). 1965.
39. Maksimov, I. V. Solnechnaya aktivnost' i dvizhenie polyusa Zemli (Solar Activity and the Motion of the Terrestrial Pole). — Uchenye zapiski LVIMU imeni admirala Makarova, No. 5. 1954.
40. Maksimov, I. V. O mnogoletnikh prilivnykh yavleniyakh v more i v atmosfere Zemli (Long-Term Tidal Phenomena in the Sea and in the Atmosphere of the Earth). — Trudy Instituta okeanologii, Vol. 8. 1954.

41. Maksimov, I. V. Nutatsionnaya stochnaya volna v Mirovom okeane i ee geograficheskie sledstviya (The Nutational Discharge Wave in the World Ocean and its Geographic Effects). — Izvestiya AN SSSR, seriya geograficheskaya, No. 1. 1956.
42. Panov, B. P. Zimnii rezhim rek SSSR (The Winter Regime of USSR Rivers). — Izdatel'stvo LGU. 1960.
43. Rubinshtein, E. S. and L. G. Polozova. Sovremennye izmeneniya klimata (Present-Day Changes in Climate). — Izvestiya AN SSSR, seriya geograficheskaya, No. 5. 1963.
44. Sleptsov-Shevlevich, B. A. Ob atmosfernoi obuslovlennosti mnogoletnikh kolebanii urovennoi poverkhnosti morya (On the Atmospheric Causes of Long-Term Variations in the Level of the Sea Surface). — Uchenye zapiski LVIMU imeni admirala Makarova. No. 13. 1959.
45. Tsoi Ik-Son. Rezhim pogody nad Koreei pri osnovnykh formakh atmosfernoi tsirkulyatsii (The Weather Regime Over Korea Corresponding to the Principal Atmospheric Circulation Forms). Author's abstract of dissertation. Gidrometeoizdat. 1957.
46. Chang Chi-Chia. Rezhim pogody nad Kitaem pri osnovnykh formakh atmosfernoi tsirkulyatsii v letnee vremya (The Weather Regime Over China Corresponding to the Principal Forms of Atmospheric Circulation in Summer). — Trudy GGO, Vol. 90. 1960. See also Acta meteorol. sinica, 29 (2). 1958.
47. Shklyaev, A. S. Osobennosti raspredeleniya osadkov i stoka na Srednem i Yuzhnom Urale i ikh svyaz' s atmosfernoi tsirkulyatsiei (Features in the Distribution of Precipitation and Runoff of the Central and South Urals and their Relationship with Atmospheric Circulation). — Uchenye zapiski Permskogo gosudarstvennogo universiteta imeni A. M. Gor'kogo. No. 112. 1964.
48. Rossby, C. G. Relation Between the Variations in the Intensity of Zonal Circulation of the Atmosphere and the Displacements of the Semipermanent Centres of Action. — J. Marine Res., Vol. 2, pp. 38—55. 1939.

D. L. Laikhtman, B. A. Kagan, and V. V. Timonov

METHODS OF STUDYING OCEAN-ATMOSPHERE INTERACTION

The ocean and atmosphere are treated as a single system consisting of four interrelated layers: a) the free atmosphere; b) boundary layer of the atmosphere; c) boundary layer of the ocean; d) deep ocean layers. Hydrothermodynamic equations are written out for each layer and conditions are formulated relating the solutions of these equations. Examples are presented of the solution of model equations with individual simplifying assumptions. In conclusion possible ways of study of the ocean-atmosphere system are indicated, these being: a) treating the problem of atmosphere-ocean interaction as that of the physics of a single ocean-atmosphere system, paying particular attention to zones where they are in direct contact, and to centers of intensive interactions between the ocean and atmosphere both in space and in time; b) developing the physical theory of ocean-atmosphere interaction by solving the two-layer problem and by subsequently checking the theory; c) studying the ocean-atmosphere interaction with a view to establishing cause-and-effect relationships; d) developing experimental methods of studying the ocean-atmosphere interaction.

Effect of the sea on the thermal regime of the atmosphere. As a result of intensive mixing of water in the upper layer of the sea and its high specific heat, the sea participates far more actively in heat exchange with the atmosphere than the land. This is of particular importance in transient processes during which large quantities of heat are exchanged between the sea and atmosphere.

Heat exchange between the sea and atmosphere at any given instant depends to a large extent on the preceding thermal regime; the sea has a large "thermal memory." Features peculiar to the sea become very evident when introducing a quantitative description of the "thermal memory" of various media. Let us assume that the isothermal active soil or sea layer is infinitely thick and is initially at temperature T_0. Then the temperature changes instantaneously to a value $T_1 \neq T_0$.

It is natural that this results in heat flux to the surface ($T_1 > T_0$) or from it ($T_1 < T_0$). The temperature of the active soil or sea layer starts changing and approaches the value T_1 over its entire thickness. The heat flux to the surface is

$$B = (T_1 - T_0)\frac{\sqrt{a\widetilde{\rho c}}}{\sqrt{\pi t}},$$

where a is the thermal diffusivity; $\widetilde{\rho c}$ is the specific heat per unit volume; and t is the time.

We shall apply the term "thermal memory" to the time during which the heat flux at the surface decreases to 0.05 cal/cm^2·min for an initial temperature difference of 5°. Then the thermal memory (in hours) can be calculated from the expression

$$t=\frac{(T_1-T_0)^2}{\pi}\left(\frac{\sqrt{a\rho}\,\tilde{c}}{B}\right)^2=\frac{10^4}{3.14}(\sqrt{a}\,\tilde{\rho c})^2.$$

Values of the thermal memory for various media are listed in the table.

TABLE. The thermal memory of various media

Medium	$\sqrt{a}\,\tilde{\rho}\tilde{c}$, cgs units	$(\sqrt{a}\,\tilde{\rho}\tilde{c})^2$, cgs units	t, hours
Fresh snow	0.002	$4\cdot10^{-6}$	$1.27\cdot10^{-2}$
Old snow	0.012	$1.44\cdot10^{-4}$	0.459
Dry sand	0.011	$1.21\cdot10^{-4}$	0.386
Wet sand	0.04	$1.6\cdot10^{-3}$	5.10
Organic soil	0.04	$1.6\cdot10^{-3}$	5.10
Wet marshy soil.	0.038	$1.444\cdot10^{-3}$	4.60
Ice	0.05	$2.5\cdot10^{-3}$	7.96
Standing water , ...	0.039	$1.521\cdot10^{-3}$	4.85
Water under highly stable conditions, moderate currents	0.32	0.1024	326
Air under highly stable stratification . . .	0.01	$1\cdot10^{-4}$	0.318
Air under equilibrium stratification . . .	0.1	$1\cdot10^{-2}$	31.8

On the physical theory of ocean-atmosphere interaction. The atmosphere and the ocean are not isolated media; on the other hand, continuous exchange of momentum, heat, moisture and salts exists between them. Strictly speaking the atmosphere and ocean comprise a single system, stratified in a certain manner. The type of exchange between the sea and atmosphere is determined by the turbulence in the boundary layers of both media. The chief cause for the appearance of boundary layers is a stepwise change in densities at the water-air interface. It is precisely this change in the atmosphere at the sea surface which is responsible for large vertical velocity gradients and their related instabilities of air flow and turbulence. The turbulent tangential stress which is exerted by the moving air flow at the sea surface sets the upper water layer into motion, carrying along with it lower-lying water layers. As one moves away from the surface, the velocity of this drift current decreases, approaching zero. Due to this, marked vertical velocity gradients may arise in the upper layer of the sea and this in turn may produce turbulence.

The turbulence intensity in both media depends on the vertical velocity and density gradients which, in turn, are acted upon by turbulence. Hence the velocity and density distributions and the turbulence intensity in the boundary layers of the atmosphere and sea, as well as the eddy momentum, heat and moisture fluxes, characterizing the energy exchange between the sea and atmosphere produced by them, cannot be regarded as independent variables. They must be sought simultaneously from specified external parameters.

For the system formed by the two boundary layers of the atmosphere and sea, these parameters are: meteorological elements at the upper boundary of the atmospheric boundary layer (geostrophic wind, temperature, moisture and radiation balance) and hydrological elements at the lower boundary of the boundary layer of the sea (geostrophic current, temperature and salinity). It is precisely these factors which, forming boundary layers, determine the given structure (average profiles of meteorological and hydrological elements, turbulence parameters and eddy fluxes of these layers), as well as the water-air interface and internal relationships between individual elements.

The external parameters of the two boundary layers in their turn are determined by the interaction of the boundary layer of the atmosphere with the free atmosphere and the boundary layer of the sea with the deep layers. However, it is known that as one moves away from the interface, the eddy momentum, heat, moisture and salt fluxes decrease and vanish at the outer boundary of the boundary layer. It is obvious that the relationship between these layers should not be implemented by means of eddy momentum, heat, moisture and salt transport, but by means of processes on an entirely different scale (of the order of magnitude of the scale of the general circulation of the atmosphere).

The connecting links in the boundary layer of the atmosphere (free atmosphere system) and in the boundary layer of the sea (deep layers system) are vertical friction currents arising in these boundary layers.

Exchange between properties of the boundary layers and the external medium occurs by means of these very vertical motions. Thus, if as a result of operation of some factors (for example, changes in baric formation) the external parameters of the free atmosphere change, then in the process of interaction these changes are transmitted to the sea and atmosphere, i. e., a mutual reconstruction of the water and air layers arises. This reconstruction extends to the boundary layers of the atmosphere and of the sea by means of eddy exchange, and is then transmitted to the free atmosphere and deep layers of the sea by means of vertical friction currents. This is the mechanism of the back effect of boundary layers on the free atmosphere and on the deep layers of the sea.

O n t h e i n t e r a c t i o n i n t h e s e a - a t m o s p h e r e s y s t e m. The entire atmosphere-sea system can be subdivided into four layers: a) free atmosphere; b) boundary layer of the atmosphere; c) boundary layer of the sea; d) deep layers of the sea.

We shall utilize the fact that the terms in the equations describing the structure of the boundary layer of the atmosphere, which reflect the effect of unsteadiness and horizontal inhomogeneity, are relatively small and may be disregarded relative to the other terms (at least in the first approximation). By virtue of this it may be assumed that the meteorological elements in the atmospheric boundary layer do not vary with time and with horizontal coordinates. The same assumptions can also be made for the boundary layer equations of the sea. Here, however, it is expedient to retain the partial time derivative of the temperature. Then the equations of hydrodynamics for each of the aforementioned layers can be written in the following manner.

Layer a:

$$\frac{\partial u_i}{\partial t} + u_\alpha \frac{\partial u_i}{\partial x_\alpha} = -\frac{1}{\rho}\frac{\partial p}{\partial x_i} + \frac{1}{\rho} K_i, \qquad (1)$$

$$\frac{\partial \rho}{\partial t} + \frac{\partial \rho u_\alpha}{\partial x_\alpha} = 0, \qquad (2)$$

$$\frac{\partial T}{\partial t} + u_\alpha \frac{\partial T}{\partial x_\alpha} - \frac{\gamma_a}{\rho g}\left(\frac{\partial p}{\partial t} + u_\alpha \frac{\partial p}{\partial x_\alpha}\right) = 0, \tag{3}$$

$$\frac{\partial q}{\partial t} + u_\alpha \frac{\partial q}{\partial x_\alpha} = 0, \tag{4}$$

$$\frac{\partial p}{\partial x_3} = -\rho g, \tag{5}$$

$$p = \rho R T. \tag{6}$$

Layer d:

$$\frac{\partial \tilde{u}_i}{\partial t} + \tilde{u}_\alpha \frac{\partial \tilde{u}_i}{\partial x_\alpha} = -\frac{1}{\tilde{\rho}} \frac{\partial \tilde{p}}{\partial x_i} + \frac{1}{\tilde{\rho}} \tilde{K}_i, \tag{7}$$

$$\frac{\partial \tilde{\rho}}{\partial t} + \frac{\partial \tilde{\rho} \tilde{u}_\alpha}{\partial x_\alpha} = 0, \tag{8}$$

$$\frac{\partial \tilde{T}}{\partial t} + \tilde{u}_\alpha \frac{\partial \tilde{T}}{\partial x_\alpha} = 0, \tag{9}$$

$$\frac{\partial S}{\partial t} + \tilde{u}_\alpha \frac{\partial S}{\partial x_\alpha} = 0, \tag{10}$$

$$\frac{\partial \tilde{p}}{\partial x_3} = \tilde{\rho} g, \tag{11}$$

$$\tilde{\rho} = \tilde{\rho}(\tilde{T}, S). \tag{12}$$

Layer b:

$$\frac{\partial}{\partial z} k \frac{\partial u_i}{\partial z} + \frac{1}{\rho} K_i = \frac{1}{\rho} \frac{\partial p}{\partial x_i}, \tag{13}$$

$$\alpha_1 \frac{\partial}{\partial z} k \frac{\partial \theta}{\partial z} + \frac{1}{\rho c_p} \frac{\partial Q}{\partial z} = 0, \tag{14}$$

$$\frac{d}{dz} k \frac{dq}{dz} = 0, \tag{15}$$

$$\frac{\partial \rho u_\alpha}{\partial x_\alpha} = 0, \tag{16}$$

$$k\left[\left(\frac{du}{dz}\right)^2 + \left(\frac{dv}{dz}\right)^2\right] - \alpha_1 k \frac{g}{\theta} \frac{d\theta}{dz} - c\frac{b^2}{k} - \alpha_2 \frac{d}{dz} k \frac{db}{dz} = 0, \tag{17}$$

$$k = l\sqrt{b}, \tag{18}$$

$$l\left[\frac{du}{dz}\frac{d^2u}{dz^2} + \frac{dv}{dz}\frac{d^2v}{dz^2} - \frac{\alpha_i}{2}\frac{g}{\theta}\frac{d^2\theta}{dz^2}\right] + \frac{\varkappa}{2}\left[\left(\frac{du}{dz}\right)^2 + \left(\frac{dv}{dz}\right)^2 - \frac{g}{\theta}\alpha_1\frac{d\theta}{dz}\right] = 0. \tag{19}$$

Layer c:

$$\frac{\partial}{\partial \xi} \tilde{k} \frac{\partial \tilde{u}_i}{\partial \xi} + \frac{1}{\tilde{\rho}} \tilde{K}_i = \frac{1}{\tilde{\rho}} \frac{\partial \tilde{p}}{\partial x_i}, \tag{20}$$

$$\frac{\partial}{\partial \xi} \tilde{\alpha}_1 \tilde{k} \frac{\partial T}{\partial \xi} = \frac{\partial T}{\partial t}, \tag{21}$$

$$\frac{\partial}{\partial \xi} \tilde{\alpha}_1 \tilde{k} \frac{\partial S}{\partial \xi} = 0, \tag{22}$$

$$\frac{\partial \tilde{\rho} \tilde{u}_\alpha}{\partial x_\alpha} = 0, \tag{23}$$

$$\tilde{k}\left[\left(\frac{du}{d\xi}\right)^2 + \left(\frac{dv}{d\xi}\right)^2\right] - \tilde{\alpha}_1 \tilde{k} \frac{g}{\tilde{\rho}} \frac{d\tilde{\rho}}{d\xi} - c\frac{\tilde{b}^2}{\tilde{k}} + \tilde{\alpha}_2 \frac{d}{d\xi} \tilde{k} \frac{d\tilde{b}}{d\xi} = 0, \tag{24}$$

$$\tilde{k} = \tilde{l}\sqrt{\tilde{b}}, \tag{25}$$

$$\tilde{l}\left[\frac{d\tilde{u}}{d\xi}\frac{d^2\tilde{u}}{d\xi^2} + \frac{d\tilde{v}}{d\xi}\frac{d^2\tilde{v}}{d\xi^2} - \frac{\tilde{\alpha}_1}{2}\frac{g}{\tilde{\rho}}\frac{d^2\tilde{\rho}}{d\xi^2}\right] + \frac{\tilde{\varkappa}}{2}\left[\left(\frac{d\tilde{u}}{d\xi}\right)^2 + \left(\frac{d\tilde{v}}{d\xi}\right)^2 - \frac{g}{\tilde{\rho}}\tilde{\alpha}_1\frac{d\tilde{p}}{d\xi}\right] = 0. \tag{26}$$

Here $i = 1, 2$ and $u_1 = u, u_2 = v$; $\alpha = 1, 2, 3$ and $x_1 = x, x_2 = y, x_3 = z$ or ξ; the z axis is directed in the atmosphere upward and the ξ axis in the sea is directed downward; $K_1 = K_x = -2\omega_z \rho v$; $K_2 = K_y = 2\rho\omega_z u$; the remaining notation is standard.

Fquations (1, 7, 13, 20) are equations of motion, (2, 8, 16, 23) are continuity equations, (3, 9, 14, 21) are heat conduction equations, (4, 15) are moisture diffusion equations, (10, 22) are salt diffusion equations, (5, 11) are equations of statics, (6, 12) are equations of state, (17, 24) are equations of eddy energy balance, (18, 25) are Kolmogorov equations relating the eddy transfer coefficient to the eddy energy, (19, 26) are the generalized representation of the Karman mixing length obtained by Laikhtman and Zilitinkevich.

System of equations (1 – 26) should satisfy the following boundary conditions:

$$\rho w\,|_{z\to\infty} \to 0; \quad \widetilde{\rho}\widetilde{w}\,|_{\xi\to\infty} \to 0,$$

$$w\,|_{z=H} = w_0 = -\int_0^H \left(\frac{\partial u}{\partial x} + \frac{\partial v}{\partial y}\right) dz,$$

$$\widetilde{w}\,|_{\xi=\widetilde{H}} = \widetilde{w}_0 = -\int_0^H \left(\frac{\partial \widetilde{u}}{\partial x} + \frac{\partial \widetilde{v}}{\partial y}\right) d\xi,$$

$$u\,|_{z=H-0} = u\,|_{z=H+0}, \quad v\,|_{z=H-0} = v\,|_{z=H+0}, \quad w\,|_{z=H-0} = w\,|_{z=H+0},$$

$$T\,|_{z=H-0} = T\,|_{z=H+0}, \quad q\,|_{z=H-0} = q\,|_{z=H+0},$$

$$\left.\frac{du}{dz}\right|_{z=H-0} = \left.\frac{du}{dz}\right|_{z=H+0}, \quad \left.\frac{dv}{dz}\right|_{z=H-0} = \left.\frac{dv}{dz}\right|_{z=H+0},$$

$$\widetilde{u}\,|_{\xi=\widetilde{H}-0} = \widetilde{u}\,|_{\xi=\widetilde{H}+0}, \quad \widetilde{v}\,|_{\xi=\widetilde{H}-0} = \widetilde{v}\,|_{\xi=H+0}, \quad \widetilde{w}\,|_{\xi=\widetilde{H}-0} = \widetilde{w}\,|_{\xi=\widetilde{H}+0},$$

$$\widetilde{T}\,|_{\xi=\widetilde{H}-0} = \widetilde{T}\,|_{\xi=H+0}, \quad S\,|_{\xi=\widetilde{H}-0} = S\,|_{\xi=H+0},$$

$$\left.\frac{d\widetilde{u}}{d\xi}\right|_{\xi=\widetilde{H}-0} = \left.\frac{d\widetilde{u}}{d\xi}\right|_{\xi=\widetilde{H}+0}, \quad \left.\frac{d\widetilde{v}}{d\xi}\right|_{\xi=\widetilde{H}-0} = \left.\frac{d\widetilde{v}}{d\xi}\right|_{\xi=H+0}.$$

Elements $u, v, T, q, \widetilde{u}, \widetilde{v}, \widetilde{T}$ and S at the lateral surfaces of the three-dimensional region under study must be specified.

It remains to express the boundary conditions at the water-air interface. It should be remembered that this interface is not plane and stationary, but undulatory with a certain translational velocity. The boundary conditions in describing profiles of meteorological elements at the water-surface layer of the atmosphere are not usually satisfied at the physical surface proper, but at the roughness surface. No consideration is given to the fact that a thin transitional layer always exists upon transition from one medium to the other. The transitional layer under water-surface conditions is the layer within which the physical water-air interface oscillates. However, since the amplitude of oscillations of the physical water surface is a variable quantity, depending on wind speed, temperature stratification in both media, etc., in other words on the internal parameters of the system proper, then, naturally, the thickness proper of the transitional layer is also not constant and should depend on the external parameters. The specific properties of this layer are associated with the fact that regularities of processes occurring within it cannot be described by averaged equations used for describing the boundary layers of the atmosphere and sea.

Since all the physical variables in this layer have large gradients, one cannot regard it as a plane and expect a continuous transition from one medium to the other. Apparently it would have been more correct to postulate a stepwise transition of some of the physical variables. Unfortunately, this important problem of the physics of the atmosphere-sea system has hardly been developed until now. Hence when specifying boundary conditions at the atmosphere-sea interface one must make the ordinary rough assumption that

$$z = z_0, \quad u = \tilde{u}, \quad v = \tilde{v},$$
$$k\rho \frac{du}{dz} = -\tilde{k}\tilde{\rho} \frac{d\tilde{u}}{d\xi}, \quad k\rho \frac{dv}{dz} = -\tilde{k}\tilde{\rho} \frac{d\tilde{v}}{d\xi},$$
$$T = \tilde{T}, \quad q = q_m(T_0), \quad S = S_0,$$
$$-\alpha_1 k\rho c_p \frac{\partial\theta}{\partial z} - \tilde{\alpha}_1 \tilde{k} \tilde{\rho} \tilde{c} \frac{\partial\tilde{T}}{\partial\xi} - Lk\rho \frac{\partial q}{\partial z} = R,$$
$$w = \tilde{w} = 0.$$

However, some conclusions on the variables in the transitional layer, i. e., on the mean wavelength λ, mean wave height h, period T^* and eddy exchange coefficient k^* averaged within this layer, can still be drawn by using dimensional analysis. Clearly, the variables of processes in the transitional layer for the quasisteady state of locally produced undulation depend on the following dimensional quantities: dynamic velocity v_*, density ρ of the air and $\tilde{\rho}$ of the water, and gravitational acceleration g. Dimensional analysis yields

$$\lambda = \frac{v_*^2}{g} f_1\left(\frac{\tilde{\rho}}{\rho}\right), \tag{27}$$

$$h = \frac{v_*^2}{g} f_2\left(\frac{\tilde{\rho}}{\rho}\right), \tag{28}$$

$$T^* = \frac{v_*}{g} f_3\left(\frac{\tilde{\rho}}{\rho}\right), \tag{29}$$

$$k^* = \frac{v_*^3}{g} f_4\left(\frac{\tilde{\rho}}{\rho}\right). \tag{30}$$

Equations (27) – (30) are in agreement with existing empirical relationships. Since the argument $\tilde{\rho}/\rho$ is practically constant, functions f_1, f_2, f_3 and f_4 can be replaced by constant coefficients determined from experimental data.

The parameters of the transitional layer were derived on the assumption of quasisteady state, i. e., it was assumed that the parameters of these layers at all times are in equilibrium with external conditions. It is easy to estimate conditions under which this assumption is justified. Obviously, the wave energy corresponding to changes in external conditions is controlled by the dissipation of this energy as eddy energy. However, then, in considering the unsteady process, one should consider two time scales: the external (t_{ext}), pertaining to the interval during which the tangential stress changes by an order of magnitude, and the internal (t_{int}), depending on parameters determining the characteristics of turbulence in the wave motion, i. e., on v_*, g, ρ and $\tilde{\rho}$. It follows from this that

$$t_{\mathrm{int}} = \frac{v_*}{g} \varphi\left(\frac{\widetilde{\rho}}{\rho}\right).$$

Consequently, the assumption of quasisteadiness is justified if

$$t_{\mathrm{ext}} \gg \frac{v_*}{g} \varphi\left(\frac{\widetilde{\rho}}{\rho}\right).$$

If $\varphi\left(\frac{\widetilde{\rho}}{\rho}\right) = 0\,(1)$, then $t_{\mathrm{ext}} \gg 0.5$ hours.

Solution of system of equations (1)—(30) with the given boundary conditions makes it possible to calculate the wind, temperature, moisture and vertical velocity distributions in the atmosphere; eddy momentum, heat and moisture fluxes; wave elements at the interface; current, temperature, salinity and vertical velocity profiles in the sea; and the distribution of eddy coefficients in the boundary layers of both media.

When the sea is ice-covered, the transitional layer in the atmosphere-sea system should be replaced by the ice layer. In this case system of equations (1)—(26) should be supplemented by equations of motion and heat conduction for ice:

$$\left.\begin{aligned} \widetilde{k}\widetilde{\rho} \frac{du}{dz}\Big|_0 + \widetilde{k}\widetilde{\rho} \frac{d\widetilde{u}}{d\xi}\Big|_0 + 2\omega_z m v_0 &= -gm \frac{\partial \zeta}{\partial x} \\ k\rho \frac{dv}{dz}\Big|_0 + \widetilde{k}\widetilde{\rho} \frac{d\widetilde{v}}{d\xi}\Big|_0 - 2\omega_z m u_0 &= -gm \frac{\partial \zeta}{\partial y} \end{aligned}\right\} \tag{31}$$

$$\frac{\partial \widetilde{\theta}}{\partial t} = k_{\mathrm{ice}} \frac{\partial^2 \widetilde{\theta}}{\partial h^2}, \tag{32}$$

where m is the mass of ice ɔer unit area; ζ is the departure of the sea surface from the horizontal; k_{ice} is the thermal diffusivity coefficient for ice; and h is a vertical coordinate, measured downward from the upper ice surface.

Then equations (1)—(26) and (31), (32) should be integrated subject to the following boundary conditions at the atmosphere-ice and ice-sea interfaces:

$$\begin{gathered} z = z_0, \quad u = u_0, \quad v = v_0, \\ -k\rho c_p \alpha_1 \frac{\partial \theta}{\partial z} - k_{\mathrm{ice}} \rho_{\mathrm{ice}} c_{\mathrm{ice}} \frac{\partial \theta}{\partial h} - Lk\rho \frac{\partial q}{\partial z} = R, \\ \theta = \widetilde{\theta}, \quad q = q_m\ (T_0), \quad w = 0, \\ \xi = \xi_0, \quad \widetilde{u} = u_0, \quad \bar{v} = v_0, \\ \widetilde{\theta} = \theta_0, \quad S = S_0, \quad \widetilde{w} = 0. \end{gathered}$$

Here z_0 and ξ_0 are roughness parameters of the upper and lower ice surfaces, respectively.

C o n s t r u c t i o n   o f   a   s i m p l i f i e d   m o d e l. It is impossible to obtain an analytic solution of system of hydrodynamic equations (1)—(30) describing the interaction between the atmosphere and sea, or of equations (1)—(26) and (31), (32) pertaining to the interaction between the atmosphere, ice cover and sea. It was hence found expedient to construct simplified models of the interaction between these media. Thus, references /2, 3/ consider the thermal and dynamic interaction in the atmosphere-sea and atmosphere-ice systems on the assumption that the external parameters of the free atmosphere and of the deep layers of the sea are specified, i. e., in the final count the problem considered is that of interaction between the boundary layers of the two media. In addition, the eddy energy balance equation for the two boundary layers was taken in these references in a simplified, integral form, which allows one to disregard unknown terms describing the diffusion of eddy energy from one layer to another. At the same time the use of the integral form of eddy energy balance equations requires specification of a model for the variation of the eddy exchange coefficients within the two boundary layers.

It was shown by a relatively large volume of gradient observations in the water-adjoining layer of the atmosphere, that the eddy coefficient in this layer increases with altitude. On the other hand, very little is known about the eddy coefficient in the remaining part of the atmospheric boundary layer.

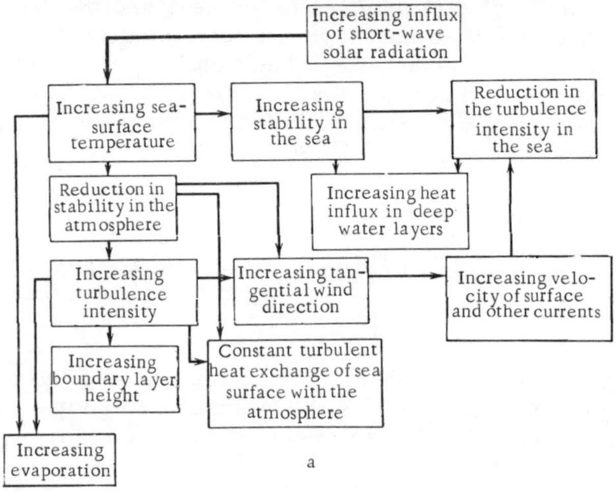

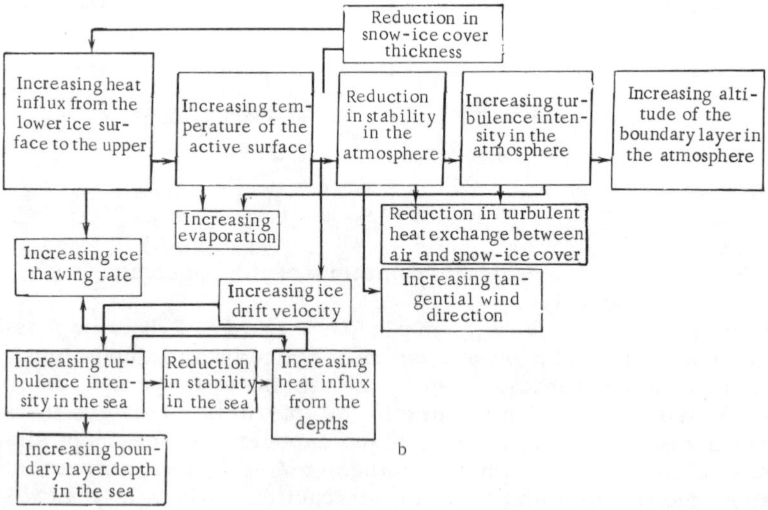

FIGURE 1. Schematic presentation of the dependence of the structure of boundary layers of the atmosphere (a) and sea (b) on external parameters.

As in the case of the atmosphere, the rate at which the eddy coefficient decreases with sea depth is unknown. It is hence expedient to replace the real distribution of the eddy coefficient in the boundary layer of the sea by its average value. It should, however, be emphasized here that such approximation of the real distribution of the eddy coefficient in the boundary layer of these media does not at all free us from the need to properly take into account the dependence of these coefficients on the geostrophic wind, current and vertical stratification, since the horizontal nonuniformity of their field produces a nonuniform field of eddies and, consequently, mesoscale phenomena of importance in short-range weather forecasting.

When approximating the real distribution of the eddy coefficient in the boundary layer of the atmosphere by a model "with an inflection" and specifying an eddy coefficient in the boundary layer of the sea which would not depend on depth, it became possible to calculate the characteristics of interaction between the sea and atmosphere under specific physico-geographical conditions /1, 4/. It was assumed in these references that conversion of eddy energy into heat is directly proportional to the transformation of energy of averaged motion into eddy energy; here the proportionality factors δ and $\bar{\delta}$ were assumed to be constant with altitude (in the atmosphere) and with depth (in the sea). This assumption makes it impossible to use the eddy energy balance equation under equilibrium stratification conditions. However, if we were to follow the idea presented in /5/, then we can dispense with this assumption, and then the eddy energy balance equation can be used for determining the magnitude of the eddy coefficient in the boundary layers of media under any stratification conditions.

Comparison of predicted with observed results. Calculations using the above formulas made it possible to estimate qualitatively the manner in which variations in individual factors change the entire structure of the system comprising the two boundary layers. Figure 1 shows the dependence of the structure of these layers on the external parameters of the problem at hand.

Figure 2 presents results of quantitative computation of the distribution of thermal and dynamic interaction characteristics between the sea and atmosphere along the tropical Atlantic — North Pole meridional section, obtained in /1/. Reference /4/ presents the calculated distribution, for the North Atlantic, in summer, of the height of the boundary and water-adjoining atmospheric layers, absolute value of the wind speed at the 1 m level in the atmosphere, components of the velocity of surface current, water temperature at different depths, eddy coefficients in the two boundary layers, boundary layer thickness of the ocean, components of the heat balance and of the tangential wind stress.

It is seen by examining the meridional and space distribution of these characteristics that the above results are in satisfactory agreement with existing physical considerations and regional features of the area under study.

Unfortunately, the heights of the boundary and water-surface atmospheric layers, as well as the depth of the boundary layer in the ocean thus obtained cannot be compared with direct measurements of these quantities, since the latter have not been made. The scattered measurements of the turbulence intensity which are presently available can be used only for comparing the

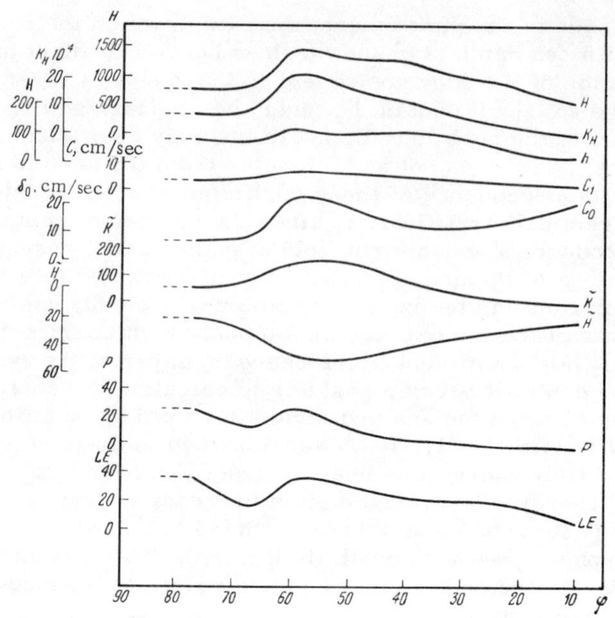

FIGURE 2. Variation in the characteristics of thermodynamic interaction betweeen the sea and atmosphere on a horizontal section in winter from the tropical Atlantic to the North Pole.

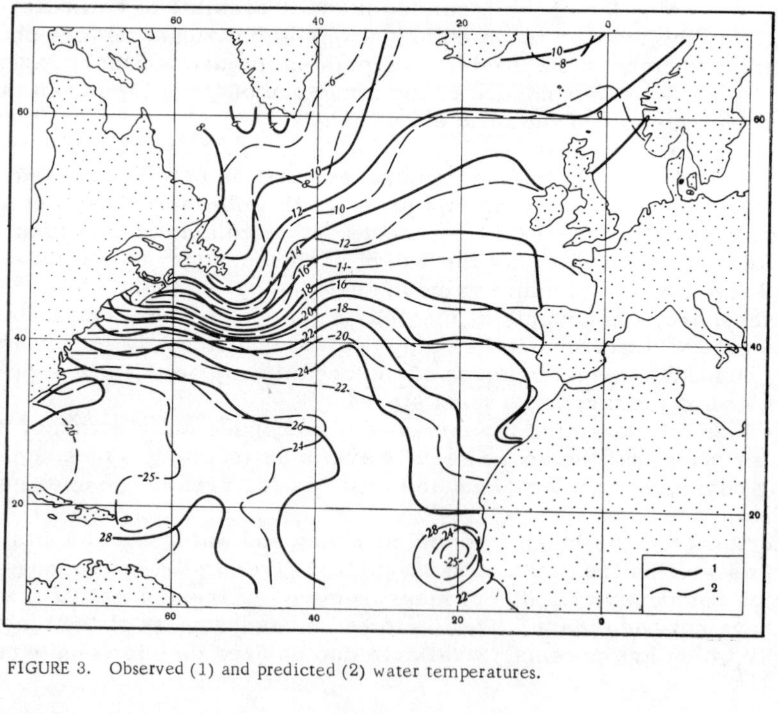

FIGURE 3. Observed (1) and predicted (2) water temperatures.

orders of magnitude of the predicted and observed magnitudes of eddy coefficients in the two boundary layers and for confirming the correctness of calculations.

At present the number of direct measurements of momentum, heat and water vapor fluxes does not suffice for comparing them with calculations. On the other hand, existing semiempirical methods for calculating the tangential wind stress, the eddy exchange between the sea surface and atmosphere and the heat expended upon evaporation in summer are suitable for comparison with theoretical calculations. Near-equilibrium conditions are established during this season in the water-surface layer of the atmosphere over the ocean. The agreement between the values of eddy heat and water vapor flux calculated in /4/ and those obtained by a method employing the heat balance equation of the sea surface /6/ was found to be quite satisfactory. It is difficult to directly compare the calculated and observed wind speed and current profiles due to the almost total unavailability of the necessary experimental data. Hence one may use the relatively numerous experimental data on wind coefficients:

Author	Wind coefficient	Author	Wind coefficient
Hela	0.0092	Rossby and Montgomery	0.0169
Palmén	0.0092	Dizer	0.0181
Thorade	0.0092	Stommel	0.0193
Wiedemann	0.0107	Chatley	0.0200
Witting	0.0107	Tickner	0.0200
Dinklage	0.0107	Yao	0.0200
Struiskii	0.0140	Kellegan	0.0221
Ekman	0.0147	Francis	0.0221
Screton	0.0134–	Hughes	0.0221
	0.0200	Van Dorn	0.0241
Stevenson	0.0134–	Tomczak	0.0294
	0.0200	Gallé	0.0295
Munk	0.0167–	Mohn	0.0328
	0.0200		

It is seen from this table that the information on the wind coefficient at present is quite contradictory. The scatter in individual values of this coefficient is usually attributed to measurement errors. Without denying the existence of these errors one must particularly emphasize the fact that, first, the above results were obtained from observations of various drifting bodies and hence do not pertain to the motion of the topmost layer, but a layer of finite thickness which was not identical in the different measurements. Second, the wind speed used in calculating the wind coefficients was measured by the three different authors at different heights above the sea surface (from 5 to 20 m). In addition, wind speed measurements proper carried out from a ship are unreliable due to the air flow distortion by the ship's superstructure. The most reliable data (those due to Hughes and Tomczak) are in satisfactory agreement with theoretical conclusions.

It is also possible to estimate the reliability of the calculated absolute values of the height of the atmospheric boundary layer, of the depth of the sea boundary layer, of the turbulence intensity in both cases, of eddy momentum, heat and water vapor fluxes indirectly, by comparing predicted and measured water temperatures. These were actually compared in Figure 3. The standard computational error was 2°. Considering the large regional differences, the temperatures may be regarded as sufficiently close. This allows us to assume that satisfactory quantitative estimates can be obtained also for the remaining variables found in the process of solving the problem. Thus, in spite of the assumptions made, the suggested model of interaction of the two boundary layers yields results which agree with available experimental data.

In connection with the above an attempt was made to use this model also for developing a scheme for short-range forecasting of the weather and hydrologic characteristics at the sea surface /5/. This work is based on the assumption that the changes in hydrologic elements in deep sea layers occur at extremely slow rates. In this case, with reference to short-range forecasts, the values of the hydrologic elements at the outer boundary of the boundary layer in the integration of the system of equations governing the free atmosphere and the two boundary layers can be replaced by their initial values. Then, specifying values of the eddy coefficients which do not vary within the two boundary layers, it is possible to find expressions for the surface-current velocity, sea-surface temperature and wind-driven wave characteristics by solving the previously presented system of equations and boundary conditions. Reference /5/ suggests a scheme for forecasting these elements for 1 — 3 days using an electronic computer.

On the problem as a whole and on ways of studying it. The construction and development of a physical theory of ocean-atmosphere interaction cannot be restricted to examination of the contact zone of these media alone. The theoretical scheme presented here considers the entire boundary layer of the atmosphere and the entire active layer of the ocean, with the latter replaced by the friction layer only in the first approximation. However, these layers too are not independent of the remaining layers of the atmosphere and hydrosphere. Hence the question of the physical interaction between the ocean and atmosphere is the question of the physics of the ocean-atmosphere system. This formulation can also be supported by the fact that both these moving envelopes of the planet, when subjected to external effects, react to them precisely as a single system which, in addition, has the ability to develop within itself self-oscillatory processes, which clearly attain substantial scales.

Phenomena which at first sight appear to be quite independent point to interactions between processes in the atmosphere and oceans. It suffices to mention the formation of properties of bottom ocean waters or the effect of the underlying surface, to become convinced once more that the contact between the two media is not damped out even at a distance of 1000 m from the sea-air interface.

Even tidal phenomena, which, as a result of their cosmic origin and of the fact that they extend over the entire thickness of both media, are usually considered separately, actually depend in both media on the interaction of the latter. In their turn, tidal phenomena form to a large extent the mixing processes and thus affect the interaction between the media. Many other such examples could be cited.

We thus conclude that neither the physics of the ocean nor the physics of the atmosphere exist separately, at least in principle.

It should also be emphasized that alongside with the attention that should be paid to the ocean and atmosphere as a single unit, particular attention should be accorded to the zone of their direct contact. In particular a more detailed study is needed of the as yet insufficiently understood mechanism of energy and mass exchange through the interface ("small-scale interaction" in American scientific terminology). Here allowance should be made for the rather important fact that the rate of interaction between the ocean and atmosphere is by far not the same in all regions of the World's ocean. Thus, zones with slow-moving surface waters and systems of strong wind currents exist in different regions; wave maps clearly show stormy and quiet waters; total heat balance maps distinctly depict limited regions of very rapid heat and moisture loss and enormous areas where the heat and moisture exchange through the ocean surface is quite weak. One may therefore refer to zones and centers of intensive ocean-atmosphere interaction. In order to understand how significant this is, it suffices to point out that approximately from 40 to 55% (according to different calculations) of the entire heat transmitted through the surface in the North Atlantic is released through the major heat transfer centers alone.

Thus, heat loss centers in oceans and seas are frequently also centers of deep water convective mixing of waters and formation of deep water masses; these are also zones of particularly strong air-mass transformations and centers of their instability, centers where heat is carried to upper atmosphere layers; in addition they are zones which have a marked effect on the formation of atmospheric centers of action and atmospheric circulation.

The study of centers and zones of intensive ocean-atmosphere interaction should thus be developed from the viewpoint of geography, by developing more rapidly the "geography of interaction," determining its "special points," as well as from the physical point of view, by the determination of general and local features of interaction between media at these points and zones. This approach should be made not only in space coordinates, but also in time, by studying the interaction between media, particularly during those months, seasons and years when it is very pronounced.

Especially great importance, however, is had by another aspect of the problem related to time: by the study of changes of processes resulting from the interaction of these media and, accordingly, of changes in the state of the ocean-atmosphere system. To understand the regularities of these variations (short-term, yearly, year-to-year, epochal) is probably the most difficult aspect of the entire problem. However, it is also the most important one, since only its solution will lead to forecasting of the state of the ocean and atmosphere, i. e., to the main applied goal of its study. In addition, it is precisely the study of changes in the ocean and atmosphere which will aid in determining the mechanism of their interaction.

The solution of this problem can be helped by all the above considerations on the approach to its study. But these should be supplemented by still another: in the study of the changes of the state of the ocean-atmosphere system one must also study the causes of these changes.

Those causes which are themselves a part of the mechanism of the ocean-atmosphere interaction, of the peculiarities in the operation of the complex heat machine which these media represent, will be discovered by themselves, in the process of development of the entire problem. The situation with respect to external, particularly extraterrestrial factors is quite different.

The above refers primarily to the rapid development and acceleration of studies regarding the effect of solar activity changes on the atmosphere and hydrosphere. It is most important here to determine the mechanism proper of transmission of solar activity oscillations and of their effect on processes in the atmosphere and ocean. This will provide a basis for understanding one of the main levers controlling changes in the state of the ocean and atmosphere, and will make it possible to put an end much faster to the forced, but not entirely normal, "blind" construction of hydrometeorological forecasts on the basis of heliogeographic relationships.

However, particular attention must be paid: to the development of a physical theory of ocean-atmosphere interaction by solving the two-layer problem, to subsequent checking of this theory, to the creation on its basis of reliable methods of theoretical calculations and their use in determining the mechanism of interaction and the role of individual factors, to the determination of its regional and planetary regularities, to the study of the change in the state of the ocean-atmosphere system under different conditions and, finally, to the development of hydrodynamic methods of short- and long-range forecasting of the state of the system.

References

1. Bogdanova, G. V. and B. A. Kagan. Opyt rascheta kharakteristik termicheskogo i dinamicheskogo vzaimodeistviya morya i atmosfery (Experience in Calculating Characteristics of the Thermal and Dynamic Interaction Between the Sea and Atmosphere). — Trudy GGO, Vol. 167. 1965.
2. Kagan, B. A. and Z. M. Utina. K teorii termodinamicheskogo vzaimodeistviya morya i atmosfery (On the Theory of the Thermodynamic Interaction Between the Sea and Atmosphere). — Okeanologiya, No. 2. 1963.
3. Kagan, B. A. O termodinamicheskom vzaimodeistvii v sisteme more-led-atmosfera (On the Thermodynamic Interaction in the Sea-Ice-Atmosphere System). — Trudy GGO i Ukrainskoi NIGMI, Vol. 144/40. 1963.
4. Kagan, B. A. O vzaimodeistvii pogranichnykh sloev atmosfery i okeana v Severnoi Atlantike (On the Interaction Between Boundary Layers of the Atmosphere and Ocean in the North Atlantic). — Trudy LGMI, Vol. 24. 1965.
5. Laikhtman, D. L. and B. A. Kagan. Skhema predvychisleniya gidrologicheskikh kharakteristik na poverkhnosti morya (A Scheme for Precalculation of Hydrologic Characteristics at the Sea Surface). — Meteorologiya i gidrologiya, No. 5. 1965.

6. Strokina, L. A. Teplovoi balans Severnoi Atlantiki (The Heat Balance of the North Atlantic). — Trudy GGO, Vol. 92. 1959.
7. Timonov, V. V. Issledovanie izmenenii gidrometeorologicheskikh uslovii Severnoi Atlantiki v svyazi s MGG i MGS (A Study of Changes in Hydrometeorological Conditions in the North Atlantic in Connection with the IGY and IQSY). — Okeanologiya, No. 2. 1961.
8. Hughes, P. A Determination of the Relation Between Wind and Sea-Surface Drift. — Quart. J. Roy. Meteorol. Soc., 82(354). 1956.
9. Tomczak, G. Neuere Untersuchungen mit Treibkörpern zur Bestimmung des Windeinflusses auf Oberflächenströmungen im Meer. — Ber. Dtsch. Wetterdienstes, 91 (2). 1963.

S. A. Kitaigorodskii and Yu. A. Volkov

SOME PROBLEMS OF THE PHYSICS OF THE WATER-ADJOINING LAYER OF THE ATMOSPHERE

(in connection with calculation of eddy momentum, heat and moisture fluxes above the sea)

The problem is considered of calculating eddy fluxes of heat, moisture and momentum in the water-adjoining layer of the atmosphere. The following models of this layer are used: 1) a layer above an aerodynamically smooth, stationary wall; 2) a layer above a rough, stationary wall; 3) a layer above a rough, moving wall.

The main difference between a liquid underlying surface and a solid rough wall is, first, that waves, as roughness elements, move with different phase velocities and, second, that turbulent flow transmits to the waves both momentum and energy. A comparison is made of various methods of calculating eddy heat, moisture and momentum fluxes over the ocean. Analysis of these methods shows that they are unreliable and that experiments are needed for refining the coefficients involved in computations.

The study of the general circulation of the atmosphere is frequently based on the energy balance of the ocean-atmosphere system. In its turn, data on this balance are obtained from calculation of heat and moisture exchange and from exchange of mechanical energy between the lower layer of the atmosphere and the surface layer of the ocean.

Development of experimentally verified schemes for such a calculation is a necessary stage in the study of processes of large-scale ocean-atmosphere interaction. At the same time the development of reliable methods of calculation of eddy momentum, heat and moisture fluxes above oceans involves the study of physics of interaction of eddy air fluxes with the upper layer of the atmosphere, and primarily the study of the structure of the water-adjoining atmospheric layer. It is the latter problem which is being discussed in this article.

SOME GENERAL PROBLEMS OF CALCULATING EDDY FLUXES IN THE WATER-ADJOINING LAYER OF THE ATMOSPHERE

At a given instant the undulatory sea surface may have quite a complicated topography. The characteristic vertical size of irregularities of this topography (mean height of visible wave "hills") is in the majority of cases much smaller than the thickness of the water-adjoining layer with constant shear stress $\tau(z) = -\rho_a \overline{u'w'} = \text{const}$ in the eddy flux above the sea surface.

In the steady state the water-adjoining layer of the atmosphere should coincide with the layer of constant vertical shear stress. Apparently, the thickness of this layer above the sea is, at a minimum, several tens of meters. Here this thickness may markedly increase with a rise in the (geostrophic) wind speed.

At the same time, the characteristic amplitude of wind waves, which can be regarded as the characteristic height of the roughness relative to a smooth wall (the undisturbed sea surface) is of the order of one meter. Since an increase in wave height with increasing wind speed is also accompanied by a rise in the thickness of the layer above the sea with constant shear stress, the stationary water hills can be regarded as roughness elements under the majority of hydrometeorological conditions. The latter is the more justified by the fact that the total thickness of the turbulent boundary layer above the sea (thickness of the planetary friction layer) always exceeds the height of wind waves by a factor of several hundred.

Hence the pattern of flow past the stationary undulatory sea surface can be identified with turbulent flow past an infinite rough solid wall (the water density ρ_w being much greater than the air density ρ_a). Then, obviously, steady flow past the stationary undulatory sea surface in the calculation of turbulent currents can be described using known relationships of "wall" turbulence, in particular, for conditions of neutral stratification, using the simplest scheme of the logarithmic boundary layer. When using this scheme one is faced with the determination of the roughness parameter for each specific underlying surface. In the most general case, including the so-called aerodynamically smooth surface, the roughness parameter of a stationary wall in the shape of the undulatory sea surface can depend on the magnitude of eddy friction proper, on the molecular viscosity of air and on the geometric characteristics of the rough-surface topography. It is of importance that even when using this idealized approach one cannot expect to obtain a universal roughness parameter "for the sea" due to the extreme variety of geometric shapes of the undulatory sea surface. The first difficulty in applying methods of the theory of the logarithmic boundary layer to the determination of the magnitude of eddy friction above the sea from data on the wind speed at one level in the self-similar region of developed wall turbulence was usually attributed to this fact.

If no simultaneous observations are made of the average wind speed, mean temperature and average specific air humidity at several levels, one must consider yet another important question: how is it possible to calculate the eddy momentum, heat and moisture fluxes above the ocean from data on average wind speed, mean temperature and average humidity at one altitude in the developed turbulence region and from data on the mean temperature of the sea surface?

In solving this problem one usually encounters the second difficulty in the use of computational methods of the theory of the logarithmic boundary layer: near the rough wall proper, i.e., near the undulatory sea surface, the regime of eddy viscosity, eddy thermal diffusivity and eddy diffusion is not described by universal theoretical relationships (as opposed to the layer with developed self-similar turbulence).

It is easy to notice that the above two difficulties can in no way be attributed to specific properties of eddy exchange processes over a stationary undulatory sea surface (also, they cannot be attributed to specific properties of eddy processes in the surface atmospheric layer). The same difficulties are encountered in engineering aerodynamics, heat engineering and diffusion kinetics. In fact, when analyzing the turbulent regime above a stationary undulatory sea surface these two difficulties are known difficulties in determining the coefficients of drag, heat exchange and diffusion for different types of rough surfaces. These coefficients may be introduced on the basis of the following relationships:

$$\tau_0 = \rho_a C_u(z) U^2(z), \tag{1}$$

$$-Q = C_p \rho_a C_\theta(z)\,[\theta(z) - \theta_w] U(z), \tag{2}$$

$$E = \rho_a C_E(z)[s(z) - s_w] U(z), \tag{3}$$

where ρ_a is the air density; C_p is its specific heat; $U(z)$, $\theta(z)$ and $s(z)$ are respectively the average wind speed, mean r temperature and average specific air humidity at altitude z above the sea surface; θ_w is the mean temperature of the sea surface proper; s_w is the average specific humidity at the sea surface ($\theta(o) = \theta_w$); τ_0 is the momentum flux; Q is the heat flux; E is the water vapor flux; C_u (z), C_θ (z) and C_E (z) are respectively the drag, heat exchange and evaporation coefficients.

It should be pointed out that within the framework of the above idealized model of the water-adjoining layer, of flow past a stationary undulatory sea surface, we encounter an excessively large "ensemble" of rough surfaces (spatial two-dimensional sea-swell spectra), if the coefficients C_u (z), C_θ (z) and $C_E(z)$ are to be found only on the basis of experimental studies (without recourse to some general considerations from the theory of similarity and from analysis of various physical hypotheses).

In continuing our examination of the idealized case of turbulent flow past a stationary undulatory sea surface, we could point out a third difficulty in calculating eddy fluxes in the water-adjoining layer of the atmosphere involving the need to allow for the effect of temperature stratification on the turbulent regime. This difficulty (as well as the first two) cannot be regarded as being related to the specific properties of turbulent exchange precisely above the sea, since determination of the effect of temperature stratification on the turbulence is the primary concern of physics of the surface layer of the atmosphere. As is known, in the case of a stratified eddy flux in a region with constant shear stress above a stationary rough surface the logarithmic boundary layer model must be replaced by a more general theory, the theory of similarity /8/, while eddy fluxes under arbitrary temperature stratification conditions should be calculated by computational methods developed in /1, 2/.

It is important to note that if observations of the average wind speed, mean temperature and average air humidity are carried out at altitudes $z \ll L$ (L being the thickness of the dynamic sublayer /8/, where the eddy-formation effects in the flow past individual roughnesses and molecular transport effects are no longer felt), then the logarithmic boundary layer model can again be used in analyzing these observations. Since the difficulties in the theoretical determination of coefficients C_u (z), C_θ (z) and C_E (z) in equations (1)—(3) may be ascribed to the need of making allowance for the two latter effects, it is possible by proper selection of the altitude in the layer $\varepsilon \ll z \ll L$ (where ε is the characteristic vertical dimension of roughness projections) to reduce the determination of $C_u(z)$, $C_\theta(z)$ and $C_E(z)$ for a stratified eddy flux to the scheme of the logarithmic boundary layer. Assuming that the altitude interval $\varepsilon \leqslant z \leqslant L$ actually exists, we shall not here consider problems related to making allowance for the effect of temperature stratification on the turbulent regime. For a treatment of these problems the reader is referred to /9, 10, 29/.

Thus, if the turbulent regime in the water-adjoining layer is considered within the framework of the idealized model, then eddy fluxes above the sea can be determined, first, by all the theoretical methods of the interpretation of gradient observations developed in physics of the surface atmosphere layer and, second, by results of theoretical and experimental determination of coefficients $C_u(z)$, $C_\theta(z)$ and $C_E(z)$ in (1)—(3) for a stationary rough solid wall, whose relief is similar to the topography of an undulatory sea surface at a given instant. It is not surprising that virtually all the results of studies on physics of the water-adjoining layer of the atmosphere and of calculations of eddy fluxes above the sea were obtained mainly by the above two methods. At the same time the question of feasibility, in principle, of carrying over known theories of the surface atmospheric layer to analysis of eddy exchange processes above the sea and the feasibility of using theoretical and experimental values of coefficients $C_u(z)$, $C_\theta(z)$ and $C_E(z)$ in (1)—(3) obtained for flow past a stationary rough wall remained uninvestigated for quite a period of time.

However, it is easy to notice that in calculating eddy fluxes above the sea, particularly in the theoretical calculation of coefficients $C_u(z)$, $C_\theta(z)$ and $C_E(z)$, we encounter a new difficulty in analysis of eddy exchange processes in the real water-adjoining layer of the atmosphere. This difficulty is due to the absence of a direct physical analogy between steady turbulent flow past a rough wall (whose relief may be similar to the topography of a stationary undulatory sea surface) and the steady regime of the mean turbulent motion over a real moving undulatory sea surface.

The first important difference between the above two cases consists in the fact that the undulatory sea surface can be regarded as a rough wall, provided it is remembered that the roughnesses proper (real wave "hills") move. Had all the waves regarded as individual roughnesses moved with the same phase velocity, then in this case, in a special reference system, the problem of describing the turbulent regime over the sea would have been similar to that of flow past a rough surface with the same relief as the undulatory sea surface at some given time.

The second difference between flow past a stationary surface and steady turbulent flow in the water-adjoining layer of the atmosphere consists in the following. When the underlying surface is a stationary rough surface, the turbulent air flux transmits to it only the momentum, while the eddy energy in the final count dissipates into heat while still in the air. In the case, however, when the underlying surface is a moving undulatory sea surface, the turbulent flux above it transmits to the waves (i.e., roughnesses) not only momentum, but also energy. This should produce in the lower air layers some region through which the energy and momentum of the fluctuating motion would be transmitted to the waves without appreciable losses. In this region the vertical profiles of the average velocity, mean temperature, average humidity, as well as the characteristics of the fluctuating motion proper could markedly differ from those observed in turbulent flow with total shear stress in the self-similar regime of developed turbulence. The last hypothesis was first formulated and suggested by Stewart /35/. The above differences are responsible for the differences in the problem of "water-adjoining" atmospheric turbulence as compared with problems of "wall" turbulence in problems of physics of the surface atmospheric layer. At the same time these differences determine also the fourth difficulty encountered in calculating eddy fluxes above the sea and, in particular, in theoretical determination of coefficients $C_u(z)$, $C_\theta(z)$ and $C_E(z)$. The solution of this latter problem requires analysis of turbulent motion processes near the very air-water interface. The complexity of this problem for the water-adjoining layer of the atmosphere is due not only to the effect of molecular and eddy-generation processes at the roughnesses (processes which are studied in analysis of heat exchange and drag at a stationary rough wall), but at the same time to the energy transfer from air to the moving roughnesses, i.e., waves. We recall that different mechanisms of such energy transmission (but without eddy generation at the waves) were presented in /11, 16/.

Thus, while only two factors may result in deviation from the self-similar developed turbulence regime in the region with constant shear stress near a solid rough stationary wall (closeness to the wall, resulting in amplified effect of molecular exchange processes and generation of eddies at the roughnesses proper), near the moving undulatory sea surface one has to deal with a third, independent factor, namely, transmission of energy from lower layers of turbulent air flow to the wave motion of the sea surface proper.

We shall attempt in this article to examine in sequence the aforementioned difficulties in formulating computational methods for determining eddy fluxes in the water-adjoining layer of the atmosphere. This will make it possible to critically evaluate the, at present, widely used methods for calculating energy exchange between the atmosphere and ocean.

SMOOTH STATIONARY WALL

In this section we shall briefly survey known results of the theory of the turbulent boundary layer at a flat plate. This should be carried out, since subsequently these results will serve as points of departure and, in addition, they may serve to illustrate the difficulties in the theoretical calculation of coefficients $C_u(z)$ and $C_\theta(z)$.

The average velocity profile, roughness parameter and drag coefficient. Consider the classical model of steady turbulent flow above a flat infinite wall. The x axis is directed along the wall, the z axis is directed upward perpendicular to it, plane $z = 0$ coincides with the wall surface. The corresponding average-velocity components are denoted by U and W, while the components of disturbance velocities are denoted by u' and w'. On the assumption of homogeneity in the x direction, the equation of motion and the conditions at the wall take the standard form

$$-\frac{d}{dz}\overline{(u'w')}+\nu\frac{d^2U}{dz^2}=0 \quad (4)$$

for

$$z=0,\quad U=0,\quad \frac{dU}{dz}=\frac{v_*^2}{\nu}, \quad (4a)$$

where $v_*=\sqrt{\frac{\tau_0}{\rho_a}}$ and ν is the kinematic coefficient of molecular viscosity.

Introducing the eddy viscosity coefficient ν_{turb}, equation (4) is frequently written in the form

$$(\nu_{turb}+\nu)\frac{dU}{dz}=v_*^2. \quad (5)$$

The universal relationships for the "wall" turbulence under study can be written as

$$\frac{U(z)}{v_*}=F_1(z^*), \quad (6)$$

$$\frac{\nu_{turb}}{\nu}=F_2(z^*), \quad (7)$$

where $z^*=\frac{zv_*}{\nu}$. The theory of the viscous sublayer and the theory of the logarithmic boundary layer can be used for determining the form of these functions in two regions /7, 15, 17/:

$$\left.\begin{aligned}F_1(z^*)&=z^*, & (8)\\ F_2(z^*)&\simeq\left(\frac{z^*}{\delta_B^*}\right)^4, & (9)\end{aligned}\right\}\quad \text{for } z^*<\delta_B^*$$

$$\left.\begin{aligned}F_1(z^*)&=\frac{1}{\varkappa}(\ln z^*+B), & (10)\\ F_2(z^*)&=\varkappa z^*, & (11)\end{aligned}\right\}\quad \text{for } z^*>\delta_T$$

where $\delta_B^*=\frac{\delta_B v_*}{\nu}$; δ_B is the upper boundary of the viscous sublayer in which the momentum transfer occurs primarily by molecular viscosity; $\delta_T^*=\frac{\delta_T v_*}{\nu}$; δ_T is the lower boundary of fully developed turbulence (where u' and w' are constant); $\varkappa$ and B are universal turbulence constants.

In the third region, the coupling zone ($\delta_B<z<\delta_T$), where momentum transfer occurs due to the fluctuating motion as well as by molecular means, function $F_1(z^*)$ is usually calculated by some interpolation formulas for ν_{turb} which, in general, should make allowance for the fact that the eddy viscosity in the coupling zone increases with distance from the wall slower than z^*, but faster than z.

Experimental data, although having a marked scatter (this is particularly true of B), yield the following results:

$$\delta_B^* = 5.0\text{—}6.7;\ \ \delta_T^* = 25\text{—}30;\ \ \varkappa = 0.39\text{—}0.41;\ \ B = 1.8\text{—}2.4. \tag{12}$$

Introducing the roughness parameter h_0, the drag coefficient $C_u(z)$, and using the expressions

$$\frac{U(z)}{v_*} = \frac{1}{\varkappa} \ln \frac{z}{h_0}, \tag{13}$$

$$C_u(z) = \left[\frac{v_*}{U(z)}\right]^2, \tag{14}$$

we derive from equation (10) that

$$h_0 = e^{-B} \frac{\nu}{v_*}; \quad \frac{h_0 v_*}{\nu} = 0.16\text{—}0.09, \tag{15}$$

$$C_u(z) = \frac{\varkappa^2}{[\ln z^* + B]^2} = \frac{\varkappa^2}{\left[\ln \frac{zU(z)}{\nu} + \frac{1}{2} \ln C_u(z) + B\right]^2}. \tag{16}$$

It is seen from equations (15) and (16) that the roughness parameter and the drag coefficient are functions of the wind speed at some altitude z (it should be remembered that equations (13)—(16) are valid, as can be seen from equation (12), for altitudes $z > 250\ h_0$).

The temperature profile and heat transfer and evaporation coefficients. Equations (4) and (4a) can similarly be written as

$$\frac{d}{dz} \overline{w'\theta'} + \chi \frac{d^2\theta}{dz^2} = 0, \tag{17}$$

for

$$z = 0, \quad \theta(z=0) = \theta_w, \quad \frac{d\theta}{dz} = -\frac{Q}{c_p \rho_a \chi}, \tag{18}$$

where χ is the molecular thermal diffusivity and Q is the heat flux to the wall.

Introducing the eddy thermal diffusivity coefficient χ_{turb}, equation (17) is frequently written in the form

$$(\chi_{turb} + \chi) \frac{d\theta}{dz} = -\frac{Q}{c_p \rho_a}. \tag{19}$$

Introducing the notation

$$\frac{\nu_{turb}}{\chi_{turb}} = \mathrm{Pr}_{turb}, \quad \frac{\nu}{\chi} = \mathrm{Pr}_\theta, \quad \theta^* = \frac{-\frac{Q}{c_p \rho_a} \cdot \mathrm{Pr}_\theta}{v_*}, \quad \widetilde{\theta} = \frac{\theta}{\theta^*} \tag{20}$$

and neglecting changes in molecular coefficients, i.e., assuming that the Prandtl number Pr_θ is constant, we can express the universal expression for the temperature profile in the form

$$\widetilde{\theta}(z) - \widetilde{\theta}_w = F_3\left(z^*, \frac{\mathrm{Pr}_{turb}}{\mathrm{Pr}_\theta}\right). \tag{21}$$

On the assumption that

$$\mathrm{Pr}_{turb} = \mathrm{Pr}_\theta = \text{const} \tag{22}$$

we obtain (equating (19) to (5))

$$F_3\left(z^*, \frac{\Pr_{\text{turb}}}{\Pr_\theta}\right) = z^* \text{ for } z^* < \delta_B^*, \tag{23}$$

$$F_3\left(z^*, \frac{\Pr_{\text{turb}}}{\Pr_\theta}\right) = \frac{1}{\varkappa}(\ln z^* + B) \text{ for } z^* > \delta_T^*, \tag{24}$$

where $\varkappa$ and B have the same meaning as above.

Introducing the heat exchange coefficient as given in equation (2):

$$-\frac{Q}{c_p\rho_a} = C_\theta(z)\, U(z)\, [\theta(z) - \theta_w],$$

there results

$$C_\theta(z) = \frac{C_u(z)}{\Pr_\theta}, \tag{25}$$

where $C_u(z)$ can be calculated from equation (16).

Clearly, only for $\Pr_\theta \approx 1$ will we derive

$$C_\theta(z) \approx C_u(z). \tag{26}$$

Equations (25) and (26) can be used for $z > 250\, h_0$, where h_0 is defined by equation (15).

If condition (22) is not used, calculation of $F_3\left(z^*, \frac{\Pr_{\text{turb}}}{\Pr_\theta}\right)$ requires detailed analysis of eddy processes in the coupling zone /7, 17/. Apparently for $\Pr_\theta \approx 1$ the use of condition (22) should result in excessive error. Since for air

$$\Pr_E = \frac{\nu}{\sigma} \approx \Pr_\theta = \frac{\nu}{\chi} \approx 1, \tag{27}$$

where $\Pr_E$ is the diffusion Prandtl number and σ is the molecular diffusion coefficient, similar expressions can also be written for evaporation (see equation (3)):

$$E = \rho_a C_E(z)\,[s(z) - s_w]\, U(z),$$
$$C_E(z) \approx C_\theta(z) \approx \frac{C_u(z)}{\Pr_{\text{mol}}}, \tag{28}$$

$$\Pr_{\text{mol}} \approx \Pr_\theta \approx \Pr_E. \tag{29}$$

We note that when $\Pr_{\text{mol}} = \Pr_{\text{turb}} = \text{const}$, the heat exchange and evaporation coefficients are uniquely related to the wind speed at a specified level. This relationship can be determined theoretically only by the assumption of condition (22). It is valid when $\Pr_{\text{mol}} = 1$. For higher values of $\Pr_{\text{mol}}$ the calculation of heat transfer and evaporation coefficients becomes very difficult, since this requires detailed analysis of the turbulence structure in the viscous sublayer and in the coupling zone ($\delta_B < z < \delta_T$). Apparently, this question is not of great significance for air, since $\Pr_{\text{mol}} \approx 1$. We note that this is not true of the sea since for water

$$\Pr_\theta \approx 10; \quad \Pr_E \approx 10^3. \tag{30}$$

ROUGH STATIONARY WALL

Here we shall consider known results of the theory of the turbulent boundary layer above rough surfaces. This will make it possible to see that the difficulties encountered in determining the coefficients $C_u(z)$, $C_\theta(z)$ and $C_E(z)$ by theoretical means become so great that in the best case the problem can be reduced to construction of some universal relationships on the basis of experimental studies.

The average velocity profile, roughness parameter and drag coefficient. Suppose we are given an ensemble of rough surfaces, whose relief is described by functions $\eta_i(x, y,)$ (subscript i denotes the number of the surface under test); the value of $\eta_i(x, y)$ is measured from some specified level. The mean height of the roughnesses (projections) of this relief, relative to an imaginary smooth wall at which the origin of measurement $z = 0$ is placed, is denoted by ε. Then, as is known in analyzing eddy exchange processes above a rough wall, one must consider three cases corresponding to different values of the parameter $\mathrm{Re}_{\mathrm{rough}} = \varepsilon^* = \varepsilon v_*/\nu$.

The first case is that of an aerodynamically smooth wall and is defined by the obvious condition

$$\varepsilon < \delta_B \tag{31}$$

or, more precisely,

$$\varepsilon^* < k_i, \tag{32}$$

where k_i is some constant, depending on the geometric characteristics $\eta_i(x, y)$ of the relief.

Inequalities (31) and (32) indicate that the mean height of roughness projections is smaller than δ_B, the viscous sublayer thickness. It is of importance that in the general case $\eta_i(x, y) \neq \eta_j(x, y)$ we have

$$k_i \neq \kappa_j \text{ for } \varepsilon_i = \varepsilon_j. \tag{33}$$

Apparently, experiments point to the existence of some limiting value $k_{\lim}$ for which, if

$$\varepsilon^* \leqslant k_{\lim}, \tag{34}$$

the wall will behave as if it were aerodynamically smooth for any i.

When conditions (31) and (32) are satisfied, the expressions for the velocity profile (8) and (10), for the drag coefficient (16) and the roughness parameter (15) remain valid.

The second case is that of a moderately rough surface. It is defined by the condition

$$\delta_B < \varepsilon < \delta_T \tag{35}$$

or, more precisely,

$$k_i < \varepsilon^* < N_i. \tag{36}$$

Inequalities (35) and (36) mean that the average height of roughness projections is greater than δ_B, the viscous sublayer thickness, and smaller than the total thickness δ_T of the layer, where the viscosity starts exerting its effect on the average motion. Here also in the general case

$$\eta_i(x, y) \neq \eta_j(x, y), \quad N_i \neq N_j \text{ for } \varepsilon_i = \varepsilon_j, \tag{37}$$

since N_i is some constant depending on $\eta_i(x, y)$.

In the case of a moderately rough wall the vertical distribution of the average velocity, unlike the universal "wall" relationship (6), should be described by expressions

$$\frac{U(z)}{v_*} = \psi(z^*, \varepsilon^*) \text{ for } 0 \leqslant \frac{z}{\varepsilon} \leqslant \zeta_i(\varepsilon^*), \tag{38}$$

$$\frac{U(z)}{v_*} = \frac{1}{\varkappa} \ln \frac{z}{h_0} \text{ for } \frac{z}{\varepsilon} \geqslant \zeta_i(\varepsilon^*). \tag{39}$$

The second of these expressions means that the logarithmic velocity distribution characteristic of the developed turbulence region may exist at altitudes where the effect of the additional eddy generation at individual projections no longer, on the average, affects the turbulence regime, i. e., at altitudes $z \gg \varepsilon$, from where it follows that $\zeta_i(\varepsilon^*) \gg 1$. Here $\zeta_i(\varepsilon^*)$ is some constant, depending on $\eta_i(x, y)$ as well as on the turbulent friction and molecular viscosity.

The roughness parameter for a moderately rough wall in equation (39) is defined, unlike (15), by an expression such as

$$h_0 = e^{-B} \frac{\nu}{v_*} f_i(\varepsilon^*), \tag{40}$$

while the drag coefficient is expressed as

$$C_u(z) = \frac{\varkappa^2}{[\ln z^* + B - \ln f_i(\varepsilon^*)]^2}. \tag{41}$$

It is important that the form of function $f_i(\varepsilon^*)$ may depend markedly on $\eta_i(x, y)$. It is shown experimentally /17/ that in the general case

$$\eta_i(x, y) \neq \eta_j(x, y), \quad f_i(\varepsilon^*) \neq C_i f(\varepsilon^*), \tag{42}$$

where C_i is a constant corresponding to subscript i.

Equations (40)—(42) explain in part why the concept of an "equivalent" roughness cannot be used for the case of moderate roughness. As is seen from equation (40), the roughness parameter may depend on v_* (or on the average wind speed) differently, depending on the underlying surface geometries.* In spite of this, function $f_i(\varepsilon^*)$ should, obviously, have certain asymptotic properties. In fact, for small ε^* we should arrive at the case

* The term underlying surface geometry denotes the "roughness relief."

of an aerodynamically smooth wall, i.e.,

$$f_i\ (\varepsilon^*) \approx 1 \ \text{for}\ \varepsilon^* \ll 1. \tag{43}$$

Conversely, for large ε^* we arrive at the other limiting regime of the so-called "developed roughness," when

$$f_i(\varepsilon^*) = C_i\ \varepsilon^* \ \text{for}\ \varepsilon^* \gg 1. \tag{44}$$

The third case is that of a wall with developed roughness. It is defined by the condition

$$\varepsilon > \delta_T \tag{45}$$

or, more precisely,

$$\varepsilon^* > N_i. \tag{46}$$

Inequalities (45) and (46) show that for a given relief $\eta_i\ (x,\ y)$ the roughness projections are on the average located in the region with developed turbulence (above the coupling zone). In the case of a wall with developed roughness, the vertical mean velocity distribution should be described by the following expressions:*

$$\frac{U\ (z)}{v_*} = \psi_i\,(z^*,\ \varepsilon^*)\ \text{for}\ 0 \leqslant \frac{z}{\varepsilon} \leqslant \zeta_i\,(\varepsilon^*), \tag{47}$$

$$\frac{U\ (z)}{v_*} = \frac{1}{\varkappa}\ln\frac{z}{h_0}\ \text{for}\ \frac{z}{\varepsilon} \geqslant \zeta_i(\varepsilon^*). \tag{48}$$

The last expression means that the logarithmic velocity distribution typical of the region with developed turbulence will exist at altitudes where the roughness no longer exerts its effect ($z \geqslant \zeta_i\,(\varepsilon^*)\,\varepsilon$). Here the roughness parameter for the developed roughness regime, precisely in the same manner as the drag coefficient, should not be an explicit function of the molecular viscosity; this yields

$$h_0 = A_i\ \varepsilon, \tag{49}$$

$$C_U\,(z) = \frac{\varkappa^2}{\left[\ln\dfrac{z}{\varepsilon} - \ln A_i\right]^2}. \tag{50}$$

Equation (49) explains why the so-called equivalent roughness can be used for the developed roughness regime. However, the use of this concept is meaningful only when the dependence of A_i on the principal geometric

* Equation (47) was examined by Levich /7/. According to his model, there should exist near the very surface of roughness a thin viscous sublayer δ'_B, within the limits of which $\psi_i\ (z^*,\ \varepsilon^*) \approx z^*$. The ratio of the thickness of this sublayer to that of the viscous sublayer at a smooth wall δ_B is $\frac{\delta'_B}{\delta_B} = \frac{1}{\sqrt{\varepsilon^*}}$. In addition, there exists a region in which the effect of viscosity is no longer substantial, but departures from the logarithmic law are substantial and are related to the separation of eddies at the roughnesses. In this region, clearly, for $\delta_B < z < \xi_i(\varepsilon^*), \varepsilon\psi_i(z^*,\ \varepsilon^*) \approx \psi_i(z^*;\ \varepsilon^*)$, while according to Levich $\psi_i(z^*/\varepsilon^*) \approx z^*/\varepsilon^*$.

characteristics of rough surfaces has been determined experimentally. This was done by Schlichting /18/, but it is clear that the ensemble of parameters used for classifying the geometry of technical rough surfaces does not suffice for classifying various complicated topographies of natural underlying surfaces (such as the undulatory sea surface). It can only be concluded from Schlichting's data that

$$A_i = F_i \sqrt{\left[\frac{\partial y(x, y)}{\partial x}\right]^2} \tag{51}$$

and equations (49) and (50) define the known square relationship for the drag. If it is used as a point of departure, then the results of these formulas are obtained as a corollary. The experimental data on roughness parameters and drag coefficients of different surfaces are not too numerous. From them one may derive that

$$\begin{gathered} k_i \approx 5{,}0; \; N_i \approx 55\text{—}100, \\ A_i \approx 10^{-1}\text{—}10^{-2}. \end{gathered} \tag{52}$$

Data obtained by Motzfeld /30/ from a study of solid wave-shaped roughnesses may be of some interest. For the case of waves with sharp crests he found that

$$A_i \approx {}^1/_{15}. \tag{53}$$

He concluded that for very shallow waves the roughness parameter is determined only by three variables: v_*, ν and the wave steepness ε/λ (λ is the wavelength). It should be emphasized that, for causes which will become clear from the subsequent discussion and which were mentioned above, one should not expect to obtain a correct idea on the resistance of a moving undulatory sea surface on the basis of experiments with stationary "wavy" roughness. Hence it is impossible to directly transfer the results from equations (52) and (53) to the case of the water-adjoining atmospheric layer (as is quite frequently done).

The temperature profile, heat transfer and evaporation coefficients. If we assume, as before, that $\mathrm{Pr}_{\mathrm{turb}}$ is constant from $z = 0$ to the altitudes of interest and we apply equation (22), then the similarity between the distribution of $U(z)$, the average velocity, and $\widetilde{\theta}(z) - \widetilde{\theta}_w$, the difference of dimensionless temperatures, will be retained, while equation (26) will be valid for heat exchange and evaporation coefficients at rough surfaces; here it should be remembered that the drag coefficient at rough surfaces will now be defined by equations (41) and (50). On the above assumption the heat exchange and evaporation coefficients for a wall with developed roughness do not depend on the friction velocity v_* (or on the wind speed $U(z)$). Latest experimental studies /17, 33/ show that this is not so and as if "aerodynamically rough surfaces,"* do not exist for heat exchange and diffusion processes of

* The term "aerodynamically rough surface" for heat transfer processes is used to denote a rough surface at which the heat transfer coefficient does not depend on the molecular viscosity (without making allowance for its dependence on Pr, which is regarded as constant). This definition is analogous to that of aerodynamically rough surfaces for the drag regime.

any given inert substance. The causes of this will become clear if it is remembered that, no matter what the contribution to the heat transfer at the wall due to turbulence produced by formation of eddies at the roughnesses proper, the difference between the mean wall temperature and the mean temperature of the fluid surrounding it (which in the final count determines the value of $C_\theta(z)$) should be controlled by molecular thermal diffusivity. Naturally, this limitation is not imposed on the momentum flux to the wall, if its roughness is sufficiently great (developed roughness). It can be assumed in addition that turbulence produced by separation of eddies at the roughnesses proper makes different contributions to the heat and momentum transfer for different rough surfaces. Within the framework of equations (5) and (19) the above circumstances result in the fact that Pr_{turb} near the rough surface may depend markedly on the vertical coordinate z, as well as on the topographical features of the rough surface, i. e.,

$$Pr_{turb} = Pr_{turb}(z^*, \varepsilon^*). \tag{54}$$

In other words, the assumption made above that Pr_{turb} = const and that $Pr_{turb}/Pr_{mol} = 1$ for an underlying surface with complex topography ($\eta(x, y)$) are not satisfied in the close vicinity of these surfaces. This makes it impossible to use equations (41) and (50) and to determine coefficients $C_\theta(z)$ and $C_E(z)$ from equation (26).

Clearly, for a wall with developed roughness we can write only that

$$\tilde{\theta}(z) - \tilde{\theta}[J_i(\varepsilon^*)\varepsilon] = \frac{1}{\varkappa}\ln\frac{z}{h_0} - \frac{1}{\varkappa}\ln\frac{J_i(\varepsilon^*)}{A_i}$$

for

$$\frac{z}{\varepsilon} \geqslant J_i(\varepsilon^*), \tag{55}$$

since equation (22) can be regarded as satisfied in the developed turbulence region, where the effect of additional eddy generation is still not felt. When

$$\frac{z}{\varepsilon} \leqslant J_i(\varepsilon^*) \quad \text{we have} \quad \tilde{\theta}(z) - \tilde{\theta}_w = F_i(z^*, \varepsilon^*, Pr_{mol}) \tag{56}$$

or

$$\bar{\theta}[J_i(\varepsilon^*)\varepsilon] - \tilde{\theta}_w = \tilde{\xi}_i(\varepsilon^*, Pr_{mol}). \tag{57}$$

It is in general impossible to calculate the form of functions F_i and $\tilde{\xi}_i$ (here it should be remembered that the form of functions (56) and (57) may depend on the underlying surface topography). Obviously, this is also valid of heat transfer and evaporation coefficients, for which in the case of rough surfaces one can only write that

$$C_\theta(z) = C_E(z) = \pi_i\left(\frac{z}{\varepsilon}, \varepsilon^*, Pr_{mol}\right) \tag{58}$$

or

$$C_\theta(z) = C_E(z) = M_i[C_u(z), \varepsilon^*, Pr_{mol}]. \tag{59}$$

In order to determine the form of functions F_i, ξ_i and π_i in equations (56)–(58), one must consider the turbulence structure in the region $z \leqslant J_i(\varepsilon^*)\varepsilon$, where both the effects of additional turbulence due to separation of eddies from individual surfaces and effects of molecular transports are already felt. It is natural that this problem (in addition to the different surface topographies) is quite complex. An example of one possible theoretical solution of this problem can be found in /7/. Apparently, the most complete experimental data on the form of function (58) are contained in /33/. In order to estimate the results presented in /33/, it is convenient to first find the relation between functions (57) and (58). For this equation (57) is substituted into (55) and coefficients $C_\theta(z)$ and $C_E(z)$ are determined. After appropriate transformations we derive for the case of developed roughness ($\varepsilon^* > N_i$):

$$C_\theta(z) = C_E(z) = \frac{C_u(z)^{1/2}}{\mathrm{Pr}_{\mathrm{mol}}\left[C_u(z)^{-\frac{1}{2}} - \frac{1}{\varkappa}\ln\frac{J_i(\varepsilon^*)}{A_i} + \xi_i(\varepsilon^*, \mathrm{Pr}_{\mathrm{mol}})\right]}, \tag{60}$$

where $C_u(z)$ is specified by equation (50).

Reference /33/ considers underlying surfaces for which when $\varepsilon^* \gg 100$ it can be assumed without much error that $J_i(\varepsilon^*) \approx 1$, while the value of $J_i(\varepsilon^*)/A_i$ for the above surface was found to change little and equal to 1.0–3.0. Experimental data in /33/ were processed on the assumption that

$$\xi_i(\varepsilon^*, \mathrm{Pr}_{\mathrm{mol}}) = \alpha_i(\varepsilon^*)^m(\mathrm{Pr}_{\mathrm{mol}})^n, \tag{61}$$

and showed that

$$m = 0.45; \quad n = 1.8; \quad \alpha_i = 0.45 - 0.7. \tag{62}$$

The quantity α_i was found to change little on transition from the case of two-dimensional to that of three-dimensional roughnesses. As is seen from equations (60)–(62), for the case of an aerodynamically rough wall ($\varepsilon^* > 100$) coefficient $C_\theta(z)$ is a function of v_*. Here it is important that as v_* for a given rough surface increases, the heat transfer coefficient decreases, i.e., it behaves as in the case of an aerodynamically smooth wall. The dependence of heat exchange on the surface topography is more complicated. This is related to changes in the effective thickness of diffusion sublayers – the viscous and thermal ones. While the thickness of the first, viscous sublayer can be virtually taken as zero (for $\varepsilon^* > 100$), in calculating the heat transfer allowance must be made for the finite effective thickness of the thermal sublayer.* In our opinion, the authors of /33/ quite correctly pointed out that there is no basis for assuming the existence of universal relationship (61) for any kind of rough surfaces. This is chiefly due to two factors: a) the nature of turbulence at the wall proper depends markedly on the surface topography (even for the same values of ε^*); b) when using the heat transfer coefficient we can determine

* The thermal sublayer, introduced by analogy with the viscous sublayer, is a region at the wall where heat transfer is implemented primarily by molecular thermal diffusivity (since it is penetrated by turbulent pulsations). For $\mathrm{Pr}_\theta \approx 1$ the thicknesses of the viscous and of the thermal sublayers should be close at a smooth wall. However, their effective thickness at a rough surface may be equal.

the heat flux per unit surface area parallel to the plane of the average motion, and the time when the true wall surface may depend quite substantially on the surface topography.

It is seen from the above how difficult it is to derive universal relationships for heat transfer coefficients from experimental studies at different rough surfaces. As was previously noted, no lesser difficulties are involved in theoretical calculations, which should be based on: 1) certain assumptions regarding the type of variation in Pr_{turb} with altitude within the limits of $0 \leqslant z \leqslant J_i(\varepsilon^*)\varepsilon$ for different surfaces; 2) some assumptions regarding the value of Pr_{turb} proper for different surfaces, i.e., on the relationship $Pr_{turb} = Pr_{turb}(\varepsilon^*)$; 3) some assumptions regarding the eddy viscosity regime in the region where eddy generation at roughnesses and molecular transport are felt (in the region $0 < z \leqslant \xi_i(\varepsilon^*)\,\varepsilon$).

Since all the above concerning heat exchange at rough surfaces applies just as well to water vapor diffusion, only calculation of eddy heat fluxes is considered further on. Calculation of evaporation and heat exchange above oceans is discussed in /12, 36, 38/.

It is found upon careful analysis of the above experimental results that all of them are based on certain hypotheses concerning the three above main problems of transport theory near stationary solid rough surfaces. Let us examine the results obtained in these references within the framework of general concepts on eddy heat exchange and eddy diffusion near stationary rough surfaces.

ON "THEORETICAL" FORMULAS FOR COMPUTING HEAT AND MOISTURE FLUXES ABOVE THE OCEAN

Formulas due to Sverdrup, Rossby and Montgomery, and Sheppard for computing eddy fluxes above oceans have come into quite extensive use. Sverdrup's theoretical conclusions /36/, discussed in detail in /12, 13/, underlie a number of mass calculations of energy exchange between the ocean and atmosphere. It hence appears pertinent to clarify the physical meaning of the hypotheses which were used by the above investigators as the basis for deriving their theoretical formulas. This is the more meaningful if it is remembered that no critical theoretical analysis of the conclusions of these investigators was made.

The Sverdrup formulas. Formulas for calculating eddy heat and moisture fluxes above the ocean were first derived by Sverdrup /36, 37/. Let us examine, following Sverdrup, a stationary undulatory sea surface. The origin for the vertical coordinate z is placed at the physical air-water interface proper.

We further assume that there exists near the surface some thin effective viscous sublayer δ_u, within which the molecular viscosity of air plays a predominant role in momentum transfer. Since at altitudes comparable with sea-wave amplitudes the vertical turbulence scale should be determined not only by the "distance from the wall," but also by the height of the roughness projections, we express it as follows:

$$l = \varkappa(z + b\varepsilon) \text{ for } z \geqslant \delta_u, \tag{63}$$

where b is some proportionality factor.

Using the Prandtl hypothesis, we get

$$\nu_{\text{turb}} = \varkappa^2 (z + b\varepsilon)^2 \frac{dU}{dz} \text{ for } z \geqslant \delta_u \tag{64}$$

and for the velocity profile

$$\frac{U(z)}{v_*} = \frac{1}{\varkappa} [\ln(z + b\varepsilon) + \text{const}] \text{ for } z \geqslant \delta_u. \tag{65}$$

The constant of integration in equation (65) is obtained from the condition

$$\frac{U(\delta_u)}{v_*} = \frac{\delta_u v_+}{\nu} \text{ for } z = \delta_u. \tag{66}$$

Then finally

$$\frac{U(z)}{v_*} = \frac{1}{\varkappa}\left\{\ln\left[\left(\frac{z + b\varepsilon}{\delta_u + b\varepsilon}\right) e^{\frac{\varkappa\delta_u v_+}{\nu}}\right]\right\}. \tag{67}$$

Since the effective viscous sublayer is very thin at the undulatory sea surface (we recall that $\varepsilon^* > 100$), we may assume

$$\frac{\varkappa\delta_u v_*}{\nu} \ll 1 \text{ for } \delta_u \ll b\varepsilon \tag{68}$$

and express equation (67) without much error in the form

$$\frac{U(z)}{v_*} = \frac{1}{\varkappa} \ln \frac{z + b\varepsilon}{b\varepsilon}. \tag{69}$$

Assuming that at altitudes far from the roughnesses the velocity profile takes the standard form (49) for the case of developed roughness, equation (69) should coincide with the following expression:

$$\frac{U(z)}{v_*} = \frac{1}{\varkappa} \ln \frac{z}{A_l\varepsilon}. \tag{70}$$

Equating equations (69) and (70), we have

$$b = A_l;\; b\varepsilon = h_0 \tag{71}$$

and finally, the known formula for the velocity profile

$$\frac{U(z)}{v_*} = \frac{1}{\varkappa} \ln \frac{z + h_0}{h_0}, \tag{72}$$

where h_0 is the roughness parameter for a wall with developed roughness. We thus consider the case when the effect of the viscous sublayer on the velocity distribution can be neglected ($\delta_u \approx 0$, i.e., the case of an aerodynamically rough wall).

Now consider hypothesis (22) developed by us for the case of an aerodynamically rough wall:

$$\tilde{\theta}(z)-\tilde{\theta}_w=\frac{1}{\varkappa}\ln\left[\left(\frac{z+b\varepsilon}{\delta_u+b\varepsilon}\right)e^{\frac{\varkappa\delta_u v_*}{\nu}}\right], \tag{73}$$

which yields

$$\tilde{\theta}(z)-\tilde{\theta}_w=\frac{1}{\varkappa}\ln\frac{z+h_0}{h_0}, \tag{74}$$

or, by definition of heat transfer and drag coefficients,

$$C_\theta(z)=\frac{C_u(z)}{\mathrm{Pr}_{\mathrm{mol}}}.$$

For $\mathrm{Pr}\approx 1$ we conclude that for rough surfaces the heat transfer coefficient is equal to the drag coefficient. The main achievement of Sverdrup consists precisely in the fact that, upon disproving this conclusion with the aid of experimental data, he suggested an improvement of the computational method introducing some diffusion or temperature sublayer of finite thickness, for which we should have

$$\delta_{\theta,E}\gg\delta_u. \tag{75}$$

Within the framework of the general equations (5) and (19) this implies that

$$\frac{\mathrm{Pr}_{\mathrm{turb}}}{\mathrm{Pr}_{\mathrm{mol}}}\gg 1. \tag{76}$$

Since $\mathrm{Pr}_{\mathrm{mol}}=1$, condition (75) yields

$$\mathrm{Pr}_{\mathrm{turb}}\gg 1,\ \nu_{\mathrm{turb}}\gg\chi_{\mathrm{turb}}. \tag{77}$$

Hence Sverdrup actually considers the following two-layer model:*

$$\tilde{\theta}(z)-\tilde{\theta}_w=z^*,\ \mathrm{Pr}_{\mathrm{turb}}\gg 1 \text{ for } z\leqslant\delta_{\theta,E}, \tag{78}$$

$$\tilde{\theta}(z)-\tilde{\theta}(\delta_{\theta,E})=\frac{1}{\varkappa}\ln\frac{z+h_0}{\delta_{\theta,E}+h_0},\ \mathrm{Pr}_{\mathrm{turb}}=1 \text{ for } z\leqslant\delta_{\theta,E}. \tag{79}$$

From equations (78) and (79) one easily finds an expression for the heat transfer coefficient at the rough wall:

$$(z)=C_E(z)=\frac{C_u(z)^{\frac{1}{2}}}{\mathrm{Pr}_{\mathrm{mol}}\left[C_u(z)^{-\frac{1}{2}}-\frac{1}{\varkappa}\ln(\delta_{\theta,E}+h_0)+\frac{1}{\varkappa}\ln h_0+\frac{\delta_{\theta,E}v_*}{\nu}\right]}. \tag{80}$$

* It was not noticed by any of the authors of /36, 12, 19, 13/ that this is a two-layer model not only for the temperature, but also for the turbulence Prandtl number. At the same time it is precisely the latter fact which is most important in starting hypotheses.

Since $\delta_{\theta,E}$ in equation (79) is the altitude up to which the logarithmic temperature distribution, characteristic of the developed turbulence region, applies, then in our more general designation we have

$$\delta_{\theta,E} = J_i(\varepsilon^*)\,\varepsilon;\;\; h_0 = A_i\varepsilon. \tag{81}$$

In accordance with equation (60), we rewrite this formula in the form

$$C_\theta(z) = C_E(z) = \frac{C_u(z)^{\frac{1}{2}}}{\mathrm{Pr}_{\mathrm{mol}}\left[C_u(z)^{-\frac{1}{2}} - \frac{1}{\varkappa}\left(\ln\frac{J_i(\varepsilon^*)}{A_i} + 1\right) + J_i(\varepsilon^*)\,\varepsilon^*\right]}. \tag{82}$$

Sverdrup used two hypotheses. In /36/ he assumed that

$$J_i(\varepsilon^*) = \frac{(0.10 - 0.15)\,\mathrm{cm}}{\varepsilon}, \tag{83}$$

i. e., that the thickness of the thermal sublayer in equation (81) is constant. As is seen from equation (82), in this case the heat transfer coefficient for an aerodynamically rough surface depends on v^* (by way of ε^*), namely, it decreases with wind speed, as is the case for a smooth wall. In /37/ he makes an assumption satisfying analogy considerations:

$$J_i(\varepsilon^*) = m_i(\varepsilon^*)^{-1},\;\; m_i \approx 27.5, \tag{84}$$

i. e., $\delta_\theta \approx \delta_T$.

Sheppard's formulas. Sheppard actually also considered a two-layer model, in which

$$\frac{\mathrm{Pr}_{\mathrm{turb}}}{\mathrm{Pr}_{\mathrm{mol}}} = 1 \text{ for } \frac{z}{\varepsilon} \geqslant J_i(\varepsilon^*), \tag{85}$$

$$\mathrm{Pr}_{\mathrm{turb}} = \mathrm{Pr}_{\mathrm{turb}}(z) \text{ for } \frac{z}{\varepsilon} \leqslant J_i(\varepsilon^*). \tag{86}$$

Hence the vertical variation of $\mathrm{Pr}_{\mathrm{turb}}$ is such that the regime of eddy thermal diffusivity in the layer $z \leqslant J_i(\varepsilon^*)\,\varepsilon$ would remain the same as in the developed turbulence layer. Hence we may write, following Sheppard, that

$$\frac{\mathrm{Pr}_{\mathrm{mol}}}{\mathrm{Pr}_{\mathrm{turb}}} = \frac{\varkappa v_* z}{\nu_{\mathrm{turb}}} \text{ for } \frac{z}{\varepsilon} \leqslant J_i(\varepsilon^*) \tag{87}$$

and

$$\tilde{\theta}(J_i\varepsilon) - \tilde{\theta}_w = \frac{1}{\varkappa}\ln\left[1 + \varkappa J_i(\varepsilon^*)\,\varepsilon^*\right] \text{ for } \frac{z}{\varepsilon} \leqslant J_i(\varepsilon^*). \tag{88}$$

Hence the heat transfer coefficient is easily obtained in the form

$$C_\theta(z) = C_E(z) = \frac{C_u(z)^{\frac{1}{2}}}{\mathrm{Pr}_{\mathrm{mol}}\left\{C_u(z)^{-\frac{1}{2}} - \frac{1}{\varkappa}\ln\frac{J_i(\varepsilon^*)}{A_i} + \frac{1}{\varkappa}\ln\left[1 + \varkappa J_i(\varepsilon^*)\,\varepsilon^*\right]\right\}}. \tag{89}$$

As we see also here the evaporation and heat transfer coefficients decrease with a rise in the friction velocity v_* (or wind speed $U(z)$).

It follows from this that both these authors made allowance for the effect of molecular transport processes at the very surface for calculating the heat transfer and evaporation processes also in the case of a very rough wall ($\varepsilon_* > 100$). However, the principal question concerning the relationship between the contribution of eddies and molecular exchange to the transfer of heat and moisture near different rough surfaces is not resolved by any of these theoretical models. The hypothetical assumptions of a sharp change in the value of Pr_{turb} near stationary rough surfaces, made implicitly by all the authors, are not based on any reliable experimental data. This also applies to shortcomings of the two-layer model for Pr_{turb}.

The use of two-layer models for the average velocity profile (Sverdrup, Sheppard) are also not substantiated physically, and the shortcomings of this approach are clear at least from the fact that the velocity profile near the roughnesses should differ from logarithmic, while according to both the above authors we obtain that near the very roughnesses ($z \approx \varepsilon$) the mean velocity and temperature profiles are the same as at the smooth wall, i. e., logarithmic. As we have seen above, such a hypothesis is inapplicable even for smooth walls and interpolation formulas must be used in the coupling zone. A more realistic model of turbulence near roughnesses was suggested by Levich /7/; the results obtained by him agree with formulas (60) – (62). As is seen, the problem of eddy heat transfer even at a stationary rough wall has not been sufficiently studied. Hence it is not surprising that actually all the calculations of heat transfer and evaporation above the sea are carried out with theoretical formulas (82) – (84), (89), derived for the case of a stationary rough wall without sufficient experimental verification. Diffusion formulas for calculating heat transfer and evaporation could have been regarded as simple empirical relationships had the form of function $C_\theta(z) = M_i[C_u(z), \varepsilon^*]$ been determined from experimental studies. However, as will be seen below, it would have been premature to use the formula

$$C_\theta(z) = C_E(z) = \pi_i\left(\frac{z}{\varepsilon}, \ \varepsilon^*, \ Pr_{mol}\right) \tag{90}$$

for analysis of experimental data on coefficients $C_\theta(z)$, $C_E(z)$ for the sea. This is due to the fact that equation (90) was derived for the case of flow around a stationary rough wall. It was already noted that actually there is no exact physical analogy between the steady turbulent flow over a moving undulatory sea surface and the flow past a stationary rough wall. The following two sections are devoted to this aspect of the problem of physics of the water-adjoining atmospheric layer.

WALLS WITH MOVING ROUGHNESSES

T h e v e l o c i t y p r o f i l e a n d r o u g h n e s s p a r a m e t e r. Suppose the relief $\eta(x, y)$ of a rough surface reproduces the topography of a stationary undulatory sea surface. Then, placing the coordinate origin with vertical coordinate z at the air-water interface proper, we may assume in the first approximation that

$$\varepsilon \approx H, \tag{91}$$

where H is the average amplitude (half-height) of real wave "hills," which compose the undulatory sea surface. To make an analogy between the flow past a stationary rough wall and the moving undulatory sea surface, it must be first assumed that the wave "hills," treated by us as roughness elements, move. We assume that all the wave "hills" move with the same speed c_i (which differs for different surfaces). Then, obviously, in the Cartesian coordinate system moving in the positive x direction (in the direction of the average wind) with velocity c_i, the pattern of turbulent flow around the undulatory sea surface will be similar to its flow past a stationary rough wall, whose relief coincides with the topography of the undulatory sea surface at a given instant of time. The average velocity in the moving coordinate system is $U(z) - c_i$. In general, at $z = 0$, i.e., at the very surface of the sea, we cannot assume, even in the moving coordinate system, that $U(z) - c_i = 0$, due to the presence of some average transport drift current. Hence the velocity should rigorously be reckoned relative to the velocity of this surface current.* However, when

$$\frac{c_i}{U_{\text{drift}}} \gg 1, \tag{92}$$

where U_{drift} is the velocity of the transport current, this effect can be neglected in the first approximation.

Introducing a relative velocity, then, by analogy with the results of the second and third sections, we have

$$\frac{U(z) - c_i}{v_*} = \frac{1}{\varkappa} \ln \frac{z}{h_0} \text{ for } z \gg h_0, \tag{93}$$

where for h_0, as well as for the region in which equation (93) is valid, one can use all the previously derived relationships. Since the average velocity $U(z)$ at several altitudes z_i is determined from gradient observations, and the roughness parameter z_0 for the sea** is determined by extrapolation to zero velocity, it is expedient to rewrite equation (93) in the form

$$\frac{U(z)}{v_*} = \frac{1}{\varkappa} \ln \frac{z}{z_0}, \tag{94}$$

where

$$z_0 = \frac{h_0}{e^{\frac{\varkappa c_i}{v_*}}}. \tag{95}$$

Using the results of the preceding sections as well as equation (91) we can thus express the mean velocity profile and the roughness parameter in the form

* This was pointed out to the authors by A. S. Monin.

** The above procedure is at the same time also the determination of the roughness parameter for the sea. It should only be remembered that it is uniquely related to the drag coefficient $C_u(z)$ (14). Accordingly, no matter what the method used for determining $C_u(z)$ (for example, fluctuation measurements), one can determine z.

$$\frac{U(z)}{v_*}=\frac{1}{\varkappa}\ln\frac{z}{z_0},\quad z_0=\frac{e^{-B}\nu}{v_* e^{\varkappa\frac{c_i}{v_*}}},$$

$$\text{for } \frac{Hv_*}{\nu}<k_i,\quad z\gg z_0 e^{\varkappa\frac{c_i}{v_*}}, \tag{96}$$

$$\frac{U(z)}{v_*}=\frac{1}{\varkappa}\ln\frac{z}{z_0},\quad z_0=\frac{e^{-B}\nu}{v_* e^{\varkappa\frac{c_i}{v_*}}}\cdot f_i\left(\frac{Hv_*}{\nu}\right),$$

$$\text{for } N_i>\frac{Hv_*}{\nu}>k_i,\ z\gg z_0 e^{\varkappa\frac{c_i}{v_*}} \tag{97}$$

and, finally,

$$\frac{U(z)}{v_*}=\frac{1}{\varkappa}\ln\frac{z}{z_0},\quad z_0=\frac{A_i H}{e^{\varkappa\frac{c_i}{v_*}}} \tag{98}$$

$$\text{for } \frac{Hv_*}{\nu}>N_i,\ z\gg z_0 e^{\varkappa\frac{c_i}{v_*}}.$$

It is seen from equations (95) — (98) that the ratio of the roughness parameter z_0 for the sea to the roughness parameter for a stationary wall h_0 (or H in the moving coordinate system) should be a function of the dimensionless velocity $\frac{c}{v_*}$. This is seen graphically in Figure 1, which presents the dependence of $\frac{z_0}{H}=f\left(\frac{c}{v_*}\right)$ using all the available and reliably analyzed experimental data. The graph shows measurements for $Re_{\text{rough}}>50$, carried out under different sea swell conditions and wind speeds, as well as laboratory data for low values of $\frac{c}{v_*}$ /26/.

The above allows the important conclusion that the roughness parameter (and consequently, the friction regime of the undulatory sea surface) depends not only on the absolute values of the spectral density of the sea swell, which determine the value of H, but to the same extent on the frequency composition of the sea swell, which determines the value of c_i. It is precisely this latter fact which sometimes escaped the attention of investigators and prevented them from a sufficiently convincing interpretation of the experimental data on z_0. The question arose as to whether sea waves can be treated as roughness elements. For an aerodynamically rough sea surface (for $Hv^*/\nu>100$, which is valid in many cases) we should have

$$z_0\sim A_i H. \tag{99}$$

Here A_i, the proportionality factor, should not be much smaller than 10^{-2} (since otherwise the waves could not be treated as roughness elements). Substituting the experimentally found values of z_0 ($10^{-2}-10^{-4}$ cm) into equation (99) and visually estimating the wave height (10^2 cm), the investigators arrived at the seemingly paradoxical result

$$A_i\approx\frac{10^{-2}-10^{-4}}{10^2}\approx 10^{-4}-10^{-6}. \tag{100}$$

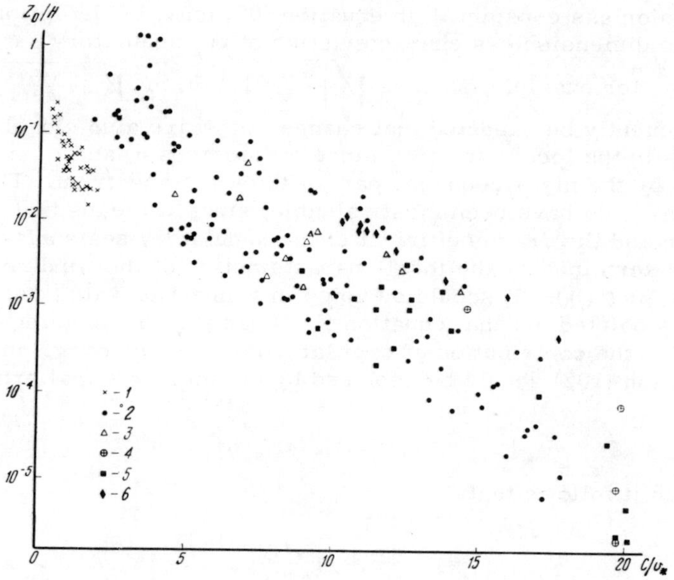

FIGURE 1. Ratio z_0/H as a function of dimensionless velocity $\frac{c}{v_*}$.
From experimental data of: 1 — Kunishi /26/; 2 — Takahashi /39/; 3 — Takeda /40/; 4 — Deardroff /20/; 5 — Snopkov /14/; 6 — joint expedition of the Institute of Oceanology of the USSR Academy of Sciences and the Institute of Atmospheric Physics (1964).

The natural reaction to this result was to conclude that the wind waves carrying the major part of the energy, i.e., making the main contribution to H, cannot be treated as roughness elements and hence only the high frequency part of the sea-swell spectrum with amplitudes of the order of 1 cm should determine the magnitude of the roughness parameter, and consequently, the resistance of the undulatory sea surface (99). This conclusion appeared quite attractive, since it led one to believe that some universal drag relationships for the undulatory sea surface could be obtained. Thus, if it is assumed that the roughness elements comprise short gravitational waves, rapidly reaching "saturation" /34/, then the amplitude of these waves should be determined only by the gravitational acceleration g and the local characteristics of the wind field (v_+ or $U(z)$). Using equation (99) one can write

$$z_0 = z_0(v_+, g) \tag{101}$$

or

$$z_0 g / v_+^2 = \text{const.} \tag{102}$$

Apparently, this is the consideration underlying the derivation of the Charnock-Ellison formula, from which it follows that the order of magnitude of the constant in equation (102) should hardly change for any regimes of wind-generated sea swell. It is of importance to note that the dimensionless constant A_l in equation (99) may, by definition, depend only on the dimensionless characteristics of the undulatory surface topography, for example, such as $\sqrt{\overline{\left[\frac{\partial\eta(x, y)}{\partial x}\right]^2}} = \sigma_\eta$ or $\sqrt{\overline{\left[\frac{\partial^2\eta(x, y)}{\partial x^2}\right]^2}} = r_\eta$. It can consequently be expected that changes in A_l are also closely related to changes in the local wind (v_+), since parameters σ_η and r_η are determined primarily by the high-frequency part of the wave spectrum. The latter conclusion would have been most valuable, since it means that the roughness parameter and the drag coefficient of the undulatory sea surface are uniquely determined by the local characteristics of the wind regime above the sea (v_+ or $U(z)$). It should be noted in connection with this, that it is sometimes pointed out that equation (102) has the shortcoming of not allowing for the contribution of capillary waves to the drag. In the latter case equation (102) should be replaced by the more general expression

$$z_0 = z_0\left(v_*, g, \frac{T}{\rho_w}\right), \tag{103}$$

from which it follows that

$$\frac{gz_0}{v_+^2} = f\left(\frac{gT}{\rho_w v_*^4}\right), \tag{104}$$

where T is the surface tension.

Equation (104) means that, when both short gravitational and capillary waves are considered in the drag regime, the roughness parameter of the undulatory sea surface may be any universal function of only the local characteristics of the wind regime above the sea (v_* or $U(z)$). In this case it is possible to construct the universal relationship $z_0 = z_0(v_*)$.

It should be noted that equation (102) as well as (104) was derived on the assumption that the roughness parameter is determined by the mean amplitude of those short gravitational and capillary waves which receive their energy directly from the given local air flow. In fact, as is shown by the latest investigations of Hasselmann /25/ and Longuet-Higgins /27/, short gravitational waves as well as capillary waves may receive energy from the low-frequency wave spectrum components due to nonlinear effects. It is, however, difficult to assume that these effects can be responsible for marked changes in z_0 with v_*, since on the assumptions made regarding the predominant role of high-frequency components of the wind wave field in the drag, the roughness parameter should be determined by the mean amplitude of the short gravitational and capillary waves, which is unlikely to change much for a given v_* due to nonlinear effects. Still one must remember that departures from the universal relationship $z_0(v_*)$ can be somewhat attributed to nonlinear effects, as a result of which the energy contained in the high-frequency region may depend markedly on the structure of the low-frequency part of the spectrum, which is not uniquely related to the local wind.

Since c and H are related to the low-frequency wave spectrum region, which depends not only on the local wind, but also on additional external variables, such as T, the wind duration, and the acceleration X /13, 4, 5/, it is difficult to expect any unique relationship between z_0 on the one hand and v_* (or the wind speed $U(z)$) on the other.

Hence, one could have expected equation (104) to have been replaced by

$$\frac{gz_0}{v_*^2} = f\left(\frac{gT}{\rho_w v_*^4}, \frac{gX}{v_*^2}, \frac{gt}{v_*}\right). \qquad (105)$$

Naturally, it is impossible to determine this relationship from individual observations of the wind profile. It should also be remembered that it is as yet impossible to formulate the problem of determining the relationship between z_0 and characteristics of the sea-wave spectrum. It may therefore be sometimes expedient to use the approach utilized in deriving equation (95). Its derivation was based on the fact that all the waves (roughness projections) move with the same phase velocity c_i. The actual spectrum of wave motion is represented as the motion of elementary wave components of different amplitude with different phase velocities.* One hence may refer only to some effective velocity of the coordinate system within which the drag of the undulatory sea surface would have been determined uniquely by some statistical characteristics of the roughness topography, in the same manner as the drag of a stationary rough wall. In general, it is rather impossible to determine theoretically the value of this effective velocity or its relationship with the parameters of the spectrum, but two facts must be borne in mind when using equations such as (95).

First, in the case of a narrow sea-swell spectrum the introduction of some phase velocity c_i into equation (95), defined as the velocity of such a spectral component whose amplitude is H, should not result in excessive error.

Second, some allowance should be made for the group nature of wind swells. If it is assumed, for example, that waves with maximum heights make the principal contribution to the drag (to the value of z_0), then the phase velocities of these waves will be close to one another, which again leads to the case when equation (95) is sufficiently exact.

The derivation of equation (95) was presented only for the purpose of pointing out the possible contribution of low and intermediate frequencies to the wave spectrum, making the main contribution to H in the drag regime. From equation (95) we can draw the following quite important conclusion: the roughness parameter of the undulatory sea surface may in general be a function of v_*, ν, wave amplitude, phase velocity, period or length of waves, as well as, by means of A_i, of certain dimensionless characteristics of the sea surface topography, such as the mean wave steepness H/λ, where λ is the wavelength. This conclusion (regarding the dependence of z_0 on v_*, ν, H, c_i and H/λ) is fully confirmed by experimental results /24/.

It follows that the use for the sea of the square law for drag in a moving coordinate system is still unreliable, as opposed to the theory of the surface air layer /8/, where the use of this relationship enables one to effectively determine roughness parameters for a limited ensemble of underlying surfaces.

The temperature profile and heat transfer coefficient. In making an analogy between the air flow above waves and the flow past a

* The fact that the high-frequency part of the wave spectrum also affects the drag regime is, in our opinion, best demonstrated by experiments in which the sea was covered by surface-active substances /24/.

stationary rough wall by taking into account the fact that the waves move, we pointed out that the drag coefficients can no longer be determined from the above formulas and that they should be obtained from equations which would reflect the motion of the individual roughness on the drag. However, in the formal approach, i. e., when introducing some relative velocity, it remains unclear in the absence of a physical analogy between the water-adjoining layer and the turbulent flow above a stationary rough surface, what are the changes to be introduced into expressions for coefficients $C_\theta(z)$ and $C_E(z)$. In order to resolve this question it is not sufficient to introduce some roughness parameter for the sea, which is a function of the phase velocity of waves. Some more precise considerations, first proposed by Stewart /35/, will be presented in the following section.

SOME FEATURES OF THE TURBULENT REGIME IN THE ATMOSPHERIC LAYER DIRECTLY ABOVE THE WATER

It must be remembered that a major difference exists between the turbulent regime in the vicinity of a stationary rough wall and that in the vicinity of moving waves. This difference was first pointed out by Stewart, whose conclusions are briefly stated as follows. A large part of the momentum transmitted by turbulent air flow is converted into the momentum of wave motions of the sea surface.* Since a simple relationship exists in irrotational [eddy-free] flow between the wave motion momentum I and the former's energy E (a relationship is considered only for an individual component of the wave field):

$$I = E/c_i, \tag{106}$$

where c_i is the phase velocity of the wave, while for air flow

$$E/I \approx U(z), \tag{107}$$

it can be easily concluded that at the altitude where reaction of the air to the wave drag is already felt, we should have

$$U(z) \geqslant c_i, \tag{108}$$

since for a given value of I the energy lost by the air should be equal to that acquired by the wave. It follows from here that an air layer (δ_w), subsequently called the interaction sublayer, in which the momentum and energy of the pulsating motion should be considered simultaneously, should exist above the sea surface. This pulsational motion does not satisfy known conditions of the self-similar regime of developed turbulence at a stationary wall. Hence Stewart /35/ suggested that, first, deviations

* No proof is presented for the validity of this conclusion by Stewart. It is only noted that this problem has not been sufficiently studied in its application to the physics of the sea.

from a logarithmic velocity profile may be observed in the lower layers of air above the sea; second, since the pulsating motion in the region $z < \delta_w$ should be ordered (so-called pseudo-turbulence after Stewart), then its contribution to the heat and moisture transport should be smaller than such a contribution of developed turbulence at a stationary wall. Hence for $z < \delta_w$ the eddy thermal diffusivity and the eddy diffusion coefficients become small, and consequently large vertical mean temperature and average specific humidity gradients can be expected in the region of the interaction sublayer.

Since a region with self-similar developed turbulence with constant eddy drag is situated above the interaction sublayer ($z > \delta_w$), the question of possible interpretation of gradient observations above the sea within the framework of known theories of the surface air layer reduces in substance to a determination of the thickness of the interaction sublayer under different hydrometeorological conditions.* Estimates show /28, 35/ that for the majority of hydrometeorological situations standard levels of observations of the mean wind and air temperature above the sea lie above the interaction sublayer. In this sense the question of extending known schemes of the surface air layer to analysis of turbulent exchange processes above the sea can be successfully resolved.

However, the situation becomes quite different when using diffusion equations (1)—(3), in particular, for theoretical calculation of the heat transfer and evaporation coefficients. Obviously, of importance here is not the absolute thickness of the interaction sublayer, but its relative magnitude as compared with the viscous diffusion sublayers arising in the flow past stationary walls.

Obviously, when $\delta_T \gg \delta_w$, the main temperature and moisture changes occur in the diffusion sublayers.** In this case it is possible to use equations for $C_\theta(z)$ and $C_E(z)$ derived for a stationary smooth wall. Since the thickness of diffusion sublayers (in particular of the viscous and thermal layers) for rough walls is smaller than its values at a smooth wall, while wave growth generally results in an increase in the interaction sublayer thickness, it should be expected that in the majority of cases we will have $\delta_w \gg \delta_T$. Then, obviously, coefficients $C_\theta(z)$ and $C_E(z)$ should depend not only on the surface relief and on the molecular coefficients, but also on the thickness δ_w of the interaction sublayer. At the same time the latter thickness is not a constant quantity. As is seen from equation (108), it should depend on v_* and on the phase velocity of the waves acquiring energy from the given quasisteady turbulent flow. Since it is difficult to find a frequency range in the wind wave spectrum which does not acquire energy from the wind, it follows from the above that air may lose its momentum and energy at very different altitudes even for the separately considered undulatory sea surface. Only in the case of a narrow wave spectrum can one expect some regularity in the velocity profiles and in characteristics of pulsating motion at altitudes of $z < \delta_w$. However, this should not be expected since, as was noted by Lighthill /28/, energy and momentum from air is transmitted not only to waves propagating in the mean-wind direction, but also to those moving at different angles to the latter. In this case

* This statement of the problem was formulated by A. S. Monin.

** Since for air $\mathrm{Pr}_{\mathrm{mol}} \approx 1$, then the viscous sublayer coincides with the temperature and diffusion sublayers.

equation (108) should be replaced by the expression

$$U(z)\cos\gamma \geqslant c_i, \tag{109}$$

where γ is the angle between the mean-wind direction and the direction of motion of an individual wave. According to Lighthill for all these reasons one should not expect universal features in the velocity profile in the interaction zone. This makes the determination of the interaction sublayer thickness even more important. This is quite difficult to do with a high accuracy, since in the final count we are not interested in δ_w, but rather in the ratio δ_w/δ_T.

To illustrate the possible effect of making allowance for the interaction sublayer, we shall now consider the known model of the so-called fully developed swell, when the steady turbulent boundary layer and a steady wind swell spectrum coexist. The entire energy received by the waves from the wind is dissipated by the former in generating eddies (or in producing drift currents). The wind-wave spectrum containing the principal amount of energy can be regarded as the spectrum of gravitational waves. The thickness of the interaction sublayer will be determined solely by two constant dynamic variables, which are available in the case of fully developed swell: v_+ and the acceleration of gravity g, which limits the vertical wave growth /34/. We thus consider a swell spectrum in which all the waves acquiring their energy from the wind reached the equilibrium stage /34/.

The condition defining the value of δ_w can be written in the form

$$\delta_w g/v_*^2 \sim 1. \tag{110}$$

At the same time the thickness of diffusion sublayers characteristic of a smooth stationary wall can be expressed as

$$\delta_T v_*/\nu \sim 1. \tag{111}$$

It is thus seen that the ratio between these thicknesses is a function of

$$\frac{\delta_w}{\delta_T} \sim \frac{v_+^3}{g\nu} = M, \tag{112}$$

i.e., the value of the dimensionless parameter M is important in analyzing the structure of the water-adjoining layer. This parameter appeared in the work of Munk /31/. Apparently, its physical meaning is as follows: at low M a greater contribution is made by the viscous regime than by the motion of the surface. On the other hand, at high M the interaction sublayer is much thicker than the viscous sublayers arising at a stationary wall. In this case the expressions for heat transfer and evaporation coefficients derived for a stationary wall lose their physical meaning. Thus, had all the observations been made under conditions of fully developed wind swell at different wind speeds, then classifying data on roughness parameters, heat transfer and evaporation coefficients with respect to values of $M(v_*)$ would have had some meaning. Actually, however, the relationship between the thickness of the interaction sublayer and viscous sublayers depends on the surface relief, and not only on v_*. Hence this problem will not be discussed in detail here, although undoubtedly it is of the greatest interest in the overall problem of the water-adjoining layer of the atmosphere. We only note that assuming $v_* = 10^1 - 10^2$ cm/sec and $g = 10^3$, we obtain $M = 10^1 - 10^4$, i.e., it is more unlikely that situations will occur when $\delta_T/\delta_w > 1$.

Modern methods of computing eddy fluxes above the ocean. The methods presently used in the physics of the sea (as well as in marine meteorology) for calculating eddy fluxes above oceans reduce to the two following approaches.

1. The values of v_* and z_0 are determined as a result of gradient observations (using the model of the logarithmic boundary layer, as well as the more general theory of the stratified ground air layer). Then one constructs an empirical relationship, such as

$$z_0 = z_0(v_*) \text{ or } z_0 = z_0(U(z)). \tag{113}$$

Since no unique relationship is found between z_0 and v_*, one must seek some average value $\bar{z}_0$ in a narrow range of wind speeds or of values of v_*, and then the empirical relationship becomes

$$\bar{z}_0 = f(v_*), \tag{114}$$

which has corresponding to it another empirical relationship

$$\overline{C_u}(z) = f(v_*). \tag{115}$$

Different forms of this relationship were suggested by many investigators, **
the differences between them being sometimes quite marked.

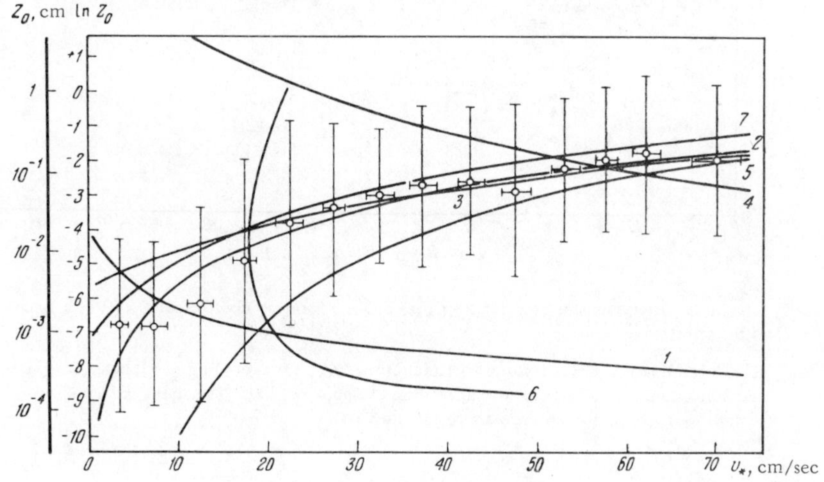

FIGURE 2. The roughness parameter $\bar{z}_0$ as a function of the friction velocity v_*:

1 — for a smooth wall; 2 — averaged values of $\bar{z}_0$ obtained by the method of least squares; 3 — according to Deacon /19/; 4 — according to Neumann /32/; 5 — according to Francis /23/; 6 — according to Kuznetsov /6/; 7 — according to Sheppard /38/.

The curves in Figure 2 illustrate quite well the feasibility of obtaining different relationships of type (114) or (115), depending on the experimental data used. We have analyzed a large number (more than 1000) of measurements of various investigators. Curve 2 was obtained from values of $\bar{z}_0$ averaged over narrow ranges of the friction velocity v_* by the method of least squares. The large standard deviations can in no way be solely attributed to imperfection of the measurements proper — they are a result of the fact that no unique physical relationship exists between z_0 and v_*.

The causes of this situation are quite clear and consist in the fact that quasisteady turbulent air flows above the sea can occur over different kinds of surfaces, the characteristics of which should include not only data on the mean altitudes, but in the first place data on the phase velocities of the wind waves (i.e., we must actually know the three-dimensional spectrum of the sea swell).

Here the principal question as to which of the wave-spectrum parameters has the greatest effect on the drag regime of the undulatory sea surface remains at present not studied either theoretically or experimentally.

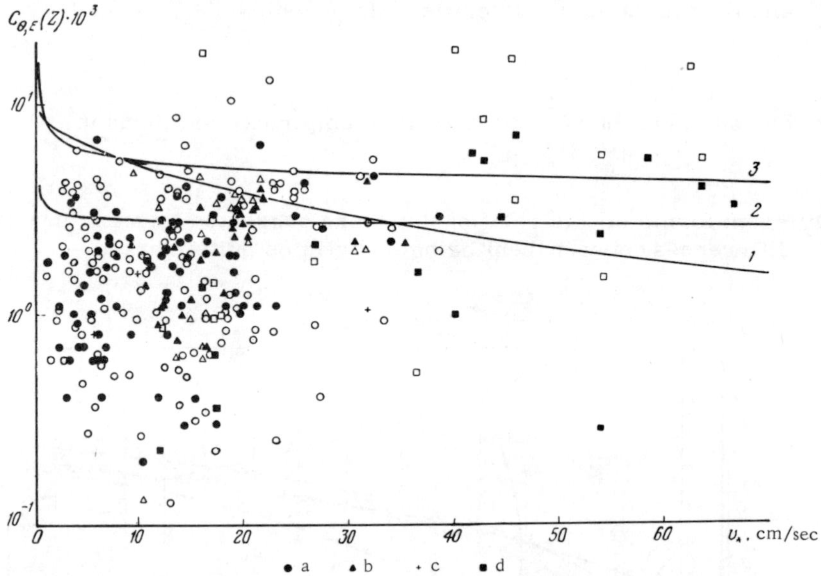

FIGURE 3. Heat transfer coefficient $C_\theta(z)$ and evaporation coefficient $C_E(z)$ as a function of the friction velocity v_*:

1, 2, 3 — see text; a — according to Takahashi /39/; b — according to Fleagle /22/; c — according to Deardroff /20/; d — according to Snopkov /14/. The darkened symbols pertain to $C_E(z)$, the open symbols pertain to $C_\theta(z)$.

2. Heat transfer and evaporation are calculated from diffusion formulas (1)—(3). However, unlike the problem of calculating the turbulent friction τ_0, sufficient experimental data are at present unavailable for calculating the eddy fluxes of Q and E in order to construct a reliable

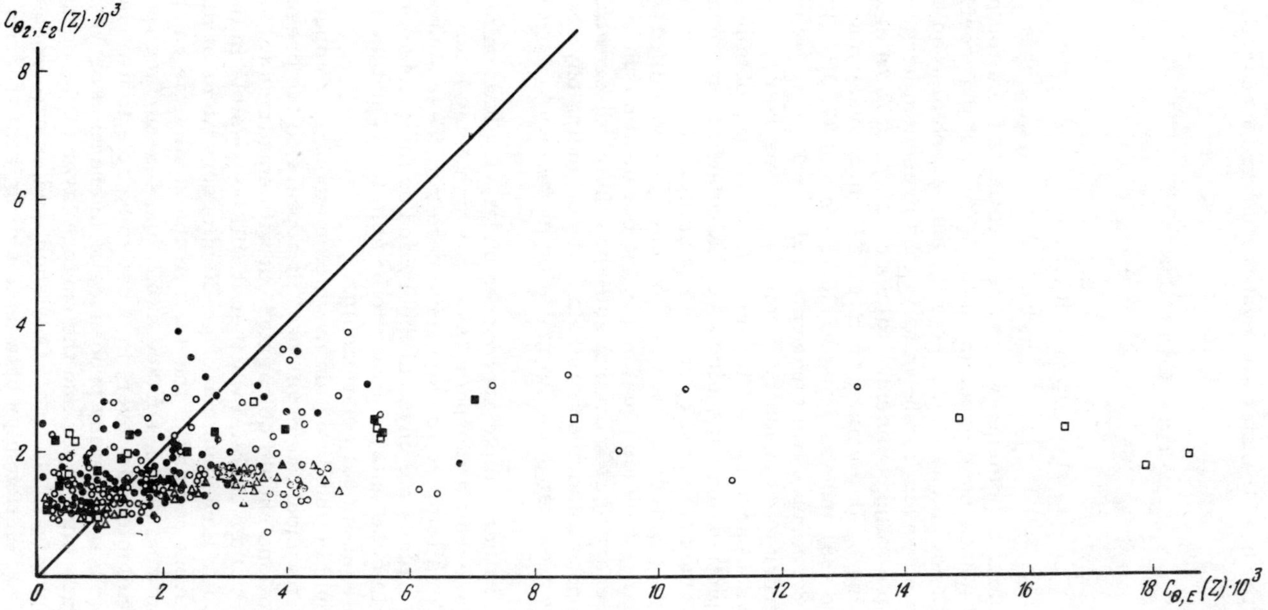

FIGURE 4. Experimental values of coefficients C_θ (z) and C_E (z) plotted against their values calculated from equation (82). For the legend see Figure 3.

empirical relationship for $C_\theta(v_*)$. One may only refer to an average statistical dependence of $C_\theta(z)$ and $C_E(z)$ on v_*, since there is no unique physical relationship between them. The unavailability of experimental data is the only reason why the heat transfer and evaporation are frequently calculated from the expressions

$$C_\theta(z) = C_E(z) = C_u(z) \tag{116}$$

or

$$C_\theta(z) = C_E(z) = M_i[C_u(z),\ v_+], \tag{117}$$

derived from various hypotheses which are unreliable from the physical point of view and not checked experimentally. The fact that coefficients $C_\theta(z)$ and $C_E(z)$ are not unique functions of v_*, but depend also on characteristics of the wind waves, is shown by graphs presented here.

Figure 3 is a plot of experimental values of $C_\theta(z)$ and $C_E(z)$ against their corresponding v_*, determined from profile measurements. Using equations (82) and (89) and assuming $\delta_{\theta,E}$ (0.12 cm and 27. 5 ν/v_*) in accordance with Sverdrup's two theorems, and $z_0 = 0.6$ cm, we calculated curves using Sverdrup and Sheppard formulas (curves 1—3).

Figure 4 shows a curve of coefficients $C_\theta(z)$ and $C_E(z)$, obtained experimentally, plotted against their values calculated from equation (82) using observed values of z_0 and v_* on the assumption that $\delta_{\theta,E} = 27.5\ \nu/v_*$.

The frequently used average values $\overline{C_\theta(z)}$ and $\overline{C_E(z)}$, obtained by the closure of the energy balance equations in the ocean-atmosphere system, first, cannot be regarded as more reliable than their "theoretical" values; and second, this method can help little in finding physical regularities in heat exchange and evaporation processes under different conditions.

These are the rather simple approaches underlying heat balance calculations in the ocean-atmosphere system which came lately into quite extensive use. We attempted to emphasize here that these approaches are not based on a plausible physical theory and are not based on sufficiently convincing experimental data, in order to be able to regard them simply as the use of reliable empirical relationships.

What then should be the direction in which methods of computing eddy fluxes above oceans should be developed? It appears to the present authors that these calculations should be carried out with equations (1)—(3). The point is that it is rather doubtful whether it will be possible in the near future to carry out a large number of gradient observations at different times and in different regions of the sea. At the same time an appreciable volume of data on wind speed, surface swell, and sea-surface temperature are at present available. However, in order to be able to use equations (1)—(3), it is necessary to study processes of interaction between the turbulent water-adjoining layer and the surface layer of the ocean on the basis of certain physical concepts. Only such studies can produce physically justified relationships between the drag, heat transfer and evaporation coefficients on the one hand, and accessible measurements of a parameter of the turbulent regime in the water-adjoining layer of the atmosphere and the upper layer of the sea on the other (in the first place, the dependence of heat exchange and evaporation on the wind speed and characteristics of the sea-surface swell).

To solve this problem in the study of the water-adjoining layer one must make extensive use of gradient and pulsation measurements, in addition to well-developed methods of interpreting these observations /8, 9/. Work of this kind, supplemented by simultaneous measurement of the sea-wave spectrum, radiation balance and temperature regime in the surface layer of the sea, will undoubtedly bear fruit and make it possible to work out methods for calculating eddy fluxes above oceans which would have a satisfactory theoretical and experimental basis.

References

1. Kazanskii, A. B. and A. S. Monin. Turbulentnost' v prizemnykh inversiyakh (Turbulence in Ground Inversions). — Izvestiya AN SSSR, seriya geofizicheskaya, No. 1. 1956.
2. Kazanskii, A. B. and A. S. Monin. O turbulentnom rezhime v prizemnom sloe vozdukha pri neustoichivoi stratifikatsii (On the Turbulent Regime in the Surface Air Layer under Unstable Stratification). — Izvestiya AN SSSR, seriya geofizicheskaya, No. 6. 1958.
3. Kitaigorodskii, S. A. Nekotorye prilozheniya metodov podobiya pri analize vetrovogo volneniya kak veroyatnostnogo protsessa (Some Applications of Analogy in Analysis of Wind Swell as a Probability Process). — Izvestiya AN SSSR, seriya geofizicheskaya, No. 1. 1962.
4. Kitaigorodskii, S. A. and S. S. Strekalov. K analizu spektrov vetrovogo volneniya, Ch. 1 (On Analyzing Spectra of Wind Swell, Part 1). — Izvestiya AN SSSR, seriya geofizicheskaya, No. 9. 1962.
5. Kitaigorodskii, S. A. and S. S. Strekalov. K analizu spektrov vetrovogo volneniya, Ch. 2 (On Analyzing Spectra of Wind Swell, Part 2). — Izvestiya AN SSSR, seriya geofizicheskaya, No. 8. 1963.
6. Kuznetsov, O. A. K voprosu ob aerodinamicheskoi sherokhovatosti poverkhnosti morya (Concerning the Aerodynamic Roughness of the Sea Surface). — Trudy IOAN SSSR, part 72. 1963.
7. Levich, V. G. Fizichesko-khimicheskaya gidrodinamika (Physicochemical Hydrodynamics). — Fizmatgiz. 1962.
8. Monin, A. S. and A. M. Obukhov. Osnovnye zakonomernosti turbulentnogo peremeshivaniya v prizemnom sloe atmosfery (Principal Regularities of Eddy Mixing in the Surface Layer of the Atmosphere). — Trudy Geofizicheskogo instituta AN SSSR, 24 (151). 1964.
9. Monin, A. S. Struktura atmosfernoi turbulentnosti (The Structure of Atmospheric Turbulence). — Teoriya veroyatnosti i ee primenenie, 3 (3). 1958.
10. Monin, A. S. O strukture polei skorosti vetra i temperatury v prizemnom sloe vozdukha (The Structure of Wind Speed and Temperature Fields in the Ground Air Layer). — Trudy IFA AN SSSR, No. 4. 1962.

11. Miles, J. W. On the Generation of Surface Waves by Shear Flows. — J. Fluid Mech., Vol. 3. 1957.
12. Samoilenko, V. S. Sovremennaya teoriya okeanicheskogo ispareniya i ee prakticheskoe primenenie (The Modern Theory of Oceanic Evaporation and its Practical Application). — Trudy GOIN, 21 (33). 1952.
13. Samoilenko, V. S. Formirovanie temperaturnogo rezhima morei (Formation of the Temperature Regime of the Seas). — Gidrometeoizdat. 1959.
14. Snopkov, V. G. O turbulentnom obmene v prizemnom sloe nad okeanom (Turbulent Exchange in the Surface Layer above the Ocean). — Trudy IOAN SSSR, Vol. 78. 1965.
15. Townsend, A. A. The Structure of Turbulent Shear Flow. — Cambridge University Press. 1956.
16. Phillips, O. M. Resonance Phenomena in Gravity Waves. — In: Gidrodinamicheskaya neustoichivost'. Nauka. 1964.
17. Hinze, J. O. Turbulence. — McGraw-Hill Book Co. 1959.
18. Schlichting, H. Lecture Series "Boundary Layer Theory." Part II. Turbulent Flows. — U. S. NACA. Technical Memorandum No. 1218. April 1949.
19. Deacon, E. L. and E. K. Webb. Interchange of Properties Between Sea and Air Small-Scale Interactions. — "The Sea," Vol. 1. 1960.
20. Deardroff, J. M. An Experimental Ocean Buoy for Air-Sea Transfer Studies. — Oceanogr. Rept. Univ. of Washington, No. 13. 1962.
21. Van Dorn, W. G. Wind Stress on an Artificial Pond. — J. Marine Res., Vol. 12. 1953.
22. Fleagle, R. G., J. W. Deardorff, and F. I. Badgley. Vertical Distribution of Wind Speed, Temperature and Humidity above a Water Surface. — J. Marine Res., Vol. 17. 1958.
23. Francis, J. R. D. Wind Stress over a Water Surface. — Quart. J. Roy. Meteorol. Soc., 80 (345). 1954.
24. Gerritsen, F. Surface Wind Stress over Water as Related to Wave Action. — Ingenieur, Bonw en water Bou Kunde, Vol. 14. 1963.
25. Hasselmann, K. On the Non-Linear Energy Transfer in a Gravity Wave Spectrum. Pt. 3. — J. Fluid Mech., Vol. 15. 1963.
26. Kunishi, H. An Experimental Study on the Generation and Growth of Wind Waves. — Disaster Prev. Res. Inst. Kyoto Univ., Bull. No. 61. 1963.
27. Longuet-Higgins, M. S. The Generation of Capillary Waves by Steep Gravity Waves. — J. Fluid Mech., Vol. 16. 1963.
28. Lighthill, M. J. Physical Interpretation of the Mathematical Theory of Wave Generation by Wind. — J. Fluid Mech., Vol. 14. 1963.
29. Monin, A. S. Turbulence in Shear Flow with Stability. — J. Geophys. Res., 64 (12). 1959.
30. Motzfeld, H. Die turbulente Strömung an welligen Wänden. — Z. angew. Math. und Mech., Vol. 17. 1937.
31. Munk, W. H. Wind Stress on Water: An Hypothesis. — Quart. J. Roy. Meteorol. Soc., 81 (349). 1965.
32. Neumann, G. Wind Stress on Water Surfaces. — Bull. Amer. Meteorol. Soc., 37 (5). 1956.

33. Owen, P. R. and W. R. Thomson. Heat Transfer across Rough Surfaces. — J. Fluid Mech., Vol. 15. 1963.
34. Phillips, O. M. The Equilibrium Range in the Spectrum of Wind-Generated Waves. — J. Fluid Mech., Vol. 4. 1958.
35. Stewart, R. W. The Wave Drag of Wind over Water. — J. Fluid Mech., Vol. 10. 1961.
36. Sverdrup, H. U. The Maritime Verdunstungsproblem. — Ann. Hydrogr., No. 3. 1936.
37. Sverdrup, H. U. On the Evaporation from the Oceans. — J. Marine Res., 1 (1). 1937.
38. Sheppard, P. A. Transfer across the Earth's Surface and through the Air Above. — Quart. J. Roy. Meteorol. Soc., 84 (361). 1958.
39. Takahashi, T. Micro-Meteorological Observations over the Sea. — Meteorol. Notes, Kyoto Univ. Meteorol. Res. Inst., Ser. 2, No. 12. 1958.
40. Takeda, A. Wind Profiles over the Sea Waves. — J. Oceanogr. Soc. Japan, 19 (3). 1963.

T. V. Bonchkovskaya

EXPERIENCE IN MODELING CERTAIN SCHEMES OF ATMOSPHERIC CIRCULATION AND HEAT EXCHANGE

The present article considers experiments with modeling certain circulation and heat exchange schemes by means of two rotating vessels, with consideration of similarity criteria. An attempt is made to model the temperature field and dynamic structure of the atmosphere over large areas. The results obtained are used to discuss the mechanism of dynamic processes of the model, i. e., the physical nature of the conventional exchange coefficients.

Modeling as a method of scientific study usually involves two work stages. First a model of the phenomenon under study is constructed, i. e., a process similar to that occurring in nature is reproduced in the laboratory, and then the model is used for solving problems of scientific and practical interest.

The concept of similarity in modeling means that the phenomena observed in nature and in the laboratory are described by the same equations with identical boundary conditions, and the relationships between controlling physical characteristics at geometrically similar points of the systems are equal to one another.

Modeling of dynamic atmospheric processes with allowance for the Coriolis force requires the use of equations of motion, expressing the equilibrium of inertia forces, pressure gradient, the Coriolis force and turbulent friction. If motions similar to those occurring in nature can be produced in the laboratory, then the effect of all the above forces must be reproduced in some way or another in the modeling medium.

Analysis of the equations of motion easily yields the following similarity parameters: $\mathrm{Ro} = u/\omega r$ is the ratio of inertia forces; $\mathrm{Re} = \omega r^2/\nu_e$ is the ratio of the inertia and "viscous" forces; and $G = \omega^2 r/g$ is the ratio of the centrifugal and gravity forces. Here ω is the angular velocity of the system, r is its characteristic dimension, ν_e is the coefficient of vertical eddy exchange, and u is the characteristic horizontal velocity. The relationship between the horizontal (r) and vertical (H) scales and the velocities, $wr = uH$, is determined from the continuity equations. The numbers Ro, Re and G in the modeling system should be of the same magnitude as in nature.

The atmospheric pressure gradient field is formed primarily due to interaction between heat sources and sinks and to transport of air masses which is accompanied by transport of heat. Hence the model must reproduce not only dynamic, but also the analogous heat processes. This means that the heat influx equation must be satisfied. For example, for a symmetrical circular heat source and steady homogeneous (over the thickness of the heat diffusion layer without allowance for sphericity) terrestrial surface, the heat fluxes can be expressed approximately in the form

$$KH\left(\frac{\partial^2 T}{\partial r^2}+\frac{1}{r}\frac{\partial T}{\partial r}\right)=\alpha(T_1-T)-\sigma T\ldots, \qquad (1)$$

where KH is the conditional heat transfer coefficient of the layer in the horizontal direction; α and σ are convective heat transfer and radiation coefficients; T is the mean temperature of the layer; and T_1 is the surface temperature.

The thermal similarity parameters for this case have the form: $\alpha r^2/KH$ and $\sigma r^2/KH$. Their numerical values in the model and nature should be the same. In addition, if analogous convective buoyancy of the masses are to be reproduced in the vicinity of heat sources, then one must introduce yet another parameter, the number $Ra=\beta\gamma g H^4/\gamma_e^2$, involving buoyancy and viscous forces. Here β is the bulk expansion coefficient and γ is the lapse rate. These are the dynamic and heat parameters which are needed for estimating whether modeling or circulation and heat exchange in a sufficiently extended layer of the atmosphere is possible in principle.

As was stated above, if the model has been obtained and its similarity to natural phenomena has been proved, then modeling will become a useful scientific tool. In particular, using this model it is possible, by varying the initial and boundary conditions, to solve some forecasting problems and it is also expedient to use them in the study of the circulation's dependence on a given factor or set of factors. And, finally, a model can be used for estimating the effect on climate of some or other changes in the thermal properties of the terrestrial surface.

These problems can, clearly, be solved by different methods, but a model enables one to solve nonlinear problems and, in addition, by visualizing the process, to follow its development at all stages without expenditure of time and labor. The far-reaching possibilities in the application of laboratory experiments to meteorology have served for a long time as an inducement to seeking methods for reproducing individual schemes of atmospheric motions under laboratory conditions.

All kinds of facilities were used, but initially all the work in this direction was done without allowance for similarity parameters and the results were basically qualitative. During the past few years it was shown in the USSR and elsewhere that atmospheric processes of different scales may in principle be reproduced in rotating vessels containing fluids and with stationary models. As a rule, the devices used consisted of water-filled vessels of various sizes and shapes. Depending on the phenomenon being modeled, the vessel was stationary or rotated about a vertical axis. The temperature and motion fields were produced by heating and cooling of individual regions of the bottom or walls of the vessels. These devices were used for reproducing various schemes of motion: interlatitude exchange and jet streams (with the periphery of the vessels heated and the vessels proper rotated), long waves, frontal waves (rotating a vessel with two layers of different density), flow past orographic obstacles, etc.

The main shortcoming of these experiments was determination of quantitative estimates using small-size vessels, which did not fully satisfy the similarity criteria. Using a larger facility Faller /2/ was able to only qualitatively reproduce a flow pattern which would be similar to that of the atmosphere. Nonetheless, the experiments carried out were quite valuable, since they made it possible to obtain parametric relationships between flow characteristics and the factors producing them. Thus, the motion produced in an annulus formed by two rotating coaxial cylinders in the field of action of parameters Ro and G was studied in /3 – 6/. The walls of the outer cylinder were heated and those of the inner were cooled.

This system can be used for reproducing three different types of motion: at high Ro (above 0.2 – 0.6) Hadley's symmetrical regime in the form of spiral heat flux from the outer to the inner walls; at Ro < 0.2 stable waves formed in the annulus, their number decreased with a reduction in Ro; and, finally, an eddy regime ensued at Ro of the order of 10^{-2}. According to my calculations, in the atmosphere Ro $= 0.06$ and $G = 10^{-3}$. Consequently, on the basis of experiments in the annulus one can expect both the symmetrical and the wave flow regimes in the atmosphere.

Subsequently Hide /4/ changed the depth and width of the annulus. However, the results then obtained have no significance in the study of atmospheric processes.

It should be noted in passing that the wave regime obtained by Rossby is easily reproduced without heating the walls. All that is needed is the production of the relative motion of two immiscible fluids of different density. Fultz /3/ in his experiments obtained this kind of relative motion by increasing or reducing the speed at which the vessel rotated. Here the circular surface at the interface between the different-density fluids becomes convex or concave, depending on whether the vessel is accelerated or decelerated, and a different number of waves is formed at the interface. These experiments clearly illustrate the process of wave generation in the presence of a large density difference and show that these waves can be quite stable under the proper conditions.

This author repeated the Hide and Fultz experiments with heating and cooling of the vessel walls. The results were similar to theirs, although the vessels used were somewhat larger. It should be noted that the inner, cooled vessel, which imposed thermal and mechanical limitations on the motion, played a large role in these experiments. In addition, the annulus width was commensurable with the dimensions of the waves produced in it. However, it is impossible to select similar conditions for wave generation in nature. Hence an attempt was made to reproduce stable waves not in an annulus, but simply in a cylindrical vessel, heating and cooling the fluid from the bottom rather than from the side, thus eliminating the inner cylinder.

These experiments were carried out in vessels of diameters 37 and 84 cm with different rotating speeds and heating /1/. It was impossible to obtain stable waves, at least for Ro $= 0.02 - 0.5$ and G of the order of 10^{-3}. Eddies formed immediately at the heated region of the bottom at the walls, with their number depending on Ro. In particular, for Ro of the order of 0.05, as under atmospheric conditions, six eddies were obtained. They shifted and were transformed with time, but their number remained the same. Moreover, the overall flow pattern in the entire vessel was similar, at first sight, to the Bjerknes general circulation model, particularly in the middle latitudes and in the equatorial zone.

In subsequent experiments a high-capacity cooling device, quite roughly modeling the cold center of the Antarctic, was placed at the center of the vessel. The cooling device's output and the rotational speed were such that Ro could be varied. The dynamic regime was held constant. Stable waves again could not be obtained, but the flow pattern was in principle similar to the general circulation in southern polar regions. Anticyclonic eddies formed above the cooling device, and eddies with cyclonic rotation were generated in regions closer to the periphery. Five or six eddies ensued around the cooling device for Ro of the order of 0.05.

The results obtained in these experiments were used for determining the extreme heating and rotation conditions needed for obtaining the required range of numbers Ro and G and of thermal regime variables corresponding to the values encountered under atmospheric conditions. In addition, these experiments, as well as those carried out in the West, showed the feasibility of obtaining models of individual dynamic systems which are similar to atmospheric phenomena (jet streams, orographic waves, convection, etc.) and led one to expect further successes in modeling of atmospheric processes. Hence an attempt was made to model the temperature field and the dynamic structure of the atmosphere over large areas, of the order of a hemisphere.

This author attempted to obtain similarity between the mean temperature field of the fluid layer in a model and the yearly average atmospheric temperature. This was done with two devices having rotating cylindrical vessels of diameters 84 and 150 cm. The first device was used for modeling the so-called "purely zonal" circulation scheme, while the second is being prepared for reproducing "nonzonal" circulation and heat exchange with allowance for the longitude distribution of heat sources and sinks.

It was assumed in modeling the "zonal" scheme that a 10 cm water layer in a vessel is analogous to the hemispherical layer of the atmosphere above the northern hemisphere, which is 5 km high. This substitution of a flat for a spherical layer was found permissible from the point of view of similarity of external forces. The relationship between external forces in the form of parameter G defines the inclination of geopotential surfaces to the terrestrial surface in nature and the inclination of the fluid level in the model. In the atmosphere $G = 3 \cdot 10^{-3}$, i.e., the geopotential and terrestrial surfaces are approximately parallel. For ω of the order of approximately 10^{-1} sec^{-1} the value of G in the model is the same as in nature and the inclination of the surfaces is very small.

The situation is not as satisfactory with respect to modeling of the latitudinal Coriolis force distribution, which is due to sphericity of the rotating earth. In the model the Coriolis force is constant along the vessel radius. Consequently, cylindrical vessels do not model its distribution and the effect of this factor on the motion has as yet not been taken into account.

Other simplifications involve the assumption that the 5 km atmospheric layer is vertically isotropic and that the momentum and heat in it are propagated by means of large-scale eddies. It was assumed that the motions in this layer are produced solely by the surplus of heat Q_n supplied in equatorial latitudes, over the heat outflow Q_u, i.e., $Q_n - Q_u > 0$, and the predominance of heat outflow over its influx at high latitudes, i. e., $Q_n - Q_u < 0$. To put it briefly, the latitudinal distribution of heat sources and sinks was considered.

The yearly average values of $Q_n - Q_u$ for each latitude of the northern hemisphere were determined from maps of heat-balance atlases. The "zonal" distribution of sources and sinks thus obtained was specified in the model as starting boundary conditions by heating the appropriate parts of the vessel's bottom and walls. Here the heating regime, the vessel rpm and the fluid layer height were selected so that all the aforementioned similarity parameters were of the same magnitude as under atmospheric conditions.

The average layer temperature T_{mod} was measured in different regions of the modeling fluid by means of thermopiles, composed of a vertical array of thermocouples. The mean layer temperature was averaged for each circulation cross section at distances $r_{1,2 \ldots n}$ from the center of the vessel, corresponding to latitudes of 10, 20, ..., 80° in nature, i.e., the "zonal" distribution of the mean layer temperature T was measured. The temperature field thus determined and averaged over 1.5 hours was compared with the latitudinal distribution of the yearly mean temperature of the atmosphere, obtained from OT_{1000}^{500} maps.

Figure 1 shows the corresponding curves of T/T_x, where the characteristic temperature T_x is the difference in the mean temperature of the layer at the equator and at the pole. In the model T_x was the difference between the temperature of the fluid layer at the walls and at the center of the vessel. The curves are in sufficiently close agreement.

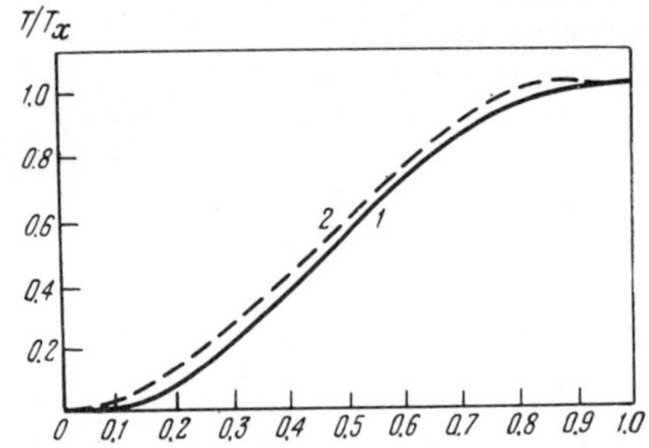

FIGURE 1. "Zonal" distribution of the average layer temperature:

1 — in nature; 2 — in the model.

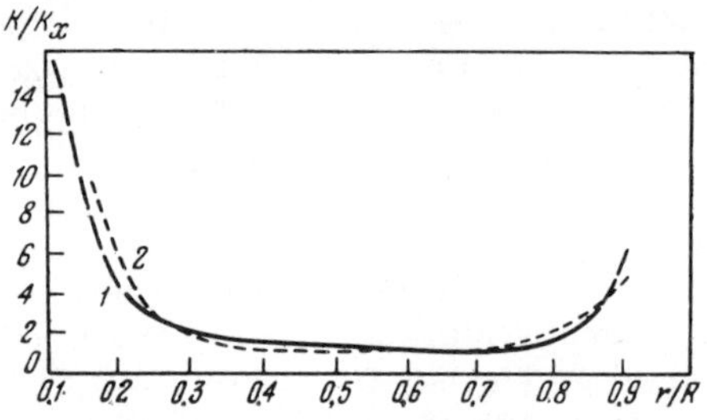

FIGURE 2. "Zonal" distribution of conventional exchange coefficients for large-scale eddies:

1 — in nature; 2 — in the model.

Since not only the temperature field, but also the horizontal heat exchange were of interest, this author has determined and compared the field of conventional horizontal exchange coefficients K_{mod} in the model and K_a in nature. In both cases K was determined from the value of the heat flux q in the equator-pole direction (in the model, from the vessel's periphery to its center) and the gradients of the mean layer temperature grad T. The calculations were carried out with the formal relationship

$$K = \frac{q}{\text{grad}\, T}.$$

The values of K thus obtained were averaged over latitude parallels for nature and by circles of corresponding radii in the model, i.e., again it was the "zonal" distribution of conventional exchange coefficients that was determined. Figure 2 shows this distribution in terms of K/K_x, where K_x is the value in nature at the 40° latitude. In the model this latitude is represented by the distance $r/R = 0.55$ along the radius from the vessel center. The agreement between these curves is also quite satisfactory.

It can thus be assumed that the principal features of the long-term interlatitude exchange in the atmospheric layer, i.e., $T(\varphi)$ and $K_a(\varphi)$, can be reproduced on a model. This means that it is correct to use this model to study the mechanism of dynamic processes, i.e., as if to bring into the open the physical nature of the conventional exchange coefficients.

It should be noted that this "zonal" model cannot have direct practical significance. However, it is of some interest for analysis of the mechanism proper of excitation of circulation processes over the earth. In particular, it can help to resolve the question of what would have been the circulation without continents and oceans and their thermal difference, but with the entire earth having a physically uniform surface, i.e., to check the validity of general circulation models of Bjerknes and others.

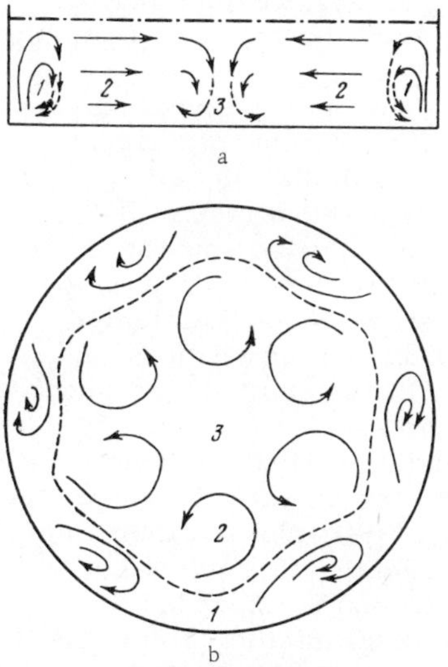

FIGURE 3. Scheme of heat exchange in a model with "zonal" distribution of heat sources:

a — direction of resulting heat transfer in radial cross section of the layer; b — flow pattern in the layer.

What kind of a mechanism is it? Figure 3 shows the flow pattern obtained with a model having a "zonal" distribution of heat sources and sinks. With respect to dynamic characteristics the entire system can be subdivided into three regions: 1) region of convective circulation; 2) region of diffusion by large-scale eddies; and 3) region of downdrafts. In this sense the experiments verify previously developed concepts. However, each of these regions has its sufficiently complicated structure. In region 1 the ascent and outflow of heated air masses in the field of rotation results in the formation of warm anticyclonic eddies. Compensating descent of air masses occurs at the central parts of these eddies. Convective circulation thus has the quite peculiar form of eddies moving slowly in a circle at the vessel walls. Temperature contrasts form at the boundary of this vorticity region. They become more pronounced and result in the formation of secondary, cyclonic-type eddies, filling region 2. These eddies further diffuse the heat toward the center of the vessel.

Consequently, had the earth surface consisted of oceans or of continents only, then a belt of anticyclonic eddies would have been observed in the 30 – 35° latitude belt, rather than a high-pressure belt. The pressure in this belt would change along the longitude due to these dynamic formations. The existence of continents and oceans, which accentuate the temperature difference along the longitude, amplifies the eddy generation and attaches eddies and baric-maximum regions to specific localities. Continents and oceans thus are not the chief, but rather the supplementary factor in the formation of tropical cyclones, in the same manner as the ruggedness of the Antarctic continent is not the cause for formation of cyclones, but only an additional factor, responsible for stationing them in certain localities.

In precisely the same manner large-scale eddies in the middle-latitude belt (region 2) are principally generated by the latitudinal heat distribution in the atmosphere. The differences between the thermal inertia of land and the sea are responsible for the nonuniform distribution of cyclonic activity in the longitudinal direction.

It is seen from this that even the most elementary model, which does not take into account the distribution of the Coriolis force, provides new insights into the mechanism of excitation of the highly complicated complex of the general circulation.

Having considered the working mechanism of the "purely zonal" scheme, let us return once more to the distribution of conventional coefficients of eddy exchange shown in Figure 2. Calculations using observations in nature and in the laboratory show a sharp rise in the value of K toward the pole and a small increase toward the equator. This means that the model is satisfactory, but provides no information on the reliability of the values of K itself. The point is that K was calculated from the formal relationship between the heat flux and the gradient of the layer's mean temperature in all the regions of the system, irrespective of the mechanism of the motions, since the latter was not clear at the start.

The experiments showed that in regions 1 and 3 vertical exchange predominates over the horizontal. The question can be raised as to whether it is correct to calculate the horizontal exchange coefficient at these points without consideration of vertical motions. Apparently, this is not so, since the principal amount of energy in these regions is expended for vertical transport, smoothing out the temperature gradients. Here this transport is characterized by large values of the vertical exchange coefficient K_z, commensurable with K. Consequently, if the topic of discussion is not the proof of the similarity of the model but of the numerical values of K, then consideration of temperature equalization due to internal vertical exchange is mandatory in the calculation of K. The values of K obtained for the region with predominant horizontal exchange by large eddies are correct.

All this applies to modeling of a simplified "zonal" model of the general circulation. Modeling has started of the yearly-average "nonzonal" circulation and heat exchange with allowance for the distribution of heat sources and sinks along the longitude. A large circulation device was used for this purpose.* It consists of a rotating cylinder 1.5 m in diameter and

* The technical specifications for the test table were compiled by A. A. Dmitriev, T. V. Bonchkovskaya and F. Ya. Klinov; V. A. Mokhov was the leading designer.

with wall height 25 cm. The number of rotations could be varied from 0.01 to 1 per second. Sectional heaters were installed beneath the metal bottom of the vessel in such a manner that each section input could be controlled separately. The unheated part of the bottom was constantly washed by cold water to produce relative cooling. In order to eliminate moving contacts in supplying signals from pickups in the rotating vessel to the recording instruments, a ring-shape platform was constructed around the vessel, which rotated synchronously with the latter. Instruments could be placed on it and it can even hold an observer. As in the smaller facility, thermocouple columns for recording the mean layer temperature at given locations were located at different points of the modeling fluid.

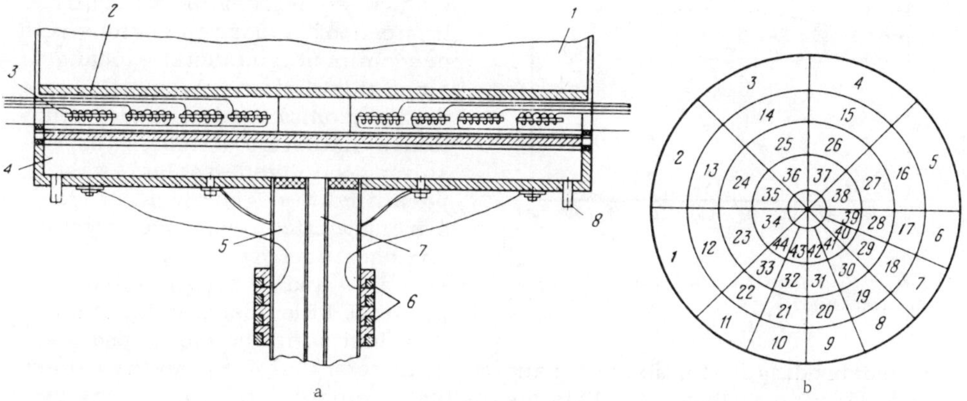

FIGURE 4. Scheme of a test vessel:

a — schematic; b — location of segments with heaters beneath the vessel bottom; 1 — modeling fluid; 2 — bottom of vessel; 3 — heaters; 4 — chamber with cooling-water circulation; 5 — rotating support column; 6 — intake manifold; 7 — tube for supplying water to the cooling device; 8 — water drain pipes.

The initial conditions specified for the model were the latitudinal and longitudinal distributions of $Q_n - Q_u$ for the northern hemisphere. This was achieved by heating the appropriate sections (from No. 44) of the vessel bottom (Figure 4).

Particular attention was paid to distortions and errors due to the slight departure from geometric similarity. An integrated projection of geographic coordinates (i. e., a projection which varies for different latitude belts) with smallest area distortion was calculated for the vessel's flat bottom.

The principal longitudinal scale for the terrestrial surface was a length of 1° of arc of the equatorial parallel (L_{nat}) with the earth radius denoted by R_{nat}. The principal scale of the model L_{mod} was taken as equal to 1° of arc at the vessel walls with radius R_{mod}. Complete geometric similarity requires that the ratios of the model of one-degree lengths $(l_{1,2 \ldots n})_{mod}$ of circles with radii $r_{1,2 \ldots n}$ to the principal scale L_{mod} be equal to the ratios of one-degree lengths of parallel circles $(l_{1,2 \ldots n})_{nat}$ at latitudes $\varphi_{1,2 \ldots n}$ to L_{nat} in nature. This means that the relative distortion of lengths at different radii in the model and at different latitudes in nature, $n = (l_{1,2\ldots n})_{\text{mod}} / L_{\text{mod}} : (l_{1,2\ldots n})_{\text{nat}} / L_{\text{nat}}$, would be unity.

The distortion in scales in the radial direction, $m = \Delta r_{\text{mod}} / R_{\text{mod}} : \Delta r_{\text{nat}} / R_{\text{nat}}$ (where Δr_{nat} is the length of 10° of latitude along the meridian, equal everywhere to approximately 1110 km; Δr_{mod} is the length between the corresponding latitude circles in the model drawn every 10°), was estimated similarly.

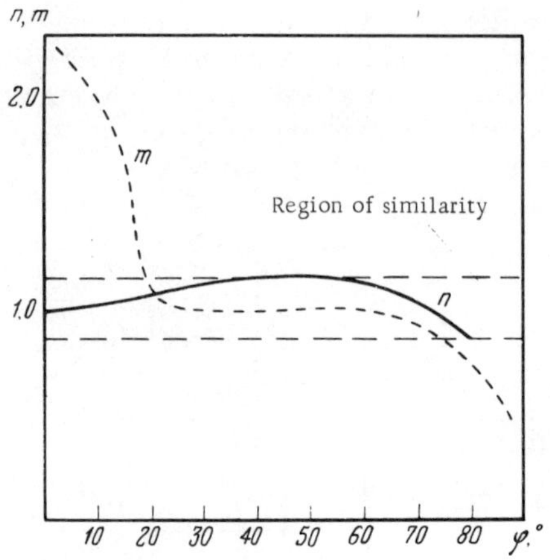

FIGURE 5. Parameters n and m at different latitudes.

A single scale cannot be maintained in projecting a spherical surface on a plane, and distortions in the horizontal n and m must exist in some region. Naturally, this cannot be done in regions with predominant horizontal exchange, since this distorts the similarity of the horizontal motions under study. Hence distortion of scales (in solving the given problem) was permitted in regions with predominant vertical exchange in polar and equatorial regions.

Below and in Figure 5 are presented the assumed latitudinal distribution in the model and the corresponding distortions of n and m. It is seen that $m = 1$ in the rather wide belt from 20 to 70°. This means that geometric similarity may be regarded as attained here.

0°	75.0	1.0	2.2	50°	41.9	1.14	1.0
10	71.3	1.03	1.8	60	33.0	1.14	1.0
20	66.8	1.05	1.0	70	24.7	1.03	0.85
30	58.4	1.11	1.0	80	15.0	0.85	0.55
40	50.2	1.15	1.0	90	00.0	—	—

In regions with distorteα scales (from 0 to 20° and from 70 to 90°), appropriate corrections were made in values of $Q_n - Q_u$ totalled over areas, in such a manner that the total amount of heat supplied by the surface to the fluid would satisfy the similarity requirements.

Preliminary experiments under the above conditions showed that the distribution of the mean temperature in the fluid layer of the model using the assumed latitudes and longitudes is similar to the mean-temperature distribution in the layer of the atmosphere, averaged over a year. However, the data obtained until now do not suffice for verifying similarity of the "nonzonal" circulation obtained in the model with that actually prevailing in the atmosphere.

References

1. Bonchkovskaya, T. V. Nekotorye rezul'taty modelirovaniya atmosfernoi konvektsii (Some Results of Modeling Atmospheric Convection). — In: Modelirovanie yavlenii v atmosfere i gidrosfere. Izdatel'stvo AN SSSR. 1962.
2. Faller, A. J. A Demonstration of Fronts and Frontal Waves in Atmospheric Models. — J. Meteorol., 13 (1). 1956.
3. Fultz, D. On the Possibility of Experimental Models of the Polar-Front Wave. — J. Meteorol., 9 (6). 1952.
4. Hide, R. An Experimental Study of Thermal Convection in a Rotating Liquid. — Philos. Trans. Roy. Soc. A, 250 (983). 1958.
5. Hide, R. Some Experiments on Thermal Convection in a Rotating Liquid. — Quart. J. Roy. Meteorol. Soc., 79 (339). 1953.
6. Riehl, H. and D. Fultz. Jet Stream and Long Wave in a Rotating Dishpan Experiment: Structure of the Circulation. — Quart. J. Meteorol. Soc., 83 (356). 1957.

G. S. Golitsyn and V. I. Moroz

PROBLEMS OF GENERAL CIRCULATION IN PLANETARY ATMOSPHERES

A brief survey is presented of the main observational data on the atmospheres of Mercury, Venus, Mars and Jupiter. The principal parameters for these planets, needed for constructing the theory of the general circulation of planetary atmospheres, are presented. In conclusion several important problems of the dynamics of atmospheres of different planets are pointed out, which apparently could be solved with the present state of knowledge about the nature of planetary atmospheres.

The study of the general circulation of planets of the solar system is of interest not only to astrophysicists. Meteorologists, who are concerned with the theory of the general circulation of the terrestrial atmosphere, can find a good object for applications, and what is more important, for checking their theoretical construction, by examining the atmospheres of other planets, the parameters of which differ from those of the terrestrial atmosphere. At the same time, the principles proper of the theory of the general circulation should be the same for all planets. In every case circulation is produced by transformation of solar energy into kinetic energy of the moving atmosphere due to nonuniform heating of the different parts of the planet. The common feature of atmospheric circulation is the fact that this motion occurs in the Coriolis force field, although the speeds at which the planets rotate about their axes vary widely. The purpose of the present survey is precisely to call the attention of meteorologists and specialists in physics of the terrestrial atmosphere to this problem. The survey presents some data (the principal parameters of planetary atmospheres) required for the construction of a theory of the general circulation and brief mention is made of some theoretical studies of this problem which have already been made. Reference is made only to planets for which most extensive studies were carried out, these being Mercury, Venus, Mars (planets of the Earth group) and Jupiter.

The only source of data which could be used for drawing any conclusions on the type of motions in atmospheres of other planets is the cloud structure observed at the planetary disks and changes in it. A large number of photographs of disks of planets, in which this structure can sometimes be seen, has been collected by Slipher /11, 12/. Frequently the cloud layer is so thick that it is virtually impossible to observe the solid surface of the planet (Venus and larger planets). Among the other planets of the solar system only Mars can be regarded as more or less similar to the earth with respect to the average thickness and structure (but not composition) of the cloud cover. It should be noted that the clouds of other planets probably do not consist of water vapor and their specific nature is frequently unclear.

It has not even been positively established that Mercury has an atmosphere at all; however, some observers found on its disk changing phenomena, which resemble clouds. Some preliminary considerations on the structure and composition of Mercury's atmosphere were recently developed by one of the authors /3/.

The Venusian disk hardly displays any details of the cloud structure. Photographs in the near-infrared region sometimes show a system of bands diverging radially from the subsolar point of the planet. The circulation fluxes can have such a direction in the case of a very slowly rotating planet, when the temperature difference between the sunlit and dark sides of the planet is greater than the temperature difference between the pole and equator. However, ultraviolet photographs show a system of parallel bands on the planet's disk. The nature and origin of these bands are unclear, but there is reason to believe that the effective level of ultraviolet radiation is markedly higher than that of the infrared.

Clouds of various types (white, yellow and blue) are observed on Mars; their directions can be used for determining the wind direction and speed. Radiometering observations in the 8 — 13 micron transparent window of the terrestrial atmosphere allowed a rough temperature distribution of the Martian disk to be obtained. It should be noted that the temperatures obtained in this manner pertain primarily to the temperature of the Martian surface proper; hence they are somewhat higher than the temperature of the relatively thin Martian atmosphere, which is treated as a meteorological parameter. An isotherm map for the Martian winter in the northern hemisphere was first published in 1950 by Hess /5/, who used results of radiometering observations carried out in 1926. This map is shown in Figure 1. References /7/, as well as Figures 2 — 4, present such maps for the winter, spring and summer (referred to the northern hemisphere). They were constructed from a much larger volume of data, collected between 1926 and 1941. It is seen by comparing Figures 1 and 2 that the data are quite representative, at least for the Martian winter. Using the temperature distribution map, as well as certain observations of cloud motion (principally of high white clouds, which projected on the limb and terminator), Hess constructed a hypothetical streamline map for the Martian winter. This map, which is shown in Figure 5, resembles a map of terrestrial weather. We note high- and low-pressure centers, cyclones and anticyclones, different fronts. The tooth-edged lines on the map denote the observed directions of cloud motion.

The most complete survey of observations of the movements of yellow clouds on Mars, together with an analysis of their possible nature, was carried out recently by Gifford /8/. Unfortunately, these movements are observed very infrequently. In analyzing observed data for the period from 1873 to 1961, Gifford was able to count only 16 examples of the movement of yellow clouds and 20 cases of movement of clouds projecting on the limb and terminator.

According to Gifford's hypothesis, the yellow clouds consist of sand or dust particles raised by wind from the Martian surface, for which reason they are determined by winds blowing in its lower atmosphere. Conversely, clouds which are projected on the limb and terminator are apparently formations of condensation origin and are observed at altitudes of the

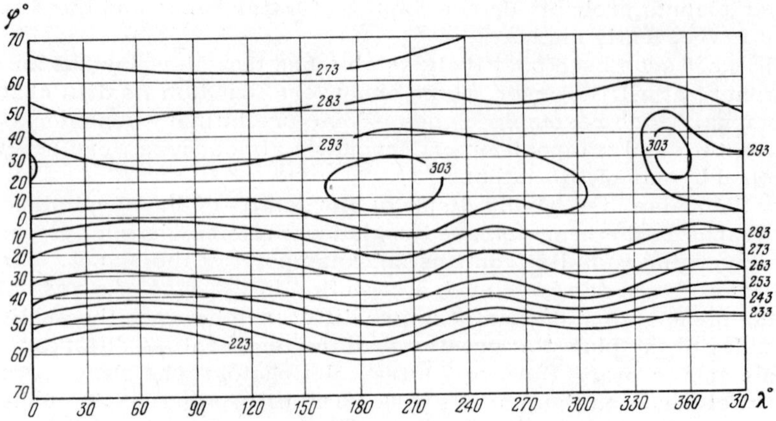

FIGURE 1. Isotherm map for the Martian winter in the northern hemisphere (after Hess).

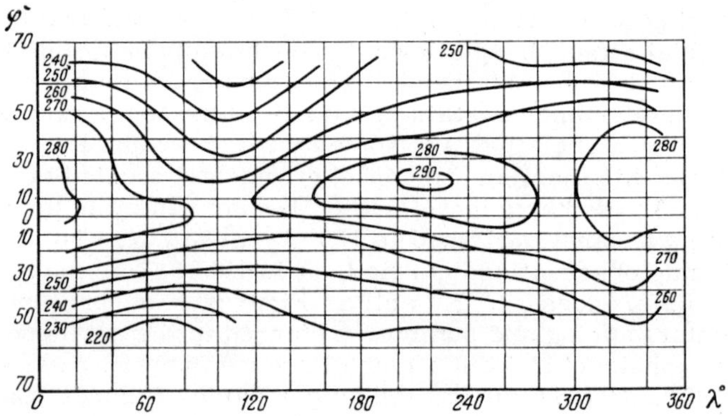

FIGURE 2. Isotherm map for the Martian winter in the northern hemisphere (after Gifford).

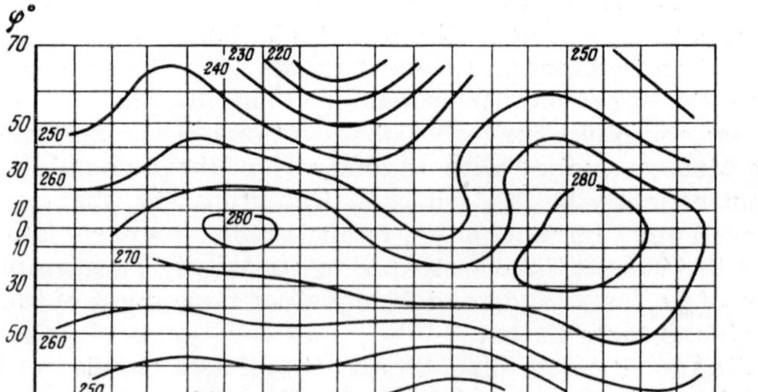

FIGURE 3. Isotherm map for the Martian spring in the northern hemisphere (after Gifford).

order of 50 km and above. The yellow clouds have a tendency to originate over hotter portions of the Martian surface and are encountered at lower latitudes, at a time when clouds projecting on the limb and terminator are encountered primarily in the middle latitudes, and when their motions are zonal and do not exhibit any clearly expressed meridional component. Yellow clouds have a general tendency to move toward the thermal equator.

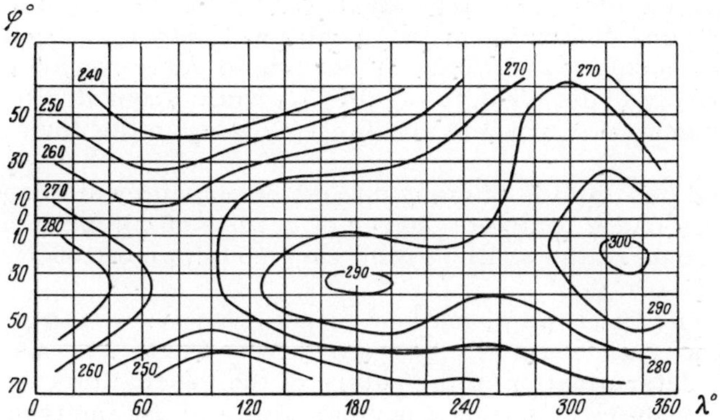

FIGURE 4. Isotherm map for the Martian summer in the northern hemisphere (after Gifford).

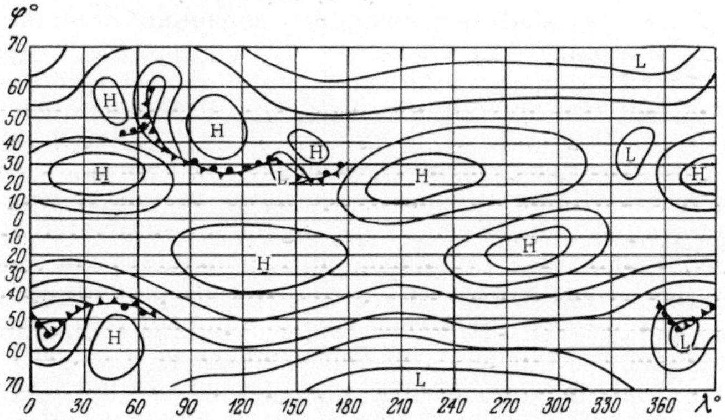

FIGURE 5. Hypothetical streamline map for the Martian winter (after Hess).

In the case of Jupiter the type of the general circulation is determined by the very form of the planetary disk, which has a number of parallel belts and zones. The belts have a lighter color and the zones a darker color. These belts and zones have a varying fine structure, and their locations proper change somewhat with time. Periodic changes with a period of 9 — 10 years are observed as well as sudden changes, when some details change their form entirely during a period of a single rotation of the planet about its axis, i. e., during a time of the order of

10 hours. Apparently, the cloud layer in the polar regions of Jupiter is lower than in the equatorial regions. It is important that the rotational velocity of the structural details and of individual belts and zones depends perceptibly on the latitude, ranging from 9 hours 50 min to 9 hours 56 min. In general the period of rotation becomes greater the farther away from the equator, which is also observed on the sun, but the angular velocity of Jupiter does not change smoothly, but rather discontinuously, at boundaries between belts and zones. The most interesting detail is the "Great Red Spot," consisting of a very large stable formation at the planet's surface. The intensity and coloration of the spot as well as its location on the planet's disk undergo secular variations. According to some considerations this spot is of purely hydrodynamic origin /6/, while according to others this is a stable accumulation of some substance floating in the Jovian atmosphere /10/.

A detailed description of many studies of planetary atmospheres is presented in /4/. A complete survey of results of comprehensive studies of all the planets of the solar system carried out as of 1962 is presented in /9/.

The principal parameters for constructing a theory of the general circulation in planetary atmospheres. Theoretical determination of the nature of the general circulation at a given planet requires knowing a number of initial parameters pertaining to the general properties of the planet's atmosphere. These parameters are not always well known and not all presented values can be regarded as finally established, but still they can serve as guidelines. These data are listed in the table, which also includes for comparison data on the terrestrial atmosphere. We shall now enumerate these characteristics.

The radius of a planet is a characteristic scale of the largest motions. The period of revolution about the sun gives information on the length of year on the planet. The period of rotation about the axis and the direction of this rotation typify the duration of the day, i. e., the time during which a given point of the planet is sunlit. We note that for all the planets, with the exception of Venus, the direction of rotation coincides with the direction of revolution about the sun. In the case of Venus these directions are opposite. The inclination of the equatorial plane relative to the ecliptic plane characterizes the change of seasons on the planet. The amount of solar energy and its "utilization" are determined for each planet by the solar constant and the integral albedo of the planet.

The composition and amount of a substance in the atmosphere also differ from one planet to another. The table lists the gravitational constants, total atmospheric pressure at the surface of the planet, partial pressure of the principal components absorbing long-wave radiation, as well as the approximate average molecular weight of the atmosphere. Then data are presented on the temperature of the planet's surface at the equator and poles, or on the sunlit and dark sides of the planet. Knowing the temperature, gravitational constant, and molecular weight one may determine the height of the homogeneous atmosphere, and estimate the vertical scale of atmospheric motions, as well as the speed of sound. Some idea can be obtained on the nature of the general circulation in planetary atmospheres, in particular, on the effect of the Coriolis force, by calculating the ratio $a\omega/c$, where c is the speed of sound, a is the radius

TABLE. Principal characteristics of planetary atmospheres

Characteristic	Mercury	Venus	Earth	Mars	Jupiter	Remarks
a, radius of planet, km	2334	6056	6371	3370	71,280 (equatorial) 66,770 (polar)	
Period of revolution, days	88	225	365	687	4339	
l, period of rotation, days	88	−243	1	1.02	0.41	The direction of rotation of Venus is opposite with a period of 250 days
Angle of inclination of the equatorial plane to the ecliptic plane, deg	<28	1.2	23	24.8	3.7	
Solar constant, erg/cm^2 · sec	$9.15 \cdot 10^6$	$2.62 \cdot 10^6$	$1.37 \cdot 10^6$	$5.9 \cdot 10^5$	$5.1 \cdot 10^4$	
Albedo	0.09	0.75 ± 0.10	0.38	0.22	0.48	
g, gravitational acceleration (ratio to 980 g/cm^2 · sec)	0.39	0.90	1.00	0.38	2.64	
p_0, total atmospheric pressure, atm	<0.001	17–20	1	0.01	2	The pressure for Jupiter is that above the cloud layer
p_{CO_2}/p_0	0.1–1	0.8–1.0	$3 \cdot 10^{-4}$	0.8–1.0	?	
p_{H_2O}/p_0	—	10^{-3}–10^{-2}	$2 \cdot 10^{-3}$	$(5 \pm 3) \cdot 10^{-5}$	10^{-3}–10^{-2}	The partial pressure of CH_4 is given for Jupiter
Molecular weight	30–44	44	29	40–44	2.6	
T_1, temperature of the sunlit part, °K	620	600–700	280	280	130	For Venus it depends on the assumed model of the atmosphere. For Jupiter this is the brightness temperature in the 8–13 micron atmospheric transparency window
T_2, temperature of the dark part, or the temperature at the poles, °K	?	550	240	200	?	These temperatures for Mercury and Venus are for the antisolar point
γ_a, adiabatic lapse rate, deg/km	3–4.5	8.5–11	10	4.5	3.2	
c, speed of sound, m/sec	440–540	370–410	340	270	650	
$a\omega/c$	$6 \cdot 10^{-3}$	$(4-6) \cdot 10^{-3}$	1.4	0.83	20	
H, height of homogeneous atmosphere, km	30–45	6–12	8	14	12	

of the planet, and ω is the angular velocity of the planet's rotation. If this ratio is small as compared to unity, then the rotation of the planet has almost no effect on the type of motions in the atmosphere (Mercury and Venus); if $a\omega/c \gg 1$, then the rotation has a decisive effect (Jupiter). In the intermediate cases (Earth, Mars) a large role, alongside rotation, is played by other factors (nonuniform heating of the atmosphere, turbulent mixing and friction, etc.). It should also be noted that by virtue of the slow rotation, the circulation in the atmospheres of Mercury and Venus should not be determined by the temperature difference between the poles and equator, but between the subsolar and antisolar points of the planet.

We note one factor, which facilitates construction of a theory of general atmospheric circulation on other planets, namely, the absence of oceans. Hence there is no need to make allowance for the complex and little-investigated mechanism of interaction between the atmosphere and ocean, which is so important for the earth; one may regard the surface of the planet as homogeneous with greater justification than for the earth.

Theoretical studies and problems of the theory of the general circulation in planetary atmospheres. An attempt to develop a theory of the general circulation of atmospheres of other planets was made by Mintz /1/. He has developed in most detail the problem of Martian circulation. The quantity of heat which the atmosphere, as a heat machine, transports from the equator to the poles on Mars is smaller than on earth. According to Mintz's calculations, the Martian atmosphere on the average should be in the symmetrical or Hadley regime, with one circulation cell from the pole to the equator. However, winter heat transport in the meridional direction may exceed the critical value and the circulation may be transformed into a wave regime with three circulation cells, situated along the latitude circle (according to Mintz, six cells should exist on earth, since the radius of the earth is twice the radius of Mars). It should be noted that as yet too little observational data are available on motions of the Martian atmosphere to substantiate or negate this model.

Mintz's theory is inapplicable to deep atmospheres (Venus and, the more so, Jupiter). The depth of the Jovian atmosphere is unknown. The radiative properties of its atmosphere are also very little known, so it is impossible to develop any rigorous theory.

In spite of gaps in our knowledge of the principal characteristics of planets, it is possible to point out certain important problems which do not appear insolvable at present: 1) to what extent can the atmospheric circulation heat the cold side of Mercury, if the planet has an atmosphere defined by the parameters listed in the table? An observational check of the predicted temperature could serve as an additional verification of the presence of atmosphere on this planet; 2) what should be the circulation system and the total mass of the atmosphere on Venus in order to account for virtually equal temperatures of the sunlit and dark sides of the planet? 3) how should one explain the visible separation of the Jovian atmosphere into different belts and zones rotating at different speeds? A theory which would provide an answer to this problem could also be used for drawing some conclusions on the depth of the Jovian atmosphere.

The current state of knowledge on planetary atmospheres is described in /1, 2, 4, 9/.

Reference

1. Kellogg, W. and C. Sagan. The Atmospheres of Mars and Venus. — NAS-NRC Publ. 944. Washington. 1961.
2. Kellogg, W. and C. Sagan. The Terrestrial Planets. — In: Annual Rev. of Astronomy and Astrophysics, Vol. 1, p. 235. 1963.
3. Moroz, V. I. Infrakrasnyi spektr Merkuriya (1.0 — 3.9 mk) (The Infrared Spectrum of Mercury (1.0 — 3.9 microns)). — Astronomicheskii zhurnal, 41 (6). 1964.
4. Sharonov, V. V. Priroda planet (The Nature of Planets). — Fizmatgiz. 1958.
5. Hess, S. L. Some Aspects of the Meteorology of Mars. — Meteorol., 7 (1). 1950.
6. Hide, R. On the Hydrodynamics of Jupiter's Atmosphere.
7. Gifford, F. A. The Surface-Temperature Climate of Mars. — Astrophys. J., 123 (1). 1956.
8. Gifford, F. A. A Study of Martian Yellow Clouds that Display Movements. — Monthly Weather Rev., 92 (10). 1964.
9. La physique des Planètes. — Mom. Soc. roy. sci. Liège, sér. 5, VII. 1962.
10. Sagan, C. On the Nature of the Jovian Red Spot.
11. Slipher, E. C. A Photographic Study of Bright Planets. — Arizona Univ. Press. 1964.
12. Slipher, E. C. Mars, the Photographic Story. — Cambridge, Sky Publ. Co. 1962.

A. A. Dmitriev

OBJECTIVE DETERMINATION OF CIRCULATION TYPES

Expansion of atmospheric fields in terms of coordinate functions of a canonical expansion usually converges very rapidly, which is its undoubted advantage. In the process of obtaining coordinate functions all the circulation types are treated as a homogeneous statistical population. This, together with the large amount of work needed for computations, is in a number of cases a shortcoming, since averaging may obliterate infrequently encountered fields. In fact autocorrelation functions for individual days differ markedly from one another /1/. For example, for OT^{500}_{1000} for a 180° shift on 20 December 1949 and 8, 9 and 14 January 1950 one observes a clearly expressed maximum which is by 50% higher than the zero-shift maximum, while on 23 and 24 December 1949 the autocorrelation function for the 180° shift was negative. The point where the correlation function as a function of the shift passes through zero also changes markedly. The power spectrum (amplitude of harmonics of the autocorrelation function) is a good objective characteristic of the circulation type, since it reflects the wave structure of the atmospheric processes. If a classification is carried out on the basis of the number of the largest and subsequent harmonics in the power spectrum, then the problem of subjective selection of graduations of a monotonically varying parameter is removed. At the same time it becomes possible to follow the day-to-day shift of individual wave components, the accuracy here being the greater the longer the wave.

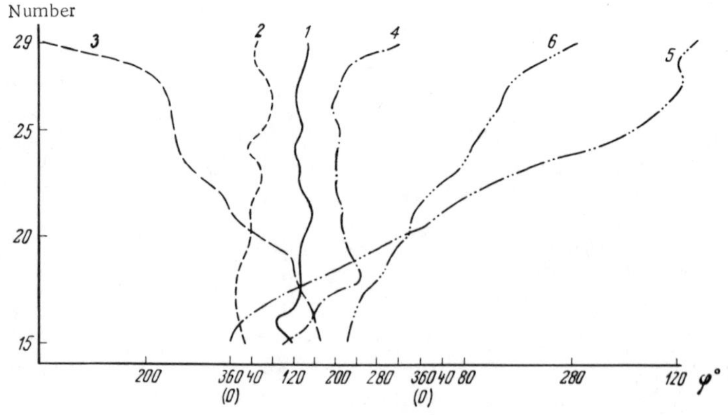

FIGURE. Phase shift of the maximum of harmonics 1 through 6 of OT^{500}_{1000} on 15—29 December 1954.

To illustrate the latter two facts let us supplement examples from /1/ by some data obtained by my student, S. V. Shpironova, during an assignment. Below are presented numbers of the first two largest harmonics obtained for a segment of the 50°N circle lying in the eastern hemisphere with Kats circulation indexes I_z /2/ for the period 17 – 29 December 1964 (see the table).

TABLE. Harmonics at 50° N and zonal circulation indexes

Date	Number of harmonics with largest amplitude	Number of second largest harmonic	I_z for 35–70°N
17	1	6	1.42
18	1	6	
19	1	6	
20	2	5	
21	2	5	1.48
22	2	6	
23	2	6	
24	3	2	
25	2	3	
26	2	4	0.82
27	2	4	
28	2	5	
29	2	6	

It is thus seen that the principal harmonic for $I_z = 1.42$ was 1, and for $I_z = 0.82$ it was 2. The period for $I_z = 1.48$, unlike the latter, has a higher value for the second largest characteristic. Analysis of the tabulated data creates the impression that the third period does not start on 26 December, as is assumed by Kats, but on the 25th, with 24 December being the reconstruction day. This is verified by the power spectrum for the entire 50° latitude circles, where harmonic 1 predominated during 17 – 23 December, harmonic 2 during 25 – 29 December and harmonic 4 from 24 December.

The figure shows the phase shift of the first six harmonics for the period under study. The phases were determined from the coefficients of $\cos n\lambda$ and $\sin n\lambda$ as $\varphi = \arctan \frac{b_n}{a_n}$. It is interesting to note that harmonics 1 and 2 change their location little, being apparently related to the underlying surface. The following, relatively long wave 3 moves from east to west, and the shorter waves 5 and 6 move to the east, which could have been expected on the basis of the theory of long waves.

It is interesting to note that calculations carried out separately for the eastern and western hemispheres showed virtually simultaneous (within 1 – 2 days) changes in circulation types, although the distribution of harmonics in different hemispheres was found to vary.

References

1. Dmitriev, A. A. Opyt ob"ektivnogo vydeleniya komponent voln OT_{1000}^{500} s primeneniem korrelyatsionnogo analiza (Experience in Objective Selection of Components of OT_{1000}^{500} Waves Using Correlation Analysis). — Trudy NIIAK, No. 28. 1964.
2. Kats, A. L. Sezonnye izmeneniya obshchei tsirkulyatsii atmosfery i dolgosrochnye prognozy (Seasonal Variations in the General Circulation of the Atmosphere and Long-Range Forecasts). — Gidrometeoizdat. 1960.

N. A. Bulinskaya

ANALYSIS OF ATMOSPHERIC CIRCULATION FOR FORECASTING PURPOSES

Atmospheric circulation is a complex phenomenon. The same data (meteorological observations, synoptic maps, maps of constant pressure surfaces) are used both for the study of climate and for weather forecasting. Hence in approaching the study of atmospheric circulation one must clearly formulate the problem corresponding to the goal to be reached.

In our case the goal is compilation of long-range weather forecasts, while the problem is to determine regularities in the development of circulation components with time. The day-to-day change in meteorological elements during this time characterizes the form of atmospheric variations corresponding to this law, and this means that any of its phases can be forecast for the future. Weather forecasting requires the development of appropriate techniques.

Regular changes in atmospheric-circulation components in time and space can be divided into two groups due to the differences in the scales corresponding to the duration of the natural process. The first (climatic) group includes processes, the thermal inconsistencies of which are produced by nonuniform heating of the surface. They produce the climate and have a periodic annual variation. The phase of the periodic variation of the atmosphere is indicated by climatic centers of action of the atmosphere (Icelandic, Siberian, etc.). The second (synoptic or circulation) group includes an ensemble of regularities, the thermal inconsistencies of which do not directly depend on the surface, but are produced by synoptic atmospheric centers of action (hot and cold sources, respectively) at high and low air pressure in the troposphere, irrespective of geographic location.

The main component of this latter group, which directly determines the weather, is a periodic fluctuation with a wavelength of 90° along the 50—60° latitude circle. The so-called signal process, produced by the special location of atmospheric centers of action, is observed 2—4 times during a season.

Our problem is to find a method for determining the phase of atmospheric centers of action which determine the wave producing the weather of each individual day. The state of the atmospheric centers of action is masked by daily changes which, similar to "noise" in information theory, are superimposed on the periodic fluctuations.

Using maps of relative constant-pressure surfaces Dmitriev calculated autocorrelation functions for one latitude circle. He obtained two waves on the globe. He established by Fourier expansion of the autocorrelation

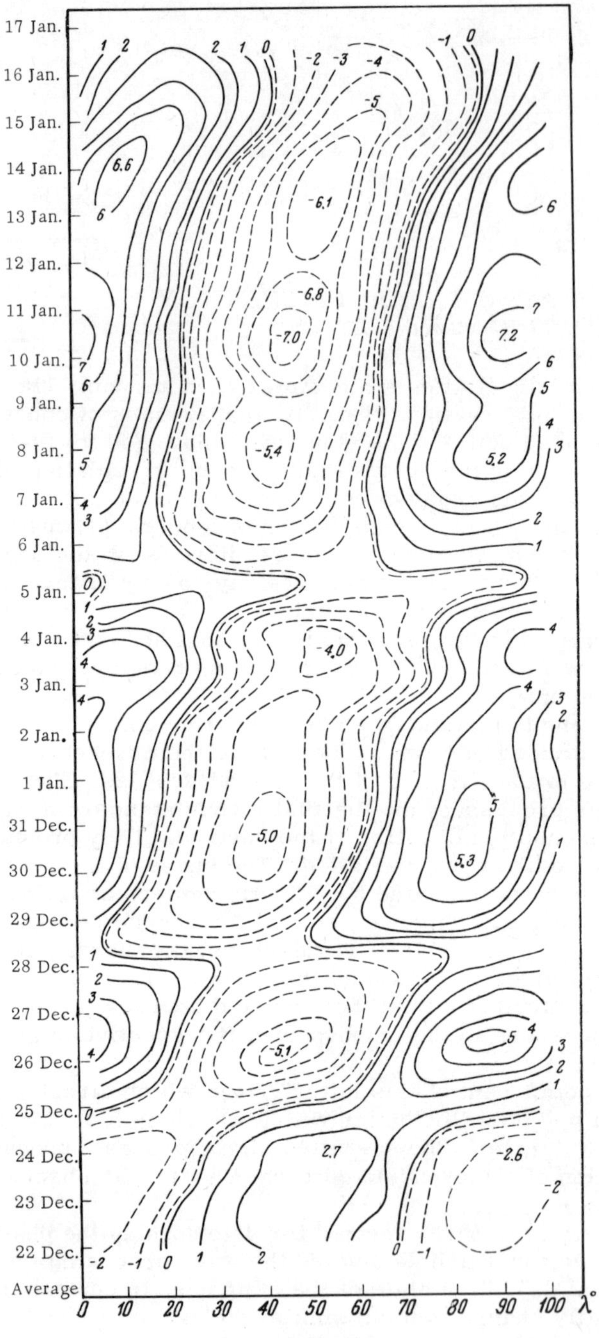

FIGURE. $\Phi_H\ h\ (\Delta)$ as a function of time.

function that the 2nd harmonic has the highest oscillatory amplitude and concluded that it predominates. In the meantime, it is easy to see from the yearly air temperature variation that a small departure from average produced by the second group of circulation components has a marked effect (hot or cold weather, etc.). Hence this author used a different approach.

The signal process was used for isolating natural processes, a catalog for which is available for 15 years. The period from 28 December 1949 to 5 February 1950 was selected from this catalog. The implementation of a random process $H(\lambda)$ was defined as a sequence of values of the relative constant-pressure surface (OT^{500}_{1000}) along a constant latitude circle at each 5° of longitude (λ). The above data were used for calculating the autocorrelation function using the standard expression

$$\Phi_H(\Delta) = \frac{1}{2\pi} \int_{-\pi}^{+\pi} H(\lambda)\, H(\lambda + \Delta)\, d\lambda. \tag{1}$$

In the case of a periodic process the autocorrelation function is periodic and contains the fundamental frequency and harmonics. The values of the cross-correlation function between $H(\lambda)$ and the n-th harmonic of the autocorrelation function were calculated for all the days. The former function has the form

$$\Phi_{H,\,n}(\Delta) = \frac{1}{2\pi} \int_{-\pi}^{+\pi} (H(\lambda) \cos[n(\lambda + \Delta)]d\lambda + \frac{A_n}{2} \cos(n\Delta - \theta_n)\ldots, \tag{2}$$

where Δ is the shift along the latitude circle, θ_n is the phase angle, and $\Phi_{H,n}$ is the sum of the periodic and random parts. Values of the cross-correlation functions were obtained for latitudes from 40 to 65° for the period from 22 December 1949 to 17 January 1950.

The distribution of atmospheric centers of action in the opposite phase in the 55 — 60° belt was fully verified. Isoplets were constructed for determining the time dependence of $\Phi_{H,n}(\Delta)$. Latitudes are represented along the abscissa with the days of the month plotted on the ordinate. It was found that one distribution prevailed during 22 — 25 December, and starting with 25 December, negative values of the cross-correlation functions prevailed for a long time between 30 and 60° of longitude.

It appears to this author that the above conclusion is of great significance since if meteorological processes are regarded as stationary random processes, then one can judge about the start of a natural process by the implementation of the process at one longitude. On the other hand, weather forecasting requires refining the boundaries of seasons obtained from the signal process and determining the duration of one phase and the regularity with which it changes in time within the cycle of fluctuations.

S. T. Serlapov

COMPARISON OF MAPS OF MAXIMUM WIND STRENGTH IN THE UPPER TROPOSPHERE WITH GROUND-SURFACE SYNOPTIC MAPS

This author studied regularities in the general circulation of the atmosphere by comparative analysis of the maximum-wind maps of the upper and lower atmosphere. Particular attention was paid to characteristics of the structure of the wind and jet-stream fields. The maps which were obtained show stagnant zones and zones with maximum wind disturbances; they also display a high planetary frontal zone and the location of surface-level cyclonic and anticyclonic formations.

Preliminary analysis of comparison of maximum-wind and surface-level maps shows that the intensity of wind disturbance zones is of great importance for forecasting purposes, and that these phenomena are frequently associated with deep cyclones and with an above-average wind intensity at the surface.

Monthly average maximum-wind and jet-stream maps are of particular importance. These are composed as follows. Centers of greatest disturbance are taken from daily maximum-wind maps and plotted with indication of wind direction and speed. When this is done one obtains the monthly and seasonal average structure of the wind state of the troposphere, which exhibits regions of geographic localization of maximum wind and of stagnation zones.

Further comparison of monthly maps compiled over a number of years can yield the corresponding "averages" of the wind situation in the troposphere and stratosphere. Departure from average will yield information for weather forecasting and for more complete classification of wind-field structure as one of the major elements of atmospheric circulation.

M. Kh. Baidal

DYNAMICS OF THE "CIRCULATION POLE" AND THEIR SYNOPTICOCLIMATOLOGICAL SIGNIFICANCE

Experience in long-range weather forecasting shows that the extreme value of anomalies is not yet forecasted, and any study of factors affecting the anomalous state of the atmosphere should be welcome.

This author's attention was attracted, in the study of the location of high cyclones in polar regions, by the fact that the former are concentrated in specific places. One can subdivide the Arctic region into six or seven circumpolar sectors, at the boundaries of which appear high anticyclones which prevail for an entire season. It was found from statistical analysis of data that these polar anticyclones are located in 30% of the cases near the pole and in the remaining 70% along the periphery; the concentration is not uniform in different sectors: 17% in the Asiatic sector, 7% in the Taimyr sector and 9% in the European sector.

The concentration of anticyclones in any sector serves as a source for the formation of cold Arctic masses. The anticyclones move from east to west without losing strength, but as soon as their strength decreases they veer southeast. Hence the concentration moves from season to season into a more westerly sector.

What hypothesis can be put forward to explain the existence of this accumulation of anticyclones and the appearance of a "circulation pole?" This question can be resolved by analysis of the secular march of circulation.

First, one can note the relationship between these data and the march of meteorological elements. For example, the temperature drops sharply in Kazalinsk when a strong anticyclone forms in the Barents Sea. This combination occurs very frequently. Such comparisons, when made over a number of years, should yield interesting regularities. Apparently there exists a common geophysical cause of the above concentration of anticyclones and of the "circulation pole," it being the relationship with the magnetic inclination.

I. L. Karol'

THE USE OF TRACERS IN THE STUDY OF THE GENERAL CIRCULATION OF THE ATMOSPHERE

The use of tracers in the study of the general circulation started recently. These methods are particularly successful when used for locations where the volume of standard aerological data is as yet insufficient for obtaining qualitative and quantitative ideas on the nature and mechanism of exchange, i.e., in high layers of the stratosphere, in polar and equatorial regions. But even when the volume of meteorological data suffices, the use of tracers is an important supplement to existing methods of study of the general circulation.

Many substances can be used as atmospheric tracers /1/, but a special position is held by natural and artificial radioactive isotopes. Unlike ozone, water vapor and carbon dioxide, which have for a long time been used in meteorology as tracers, radioactive isotopes in the atmosphere are an inert admixture, they do not affect the transport process, they have negligible mass, but one which lends itself to precise quantitative measurement. The quantitative characteristics of their sources and sinks in the atmosphere have been quite well determined recently, and the regular radioactive decay of various isotopes with different half-lives determines the time scales of the atmospheric processes under study.

The use of natural radioisotopes (such as radon and its fission products, supplied to the atmosphere from the land or forming in the former due to cosmic radiation (such as ^{14}C, ^{3}H and ^{7}Be)) as tracers has also other advantages, since their sources are sufficiently steady and are located so that the isotopes move in the atmosphere from the bottom upward (radon) as well as from the top downward (isotopes formed by cosmic rays).

Studies of meteorological processes using radioactive tracers were carried out in the Institute of Applied Geophysics. ^{210}Pb, which is a fission product of radon with a 22-year half-life, was used to determine the parameters of vertical exchange in the troposphere and stratosphere through the tropopause, as well as between the northern and southern hemispheres in the stratosphere.

^{210}Pb produced in nuclear bursts was utilized for the study of synoptic conditions of large-scale penetration of stratospheric air masses into the troposphere and of tropospheric exchange in the tropical convergence zone. These results were reported at the 3rd International Conference on Peaceful Uses of Atomic Energy in Geneva in September 1964.

The use of radioactive tracers in meteorology was also discussed at the conference on nuclear meteorology, which took place in February 1964 in Obninsk. It can be claimed on the basis of results of the conference that the method of radioactive tracers is quite promising and should be accorded its due place in meteorology in general, and in the study of the general circulation of the atmosphere in particular.

Reference

1. Karol', I. L. and S. G. Malakhov. Voprosy yadernoi meteorologii (Problems of Nuclear Meteorology). — Gosatomizdat. 1962.

Yu. B. Khrabrov

MODELING OF TYPHOONS AND THUNDERCLOUDS

This author had personal acquaintance with work done in Japan on modeling typhoons and thunderclouds; the work was carried out there under the direction of Takahashi.

All the modeling devices are placed on a specially equipped rotating platform. An analogy of the Coriolis force in the course of a typhoon's passage from low to high latitudes is obtained by changing the platform rpm, i. e., the magnitude of the centrifugal force. Takahashi uses colored gas in his models. The heat souces and sinks are imitated by the gas direction, electric heating and cooling with dry ice. The experiments are carried out in round and rectangular Plexiglas vessels which, in their turn, can be made to rotate about their axes. Takahashi obtained a graphic model of the development of a typhoon from its genesis in lower latitudes to its transformation into an ordinary polar-front cyclone. The eye of the hurricane and other features of typhoon structure are quite clearly seen. This author is in the process of applying similarity theory for application of results thus obtained to real atmospheric processes.

Studies of thundercloud models as a function of the spatial structure and rate of heat influx into different regions of the cloud are also quite remarkable.

Heat influx at different levels is obtained by electric heating coils which can be uniformly inserted into various regions of the cloud. The most interesting result of this experiment is the graphic confirmation of the known property of thunderclouds, consisting in their movement in the direction of the warmest region of the surface. Under experimental conditions the motion of the cloud most effectively "controls" the motion of the lowermost electric heater. It is assumed by Japanese meteorologists that according to these experiments it is best to compile forecasts of the movement of an actual thundercloud not on the basis of the free-atmospheric wind, but in the direction of the surface-level wind which carries some quantities of air heated at the earth surface. At present this method is used by Japanese synoptitians for forecasting the movement of thunderclouds.

R. F. Usmanov

UNDERWATER TRENCHES IN OCEANS AS FACTORS AMPLIFYING ATMOSPHERIC PROCESSES

The bottom relief of seas and oceans serves as a good morphological indicator of anomalies in the force of gravity. In the overwhelming number of cases negative anomalies of gravity are observed in the region of underwater trenches and troughs, with positive anomalies occurring above elevations /1, 2/.

Space and time disturbances in the force of gravity at the earth are usually ignored in analysis of weather maps. In the meantime, many other data determined in analysis of climatic maps and weather charts point to the marked effect of variations in the force of gravity at the surface on the development of atmospheric processes. This author pointed out, as early as in 1959, the extraordinary stability of cyclone formation over the Arctic zone of underwater trenches /3/; frequent cases were also detected of genesis and deepening of typhoons over underwater trenches and troughs. Subsequently these phenomena were confirmed on the basis of more complete information on typhoons in analyzing the locations of the genesis and deepening of the 1964 typhoons, as well as in the analysis of 56 cyclones intersecting the underwater Emperor Seamounts in the northwestern Pacific during the period from November 1963 to April 1964. Seventy-five percent of these cyclones deepened past this ridge, 20% remained unchanged and 5% were somewhat filled.

Assuming that departures from geostrophic motion and static state of the atmosphere are determined by disturbances in the force of gravity, let us introduce anomalies and time variations in the force of gravity into the equations of motion (without dissipative forces) generally used in meteorology. Then the first terms of these equations, reflecting the geostrophic motion and static state, respectively, will be equal to zero; the order of magnitude of velocities arising while an air particle is subjected to accelerations associated with disturbances in the force of gravity will be fully commensurable with the actual changes in wind and vertical velocities.

Apparently, this is the reason for the appearance, in regions of underwater trenches associated with negative anomalies of the force of gravity, of very favorable conditions for appreciable vertical velocities and deviations from the geostrophic wind, which can be regarded as the reason for frequent formation and deepening of cyclones over underwater trenches.

References

1. Vening-Meines, F.A. Gravimetricheskie nablyudeniya na more (Gravimetric Observations in the Sea). — Geodezizdat. 1940.
2. Morskie gravimetricheskie issledovaniya (Marine Gravimetric Studies). — Collection of articles, Nos. 1—2. Izdatel'stvo MGU. 1961, 1963.
3. Usmanov, R.F. O sovpadenii antarkticheskoi zony nizkogo davleniya s podvodnymi antarkticheskimi kotlovinami (On the Coincidence of the Antarctic Low-Pressure Zone with Antarctic Underwater Trenches). — Informatsionnyi byulleten' SAE, No. 24. 1960.